Introduction to Algebraic Quantum Field Theory

Mathematics and Its Applications (*Soviet Series*)

Volume 19

Introduction to Algebraic Quantum Field Theory

by

S. S. Horuzhy

Steklov Mathematical Institute,
Academy of Sciences of the U.S.S.R.,
Moscow, U.S.S.R.

Kluwer Academic Publishers

Dordrecht / Boston / London

Library of Congress Cataloging in Publication Data

CIP

Khoruzhiĭ, S. S. (Sergeĭ Sergeevich)
Introduction to algebraic quantum field theory.

Translation of: Vvedenie v algebraicheskufu kvantovufu teorifu polfa.
Bibliography: p.
Includes index.
1. Quantum field theory. 2. Algebra. 3. Axiomatic set theory. I. Title.
QC174.45.K4913 1988 530.1'4 88-4377

ISBN 90-277-2722-8

Published by Kluwer Academic Publishers,
P.O. Box 17, 3300 AA Dordrecht, The Netherlands.

Kluwer Academic Publishers incorporates
the publishing programmes of
D. Reidel, Martinus Nijhoff, Dr W. Junk and MTP Press.

Sold and distributed in the U.S.A. and Canada
by Kluwer Academic Publishers,
101 Philip Drive, Norwell, MA 02061, U.S.A.

In all other countries, sold and distributed
by Kluwer Academic Publishers Group,
P.O. Box 322, 3300 AH Dordrecht, The Netherlands.

Orginal title: Vvedenie v algebraicheskufu kvantovufu teorifu polfa.
Published by Nauka Publishers, Moscow, © 1986
Translated from the Russian by K.M. Cook

Contents

'Et moi, ..., si j'avait su comment en revenir, je n'y serais point allé.'

Jules Verne

The series is divergent; therefore we may be able to do something with it.

O. Heaviside

One service mathematics has rendered the human race. It has put common sense back where it belongs, on the topmost shelf next to the dusty canister labelled 'discarded non-sense'.

Eric T. Bell

Mathematics is a tool for thought. A highly necessary tool in a world where both feedback and non-linearities abound. Similarly, all kinds of parts of mathematics serve as tools for other parts and for other sciences.

Applying a simple rewriting rule to the quote on the right above one finds such statements as: 'One service topology has rendered mathematical physics ...'; 'One service logic has rendered computer science ...'; 'One service category theory has rendered mathematics ...'. All arguably true. And all statements obtainable this way form part of the raison d'être of this series.

This series, *Mathematics and Its Applications*, started in 1977. Now that over one hundred volumes have appeared it seems opportune to reexamine its scope. At the time I wrote

> "Growing specialization and diversification have brought a host of monographs and textbooks on increasingly specialized topics. However, the 'tree' of knowledge of mathematics and related fields does not grow only by putting forth new branches. It also happens, quite often in fact, that branches which were thought to be completely disparate are suddenly seen to be related. Further, the kind and level of sophistication of mathematics applied in various sciences has changed drastically in recent years: measure theory is used (non-trivially) in regional and theoretical economics; algebraic geometry interacts with physics; the Minkowsky lemma, coding theory and the structure of water meet one another in packing and covering theory; quantum fields, crystal defects and mathematical programming profit from homotopy theory; Lie algebras are relevant to filtering; and prediction and electrical engineering can use Stein spaces. And in addition to this there are such new emerging subdisciplines as 'experimental mathematics', 'CFD', 'completely integrable systems', 'chaos, synergetics and large-scale order', which are almost impossible to fit into the existing classification schemes. They draw upon widely different sections of mathematics."

By and large, all this still applies today. It is still true that at first sight mathematics seems rather fragmented and that to find, see, and exploit the deeper underlying interrelations more effort is needed and so are books that can help mathematicians and scientists do so. Accordingly MIA will continue to try to make such books available.

If anything, the description I gave in 1977 is now an understatement. To the examples of interaction areas one should add string theory where Riemann surfaces, algebraic geometry, modular functions, knots, quantum field theory, Kac-Moody algebras, monstrous moonshine (and more) all come together. And to the examples of things which can be usefully applied let me add the topic 'finite geometry'; a combination of words which sounds like it might not even exist, let alone be applicable. And yet it is being applied: to statistics via designs, to radar/sonar detection arrays (via finite projective planes), and to bus connections of VLSI chips (via difference sets). There seems to be no part of (so-called pure) mathematics that is not in immediate danger of being applied. And, accordingly, the applied mathematician needs to be aware of much more. Besides analysis and numerics, the traditional workhorses, he may need all kinds of combinatorics, algebra, probability, and so on.

In addition, the applied scientist needs to cope increasingly with the nonlinear world and the extra mathematical sophistication that this requires. For that is where the rewards are. Linear models are honest and a bit sad and depressing: proportional efforts and results. It is in the non-linear world that infinitesimal inputs may result in macroscopic outputs (or vice versa). To appreciate what I am hinting at: if electronics were linear we would have no fun with transistors and computers; we would have no TV; in fact you would not be reading these lines.

There is also no safety in ignoring such outlandish things as nonstandard analysis, superspace and anticommuting integration, p-adic and ultrametric space. All three have applications in both electrical engineering and physics. Once, complex numbers were equally outlandish, but they frequently proved the shortest path between 'real' results. Similarly, the first two topics named have already provided a number of 'wormhole' paths. There is no telling where all this is leading - fortunately.

Thus the original scope of the series, which for various (sound) reasons now comprises five subseries: white (Japan), yellow (China), red (USSR), blue (Eastern Europe), and green (everything else), still applies. It has been enlarged a bit to include books treating of the tools from one subdiscipline which are used in others. Thus the series still aims at books dealing with:

- a central concept which plays an important role in several different mathematical and/or scientific specialization areas;
- new applications of the results and ideas from one area of scientific endeavour into another;
- influences which the results, problems and concepts of one field of enquiry have, and have had, on the development of another.

One pervasive trend in modern mathematical times is the increasing penetration of algebraic-type thinking and algebras into various domains of thought. Here is another example, the algebraic approach to quantum field theory. It is based on the idea of taking the algebra of local observables as the point of departure and, as such, the author states the year of birth of this approach to be 1957, when the concept of an algebra of local observables first appeared in a paper by Haag; perhaps the conception of the theory can be placed in 1955 when the same Haag published a solid mathematical theorem indicating that Fock space might not be a wide enough context.

To an algebraist, there is something conceptually very appealing in the observables algebra approach to quantization matters. To him, a classical dynamical system is basically a commutative algebra (of functions) together with a Poisson bracket which is used to define the dynamics; and a quantum version is something even simpler: namely an associative algebra (of operators). The (semi-)classical limit relation is expressed by saying that the associative algebra is commutative modulo h, Planck's constant, and the Poisson algebra structure results from considering the first order part in h of the associative algebra's commutator difference Lie product. This makes a nice simple conceptual framework placing quantization within the framework of the deformation theory of algebras, concerning which there is a good deal of theory. Just how fresh this point of view is is hard to say; probably not particularly, but currently it has added attractiveness within the framework of quantum groups (=Hopf algebras). To what extent one can get away with such a simple basic underlying picture - even as only the shadow of an underlying more general scheme of axiomatically constrained correspondences - is yet another matter; there is, in any case, an awful lot more involved, particularly in the case of infinitely many degrees of freedom. Indeed, to find out just how much more is involved, this is the book to consult, the first of its kind written by a scientist who has contributed greatly to the field.

The shortest path between two truths in the real domain passes through the complex domain.

J. Hadamard

La physique ne nous donne pas seulement l'occasion de résoudre des problèmes ... elle nous fait pressentir la solution.

H. Poincaré

Never lend books, for no one ever returns them; the only books I have in my library are books that other folk have lent me.

Anatole France

The function of an expert is not to be more right than other people, but to be wrong for more sophisticated reasons.

David Butler

Bussum, September 1989

Michiel Hazewinkel

Foreword

The algebraic approach to quantum field theory is now about thirty years old if we assume the report by R. Haag in 1957 Colloque de Lille to be its birth. It was in this report that the concept of algebra of local observables first appeared. In 1961–2 the new approach was given its first systematic exposition in the Zürich lectures by H. Araki and since then it has become firmly established as one of the axiomatic approaches to quantum field theory, alongside the field axiomatics by Wightman and S-matrix axiomatics by Bogoliubov. The common aim of all these approaches was a rigorous analysis of the structure of relativistic quantum systems without referring to any approximations or model assumptions. The algebraic approach advanced the locality principle as the basis for such an analysis along with the notion of a local observable, i.e. an arbitrary physical quantity which can be measured in some finite region of space-time. Accordingly, it became known as local quantum theory.

The sixties, which were a period of flourishing axiomatic studies, produced a vast set of results which were rigorous corollaries of the Haag–Araki or Haag–Kastler systems of axioms (according to the two variants of the algebraic approach, 'concrete' and 'abstract'). But the axioms certainly do not comprise even the most important features of quantum field systems: it is hardly possible, even in principle, to attain a description of such a system which would be strictly axiomatic and yet reasonably complete, i.e. displaying main dynamical mechanisms. For this reason, in the next stage the rigorous methods of relativistic quantum theory were complemented by new elements and extended beyond the sphere of general structural studies to the analysis of concrete systems and models; due to this, however, they ceased to be strictly axiomatic. In the case of the Wightman approach, this stage was achieved in the constructive field theory which was subsequently combined with Euclidean and Markovian theories. As for the algebraic approach, the central problem was to transform it into a realistic field theory: its formalism did not initially include the notion of quantized field and, hence, it could not describe many important properties connected with this notion, such as antiparticles, charge conjugation, and types of statistics. A method enabling one to introduce all these notions and properties into the theory of local observables was developed (in the cycle of papers by Doplicher, Haag and Roberts in 1969–74) at first for a class of systems not including long-range forces, zero-mass particles and gauge fields. The core of the method is a special technique by means of which, starting with the ground state of a system and the set of its local observables, one constructs physical states with nontrivial values of field-like characteristics, such as charges and internal quantum numbers ('charged states', according to a generalized concept of charge precisely defined in Section 2.3). It proves possible, basing oneself on such states, to define all the physical quantities characterizing a system of quantum fields, to verify that the basic properties of such systems are fulfilled (connection between spin and statistics, existence of antiparticles for each type of particle, etc.) and even, in principle, to reconstruct field operators, although this reconstruction is very intricate and has not yet been worked out thoroughly in the general case. Finally, in the following stage it was discovered after all that the said method could be made much more general and versatile so that it encompassed a very wide range of systems including those which are now thought to play the leading role in interactions of elementary particles: systems with topological charges, phase transitions, massive and massless, abelian and nonabelian gauge fields. For each type of system the method can be modified in such a way that the structure and dynamics of the particular type are taken properly into account.

The development of methods providing a rigorous and yet reasonably detailed description of quantum field systems on the basis of their local observables, can be considered as an important achievement of the algebraic approach. Due to these methods, a new point of view on the structure of a relativistic quantum system is emerging; field-like objects and field-theoretic properties of such systems are obtaining constructive definitions in terms of local observables; and relativistic quantum theory as a whole is acquiring new means of studying many topical models. Thus, a new stage of local quantum theory, usually called algebraic quantum field theory, is taking shape as an integral part of the complex of rigorous methods and formalisms of modern quantum field theory. The base of this theory is constituted by theorems of the axiomatic algebraic approach, methods of reconstructing quantum field models from sets of their local observables and moreover – it should be added here – the formalism of field algebras which describes the algebraic structure of systems defined by a certain family of quantized fields.

The structure of this book follows exactly that of its subject: Chapter 1 expounds the axiomatic approach, Chapter 2 discusses the transition from the formalism of observables to field-theoretic description and Chapter 3 presents the formalism of field algebras. But the material available is too extensive for a one-volume book, especially if we recall that so far there has not been a single monograph on algebraic quantum field theory. For this reason, many results concerning the subjects of Chapters 2 and 3 have been left out. For example, in Chapter 2 we give only an introductory treatment of the methods of reconstructing field-theoretic properties and do not examine concrete models. Chapter 3 does not include algebraic methods of constructive field theory and does not analyze particular kinds of field algebras such as field algebras on manifolds, on null planes, of massless and conformal-invariant systems (hence, the Buchholz scattering theory for massless particles is also omitted though it is a beautiful and efficient application of field algebras). Finally, there are some very recent results, mainly concerning the split property and nuclearity condition, which we have only had time to mention briefly in spite of their value for the structural topics of Chapter 1. Proper discussion of the subjects listed deserves a separate book which will hopefully be written.

The mathematical base of algebraic quantum field theory, as its name suggests, is the theory of algebras. Physical books drawing heavily upon this theory traditionally include special sections with the relevant mathematical elements. In our case, however, space made this impossible, and so the subsidiary information has been restricted to a system of basic definitions (Section 1.1) and brief comments on a few subjects (Arveson spectral theory and like) which seemed to be less known to physicists. Only the theory of unbounded operator algebras is presented more systematically (Section 3.1.1) because it has been absent in mathematical monographs so far. On the other hand, almost all the results are supplied with complete proofs and references to used mathematical facts. The algebraic part of the latter can, as a rule, be found in a few well known works, e.g. Naimark [18], Dixmier [7], Takesaki [28], Bratteli and Robinson [54]. As for physics, the necessary information – apart from general quantum theory, of course – is essentially restricted to Wightman axiomatic theory on which our basic reference book is that by Bogoliubov *et al.* [4]. The bibliography at the end of the book is designed as a bibliography of algebraic quantum field theory comprising most of the original papers devoted to this approach (therefore, materials of Conferences and Schools are usually not included).

This book was planned and discussed several years ago together with V.N. Sushko whose close working cooperation of many years was of great value to me. Subsequently ill health made it impossible for V.N. Sushko to take part in the writing, however. Deeply regretting this, I express my utmost gratitude to him. I also wish to thank Drs. A.V. Voronin, A.S. Holevo, and Yu.M. Zinoviev for helpful discussions of several parts of the book.

S.S. HORUZHY

CHAPTER 1

Axiomatic Formalism

1.1. Introduction. The Algebraic Approach as a Local Quantum Theory

The subject of the present book is traditionally referred to as the algebraic approach to quantum field theory. This name indicates only the mathematical formalism employed, however, and does not, therefore, express the essence of the method completely. The algebraic approach to quantum field theory is not just a special mathematical formalism; it is a particular school of thought in physics. It can be thought of as a trend which aims at describing relativistic quantum systems (i.e. systems of quantized fields, as we understand them today) by means of the theory of algebras and in terms of observables and states treated as fundamental physical objects. In its underlying principles, this approach originates from and is a direct development of, the algebraic formulations of nonrelativistic quantum mechanics proposed in the late 1920s by von Neumann, Dirac, Jordan, and others. Initially, it was stated as an axiomatic formalism. Later on, as already mentioned, it ceased to be purely axiomatic. Today the algebraic approach is no longer an abstract axiomatic construction, but a modern dynamical method applied to description of concrete quantum field systems. However, since the results of the preceding stage were mainly general and rigorous theorems, they are incorporated as a firm basis in new investigations. And observables and states are invariably the basic physical concepts.

Thus, to present the algebraic approach, one has to begin by defining and analyzing these concepts, in order to demonstrate that the observable-state language provides an exact and complete description of relativistic quantum systems. The logical way to demonstrate this is, first of all, to define what is the real experimental situation in relativistic quantum physics; next, to show that the situation does actually lead us to the concepts of observables and states and can be described adequately in terms of these concepts; and finally, to represent the latter by appropriate mathematical objects. The mathematical structure arising from this representation will provide a proper foundation for the algebraic approach. We shall not describe here this deductive procedure (which is more relevant to the interpretation and conceptual analysis of quantum theory), but shall indicate the most important steps; more detailed expositions may be found in [10, 37, 235, 334, 368].

The basic components of the experimental situation are a microscopical system under study (which is, as a rule, a configuration of elementary particles) and a macroscopical instrument interacting with the system which brings about certain detectable changes in it. Conclusions on the behaviour and characteristics of the system are then drawn from the changes detected in the instrument states. As all statements in microscopical physics are statistical (probabilistic), numerous copies of the system investigated (a statistical ensemble) must be available. It is this ensemble that is usually called the *system state*, according to Segal [379]. Clearly, the ensemble and its behaviour depend essentially on the preparation procedure which produced the former. Therefore, to be more general, we can say that any state includes the procedure which prepares the ensemble and the ensemble itself which results from the procedure. Alternatively, one says that a state is an initial condition for the quantal system before the experimental event. For instance, in a typical quantum mechanical experiment on electron diffraction, the physical system is an electron, and the system state is a statistical ensemble determined by an electron source which is a part of the experimental set-up. In the course of an experimental event, the system is subject to various external perturbations affecting the system states in a certain way. A perturbation acting during a finite time interval and producing a system transition from an initial state to a final state is usually called an *operation*. A (finite or infinite) sequence of operations resulting in the measurement of a certain physical quantity (as well as the quantity itself in the usual terms) is called an *observable**. In the example of electron diffraction, the system observables are given by a screen which detects the electrons. If the screen is considered to be a collection of elementary detectors capable of producing only yes or no signals, then the observable value is determined by a set of affirmative and negative responses generated by the system.

It is evident from the above definitions that one can reconstruct the totality of experimental data on a quantum system, if the total set of the system states and the set of all its observables are given, as well as probability distributions of values allowed for each observable in each state. In this respect, the formalism based upon observables and states provides a complete description of the physical system. Another and stronger concept of complete description is present, however, in physical theory; besides a description of experimental data which are external manifestations of the system, one should be aware of its internal dynamics, i.e. the mechanisms which are responsible for interactions between its constituents. Theories which provide us with a complete description of both external and internal properties are called dynamical: they incorporate a dynamical apparatus which generally includes some non-observable quantities in addition to observable ones (for instance, the dynamical apparatus of quantum field theory includes non-observable fermion fields). Therefore, an important question was whether the observable-state formalism could produce a dynamical theory of relativistic quantum phenomena. The answer could only be given in the course of the development of the algebraic approach. At its initial stage as an axiomatic formalism (Chapter 1), this

* In this conventional treatment, the concept of operation is more elementary than that of observable and relies on more general assumptions about measuring devices. For this reason, there were attempts to develop the algebraic formalism based upon the concept of operation (e.g. [235, 368]). However, properties of locality and localizability, fundamental for relativistic quantum theory, are more naturally expressed in terms of observables.

approach was definitely not a dynamical theory; subsequently, however, it gradually began to be transformed into a dynamical theory. This transformation is the main subject of Chapter 2.

The above discussion is relevant to quantum theory in general, both relativistic and nonrelativistic. All the specific features of relativistic systems manifest themselves gradually, of course, yet there is a fundamental property which appears already at the early stage and deeply affects the conceptual (and, consequently, the mathematical) structure of algebraic quantum field theory. This property is locality, which is, in its turn, a combination of two properties: localization and causality. The former means that since any physical experiment takes place in a finite spacetime region, each physical quantity determined directly from the experiment is also associated with some region (localized in it). As to the latter property, one should keep in mind that no signal velocity can exceed the velocity of light, and no processes taking place in spacetime regions separated by spacelike intervals can affect each other (Einstein's causality principle). Consequently, each observable must also be causal, i.e. compatible with any other observable if their localization regions are mutually spacelike. Observables having the described properties of localization and causality are called *local observables*. This concept is the real cornerstone of the algebraic approach in quantum field theory.

Thus the whole set $\mathcal{A}_{\text{obs}}$ of observables for a given relativistic quantum system is the union of the sets $\mathcal{A}_{\text{obs}}(O)$, elements of which are observables belonging to finite spacetime regions* O. Each set of this kind includes all the observables which can be measured in the region O and are compatible with the observables from all the sets $\mathcal{A}_{\text{obs}}(O_1)$, associated with the regions O_1 separated from O by spacelike intervals. In other words, the partition $\mathcal{A}_{\text{obs}} = \cup_O \mathcal{A}_{\text{obs}}(O)$ is the distribution of the set of all observables with respect to their localizations. In general, however, the observables may also have other physical attributes, besides localization, and the observables belonging to the same set $\mathcal{A}_{\text{obs}}(O)$ may be distinguished by these attributes. Consequently, a complete determination of the set of observables for a given quantum system must include, in principle, not only the distribution in localizations, but also distributions in all other characteristics. At this point, the conceptual structure of the algebraic approach is supplemented by another important assumption: no other characteristics of observables, besides their localizations, are necessary for a complete description. It is sufficient to know solely the local structure of the set of observables, namely, the partition $\mathcal{A}_{\text{obs}} = \cup_O \mathcal{A}_{\text{obs}}(O)$. In other words, the whole content of the physical theory is encoded in the mapping $O \to \mathcal{A}_{\text{obs}}(O)$ which determines the distribution of observables in their spacetime localizations. This means that at its initial stage the algebraic approach does not discriminate between observables belonging to the same

* It is convenient to treat as observables also global, integral characteristics of physical systems like total charge, total energy momentum vector, etc., obtained from local quantities (the only ones which are actually measured in experiments) by means of various limiting operations. In this sense, one can talk about 'quasilocal' or 'global' observables, meaning certain limits of sequences consisting of true (that is local) observables.

region O *. Thus, the genuine object of study is not a separate observable, but the set of all observables localized in a given region.

In its essence, this assumption, which ignores all other attributes of an individual observable, except localization, is just another idealization in the course of formulating the algebraic approach. Naturally, one cannot do without certain idealizations and abstractions in constructing any physical theory, and some of them have already been used implicitly (for instance, an obvious abstraction was the concept of a 'physical system' as a part of the universe isolated from its environment); but a good justification is always required. To justify the assumption in question, one should turn again to the typical experimental situation in quantum physics. Experimental data in their original form always refer to the localization properties of microscopical objects. What is registered by various detectors, counters or chambers is always the fact that a particle has appeared at a certain spacetime region. A blackening in a photoemulsion, a track in a bubble chamber, an interference pattern on a screen are all geometrical facts, and no experiment in elementary particle physics can yield facts of a different nature**. As stated in a pioneering work by Haag and Kastler [235], "ultimately all physical processes are analyzed in terms of geometrical relations of (unresolved) phenomena". The experience of the algebraic approach confirms this statement and demonstrates that the set of local observables is indeed sufficient to reconstruct all the physically relevant information on a system. As we show later, the set of all local observables not only determines the results of all possible experiments, but also makes it possible to gain a deeper knowledge of the system dynamics, including interaction mechanisms and quantized fields, even though the latter are not always observable.

Thus we have defined observables and states, the basic physical quantities of the algebraic approach, and found that locality is the main specific principle in the observable-state formalism for relativistic quantum theory. By virtue of this principle, all observables are localized to bounded regions of the Minkowski spacetime and the basic physical object is not a separate observable, but something which might be called a local structure, namely, the set of all observables for a given region O, with a number of fundamental physical conditions (axioms) imposed on this set.

Therefore, fundamentally the algebraic approach was conceived as a general algebraic theory of local structures for relativistic quantum physics or, as it is called more briefly, local quantum theory. Correspondingly, its framework is given by fixing the whole set of local structures or, equivalently, by the mapping $O \to \mathcal{A}_{\text{obs}}(O)$ associating with any

* This does not mean, of course, that the algebraic approach denies any differences between observables of the same region, or restricts itself to the systems where such differences are absent (excluding, in particular, all fields and particles having internal degrees of freedom). The point is that the approach assumes from the start (and proves it later) that it is possible to develop a complete description of a physical system without including the differences in question in the input information.

** It should be stressed that we mean here only the character of original manifestations of microscopic objects. Naturally, these manifestations may provide unambiguous information on those properties of the objects which are not of a spatial nature anymore: for example, on projections of their spins, as is the case in the classical Stern–Gerlach experiment. It is legitimate to consider such properties also as experimentally measurable.

bounded spacetime region the set of all observables which can be measured by means of experiments in this region.

Now we are ready to a mathematical formulation of the theory. To this end we start from the following.

Basic Definitions

A^*. Algebra is a linear space $\mathcal{A}$ over the field of complex numbers $\mathbf{C}$, provided with an associative and distributive multiplication: $A, B \in \mathcal{A} \to AB \in \mathcal{A}$. An algebra $\mathcal{A}$ is called commutative (or abelian) if $AB = BA$. An algebra is called involutive (or $*$-algebra), if an involution is introduced in $\mathcal{A}$, which is an antilinear mapping $A \to A^* \; \forall A \in \mathcal{A}$, with the properties $A^{**} = A$, $(AB)^* = B^*A^* \; \forall B \in \mathcal{A}$. An algebra $\mathcal{A}$ is called a $*$-subalgebra of a $*$-algebra $\mathcal{B}$, if $\mathcal{A} \subset \mathcal{B}$, and $\mathcal{A}$ is a $*$-algebra with respect to the operations defined in $\mathcal{B}$. An element $A \in \mathcal{A}$ is called an identity if $AB = BA = B \; \forall B \in \mathcal{A}$, Hermitian if $A^* = A$, and positive $(A \geq 0)$ if $A = B^*B$ for some $B \in \mathcal{A}$ (cf. however p. 168). The set $\mathcal{A}^+$ of all positive elements is called the positive cone of $\mathcal{A}$. A subspace $\mathcal{J} \subset \mathcal{A}$ is called a left (or right, or two-sided) ideal, if $A \in \mathcal{A}$, $B \in \mathcal{J}$ imply $AB \in \mathcal{J}$ (respectively, $BA \in \mathcal{J}$ or $AB \in \mathcal{J}$ and $BA \in \mathcal{J}$). An ideal is called maximal (or minimal) if it is not embedded in (respectively, does not contain) another ideal of the same kind, except $\mathcal{A}$ itself (respectively, $\{0\}$). Let $\mathcal{A}$ be a $*$-algebra and a Banach space with a norm $\| \cdot \|$. Then $\mathcal{A}$ is called a B^*-algebra, if $\| A^* \| = \| A \| \; \forall A \in \mathcal{A}$; it is called a C^*-algebra, if $\| A^*A \| = \| A \|^2$; and a W^*-algebra if $\mathcal{A} = (\mathcal{A}_*)^*$. (Here $\mathcal{A}_*$ is a Banach space, and $(\mathcal{A}_*)^*$ its topological dual space; $\mathcal{A}_*$ is called a predual of $\mathcal{A}$.) Each W^*-algebra is also a C^*-algebra; each C^*-algebra is also a B^*-algebra. An N^*-algebra $\mathcal{A}$ (where $N = B, C$ or W) is called simple, if it has no closed two-sided ideals, except $\{0\}$ and $\mathcal{A}$. Elements of a dual space $\mathcal{A}^*$ of an N^*-algebra $\mathcal{A}$ are linear continuous functionals on $\mathcal{A}$. A functional $\omega \in \mathcal{A}^*$ is called positive, if $A \geq 0$ implies $\omega(A) \geq 0$ and faithful if $A \geq 0, \omega(A) = 0$ imply $A = 0$. A positive functional is called a state, if its norm equals unity. We shall denote by $\mathcal{A}^{*+}$ the set of all positive functionals and by $\mathcal{A}_1^{*+}$ the set of all states. If $\omega = \lambda\omega_1 + (1 - \lambda)\,\omega_2$ (*), where $\lambda \in (0, 1)$ and $\omega, \omega_1, \omega_2 \in \mathcal{A}_1^{*+}$, one says that ω_1 and ω_2 are majorized by ω (the symbol we use is $\omega_1, \omega_2 \prec \omega$). If equality (*) implies $\omega_1 = \omega_2 = \omega$, the state ω is called pure (or non-decomposable); otherwise, the state is called mixed (a mixture).

W^*. Let $\mathcal{B}(\mathcal{H})$ be the algebra of all bounded operators in the Hilbert space $\mathcal{H}$ with a scalar product $(\cdot, \cdot)$, and $C(\mathcal{H})$ be the algebra of all scalar operators (multiples of the identity operator I). The main topologies in $\mathcal{B}(\mathcal{H})$ we shall use are uniform, weak, strong, ultra-weak and ultra-strong; they are defined, respectively, by the operator norm and the semi-norms

$$|(\xi, A\eta)|, \quad \|A\xi\|, \quad \sum_{i=1}^{\infty} |(\xi_i, A\eta_i)|, \quad \left(\sum_{i=1}^{\infty} \|A\xi_i\|^2\right)^{1/2},$$

where

$$A \in \mathcal{B}(\mathcal{H}), \quad \xi, \eta, \xi_i, \eta_i \in \mathcal{H}, \quad \sum_{i=1}^{\infty} \| \xi_i \|^2 < \infty, \quad \sum_{i=1}^{\infty} \| \eta_i \|^2 < \infty,$$

$\| \cdot \|$ is the norm of a vector or an operator in $\mathcal{H}$. For the set $\mathcal{M} \subset \mathcal{B}(\mathcal{H})$, the set $\mathcal{M}'$ of all elements of $\mathcal{B}(\mathcal{H})$ commuting with all elements of $\mathcal{M}$ is called the commutant of $\mathcal{M}$. A *-algebra $R \subset \mathcal{B}(\mathcal{H})$, satisfying the condition $R = R''$, is called a von Neumann algebra. Each W^*-algebra is *-isomorphic to a von Neumann algebra; we use the both terms as synonyms. A set $Z = R \cap R'$ is called the centre of R; if $Z = \mathbb{C}(\mathcal{H})$, R is called a factor. If $\mathcal{A}$ is a *-subalgebra of $\mathcal{B}(\mathcal{H})$ and $\mathfrak{M}$ is a lineal (linear set) in $\mathcal{H}$, let $P^{\mathcal{A}}_{\mathfrak{M}}$ be a projection onto the subspace $\mathcal{H}^{\mathcal{A}}_{\mathfrak{M}} = \overline{L\{\mathcal{A}\mathfrak{M}\}}$ (the bar stands for the closure in $\mathcal{H}$, $L\{X\}$ means the linear span of a set X). A vector $\Phi \in \mathcal{H}$ is called a cyclic (respectively, separating) vector for R, if $\mathcal{H}^{R}_{\Phi} = \mathcal{H}$ (respectively, if $A\Phi = 0$ for $A \in R$ implies $A = 0$); it is called bicyclic, if it is both cyclic and separating. By definition, a closed operator T is affiliated with R (the notation is $T_\eta R$, if it commutes with all unitary operators from R'. For a projection $P \in R$ (or R'), the set R_P of all restrictions $PAP|_{P\mathcal{H}}$, where $A \in R$, is a von Neumann algebra called the reduction (resp., induction) of R; the standard notation is $R'_P \equiv (R_P)'$. The projection $P^{Z'}_{P\mathcal{H}} \in Z$ is called the central support of a projection $P \in R$. Let E, F, F' be projections in R. By definition, E is equivalent to F (the notation used is $E \sim F \bmod R$), if $E = W^*W$ and $F = WW^*$ for some $W \in R$; E majorizes F ($E \succ F \bmod R$), if $E \sim F' \geq F$ (i.e. $F'\mathcal{H} \supset F\mathcal{H}$) for some F'. A projection E is called finite, if $E \sim F \geq E$ implies $F = E$; otherwise, it is called infinite; it is called purely infinite, if it does not majorize any finite projection. A projection E is called abelian, if the reduction R_E is abelian, and minimal if $R_E = \mathbb{C}(\mathcal{H})$. An algebra R is called finite, infinite, or purely infinite, if the identical projection $I \in R$ has the corresponding property; the algebra is called σ-finite, if any system of its mutually orthogonal projections is at most countable; the algebra is called properly infinite, if all non-zero projections from Z are infinite. An algebra is called continuous, if any of its projections is a sum of two equivalent and mutually orthogonal projections. An algebra is called a discrete or type-I algebra if it is isomorphic to a W^*-algebra with the abelian commutant; it is called a type-II algebra if it is continuous and not purely infinite; it is called a type-I_n (or I_∞) algebra if it belongs to type I and is finite (resp., infinite); it is called type II_1 (or II_∞) algebra if it belongs to type II and is finite (resp., infinite). An algebra is called semi-finite if it is of type I or type II; it is called type-III algebra if it is purely infinite. If Ω is a bicyclic vector of R, the set $\mathcal{U} = R\,\Omega$, where the product operation, $\xi_i = A_i\Omega$ $(A_i \in R, i = 1, 2) \to \xi_1 \xi_2 = A_1 A_2 \Omega$, and the involution $\xi = A\,\Omega \to \xi^\# = A^*\,\Omega$ are introduced, is called an achieved generalized Hilbert algebra (AGHA). The involution $\xi \to \xi^\#$ can be extended to a closed antilinear operator S whose definition domain completed with respect to the scalar product $(\xi_1, \xi_2)_\# = (\xi_1, \xi_2) + (S\,\xi_2, S\,\xi_1)$ is the Hilbert space $\mathcal{D}^\#$. Similarly, the operator $S^* \equiv F$ determines the scalar product $(\cdot, \cdot)_\flat$, the Hilbert space $\mathcal{D}^\flat$, and the involution $\eta = T\,\Omega \to \eta^b = F\eta = T^*\,\Omega$, $T \in R'$, in AGHA $\mathcal{U}' = R'\,\Omega$. Given the polar decomposition $S = J\,\Delta^{1/2}$, the antiunitary operator J is called the modular involution for the pair (R, Ω); its cardinal property is $J\,R\,J = R'$. The positive self-adjoint operator Δ is called the modular operator for the pair (R, Ω); $\Delta^{it} \cdot \Delta^{-it}$ (where $t \in \mathbf{R} = (-\infty, \infty)$) is an ultraweakly continuous automorphism group of R, leaving Ω invariant; this group is called the modular group for the pair $(R\ \Omega)$ [28, 5, Section 2.5]. Ultraweakly continuous functionals on R are called normal; the set of these functionals coincides with the predual $R_* \subset R^*$. The states on R, having the form $\omega_\Phi = (\Phi, \cdot\,\Phi)$ where $\|\Phi\| = 1$ (i.e. $\Phi \in S_1$, S_1 being the unit sphere in $\mathcal{H}$) are called vector states; the projection $P^{R'}_\Phi$ is called the support of ω_Φ, and the set $\mathfrak{M}^R_\Phi = \{\chi \in S_1 \mid (\chi, A\,\chi) = (\Phi, A\,\Phi)\ \forall\, A \in R\}$ is called the $\mathcal{H}$-image of ω_Φ. A function $\omega : R^+ \to \mathbf{R}_+ = [0, \infty)$ is called a weight on R, if $\omega\,(A + B) = \omega\,(A) + \omega\,(B)$, $\omega\,(\alpha\,A) = \alpha\omega\,(A)$, $A, B \in R^+$, $\alpha \in \mathbf{R}_+$ (we put $0 \cdot \infty = 0$); if, in addition, $\omega\,(A^*A) = \omega\,(AA^*)\ \forall\, A \in R$, the function ω is called a trace on R. A weight ω is called normal, if $A = \Sigma_{i=1}^{\infty} A_i$ for $A,\ A_i \in R^+$ implies $\omega\,(A) = \Sigma_{i=1}^{\infty}\,\omega\,(A_i)$; it is called semi-finite, if the linear span of all $A \in R^+$ satisfying the condition $\omega\,(A) < \infty$ is ultraweakly dense in R; it is called faithful, if $\omega\,(A) = 0$ implies $A = 0$. By means of an extension of the GNS construction (see Section C*), the faithful normal semi-finite weight ω on R determines a Hilbert space $\mathcal{H}_\omega$ and a von Neumann algebra R_ω which is isomorphic to R and has a bicyclic vector $\Omega_\omega \in \mathcal{H}_\omega$; the pair $(R_\omega, \Omega_\omega)$ determines the modular operator Δ_ω with the spectrum $\sigma(\Delta_\omega)$. The set $S(R) = \cap_{\omega \in \Gamma}\,\sigma\,(\Delta_\omega)$ (where Γ is the set of all faithful normal semi-finite weights on R) is put into correspondence to every factor R, and one says that R is of type III_1 if $S(R) = \mathbf{R}_+$, type III_λ if $S(R) = \{0\} + \lambda^n$, where $\lambda \in (0, 1)$, and $n = 1, 2, \ldots$; and type III_0 if $S(R) = \{0, 1\}$. Another invariant of a factor R is the set $T\,(R)$ of all $t \in \mathbf{R}$ such that the modular automorphism $\Delta^{it}_\omega \cdot \Delta^{-it}_\omega$ is inner for some faithful normal semi-finite weight ω. S and T are called the Connes invariants, and a factor R is semi-finite if and only if $S(R) = \{1\}$, $T(R) = \mathbf{R}$. A von Neumann algebra R is called hyperfinite, if there is a monotone increasing sequence of von Neumann algebras R_k, $k = 1, 2, \ldots$ which are finite-dimensional (i.e. R_k is *-isomorphic to $\mathcal{B}(\mathcal{H}_k)$ in some finite-dimensional $\mathcal{H}_k$) and $R = \overline{\cup_{k=1}^{\infty} R_k}^{w}$. If R is a factor, then R_k are factors of types I_{n_k}, $n_k < \infty$. Hyperfinite factors of types II_1, II_∞, III_λ and III_1 are unique up to a *-isomorphism [122].

*C**. A mapping $\rho : \mathcal{A} \to \mathcal{B}$ of two C^*-algebras $\mathcal{A}$ and $\mathcal{B}$ is called a *-morphism if it is linear and preserves the involution and the multiplication law; such a mapping is called a *-isomorphism if it is one-to-one, and a *-automorphism if, in addition, $\mathcal{B} = \rho\,(\mathcal{A}) = \mathcal{A}$. A representation of A is a pair $(\pi, \mathcal{H}_\pi)$, where $\mathcal{H}_\pi$ is a Hilbert space and π is a *-morphism of $\mathcal{A}$ into $\mathcal{B}(\mathcal{H}_\pi)$; $\mathcal{H}_\pi$ is called the representation space, the algebra $\pi(\mathcal{A})$ is the representation image, the set $\ker \pi = \{A \in \mathcal{A} \mid \pi\,(A) = 0\}$ is the representation kernel. A representation is called faithful if π is a *-isomorphism, nondegenerate if $\{\chi \in \mathcal{H}_\pi \mid \pi\,(\mathcal{A})\,\chi = \{0\}\} = \{0\}$, cyclic if the algebra $\pi\,(\mathcal{A})$ has a cyclic vector, irreducible if $\pi(\mathcal{A})' = C(\mathcal{H})$, factorial if $\pi(\mathcal{A})''$ is a factor, and normal if $\mathcal{A}$ is a W^*-algebra and π is ultraweakly continuous. A representation $(\pi, \mathcal{H})$ is called the direct sum of representations $(\pi_\alpha, \mathcal{H}_\alpha)$ (the standard notation is $\pi = \oplus_\alpha \pi_\alpha$) if $\mathcal{H} = \oplus_\alpha \mathcal{H}_\alpha, \pi(A)|_{\mathcal{H}_\alpha} = \pi_\alpha(A)\ \forall\, \alpha, \forall\, A \in \mathcal{A}$; the direct sum $\oplus_\alpha R_\alpha$ of von Neumann algebras R_α is defined similarly. The canonical correspondence $\omega \to \pi_\omega$ between states and cyclic representations of a C^*-algebra [8, Section 2.4] is called the GNS (Gelfand–Naimark–Segal) construction. Let $(\pi_i, \mathcal{H}_i)$, $i = 1, 2$, be some representations of $\mathcal{A}$. A bounded operator $T: \mathcal{H}_1 \to \mathcal{H}_2$ such that $T\pi_1(A) = \pi_2(A)\, T\ \forall\, A \in \mathcal{A}$, is called an intertwining operator between π_1 and π_2. The projection $P_i = P^{\mathcal{Z}}_{\mathcal{H}_i}$ in $\mathcal{H}_1 \oplus \mathcal{H}_2$, where $\mathcal{Z}$ is the centre of $\{(\pi_1 \oplus \pi_2)(\mathcal{A})\}''$, is called the central support of π_i, $i = 1, 2$. Representations π_1 and π_2 are called unitarily equivalent (the notation is $\pi_1 \simeq \pi_2$) if there is an isometric intertwining operator between π_1 and π_2; they are quasi-equivalent $(\pi_1 \approx \pi_2)$, if $P_1 = P_2$, disjoint $(\pi_1 \between \pi_2)$ if $P_1 P_2 = 0$, and weakly (or physically) equivalent $(\pi_1 \sim \pi_2)$, if $\ker \pi_1 = \ker \pi_2$. Let $\mathcal{B} \subset \pi(\mathcal{A})'$ be an abelian von Neumann algebra. There exist a Borel space Z and a positive measure ν on Z such that the space $\mathcal{H}$ and the algebra $\mathcal{B}'$ as well as all its von Neumann subalgebras (in particular, $\pi(\mathcal{A})''$) can be represented in the form of integral decompositions

$$\mathcal{H} \simeq \int^{\oplus} \mathcal{H}_\xi \, d\nu(\xi), \quad \mathcal{B}' \simeq \int^{\oplus} \mathcal{B}'_\xi \, d\nu(\xi), \quad \pi(\mathcal{A})'' \simeq \int^{\oplus} \pi(\mathcal{A})''_\xi \, d\nu(\xi), \tag{1.1}$$

where the sets $\{\mathcal{H}_\xi\}$ and $\{\mathcal{B}'_\xi\}$, $\xi \in Z$, are called the field of Hilbert spaces and the field of von Neumann algebras, respectively. The decompositions are called the direct integrals of these fields; their rigorous definition [7, Chapter 2; 29, Section IV.8] needs certain separability conditions. The operators belonging to $\mathcal{B}$ are scalar in each Hilbert space $\mathcal{H}_\xi$; they are called diagonalizable operators for decompositions (1.1), and the operators belonging to $\mathcal{B}'$ are called decomposable. If $\mathcal{B}$ is a maximal abelian subalgebra of $\pi(\mathcal{A})'$, (1.1) is the decomposition of the representation π (the algebra $\pi(\mathcal{A})''$) into irreducible representations (von Neumann algebras): $\pi(\mathcal{A})''_\xi = \mathcal{B}(\mathcal{H}_\xi)$ almost everywhere in the measure ν. If $\mathcal{B}$ is the centre of $\pi(\mathcal{A})''$, (1.1) is the decomposition of the representation π (the algebra $\pi(\mathcal{A})''$) into factorial representations (factors): $\pi(\mathcal{A})''_\xi$ is a factor almost everywhere in the measure ν; in this case (1.1) is called the central decomposition of π or $\pi(\mathcal{A})''$. In the particular case when Z is a discrete set the decompositions in (1.1) are direct sums.

1.2. Axioms of the Algebraic Approach

We have outlined the physical picture of the algebraic approach as a theory of observables and states. The next step is the mathematical identification of the picture: its basic elements must be associated with certain mathematical objects. The identification adopted in the algebraic approach is quite simple and we could have stated it from the outset: observables are represented by Hermitian elements of a C^*-algebra (or a von Neumann algebra), physical states are represented by states on this algebra. There is, however, a natural question: is this identification only one of convenience or can it be deduced from some general properties of quantal processes and quantum measurements? In quantum mechanics such a deduction of the conventional mathematical formalism is the central problem of a special approach known as axiomatic quantum mechanics. Several sets of axioms have been proposed which specified the main features of quantum measurements and enabled one to derive rigorously the traditional Hilbert space formalism (see e.g. [14]). For the algebraic approach to quantum field theory no similar program has been performed and we do not wish to discuss partial results obtained so far [232, 45, 242, 59, 331, 334]. But we note instead that it is not difficult to point out a few simple properties of observables and states, the combination of which brings us very close to the mathematical structure desired, even though it does not provide for a straightforward deduction.

The basic physical elements of the formalism have already been introduced. As we have seen, they include the set $\mathcal{A}$ of all observables for a quantum system, the set Σ of all its states and the probability distributions for all the observables in all the states. Which restrictions are imposed on the mathematical identification of these elements by their physical properties? First of all, the statistical interpretation of quantum theory suggests that the probability distribution plays the role of a link between observables and states: to each pair consisting of an observable and a state it puts into correspondence a real number, the expectation of this observable in this state. In other words, the probability distribution establishes a certain duality between the set of observables and the set of states. Therefore, given a mathematical identification of the set $\mathcal{A}$, the set Σ will have to be the dual object, in the sense of the duality relation set up by the probabilistic interpretation of the theory. Next, if states are treated as statistical ensembles, one has to assume that they are normalizable and that the superposition principle holds in the set of states; namely, all linear normed combinations of states with nonnegative coefficients ('mixtures') must also belong to the set of states. That is to say, the set of all states must be convex, and the set of their linear combinations with nonnegative coefficients must be a convex cone. Finally, a composition must be introduced in the set of all observables (since a successive determination of an observable B and then an observable A is again an observable) and this composition must be noncommutative, in general (since quantum observables may be incompatible). All these conditions taken together restrict substantially the set of allowed identifications and show clearly that the structure of a noncommutative algebra is most natural for the set of observables. The conditions do not suggest, however, which class of algebras would fit, i.e. whether the algebras should be associative, involutive, normed, etc. To proceed further, one has to use more intricate properties of quantum measurements; but even employing such information one cannot eliminate all the ambiguity in the identification. Thus the final choice of the latter takes into account also the technical convenience arguments. The conclusion by R. Haag in his recent review of the problem [232] is as follows: "The mathematical structure of orthodox quantum theory results if one adds to the basic conceptual starting point (statistical description of idealized experiments) a principle of time reversal symmetry and some assumptions of maximal simplicity''.

Thus the physical arguments combined with those of convenience and simplicity produce the standard identification postulate: each observable corresponds to an Hermitian element of a normed *-algebra called the *algebra of observables*, and each physical state corresponds to a state on this algebra; values taken by the states on elements of the algebra are expectation values of the observables. In essence, this postulate is the same as that of the traditional quantum mechanics based by Dirac and von Neumann on the idea that the set of observables has a natural algebraic structure and each separate observable is represented by a self-adjoint (as all measured values of the observables are real numbers) operator in Hilbert space. It was also noted in the early works that taking the subset $\mathcal{A}_H$ of all Hermitian elements of an observable algebra $\mathcal{A}$ and introducing there a product operation, $A, B \in \mathcal{A}_H \rightarrow A \circ B = 1/2\,(AB + BA)$, one can consider directly $\mathcal{A}_H$ as the algebra of observables without appealing to A. The set $\mathcal{A}_H$ with the product operation $A \circ B$ is a commutative, non-involutive and nonassociative algebra, which is called the *Jordan algebra.* Such noncanonical schemes have not been developed very far, however, as they involve a lot of mathematical difficulties which are not balanced by noticeable physical gains. (Recent progress in the theory of Jordan algebras might lead to a revival of this approach, cf. [10, 24].) Another remark concerns the elimination of unbounded operators from the mathematical formalism. Obviously, in choosing observable algebras from the class of normed algebras, one has to admit only bounded operators to represent observables. But it is well known that many important observables in quantum theory correspond to unbounded operators. In particular, the coordinate and momentum observables satisfying the canonical commutation relations can never be represented by bounded operators. This fact does not cause any trouble if all observable operators are self-adjoint; in this case any unbounded observable operator is in a one-to-one correspondence with a family of bounded operators which are its spectral projections. The latter generate the von Neumann algebra of observables, while the original 'unbounded observables' are represented by operators affiliated with this algebra. Most frequently, however, operators of observables are given originally only as symmetric operators; it is true that usually they possess self-adjoint extensions, but these are not unique, as a rule. Thus one can conclude that the decision to avoid unbounded observable operators and take algebras of observables for quantum systems from the class of normed *-algebras does not make the algebraic approach less general, but it does spoil the uniqueness of the description, in the general case.

Let us now consider which complements this scheme needs in the relativistic region. First, in view of the fundamental principle of locality, the identification postulate is now related, more accurately, not to a separate observable, but rather to the set $A_{\text{obs}}(O)$ of all observables belonging to a finite region O in the Minkowski space M. Having neither specified any other properties of the observable besides its localization, nor introduced concepts discriminating between observables from the same set $A_{\text{obs}}(O)$, we cannot carry out the identification for each observable taken separately. Instead of this we associate with every set $A_{\text{obs}}(O)$ an involutive algebra $A(O)$ and only Hermitian elements of the algebra are considered as representatives of physical observables. Next, we must define more precisely the class of regions $O \subset M$. Minkowski space is provided with the natural topology of $\boldsymbol{R}^4$ given by the Euclidean distance. Since it is reasonable to assume that the observables of any open region O determine all the observables of its closure $\bar{O}$, it is sufficient to consider only open regions, or only closed regions. The choice in the algebraic approach is made in favour of open regions (although there are some works based upon closed regions too).

Thus the mathematical formalism of the algebraic approach is determined by the fundamental correspondence, $O \to A(O), \forall\, O \in B(M)$, where $B(M)$ is the set of all bounded open regions in the Minkowski space. The algebra $A(O)$ is called the *algebra of local observables* (or *local algebra)* of the region O. As explained above we do not assume that the set of all observable algebras $\{A(O)\}_{O \in B(M)}$, is specified uniquely for a given physical system (although the original set of observables, $A_{\rm obs}(O)$, was, of course, unique for a given system). It will be seen that different sets of local algebras corresponding to the same system can have different properties; in particular, a special role is played by the sets of maximal algebras which cannot be embedded into other possible algebras of observables.

Moreover, it is not necessary now for the algebras of observables to be concrete operator algebras; they may be abstract C^*-algebras as well. This generalization of the algebraic formalism was proposed in 1964 by Haag and Kastler [235]. Their basic argument was that the state of a physical system cannot be determined with absolute precision from experimental data. Any combination of measurements in a state ω involves some finite errors, so one does not get precise expectation values for observables A_i in this state, $\lambda_i = \omega(A_i)$, but only finite intervals for those values $(\lambda_i - \varepsilon_i, \lambda_i + \varepsilon_i)$. Therefore one cannot decide whether the system is in a state ω or ω', if for any observable A_i its expectation values in these states differ within the error limits ε_i. States which cannot be distinguished experimentally from each other are called *physically equivalent.* It is not difficult to show [235, 368] that the physical equivalence of states ω and ω' corresponds exactly to the weak equivalence (Section 1.1) of the representations determined by these states, so the terms *physical equivalence* and *weak equivalence* are often used as synonyms in the physical literature. Introducing the algebra of observables as an abstract C^*-algebra, we do not fix any particular representation of it, but retain the possibility of various physically equivalent representations. It is noteworthy, however, that the whole physical information contained in the abstract algebraic description can be expressed by a concrete description as well, say, if we choose the universal representation of the abstract observable algebra [8, Section 2.7.6]. The choice between concrete and abstract algebras of observables is rather a matter of convenience which hardly affects the physical content of the formalism. It can make a difference only if one employs a representation of the observable algebra having some special properties (for instance, the existence of a cyclic vacuum vector, see below).

Thus the algebraic approach exists in two versions: the concrete approach (or the *Haag–Araki theory*), where the local algebras are von Neumann algebras, $R(O)$, in a Hilbert space $\mathcal{H}$, and the abstract approach (or the *Haag–Kastler theory*), where the local algebras are abstract C^*-algebras, $\mathfrak{A}(O)$. They are given by the fundamental mapping, $O \to R(O)$ or $O \to \mathfrak{A}(O)$, for all regions $O \in B(M)$, and the mapping is subject to the following set of fundamental physical conditions (axioms of local structures).

AXIOM I. Isotony

$$O_1 \subset O_2 \Rightarrow R(O_1) \subset R(O_2) \text{ (respectively, } \mathfrak{A}(O_1) \subset \mathfrak{A}(O_2)).$$

Since this axiom is adopted, the set $\{R(O)\}_{O \in B(M)}$ or $\{\mathfrak{A}(O)\}_{O \in B(M)}$ gets the structure of a net [5, Section 2.2.3]. As the inductive limit operation is defined for nets of C^*-algebras [23, Section 1.23], approximating an arbitrary unbounded region $\tilde{O}$ with a sequence of monotonously expanding bounded regions, one can associate the algebra of observables with any region $\tilde{O}$, in particular, with the whole space $\tilde{O} = M$. The C^*-inductive limit of the net of the C^*-algebras $\mathfrak{A}(O)$, $B(M) \ni O \subset \tilde{O}$, is the C^*-algebra $\mathfrak{A}(\tilde{O})$. As to the von Neumann algebras $R(O)$, $B(M) \ni O \subset \tilde{O}$, both the C^*-inductive limit $\mathfrak{A}(\tilde{O})$ and the W^*-inductive limit, the von Neumann algebra $R(\tilde{O})$, exist, and $\mathfrak{A}(\tilde{O})'' = R(O)$. In the case $\tilde{O} = M$, the (concrete or abstract) algebra $\mathfrak{A}(M) \equiv \mathfrak{A}$ is called the *quasilocal C^*-algebra*, the algebra $R(M) \equiv R$ is called the *global algebra*. A nonclosed $*$-algebra $\mathcal{A} \equiv \cup_{O \in B(M)} R(O)$ defined as a set-theoretical union of all algebras $R(O)$, is also considered sometimes; it is called the *algebra of all local observables*. The states of the system are most conveniently represented by elements of dual space $\mathfrak{A}^*$ of the quasilocal algebra. However, the set S_{phys} of 'physical' states which are actually realizable for our system, may not coincide with the set $\mathfrak{A}_1^{*+}$ of all states on $\mathfrak{A}$, but may be a part of this set characterized by physical conditions (like symmetry properties, etc.).

AXIOM Ia. Continuous Isotony

Let $\{O_i\}_{i=1}^{\infty}$ be a sequence of regions (open but not necessarily bounded) converging to a region O in the following sense:

$$\liminf O_i \equiv \bigcup_{k=1}^{\infty} \bigcap_{i=1}^{\infty} O_i = \limsup O_i \equiv \bigcap_{k=1}^{\infty} \bigcup_{i=1}^{\infty} O_i = O \ . \tag{2.1}$$

Then for a monotonously increasing sequence $(O_i \subset O_{i+1})$ one has

$$R(O) = \{R(O_i),\ i = 1, 2, \ldots\}'' ,$$

and for a monotonously decreasing sequence $(O_i \supset O_{i+1})$ one has

$$R(O) = \bigcap_{i=1}^{\infty} R(O_i) \ .$$

(In the first case the region O is called the *inductive limit*, in the second case it is called the *projective limit* of the sequence O_i [16, Section 1.1.2]. The condition for the projective limit to be an open region puts, in general, a restriction on the sequence.)

This enhancement of isotony (validity of the usual form of the latter is supposed here) is, obviously, a certain requirement of continuity for the mapping $O \to R(O)$. Precisely, the conditions referring to the representation of O in the form of the inductive and projective limit, mean that this mapping is 'continuous from the inside' and 'from the outside', respectively. In the general formalism of observables and states such requirements can hardly be justified, and condition Ia is considered here as a technical assumption rather than a physical axiom. It was suggested in 1964 by Kraus [287] and used very rarely since then, although its use makes it easier to derive many structural properties, in particular, factoriality and hyperfiniteness of $R(O)$ (see Section 1.3.1).

On the other hand, for nets of field W^*-algebras $F(O)$ generated by Wightman fields $A(f)$ (Section 3.2) continuity properties can often be verified. For example, in the typical case of essentially self-adjoint fields

$$\overline{A(f)} = A(f)^*, F(O) = \left\{ \exp\left(i\overline{A(f)}\right) \middle| \operatorname{supp} f \subset O \right\}$$

the net $F(O)$ is continuous from the inside, due to the fact that the support of a test function is, by definition, a closed set. For the same reason, this continuity holds in all free, generalized free and constructive models, described by canonical fields $\varphi(h)$, $\pi(g)$ and local nets generated by their Weyl forms $U(h) = \exp(i\varphi(h))$, $V(g) = \exp(i\pi(g))$ (see Section 3.3, formula (3.16 a, b)). On the contrary, the net $\{F^w(D)\}$ of double cones algebras defined by means of wedge-like regions (see Section 3.2, formula (2.51)) is, by construction, continuous from the outside. (Note that in both these cases only one out of the two continuity conditions in Ia is generally fulfilled.)

AXIOM II. Weak Additivity

$$\forall_{O \in B(M)} \quad \overline{\bigvee_{a \in M} R(O+a)}^{u} = \mathfrak{A} \left(\text{respectively,} \quad \overline{\bigvee_{a \in M} \mathfrak{A}(O+a)}^{u} = \mathfrak{A}\right),$$

where $O + a$ is the region O translated by the 4-vector a, the closure is in the C^*-norm and all the algebras $\mathfrak{A}(O+a)$ are considered as subalgebras of $\mathfrak{A}$. The symbol $\vee_i \mathfrak{A}_i$ means the *-algebra generated by *-algebras $\mathfrak{A}_i$.

Since the region can be taken however small, this axiom is essentially the assumption that the spacetime continuum is completely homogeneous, and no subquantal phenomena like minimal length, etc., are present.

AXIOM IIa. Strong Additivity

For any family of regions $\{O_i\}_{i \in J}, O_i \in B(M)$, one has $R(\cup_{i \in J} O_i) = \{\cup_{i \in J} R(O_i)\}''$ (respectively, $\mathfrak{A}(\cup_{i \in J} O_i) = \overline{\vee_{i \in J} \mathfrak{A}(O_i)}^{u}$). Evidently, weak additivity (Axiom II) and isotony (Axiom I) are both implied by this condition. But, in spite of being so strong, it is quite realistic for field systems. Indeed, as soon as in the Schwartz space and in other typical testfunction spaces there is the resolution of unity (i.e. for any local finite covering $\{O_i\}_{i\in I}$ of Minkowski space and for any testfunction f there exists a representation $f = \Sigma_{i\in \mathcal{J}} f_i$, where the testfunctions f_i have the supports in O_i), it is natural to expect that the strong additivity holds for nets of algebras generated by the field operators $A(f)$. On the other hand, examples are known where Axioms IIa and II are violated. A typical example are the observable algebras of Fermi systems generated by bilinear combinations of bounded fermion field operators, like $\bar{\psi}(f)\,\psi(g)$: the algebra $R(O_1 \cup O_2)$ includes operators of the form $\bar{\psi}(f)\,\psi(g)$, $\operatorname{supp} f \subset O_1$, $\operatorname{supp} g \subset O_2$, while the algebra $R(O_1) \vee R(O_2)$ does not include them, in general. (For free fermion fields Axiom II is, however, valid [125].) It was also shown [345] that if the net $\{R(O)\}$ does not satisfy Axiom IIa, one can construct some new algebras $R_1(O) \subset R(O)$, which satisfy it.

AXIOM III. Causality (locality)

$$\forall_{O_1, O_2 \in B(M)} \quad O_1 \subset O'_2 \Rightarrow R(O_1) \subset R(O_2)' \text{ (respectively, } [\mathfrak{A}(O_1), \mathfrak{A}(O_2)] = O),$$

where $O' = \operatorname{int}\{x \in M \mid (x-y)^2 < 0, \forall\, y \in O\}$. (Here int stands for the set of all internal points of a set, and the set in the brackets is called the *spacelike complement* of the region O.) Like the set-theoretical complement the spacelike complement of an open region O is closed, and

$$\{x \in M \mid (x-y)^2 < 0, \forall\, y \in O\} = \{x \in M \mid (x-y)^2 \leq 0, \forall\, y \in O\}.$$

Respectively, the spacelike complement of a closed set is open. A notation $O_1 \times O_2$ is often used for the relation $O_1 \subset O'_2$; if $O' = \emptyset$ the region O is called *spatially complete*.

Axiom III can also be stated in another form: for any $O \subset M$, one has $R(O') \subset R(O)'$. One can easily verify that the two formulations are equivalent.

As we have already mentioned, the latter axiom (which is also called the *Einstein causality*) represents the most important specific feature of relativistic systems: the compatibility of any two observables belonging to regions separated by spacelike distance. In general, this is true for classical, as well as quantal relativistic systems, but classical systems have no incompatible observables at all and so the Einstein causality is not meaningful for them. In quantum relativistic theory the situation is quite the opposite: here Axiom III is the single geometrical condition which ensures compatibility of observables. One should bear in mind, however, that its mathematical formulation given above is not unique. Many modifications of it have been proposed; some of them will be considered below in this Section.

AXIOM IV. Primitive causality

For any 'complete spacelike slice',

$$S_\varepsilon^{(n)} = \{x \in M \mid |n_\alpha x^\alpha| < \varepsilon\},$$

given by a timelike normal 4-vector n_α, $(n_0)^2 - n^2 = 1$, $(n_0)^2 - \mathbf{n}^2 > 0$, one has

$$R\,(S_\varepsilon^{(n)}) = R \text{ (respectively } \mathfrak{A}(S_\varepsilon) = \mathfrak{A}).$$

Often this axiom is formulated only for a 'time-slice' S_ε, corresponding to $n_\alpha = (1, 0, 0, 0)$. Taking into account the following Axiom V, the two formulations are equivalent.

This axiom was suggested by the intuitive picture of causal propagation inherent in any physical process. It is natural to suppose that quantities describing a physical system satisfy certain dynamical equations (even if the latter are too complicated to be formulated explicitly). Then their values at any time moment are completely determined by the initial data. As the values of physical quantities in quantum theory (say, field operators) usually are not well defined at fixed time moments and therefore can be defined only in a neighbourhood of the hyperplane $t = t_0$, this brings us right to primitive causality. Thus it reveals another aspect of causality in physics, different from the Einstein causality and inherent to nonrelativistic as well as relativistic systems. As in the case of the locality axiom, however, we cannot be sure that the physical arguments behind this axiom give us the formulation above as the only possibility. Various modifications were suggested here too and we shall now reproduce the most important ones.

AXIOM IVa.

For any family of regions $\{O_i\}_{i\in I}$, such that $(\cup_{i\in I} O_i)' = \emptyset$, one has

$$\Big\{\bigcup_{i\in \mathcal{I}} R\,(O_i)\Big\}'' = R \;\; \Big(\text{respectively, } \overline{\bigvee_{i\in \mathcal{I}} \mathfrak{A}(O_i)}^{\,u} = \mathfrak{A}\Big).$$

Since $S_\varepsilon^{(n)\prime} = \emptyset$, this formulation enhances Axiom IV directly.

AXIOM IVb.

Let us define the following extension of an arbitrary region $O \subset M$:

$$O_c = \{x \in M \mid \lambda_{x,k} \cap O \neq \emptyset, \forall\, k \geq 1\},$$

$$\lambda_{x,k} = \{y \in M \mid (y^0 - x^0)^2 = K^2\,(\mathbf{y} - \mathbf{x})^2\}.$$

(This extension is called the *causal shadow* of O, or also its *causal hull* or *causally dependent* region. Clearly, it is the set of all points x in M such that every timelike or isotropic straight line $\lambda_{x,k}$, which includes x, crosses the region O.) Then

$$R\,(O) = R\,(O_c).$$

Since, obviously, $(S_\varepsilon^{(n)})_c = M$, this condition enhances Axiom IV. It can be interpreted as a 'local form' of the primitive causality: if Axiom IV is considered as the condition $R\,(O) = R\,(O_c)$ for the complete spacelike slice $O = S_\varepsilon^{(n)}$ with $O_c = M$ ('global primitive causality'), then Axiom IVb is the extension of this condition to arbitrarily small regions. This 'local' primitive causality already incorporates certain features of relativistic dynamics: obviously Axiom IVb implies that the physical process is propagating along characteristics lying within light cones (in other words, the propagation is a hyperbolic problem). A simple example of region O_c is shown in Figure 1.

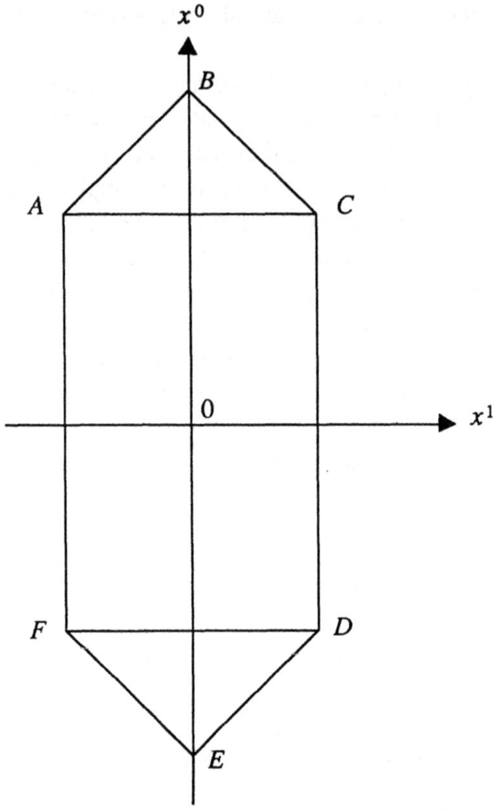

Fig. 1. Causal shadow $Z_C = ABCDEF$ of the cylinder $Z = ACDF = \{x \in M \mid |x^0| < T, |\mathbf{x}| < r\}$.

AXIOM V. Relativistic Covariance

In the concrete approach: in the Hilbert space $\mathcal{H}$ there exists a unitary strongly continuous representation $U(a, \Lambda)$ of the connected component $\mathcal{P}_+^\uparrow$ of the Poincaré group such that for every $O \in B(M)$,

$$U(a, \Lambda)\, R(O)\, U(a, \Lambda)^{-1} = R(\Lambda^{-1}(O - a)) .$$

In the abstract approach: there exists a representation of the Poincaré group $\mathcal{P}_+^\uparrow$ by the automorphisms $\alpha_{(a, \Lambda)}$ of the quasilocal algebra $\mathfrak{A}$, such that

$$\alpha_{(a,\Lambda)}(\mathfrak{A}(O)) = \mathfrak{A}(\Lambda^{-1}(O - a)).$$

This axiom introduces a group structure in the algebraic quantum theory, in addition to the algebraic structure given by the net $R(O)$. In physical terms, the axiom means that for any relativistic quantum system elements of the Poincaré group are symmetry transformations. General formalism for describing physical symmetries in the algebraic approach will be outlined below, in Section 1.3. The Poincaré symmetry plays a special role, however, since it is relevant not to properties of a particular physical system, but to the structure of the spacetime itself. In other words, the Poincaré symmetry stems from the principle of relativity, stating that the results of any experiment must be independent of the choice of an inertial reference frame employed in the experiment. For this reason, it is the sole symmetry which is adopted as a fundamental axiom in any formulation of relativistic quantum theory.

We shall see below (Section 1.3) that the presence of a physical symmetry implies, in the general case, only the existence of a certain pair of mappings in the set of observables and in the set of states, and these mappings may easily not result in the existence of a unitary representation of the corresponding group. Therefore, the formulation of Axiom V in the Haag–Araki theory is, in general, an enhancement of the physical condition. Nevertheless, in a fairly wide class of theories, where the global algebra R is a direct sum of discrete factors, this formulation can be derived from the general definition of symmetry [447].

AXIOM VI. Spectrum Condition

In the concrete approach: the spectrum of the energy-momentum or, which is the same, the support of the spectral measure $E(p)$ of the representation $T(a) \equiv U(a, 1)$ of the translation group $\hat{M}$* , is contained in the closed forward light cone,

$$T(a) = \int \exp(ipa)\, \mathrm{d}E(p), \ \operatorname{supp} E(p) \subset \overline{V^+} \equiv \{p \in M \mid p^2 \geq 0, p_0 \geq 0\} .$$

AXIOM VIa. Strong Spectrum Condition (existence of a 'mass gap'):

$$\operatorname{supp} E(p) \subset \{p = 0\} \cup V^+_{(m)}, \ V^+_{(m)} \equiv \{p \in M \mid (p^2)^{1/2} = m > 0\} .$$

This enhancement of Axiom VI can hold, evidently, only in the absence of massless particles.

In the abstract approach formulating the spectrum condition presents a problem because in the absence of the unitary group $T(a)$ we do not know how to define energy-momentum and mass operators. Various solutions of this problem have been proposed [142, 326, 289], but the underlying idea has always been the same: what should be taken as a spectrum condition in the abstract approach is some sufficient condition of the existence of such quasilocal algebra representation which is translation covariant (i.e. translations in M are implemented by unitary operators in the representation space) and satisfies the spectrum condition in the form of Axiom VI (or VIa). Representations of this type are usually called *positive energy* (or just *positive*) *representations*. Thus the 'abstract spectrum condition' must be a requirement on the algebra $\mathfrak{A}$ with the abelian automorphism group α_a, ensuring the existence of a positive representation of $\mathfrak{A}$. (Note that the existence of translation covariant representations is ensured automatically: any representation corresponding to a translation invariant state on $\mathfrak{A}$ would have this

* The symbol $\hat{M}$ stands for the translation group of M, i.e. M considered as an affine space.

property, while the existence of such states readily follows from the Markov–Kakutani theorem [9, Chapter 5, Section 10]. The desired condition was first found by Doplicher [142]: the quasilocal algebra $\mathfrak{A}$ has at least one positive representation if and only if

$$\mathcal{L} \subsetneq \mathfrak{A} , \tag{2.2}$$

where $\mathcal{L}$ is the translation invariant left ideal in $\mathfrak{A}$, constructed explicitly as follows:

$$\mathcal{L} = L\,\{\mathfrak{A}\ \mathfrak{B}\}\ ,\ \mathfrak{B} \equiv \{B \in \mathfrak{A} \mid B = \int d^4x f(x)\, \alpha_x (A), A \in \mathfrak{A}\ ,$$

$$f \in L^1 (\mathbf{R}^4)\ ,\ \tilde{f}(p) = 0\ \forall\ p \in V^+\}\ . \tag{2.3}$$

(Here the existence of the integral and the condition $B \in \mathfrak{A}$ result from the fact that the automorphisms α_x preserve the norm in $\mathfrak{A}$.) Moreover, if the condition

$$L\,\{\mathfrak{A}\ \mathfrak{B}\ \mathfrak{A}\} = \mathfrak{A} \tag{2.4}$$

is valid, there exists a nontrivial positive representation of $\mathfrak{A}$. Let us prove this criterion.

Necessity. Let a state ω on $\mathfrak{A}$ correspond to a positive representation π with a translation invariant (vacuum) vector Ω and translation operators $T(a)$. Then for any $B \in \mathfrak{B}$ one has

$$\pi (B)\, \Omega = \int d^4x f(x)\, T(x)\, \pi (A)\, T(x)^{-1}\, \Omega = \int d^4x f\ (x)\, T(x)\, \pi (A)\, \Omega$$

$$= E(\tilde{f})\pi (A)\, \Omega = 0\ , \tag{2.5}$$

where E is the spectral measure of the group $T(a)$ (because of the spectral condition for $T(a)$, $\operatorname{supp} E \cap \operatorname{supp} \tilde{f} = \emptyset$). Hence

$$\omega (B^*B) = \| \pi (B)\, \Omega \|^2 = 0\ . \tag{2.6}$$

Evidently, equation (2.6) holds for every $B \in \mathcal{L}$, as well as $B \in \bar{\mathcal{L}}^u$ and by virtue of the Schwartz inequality, $\omega (B) = 0$ for all $B \in \bar{\mathcal{L}}^u$. As $\omega \neq 0$, one gets $\mathcal{L} \neq \mathfrak{A}$.

Sufficiency. If $\mathcal{L} \neq \mathfrak{A}$, one has $\bar{\mathcal{L}}^u \neq \mathfrak{A}$ [18, Section 2.8.3], and the set S_{sp} of all states ω on $\mathfrak{A}$ such that $\omega (B^*B) = 0$, $\forall\ B \in \mathcal{L}$, is not empty. The set S_{sp} is convex and closed in the weak topology of the dual space, i.e. it is weakly *-compact * (for $\omega_\alpha \in S_{sp}$, $\omega_\alpha(A) \to \omega (A)$, $\forall\ A \in \mathfrak{A}$, implies $\omega \in S_{sp}$). The translation invariance of $\bar{\mathcal{L}}^u$ (resulting from that of $\mathcal{L}$) implies that S_{sp} is also translation invariant. Thus the conditions of the Markov–Kakutani theorem are valid for S_{sp} , and there is a translation invariant state $\omega_0 \in S_{sp}$. By definition of S_{sp} , equation (2.6) holds for ω_0 , and as the vacuum vector is cyclic in the representation π_{ω_0} , for the spectral measure E_0 of the unitary translation group in $\mathcal{H}_{\omega_0}$ one gets from (2.5):

* The existence of the identity element $I \in \mathfrak{A}$ is assumed.

$E_0\ (\tilde{f}) = 0, \forall f \in L^1(\boldsymbol{R}^4)$, $\mathrm{supp}\ \tilde{f} \cap \overline{V^+} \neq \emptyset$. This means that π_{ω_0} is a positive representation. Finally, the condition (2.4) is added to exclude the possibility that the representation π_{ω_0} is trivial; this possibility appears if $\mathcal{L}$ is embedded in a proper two-sided ideal of $\mathfrak{A}$, but such an embedding is incompatible with (2.4).

Thus the conditions (2.2) and (2.4) can be accepted as a formulation of Axiom VI in the abstract approach. The properties of positive representations and criteria of the positivity were studied in a number of works (e.g. [81, 338]). In particular, Borchers [81] obtained a result which is in a sense complementary to the Doplicher condition: a necessary and sufficient condition, under which a given translation covariant representation π_0 is positive (the notation is $\pi_0 > 0$). In order to formulate this condition, let us introduce a generalization of the ideal $\mathcal{L}$: for any $a \in V^+$ we define an ideal $\mathcal{L}_a \subset \mathfrak{A}$ by equation (2.3) with functions $f \in L^1(\boldsymbol{R}^4)$, $\tilde{f}\ (p) = 0, \forall\ p \in -a + V^+$. Let us also introduce in the space $\mathcal{H}_0$ of the representation π_0 the subspace

$$\mathcal{H}_a = \{\xi \in \mathcal{H}_0 \mid \pi_0\,(B)\,\xi = 0, \forall\ B \in \mathcal{L}_a\}\ .$$

Then the Borchers criterion is

$$\pi_0 > 0 \Leftrightarrow \overline{L\left\{\ \bigcup_{a \in V^+} \mathcal{H}_a\right\}} = \mathcal{H}.$$

Evidently, the physical meaning of the spectrum condition is the same here as in other approaches to quantum field theory: for any relativistic quantum system the spectra of energy, $\hat{p}_0$, and mass, $m = (\hat{p}^2)^{1/2}$, must be nonnegative. Mathematically this axiom ensures, exactly as it did in the Wightman theory, the possibility of analytic continuation in variables of the configuration space. As a result, the mathematical tools of the algebraic approach include the technique of analytic functions exploited extensively in lots of proofs (as we shall see very soon).

The formulation of the set of axioms is now complete. The most noticeable difference between the Axioms I–VI and those employed in the Wightman theory is the absence of a postulate declaring the existence of a vacuum state. This is because the algebraic apparatus is more general. As we have seen, in the abstract approach the quasilocal algebra $\mathfrak{A}$ always possesses a translation invariant state, corresponding to a representation with a cyclic vacuum vector. In the concrete approach, the Hilbert space may have no vacuum vector, but the Haag–Araki theories of this type are not at all meaningless from the physical point of view; they appear in fact in many cases. In the theory of canonical systems it is well known that a physical model may need to be described by a non-Fock representation of canonical (anti) commutation relations, having no vacuum state (cf., for instance, [10, 350, 290]). In the general formalism we shall meet such theories in Chapter 2 and we shall see that they correspond to charged sectors of systems with superselection rules. In this case, for the vacuumless theory there is an associated theory with a vacuum ('vacuum sector') and the way in which both theories are related to each other is one of the most important dynamical characteristics (Sections 2.2 and 2.3)

All the experience of the algebraic approach shows that the axioms included in the present system (at least, in their basic versions I–VI) are fairly general and well justified physical requirements. They are all compatible, since there are models where they are all satisfied, e.g. the model of free neutral scalar field (Section 3.3). None of them is a consequence of others; all of them are independent physical conditions. This important fact has been established by finding model examples satisfying all the axioms except one. For instance, for the primitive causality axiom such an example is a certain class of generalized free fields (Section 3.3). As to the spectrum condition, there were some suggestions at the early stage that it can be derived from relativistic covariance and locality (see, e.g., [368]) These suggestions were not confirmed. The independence of the spectrum condition was proved by a geometrical model constructed by Doplicher, Regge, and Singer [157]. The model uses a net of local C^*-subalgebras $\mathfrak{A}(O)$ of the C^*-algebra $\mathfrak{A}$ of canonical anticommutation relations* over a special Hilbert space $K = L^2(F, du)$, where F is the set of all timelike and spacelike straight lines in Minkowski space, du is the Lebesque measure on F. The algebra $\mathfrak{A}$ is generated by all even polynomials in fermion fields $\psi(f)$, $f \in K$. When treated as the quasilocal algebra of the abstract approach, it satisfies the axioms of relativistic covariance and locality, but has no positive representations [157].

On the other hand, the algebraic approach, as well as other axiomatic approaches to quantum field theory, has never asserted that its axioms present a unique possible system, the only one providing a physically meaningful description of relativistic quantum objects. It is not excluded that in future some new axioms will be adopted and the existing ones might well be subject to modifications. We have already presented some frequently used modifications of isotony, weak additivity and primitive causality (Axioms Ia, IIa, IVa, IVb). In all those cases, the modifications enhance the basic formulations and are adopted usually to help in proving some structural theorems. Numerous existing modifications of the causality (locality) axiom are of a different kind. The principle of relativistic causality can be put into various mathematical forms and in terms of various objects: observables, states, elementary observables, corresponding to projection operators. And it would be hard to say that any particular formulation includes the total contents of this broad physical principle. Essentially, any condition stating that the sets of objects belonging to regions separated by spacelike intervals are mutually independent in some sense, can be considered as an expression of relativistic causality. Here we give the most important of these 'causal properties'.

(1) *Causal Independence* [235]. Local algebras $R(O_1)$ and $R(O_2)$ $(O_1 \times O_2)$ are called causally independent, if for any pair of local states, $\omega_1 \in R(O_1)^{*+}_1$, $\omega_2 \in R(O_2)^{*+}_1$, there is a global state $\omega_1 \in \mathfrak{A}^{*+}_1$, such that $\omega|_{R(O_k)} = \omega_k$, $k = 1, 2$.

* Algebras of this kind have been described in [10, Chapter 3].

Clearly, this condition by Haag and Kastler is quite justified from the physical point of view, being a direct expression of relativistic causality: if two regions, O_1 and O_2, are separated by spacelike intervals, then any state over O_1 must be compatible with any state over O_2.

(2) *Strict Locality* [287]. If $O_1 \times O_2$, then for any nontrivial projection $P_1 \in R(O_1)$ and any vector $\Phi \in \mathcal{H}$ there is a vector $\Psi \in P_1 \mathcal{H}$ such that $(\Phi, A\,\Phi) = (\Psi, A\,\Psi)$, $\forall A \in R(O_2)$.

(3) *The Schlieder Property* [376]. If $O_1 \times O_2$, then for any pair of nonzero operators $A \in R(O_1)$ and $B \in R(O_2)$ one has $AB \neq 0$.

(4) *Extended Locality* [377]. If $O_1, O_2 \in B(M)$ and $O_1 \times O_2$, then $R(O_1) \cap R(O_2) = C(\mathcal{H})$.

(5) *Duality*. (A direct enhancement of Axiom III): $R(O)' = R(O')$.

All these conditions, as well as Axiom III itself, have similar physical contents, expressing the Einstein causality in various ways. Thus each condition might with some reason play the role of the causality axiom, instead of Axiom III. Nevertheless, the relations between them are far from being evident; in some cases these relations are not clarified even now. A special analysis of each condition is required for the complete description of the causal structure of our theory; it will be carried out in Sections 1.3.3 and 1.3.4.

As to the possibility that some completely new axioms will appear, it seems reasonable to find and postulate a condition which would ensure the existence of a corpuscular interpretation and the scattering theory in the algebraic approach. The actual need for an additional condition to ensure these properties is verified, as usual, by a counter-example: a generalized free field with a rapidly decreasing spectral measure satisfies Axioms I–VI (see Section 3.3), but has no corpuscular interpretation. To provide this interpretation in field theory, irreducible representations $[m, s]$ of the Poincaré group are associated with particles. This concept of a particle is, however, not the most general and may be invalid in the presence of massless particles. In local quantum theory, the search for the corpuscular interpretation criterion used a different, more general concept of a particle as an 'asymptotically stable centre of localization'. This is not a simple way, because the mathematical formulation of such a concept is not self-evident and, probably, nonunique. So it took a long time to develop this approach to the problem. The idea was first suggested as far back as 1965 [238, 374], but it did not begin to yield substantial results until recently. The following formulation of the corpuscular interpretation criterion is now generally accepted.

Nuclearity condition [114, 105]. For every region $O \in B(M)$ and every $\beta > 0$, the mapping

$$\theta_\beta : A \in R(O) \rightarrow \theta_\beta(A) = e^{-\beta \hat{p}_0} A\,\Omega \in \mathcal{H}$$

($\hat{p}_0$ is the generator of time translations and Ω a vacuum vector) is nuclear, which means that there exists a sequence of vectors $\Phi_i \in \mathcal{H}$ and a sequence of functionals $\omega_i \in R(O)^*$ such that

$$\theta_\beta (A) = \sum_{i=1}^{\infty} \omega_i(A)\, \Phi_i \, ; \quad \sum_{i=1}^{\infty} \| \omega_i \| \cdot \| \Phi_i \| < \infty \, . \tag{2.7}$$

Moreover, the following bound holds:

$$\inf \sum_{i=1}^{\infty} \| \omega_i \| \cdot \| \Phi_i \| \le \exp\, [(\beta_0 / \beta)^n] \, , \tag{2.8}$$

where the infimum is to be taken over all ω_i, Φ_i satisfying (2.7), and β_0, n are positive constants depending only on O.

Mathematically, (2.7) means that for any θ_β, the image of its restriction to the set $R(O)^{is}$ of all partial isometries from $R(O)$, i.e. the set

$$\mathcal{N}(O\, ; \beta) = \exp\,(-\beta\, \hat{p}_0)\, R\, (O)^{is}\, \Omega \, ,$$

is a nuclear set in the sense of Grothendieck (see [114, Section 2]) and (2.8) is a bound on the 'nuclearity index' of this set, which is defined as the expression in the left-hand side of (2.8). As for the physical meaning, the nuclearity condition is suggested by the parallel with statistical mechanics and the fact that $\mathcal{N}(O\, ; \beta)$ is an analogue of the set of states contained in a finite volume of phase space: its vectors are strictly localized in O in the sense of Knight–Licht(see p. 148)and lie in a bounded region in the momentum space, due to the energy cut-off. According to heuristic arguments by Buchholz and Wichmann [114], the nuclearity of θ_β is connected with the interpretation of $\exp\,(-\beta\, \hat{p}_0^{(V)})$ ($\hat{p}_0^{(V)}$ is the operator of energy in finite volume V) as a density matrix of a Gibbs state and with the fact that in a majority of physical theories this density matrix is a trace operator. Bound (2.8) is based on the analogy between the nuclearity index and statistical sum $\mathrm{Tr}\, \exp\,(-\beta\, \hat{p}_0^{(V)})$ and is rooted in the condition of the finiteness of pressure in thermodynamical limit at fixed temperature β^{-1}. (The value of n, at least, in free and asymptotically free theories, can be taken equal to the dimension of the localization region.) Pressure being directly related with the density of energy levels of a quantum system, it is (2.8) in the first place, that ensures the corpuscular interpretation: it demands that the number of states of a system in a bounded region should not increase with energy much faster than the corresponding number for a system of identical particles in a finite volume. In other words, (2.8) restricts the number of 'local degrees of freedom'of a system in a way, typical for systems of particles. Continuing this line of heuristic arguments, Buchholz and Wichmann show [114, Section 4] that the nuclearity condition must, indeed, result in the presence of corpuscular states.

Study of the nuclearity condition is now one of the main topics in structural studies in the algebraic approach (see [114, 105, 112, 113]; cf. [182, 233] for the discussion of the early variants). Its fulfillment was proved in the model of free massive [114, Appendix] or massless [112] scalar field. As for general theory, the most important result [105, Theorem 2.1] is that the nuclearity condition implies the following structural property of a local net $\{R(O)\}$ called the 'split property': for any $O_1, O_2 \in B(M)$ with $O_1 \subset \text{int}\, O_2$ (which will be denoted as $O_1 << O_2$), a discrete factor M can be found such that $R(O_1) \subset M \subset R(O_2)$. This property is simpler to check than the nuclearity condition and so it is used for selecting models which do not admit of the corpuscular interpretation. Its violation was proved, for example, for theories with noncompact internal symmetry groups and generalized free fields with the continuous distribution of mass (Section 3.3). On the other hand, its fulfillment (not guaranteeing, in general, the nuclearity condition and corpuscular interpretation) was proved for free fields of various spins, constructive field models $P(\varphi)_2$ and $(Y)_2$ and some other models.

In contrast to the nuclearity condition, the split property was known in the algebraic approach for a long time, though it was called either the 'Borchers property' or 'F-property' (the latter, alluding to 'funnel', is due to the fact that the fulfillment of the split property turns the quasilocal algebra into a 'funnel of type I', i.e. a C^*-algebra with a net of subalgebras of type I, see [234]). It will appear repeatedly in this book, being connected with many structural properties studied below: hyperfiniteness (Section 1.3.1), duality (Section 1.3.3), causal independence (Section 1.3.4). It was also applied to the problem of constructing operators of local charges in local quantum theory [147, 155, 191, 129] and studied intensely from a purely mathematical point of view: the theory of 'split W^*-inclusions' (i.e. the triplets of von Neumann algebras of the form $R_1 \subset M \subset R_2$, M is a discrete factor) developed in [156, 131, 312, 129] turns out to be a valuable new tool in the theory of W^*-algebras.

1.3. Structure of the Local Quantum Theory: Theorems Derived from the Axioms

The set of algebras $\{R(O)\}_{O\in B(M)}$ (or $\{\mathfrak{A}(O)\}_{O\in B(M)}$), satisfying Axioms I–VI, is called the *net of local algebras*. We have to find the main properties of this net. We have got an algebraic structure presented in an axiomatic form and so the problem is to derive rigorous implications from Axioms I–VI. Thus this stage in the development of the algebraic approach is a purely axiomatic stage. Its principal results were achieved during the sixties.

The algebraic structure, determined by the net $R(O)$ and Axioms I–VI, has three kinds of properties: local properties referring directly to the algebras $R(O)$, global properties, which refer to unions of $R(O)$ over all Minkowski space, i.e. the algebras $\mathcal{A}$, $\mathfrak{A}$, R, and finally, group properties, which refer to transformation groups acting in the net (these may be internal symmetry groups, besides the Poincaré group and its subgroups). The local properties are, in their turn, of two types: properties of an individual algebra $R(O)$ and relations between algebras of different regions. It is natural to start the study of all these properties by analyzing the central object of the algebraic approach, the individual local observable algebra.

1.3.1. LOCAL ALGEBRA

What kind of object is the algebra $R(O)$? In the framework of the Haag–Araki (concrete) approach it belongs to the category of W^*-algebras and so its study means analyzing the structural properties of a certain von Neumann algebra. The most basic of these properties include the structure of the sets of cyclic and separating vectors of $R(O)$, the structure of the centre of $R(O)$ (to begin with, one needs to find out whether $R(O)$ is a factor) and the type of $R(O)$. Most of the results obtained here use not only Axioms I–VI, but also a global condition, an assumption about the structure of the global algebra R. Usually, this is the condition that R is irreducible, or is a factor of type I_∞. As we shall see in Chapter 2, this condition means physically that the system has no superselection rules or its description is carried out in a single coherent superselection sector. In many cases a weaker form of the global condition can be used, which takes R from fairly general class of global algebras of systems with superselection rules: $R = \oplus_{\gamma\in\sigma} R_\gamma$, where σ is an arbitrary set, and R_γ are I_∞-type factors.

Besides the global condition, some restrictions on the class of regions $O \in B(M)$ are often adopted in studies of local algebras. The idea is usually to simplify the proofs of structural properties, because an arbitrary region O may be of quite involved geometry and topology, which can make these proofs much more difficult. On the other hand, there are many classes of regions O with various special properties facilitating the analysis of $R(O)$. Clearly, the optimal choice would be a class convenient for deriving structural properties, and large enough (say, enabling us to cover or approximate any region of $B(M)$ and thus completely determining the whole net $\{R(O)\}_{O\in B(M)}$). Both conditions are perfectly met by the class of regions known as 'double cones'. An arbitrary double cone D (Figure 2) is given by a pair of timelike separated points $x, y \in M$; if $x^0 > y^0$, it is the intersection of the open forward light cone V_y^+ with its apex at y and the open backward light cone V_x^- with its apex at x:

$$D = D_{x,y} = V_y^+ \cap V_x^- = \{z \in M \mid x - z \in V^+ , z - y \in V^+\}.$$

The segment joining x and y is called the *axis* of the double cone $D_{x,y}$. The intersection of the double cone with the spacelike hyperplane $S_{x,y}$, normal to its axis, is called the *base* of the double cone. For symmetrical double cones $D_{x,y}$ specified by the condition $\boldsymbol{x} - \boldsymbol{y}$, the base is a three-dimensional ball of radius $1/2\,|x^0 - y^0|$, having its centre at $1/2\,(x + y)$. Double cones form a subclass within a wider class of regions called 'diamonds' and defined by the property $O = O''$ (by our convention the domains O', O'' etc. are all open). As $O \subset O''$ implies $O' = O'''$, the regions O' and O'' are diamonds for any $O \subset M$, and for $O \in B(M)$ the diamond O' is an unbounded region. A bounded diamond $O = O'' \in B(M)$ need not be a double cone; a simple example of such a diamond is shown in Figure 2. Double cones are used extensively in the algebraic approach, serving as a standard basic set of regions, which is sufficient (together with the set of regions D') for analyzing most problems. The subclass of symmetrical double cones $D_{x,y}$, having a simple representation

$$D_{(t,0),(-t,0)} = \{x \in M \mid |x^0| + |\mathbf{x}| < t\},$$

is already sufficient in many cases.

Let us start investigating the algebra $R(O)$ with the analysis of its cyclic and separating vectors, since the existence of such vectors, as will soon be quite clear, is a necessary ingredient in obtaining most of the structural properties. The crucial result in this field is the famous theorem (or lemma) by Reeh and Schlieder. Its most general formulation [76] uses analytic vectors of the positive self-adjoint (by virtue of Axioms V and VI) energy operator $\hat{p}_0$, i.e. the generator of the time translation subgroup.

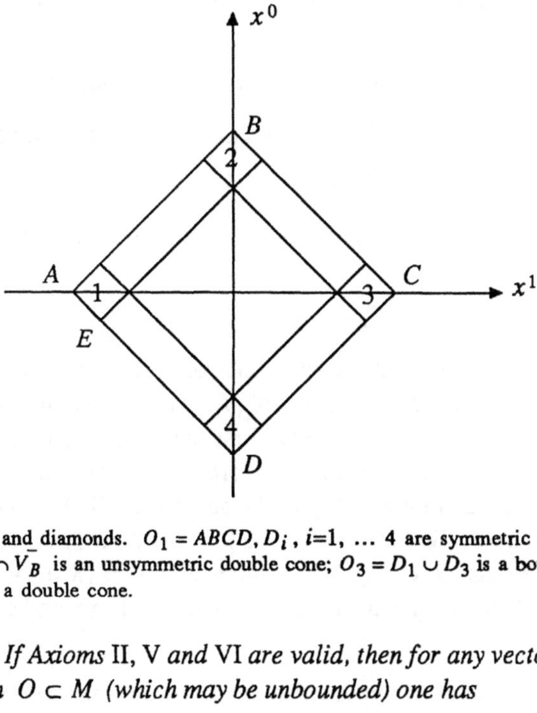

Fig. 2. Double cones and diamonds. $O_1 = ABCD, D_i$, $i=1, \ldots 4$ are symmetric double cones; $O_2 = (D_1 \cup D_2)'' = V_E^+ \cap V_B^-$ is an unsymmetric double cone; $O_3 = D_1 \cup D_3$ is a bounded disconnected diamond which is not a double cone.

THEOREM 1.3.1. *If Axioms* II, V *and* VI *are valid, then for any vector analytic* for* $\hat{p}_0$, *and any region* $O \subset M$ *(which may be unbounded) one has*

$$\mathcal{H}_\Psi^{R(O)} = \mathcal{H}_\Psi^{R}.$$

Moreover, if Axiom III *is valid,* Ψ *is separating for the centre* Z *of the global algebra* R, *and if the region* O *is spatially incomplete (i.e.* $O' \neq \emptyset$*), then* Ψ *is separating for* $R(O)$, *as well as for* $R(O)' \cap R$.

* In some texts (e.g. [433,4]) the Reeh–Schlieder theorem is formulated for so-called vectors with compact energy defined by the following property: if $E(p)$ is the spectral measure of the unitary group $T(a)$ then the vector-valued function $E(p)\Phi : \mathbf{R}^4 \to \mathcal{H}$ has a compact support in $\mathbf{R}^4$. Clearly, such vectors form a subset in the set of all vectors analytic in $\hat{p}_0$: the series $\Sigma_{n=0}^{\infty} \| \hat{p}_0^n \Phi \| z^n/n!$ is convergent for them in any circle of a finite radius in the complex z-plane, and the function $T(a)\Phi$ has an analytic continuation in the 4-vector argument a to a function holomorphic in the whole space $\mathbf{C}^4$.

Proof. By the assumption, the series $\Sigma_{n=0}^{\infty} \| \hat{p}_0^n \Psi \| z^n/n!$ is converging in a circle $|z| < c, c > 0$. Combined with Axiom VI, this implies that the vector-valued function $T(a)\Psi$ can be continued analytically to the tube $\mathcal{T} = \boldsymbol{R}^4 + i\Gamma$, where $\Gamma = \{y \in M \mid y \in -ce^0 + V^+\}$ and e^0 is the unit vector along the x^0-axis. For an arbitrary $O \subset M$ let us choose a subregion $O_1 \subset O$ such that $O_1 + a \subset O$ for each a from some neighbourhood N of the origin. Denoting $A(a) = T(a)AT(-a)$ for $A \in \mathcal{B}(\mathcal{H})$, we shall prove that for any $\Phi \in \mathcal{H}$, the equality

$$(\Phi, A_1(a_1) \dots A_n(a_n)\Psi) = 0, \ \forall A_k \in R(O_1), \ a_k \in N, \tag{3.1}$$

$(k = 1, \dots n)$ implies $(\Phi, A\Psi) = 0, \forall A \in R$. To this end, we note that due to Axiom VI the function

$$(\Phi, T_1(a_1) A_1 T(a_2 - a_1) A_2 \dots T(a_n - a_{n-1}) A_n T(-a_n)\Psi) \tag{3.2}$$

is the boundary value of an analytic function holomorphic in the tube $\mathcal{T}_n = \boldsymbol{R}^{4n} + i\Gamma_n$,

$$\Gamma_n = \{(y_1 \dots y_n) \in \boldsymbol{R}^{4n} \,|\, y_1 \in V^+, y_2 - y_1 \in V^+ \dots y_n - y_{n-1} \in V, y_n \in ce^0 + V^-\}.$$

As soon as c is strictly positive, Γ_n is not empty and includes the open neighbourhood of the origin,

$$y_k = 0, \ 0 < y_1^0 < y_2^0 \dots < y_n^0 < c, \ k = 1, \dots, n.$$

Consequently, by the analytic continuation principle, (3.1) implies that the function in (3.2) vanishes identically for all $a_1, \dots a_n$ of M. Meanwhile, by Axiom II, $\{\cup_{a \in M} R(O_1 + a)\}'' = R$, hence we get the first statement of the theorem.

The arguments presented also show that the orthogonality of the vector Φ to the set $\{R(O) \cup R'\}''$ implies its orthogonality to the set $\{R \cup R'\}''\Psi = Z'\Psi$; i.e. if Ψ is cyclic for Z' it is also cyclic for $\{R(O) \cup R'\}''$ (the third statement). Finally, if $O' \neq \emptyset$, then there exists $O_1 \in B(M)$ for which $R(O)' \supset \{R(O_1) \cup R'\}''$, and so all cyclic vectors of the latter algebra are cyclic for $R(O)'$ (the second statement of the theorem). ☐

It is clear that the proof given proves also the following simple corollary.

COROLLARY. *Let* Ψ *be an analytic vector of* $\hat{p}_0$, *and* $O_k\,(k = 1, 2, \dots)$ *be an arbitrary set of regions in* M. *Then any family of the vectors*

$$\mathcal{L} = \{\Phi_n = A_1 \dots A_n\Psi, \ A_k \in R(O_k), k = 1, 2, \dots n, \ n = 1, 2, \dots\},$$

has the 'Reeh–Schlieder property', $\overline{L\{\mathcal{L}\}} = \mathcal{H}_\Psi^R$.

Evidently, the theorem is completely based on the properties of the translation automorphisms $T(a) \cdot T(-a)$ of the global algebra R, namely, spectrum condition and weak additivity. The intimate connection between the Reeh–Schlieder property (i.e. $\overline{R(O)\Psi} = \overline{R\Psi}$ for analytic vectors of $\hat{p}_0$) and these two postulates manifests itself in a number of facts. For instance, it is not difficult to derive from Theorem 1.3.1 that the spectrum of $T(a)$ (the spectrum of the energy-momentum operator in local quantum theory) cannot be bounded; see Proposition 1.3.44. Moreover, one can show that for representations $T(a)$ satisfying the spectrum condition, weak additivity is not only sufficient, but also necessary for the Reeh–Schlieder property.

PROPOSITION 1.3.2 (the converse Reeh–Schlieder theorem). *Let a von Neumann subalgebra $R_1 \subset R$ have the Reeh–Schlieder property,*

$$\overline{\mathcal{H}_\Psi^{R_1}} = \overline{\mathcal{H}_\Psi^{R}},$$

for any vector analytic in the energy. Then R_1 has the property of weak additivity, $\{\cup_{a\in M} T(a) R_1 T(-a)\}'' = R$.

Proof is based upon another important fact which stems from the spectrum condition only: without loss of generality, one can assume $T(a)$ to be the inner automorphisms of R, i.e. $T(a) \in R, \forall a \in M$ (see Proposition 1.3.43). Let us denote $R_1^T \equiv \{\cup_{a\in M} T(a) R_1 T(a)^{-1}\}''$. As R_1^T is invariant under the action of $T(a)$, spectrum condition and the general properties of strongly continuous automorphism groups of von Neumann algebras [78, lemmas 3 and 4] imply that the group $T(a)$ can be represented in the form $T(a) = T_1(a) T_2(a)$, where $T_1(a) \in R_1^T$ and $T_2(a) \in (R_1^T)' \cap R$. Obviously if E is a spectral projection of the unitary group $T_2(a)$, E commutes both with $T(a)$ and R_1^T. Hence, it follows that for any vector Ψ analytic in the energy, E is also analytic in the energy. Thus

$$\overline{RE\Psi} = \overline{R_1 E\Psi} = \overline{ER_1\Psi} = \overline{ER\Psi}\ . \tag{3.3}$$

As the set $\mathfrak{N}$ of all vectors analytic in the energy is dense in $\mathcal{H}$, and E commutes with $\hat{p}_0$, the set of vectors $\{E\Psi, \Psi \in \mathfrak{N}\}$ is dense in $E\mathcal{H}$. Therefore equation (3.3) means that $E\mathcal{H}$ is invariant under R and $E \in R'$, so that $T_2(a) \in R' \cap R = \mathcal{Z}$, and $T(a) AT(-a) = T_1(a) AT_1(-a)$ for every $A \in R$. Hence one can assume that $T(a) \in R_1^T$. Meanwhile, if $\hat{p}_0$ is affiliated with R_1^T, and P is an arbitrary projection from $(R_1^T)'$, then the vector $P\Psi$ remains analytic in the energy for any vector $\Psi \in \mathcal{H}$ with this property. Hence, using equalities analogous to (3.3), we conclude that $P \in R'$ and, consequently, $(R_1^T)' \subset R'$. As $R_1^T \subset R$, we get $R_1^T = R$. □

The results obtained still do not answer the question about the existence and properties of cyclic vectors for $R(O)$. One more step is needed: we must analyze the set $\mathfrak{N}$ of all vectors analytic in the energy and relate it to the set C_R of all cyclic vectors of R. Clearly, all vectors from $\mathfrak{N} \cap C_R$ will be cyclic for $R(O)$. It is not difficult to establish the necessary properties of the set $\mathfrak{N}$.

PROPOSITION 1.3.3. *If Axioms* V *and* VI *are valid, the set $\mathfrak{N}$ has the following properties:*

(i) *for any projection $P \in R'$, the set $\mathfrak{N} \cap P\mathcal{H}$ is linear and dense in $P\mathcal{H}$,*

(ii) *for any $\Psi \in \mathfrak{N}$ one has $R'\Psi \subset \mathfrak{N}$,*

(iii) *if R is a type-I factor, there are vectors in $\mathfrak{N}$ corresponding to pure states on R,*

(iv) *if a vector Φ is cyclic for R, then the vector* $\exp(-\hat{p}_0)\Phi$ *is analytic in the energy and cyclic for R: $\Phi \in C_R \Rightarrow \exp(-\hat{p}_0)\Phi \in \mathfrak{N} \cap C_R$,*

(v) *if Axioms* I–III *hold and* R *is a factor, then nontrivial local observables cannot preserve the analyticity in the energy:*

$$\Psi \in \mathfrak{N},\ A \in \mathcal{A},\ A \neq \lambda I \Rightarrow A\Psi \notin \mathfrak{N}.$$

Proof. The properties (i), (ii) and (iv) can be directly checked, if the property $T(a) \in \underset{\sim}{R}$ is taken into account. The property (iii) follows from (i) due to the fact that if a vector Φ represents a pure state on a von Neumann algebra R, then all vectors from $\mathcal{H}^{R}_{\Phi}$ also represent pure states. On the other hand, the property (v), which concerns the local structure of the theory, is a consequence of a rather technical Borchers theorem [76, Theorem 6], establishing certain relations between vectors in $\mathfrak{N}$ and some families of operators in $R(O)$ and $\mathcal{Z}$. ❐

Now we are able to state a number of conditions sufficient for the existence of cyclic as well as bicyclic, in view of the second part of the Reeh–Schlieder theorem, vectors of the algebra $R(O)$: (i) if $R = \mathcal{B}(\mathcal{H})$, then all vectors $\Psi \in \mathfrak{N}$ are bicyclic for any $R(O)$ with $O' \neq \emptyset$, (ii) if a translation-invariant vector Ω (a vacuum) is cyclic for R, then Ω is bicyclic for any $R(O)$ with $O' \neq \emptyset$ (translation-invariant vectors being analytic in the energy), (iii) if a vector Φ_0 is cyclic for R, then the vector $\Psi = \exp(-\hat{p}_0)\,\Phi_0$ is bicyclic for any $R(O)$ with $O' \neq \emptyset$. The last condition is the least restrictive one: according to the well known criterion [7, p. 302], if R is infinite (which is the case in our theory, as we shall see soon), this condition is equivalent to the σ-finiteness of R' and so it always holds, for instance, in separable Hilbert spaces. The conclusion drawn we formulate as the following proposition.

PROPOSITION 1.3.4. *Let Axioms* I–III, V *and* VI *hold, and the algebra* $\mathcal{Z}'$ *have a cyclic vector. Then all algebras* $R(O)$ *with* $O' \neq \emptyset$ *have a common separating vector. If the global algebra* R *has a cyclic vector (if* R *is infinite, this is equivalent to the* σ*-finiteness of* R'*), then there exists a common bicyclic vector for all local algebras* $R(O)$ *with* $O' \neq \emptyset$.

Instead of using vectors analytic in the energy, one can get the results about bicyclic vectors of $R(O)$ by starting from general theorems on cyclic and separating vectors of von Neumann algebras. Some enhancement of the preceding results can be obtained along this line.

PROPOSITION 1.3.5 [140]. *Let Axioms* I *and* III *hold, all algebras* $R(O)$ *be infinite and their commutants* $R(O)'$ σ*-finite. Then there exists a dense* G_δ*-set* of vectors, each of which is bicyclic for all* $R(O)$, $O \in B(M)$.

Proof. Take a countable subset of $B(M)$, consisting of all open balls in M having rational values of their radii and rational coordinates of their centres. Denoting this subset $B_0(M)$, let $C \equiv \cap_{O \in B_0(M)} C_{R(O)}$. Since for any $O \in B(M)$ there exists $O_1 \in B_0(M)$ such that $O_1 \subset O$, one has $C = \cap_{O \in B(M)} C_{R(O)}$. According to the

* That is, a set which can be represented as an intersection of a countable family of open sets.

above criterion [7, p. 302], $C_{R(O)} \neq \emptyset$ and the extension of this criterion obtained in [140] gives that C is a dense G_δ-set. Moreover, for any $O \in B(M)$ there exists a domain $O_2 \in B_0(M)$ spacelike separated from it, so that by Axiom III one has $R(O_2) \subset R(O)'$. Hence any vector from C is separating for $R(O)$. □

Thus we see that the presence of cyclic and bicyclic vectors of local observable algebras is a fairly simple and general mathematical fact, which stems from the properties of translation automorphisms as well as general theorems concerning infinite von Neumann algebras (in the latter case, the axioms of the local structure are needed to ensure the infiniteness of $R(O)$). However, this fact is physically quite important: as soon as an arbitrary vector $\Phi \in \mathcal{H}$ can be approximated by vectors of the form $A\Psi$, $A \in R(O)$, performing operations localized in an arbitrarily small region O we are able to prepare (say, from the vacuum state) a state which is as close as we need to any given vector or normal state.

Now we turn to studying the centre $\mathcal{Z}(O)$ of $R(O)$. Its observables, being compatible with all observables of the system considered, behave like observables of a classical system. Hence it can be said that a system with the trivial algebra $\mathcal{Z}(O)$ behaves in the region O like a purely quantal system, having no classical characteristics. We know also that all physical systems exhibit classical behaviour at large (macroscopic) distances. Thus we can expect that local algebras should be factors for 'sufficiently small' regions, whereas the observables commuting with all other observables, if they are present, should belong to unbounded regions or be some limits of local observables. These very loose arguments will be made more precise in Chapter 2, where 'central' observables will be connected with superselection rules (cf. [236]).

Since the centre of a von Neumann algebra is invariant under unitary transformations, by axiom V the algebra $\mathcal{Z}(O)$ is the same for all regions obtainable from one another by Poincaré transformations. In particular, one has:

$$R(O) \text{ is a factor} \Rightarrow R(\Lambda(O+a)) \text{ is a factor for all } (a, \Lambda) \in \mathcal{P}_+^\uparrow .$$

The axiomatic formalism was not too successful in proving factoriality of $R(O)$. Despite the above reasoning, this property was established for bounded regions (double cones) only with the help of continuous isotony (Axiom Ia) not included in the usual list of axioms.

PROPOSITION 1.3.6 (cf. [287]). *Let Axioms* Ia, IIa, III *and* IV *hold, and D be a double cone. Then*

$$\mathcal{Z}(D) \subset \mathcal{Z}, \ \mathcal{Z}(D') \subset \mathcal{Z}. \tag{3.4}$$

In particular, if R is a factor, $R(D)$ and $R(D')$ are also factors.

Proof. The proof is based upon a trick, for the sake of which Axiom Ia was first introduced in [287]. The region $D \cup D'$ is represented as the projective limit of regions $S_k \supset S_k^{(n)}$, as $k \to \infty$, $\varepsilon_k \to 0$, where $S_{\varepsilon_k}^{(n)}$ is a complete spacelike slice determined by a four-vector n, normal to the base of the double cone,

$$D \cup D' = \text{int} \bigcap_{k=1}^{\infty} S_k \supset \text{int} \bigcap_{k=1}^{\infty} S_k^{(n)} .$$

Hence, by virtue of Axioms Ia and IV,

$$R\,(D \cup D') = \bigcap_{k=1}^{\infty} R\,(S_k) = R\ ,$$

and taking into account Axiom IIa,

$$R' = R\,(D)' \cap R\,(D')' ,$$

whence Axiom III gives immediately (3.4) ❒

It is evident from the proof that we have actually arrived at a stronger result: $R\,(O)$ is a factor (if R is a factor) for any region O, for which the closure of the union $O \cup O'$ contains the complete spacelike slice $S_\varepsilon^{(n)}$, or which is a projective limit of regions containing such a slice. Clearly, such regions are not only the double cones. Moreover, the scheme of the proof implies that $R\,(O)$ is a factor for an even larger class of regions. Indeed, if O_1 is a region such that $R\,(O_1) = R\,(O)$, and the closure of $O_1 \cup O'$ contains a slice $S_\varepsilon^{(n)}$, then by applying the proof to $O_1 \cup O'$ one gets again $\mathcal{Z}(O) \subset \mathcal{Z}$, $\mathcal{Z}(O') \subset \mathcal{Z}$. Relations of the form $R\,(O_1) = R\,(O)$ for $O_1 \neq O$ are quite common in local quantum theory; they are the so-called 'dependence relations', to be considered below. Each relation of this type produces, by our proof, a new family of regions O, for which $R\,(O)$ are factors. One well known relation of this kind is the Borchers theorem on the spacelike hull (Proposition 1.3.14), and the family of 'factorial' regions provided by this theorem (which have the special name of 'normal' regions), will be analyzed separately, see Corollary 1 to Proposition 1.3.14. Significantly, they include an important class of regions, the cylinders. Finally, the global condition put on the algebra R can be relaxed considerably: analyzing superselection rules in Section 2.2 we shall see that the algebras $R\,(D)$ remain factors in the large class of theories with an arbitrary set of physically equivalent coherent superselection sectors. Whereas algebras $R\,(D')$ are factors only in a much narrower class of theories having strongly locally quasi-equivalent coherent sectors. Without using Axiom Ia, factoriality of local algebras was established in the axiomatic approach only for a few special classes of unbounded regions.

DEFINITION 1.3.1 [164]. An open region $O \subset M$ is called *lightlike monotone*, if $O' \neq \emptyset$ and one of two following conditions holds,

$$(A_+)\ O + a \subset O + b\ , \quad \text{for } \lambda_a > \lambda_b\ ,$$

$$(A_-)\ O + a \supset O + b\ , \quad \text{for } \lambda_a > \lambda_b\ ,$$

where a and b are arbitrary 4-vectors along a fixed lightlike direction n : $a^\mu = \lambda_a n^\mu$, $b = \lambda_b n^\mu$, $\lambda_{a,b} \in \mathbf{R}$, $(n^\mu)^2 = 0$.

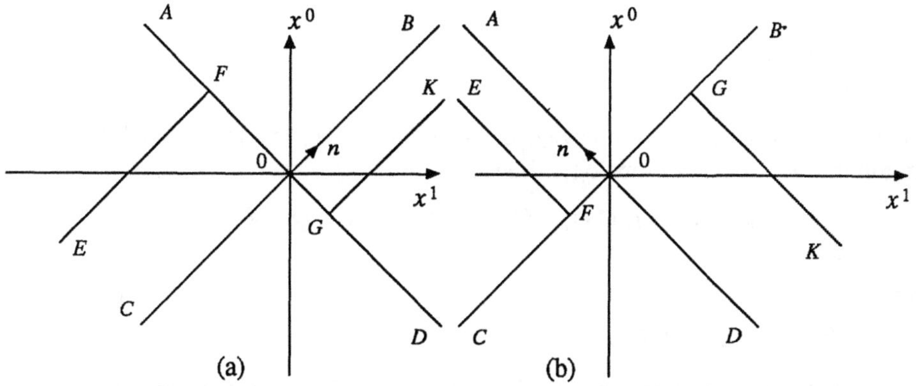

Fig. 3. Examples of lightlike monotone regions. (a) $n = 1/\sqrt{2}\,(1, 1, 0, 0)$; the regions $AOC = W_L$ and $EFOC$ satisfy condition (A_-), $BOD = W_R$ and $BOGK$ satisfy condition (A_+). (b) $n = 1/\sqrt{2}\,(1, -1, 0, 0)$ the regions W_L and $AOFE$ satisfy condition (A_+), W_R and $KFOD$ satisfy condition (A_-).

It is easily seen that the regions satisfying condition (A_+) (resp. (A_-)) must be unbounded in the n (resp. $-n$) direction. Simple examples of such regions are so-called *wedges*,

$$W_R = \{x \in M \mid x^1 > |x^0|\}, \quad W_L = \{x \in M \mid x^1 < -|x^0|\}. \tag{3.5}$$

The wedge regions and those obtained from them by the Poincaré transformations have a number of good properties, and we shall frequently use them (see esp. Section 3.2).

PROPOSITION 1.3.7 [164]. *Let Axioms* I–III, V *and* VI *hold and suppose that there is a unique translation invariant vector* Ω, *which is cyclic for* $\mathcal{Z}$. *Then* $R(O)$ *is a factor for any lightlike monotone region* O.

Proof. Let O satisfy, for instance, condition (A_+). Then for any $a = \lambda_a n$ one has $R(O+a) \subset R(O)$, if $\lambda_a > 0$, and $R(O) \subset R(O+a)$, if $\lambda_a < 0$ (the inclusions are opposite, if condition (A_-) holds). If Q is an arbitrary projection from $\mathcal{Z}(O)$, then both inclusions lead to $Q(a)\,Q = QQ(a)$, and

$$(\Omega, QT(a)\,Q\,\Omega) = (\Omega, QT(-a)\,Q\,\Omega), \ \forall\, a = \lambda_a n. \tag{3.6}$$

Due to the spectrum condition, the generator of the one-parameter translation subgroup $T(\lambda_a n)$ is a positive operator, so the l.h.s. of (3.6) can be continued analytically in λ_a into the upper half-plane as a holomorphic function bounded at infinity, while the r.h.s. can be continued into the lower half-plane. As the boundary values of these functions coincide on the real axis, we get a single analytic function, bounded in the whole complex plane, which must be a constant, according to the Liouville theorem. At $a = 0$ this function equals $\| Q\,\Omega \|^2$. We shall find its asymptotics at $|a| \to \infty$ using the asymptotic behaviour of the translation operators. As shown in Section 1.3.5, if the theory has a unique translation invariant vector Ω, then for any direction e in

Minkowski space one has

$$\underset{|\lambda| \to \infty}{w\text{-lim}} \; T(\lambda e) = P_0 , \tag{3.7}$$

where P_0 is the projection on the ray corresponding to Ω, and w-lim stands for the limit in the weak operator topology. Hence,

$$\lim_{|\lambda_a| \to \infty} (\Omega, QT(\lambda_a n)\, Q\, \Omega) = (\Omega, QP_0 Q\, \Omega) = (\Omega, Q\, \Omega)^2 ,$$

so that

$$\| Q\, \Omega \|^2 = (\Omega, Q\, \Omega)^2 .$$

This equality can be true only if $Q = 0$, or $Q\,\Omega = \Omega$. If Ω is a cyclic vector for Z', then, due to the Reeh–Schlieder theorem, Ω is separating for $R(O)$, and the equality $(Q - I)\,\Omega = 0$ for $Q \in R(O)$ implies $Q = I$. ❐

The same method, based upon the asymptotics (3.7) of the translation operators and the analytic properties owing to the spectrum condition, may be used to prove the factor property for a somewhat different class of unbounded regions, namely, infinite spacelike strips.

PROPOSITION 1.3.8. [387]. *Let all the assumptions of Proposition* 1.3.7 *hold, and the region* $O \subset M$ *with* $O' \neq \emptyset$ *be such that* $O = O + \lambda e, \forall \lambda \in \mathbf{R}$, *for some spacelike vector* e. *Then* $R(O)$ *is a factor.*

Proof. The translations $T(\lambda e)$ form a one-parameter group of automorphisms of $R(O)$, and $\mathcal{B} = R(O) \cap \{T(\lambda e)\}'_{\lambda \in \mathbf{R}}$ is the algebra of fixed points of this automorphism group. Due to (3.7), $P_0 \in \{T(\lambda e)\}''_{\lambda \in \mathbf{R}}$, and for an arbitrary projection $E \in \mathcal{B}$ one has

$$EP_0 = P_0 E \neq 0 ,$$

since Ω is separating for $R(O)$ and, consequently, $E\,\Omega \neq 0$. P_0 being a one-dimensional projection, $E\,\Omega = \Omega$, and $(E - I)\,\Omega = 0$ implies $E = I$. Thus $\mathcal{B} = C(\mathcal{H})$.

Now we get the desired result, proving that $Z(O) \subset \mathcal{B}$. Let us take an arbitrary Hermitian element $A = A^* \in Z(O)$ and a region $O_1 \subset O$ such that for a neighbourhood N of the origin in M one has $O_1 + N \subset O$. Let us consider two functions,

$$F(a) = (T(a)\, A\, \Omega, B\, \Omega), \quad G(a) = (T(-a)\, B^*\, \Omega, A\, \Omega), \quad B \in R(O_1) .$$

Because of the spectrum condition, F is a boundary value of an analytic function, holomorphic in the future tube, $\mathcal{T}^+ = \mathbf{R}^4 + iV^+$. Similarly, G is holomorphic in the past tube, $\mathcal{T}^- = \mathbf{R}^4 + iV^-$. For $a \in N$, $B(a)$ commutes with A and so one has $F(a) = G(a)$; whence, by the edge of the wedge theorem, one has $F(a) = G(a)$ for all $a \in M$. But $F(a)$ is the Fourier transform of a complex measure with the support in V^+, and $G(a)$ is the Fourier transform of a complex measure with the support in V^-.

Due to this, $F(a) = G(a)$ implies $F = G =$ constant. Hence

$$(T(a) A \Omega, B \Omega) = (A \Omega, B \Omega), \forall B \in R(O_1).$$

By the Reeh–Schlieder theorem, Ω is a cyclic vector for $R(O_1)$, so $A(a)\Omega = A\Omega$, $\forall a \in M$. In particular, for $a = \lambda e$ we have $A(\lambda e) \in R(O)$, and, Ω being a separating vector for $R(O)$, we get $A(\lambda e) = A$, $\forall \lambda \in \boldsymbol{R}$; thus $A \in B$. ❐

We shall show in Section 1.3.5 that the method we have used here is universal: the automorphism group of a W^*-algebra satisfying the spectrum condition, always leaves the centre of the algebra point-wise invariant. We shall see as well (Proposition 1.3.12) that for the regions in Propositions 1.3.7 and 1.3.8, the type of the factor $R(O)$ can also be found.

It is exactly the type of the local algebra that will be our next subject. The starting point here is the fact that all $R(O)$ as well as their commutants $R(O)'$ for all regions $O \subset M$ are infinite von Neumann algebras. It is quite easy to prove this, but the proof needs bicyclic vectors of $R(O)$, whose existence requires the global condition formulated in its weakest form in Proposition 1.3.4.

PROPOSITION 1.3.9. *Let Axiom* I–III, V, VI *hold, and the global algebra R have a cyclic vector. Then all the algebras $R(O)$ are infinite. For all regions O with $O' \neq \emptyset$ the commutants $R(O)'$ are also infinite.*

Proof. The main part of the proof is the simple, but important

LEMMA 1.3.10. (Wightman's inequality). *Let Axioms* I–III *and* V *hold, and the regions $O_1, O \subset M$ be such that $O_1 \ll O$. Then $R(O_1) \neq R(O)$.*

Proof of Lemma. Suppose that $R(O_1) = R(O)$. The property $O_1 \ll O$ is equivalent to the existence of a neighbourhood N of the origin in M such that

$$R(O_1 + a) \subset R(O) = R(O_1), \ \forall a \in N,$$

whence

$$R(O_1 + a + a') \subset R(O_1), \ \forall a, \ a' \in N.$$

Since any vector $b \in M$ can be represented as a sum of a finite number of vectors belonging to N, by virtue of Axioms II and V we have

$$R(O_1) = R, \ R(O_1 + a) = R, \ \forall a \in M.$$

Taking now $a \in M$ such that $O_1 + a \subset O'_1$, we get from these equalities and Axiom III,

$$R = R(O_1 + a) \subset R(O_1)' = R',$$

so R is an abelian algebra. This is impossible, however, since any nontrivial quantum theory must include incompatible observables. ❐

Now we turn back to the proof of Proposition 1.3.9. Due to the Reeh–Schlieder theorem, under the assumptions taken, the algebras $R(O_1)$ and $R(O)$ have a common bicyclic vector. Thus one can apply the following theorem by Kadison [268] (which he proved specially to demonstrate that $R(O)$ is infinite): if R_1, R are von Neumann algebras having a common bicyclic vector, and $R_1 \subsetneq R$, then R_1 and R are infinite. This proves that $R(O)$ is infinite; the infiniteness of $R(O)'$ for $O' \neq \emptyset$ is a consequence of the locality axiom.

COROLLARY. *Let Axioms* Ia, IIa, III–VI *hold, and* R *be a factor having a cyclic vector. Then local algebras* $R(O)$, $O \in B(M)$, *can neither be abelian, nor include compact operators.*

Proof. The first statement is obvious. To verify the second statement, note that for any region $O \in B(M)$ there is a double cone $D \supset O$, and under the present assumptions $R(D)$ is a factor. If $R(O)$ includes a compact operator, a finite-dimensional projection P is present in $R(O)$, and, consequently, in $R(D)$. The induction of the factor $R(D)'$ by that projection is an isomorphism between $R(D)'$ and the finite factor $R(D)'_P$, hence $R(D)'$ is also a finite factor, which is impossible in view of Proposition 1.3.9. (Clearly, the absence of compact operators in $R(O)$ can also be proved by means of embedding O into a wedge region of the form $\Lambda(W_R + a)$, or a strip $S = S + \lambda e$, $e^2 < 0$. In this case, instead of Axioms Ia, IIa, the proof will use weaker Axioms I, II, and the assumption of the existence of a unique vacuum.) ❐

Proceeding along the same line, one can point out explicitly large classes of infinite projections in the algebra $R(O)$. Let us introduce another type of 'strong inclusion' for regions in M: we shall use the notation $O_1 < O$, if $O_1 \subset O$, and there exists an open region O_2 such that $O'_1 \cap O \supset O_2 \neq \emptyset$. Then one can prove the following proposition.

PROPOSITION 1.3.11 [79]. *Let Axioms* I–III, V *and* VI *hold and regions* O_1, O *be such that* $O_1 < O$ or $O_1 << O$. *Then any projection* $E \in R(O_1)$ *is infinite in the algebra* $R(O)$. *Moreover, if* $O_1 < O$, *one has* $E \sim F \mod R(O)$, *where* F *is the central support of* E *in* $R(O)$; *if* $O_1 << O$, *one has, in addition,* $F \in R(O) \cap R'$.

Proof.

(1) $O_1 < O$. Evidently, due to $F \geq E$, the infiniteness of E and F would result from $E \sim F$. To prove the latter let us suppose first that R has a cyclic vector. Then, by Proposition 1.3.4, there exists $\Psi \in \mathcal{H}$ bicyclic for all local algebras of spatially incomplete regions. Let us take the projection $F' = P^{R(O)}_{E\Psi}$ and show that $F' = F$. By construction, $F' \in R(O)'$. Taking $O_2 \subset O \cap O'_1$, one gets due to the cyclicity of Ψ for $R(O_2)$,

$$F'\mathcal{H} = \overline{R(O)E\Psi} = \overline{R(O)R(O_2)E\Psi} = \overline{R(O)ER(O_2)\Psi} = \overline{R(O)E\mathcal{H}} ,$$

i.e. F' is majorizing E. In addition,

$$R(O)'F'\mathcal{H} = R(O)' \, \overline{R(O)E\mathcal{H}} = \overline{R(O)'R(O)E\mathcal{H}} = \overline{R(O)ER(O)'\mathcal{H}}$$

$$= \overline{R(O)E\mathcal{H}} = F'\mathcal{H}, \tag{3.8}$$

i.e. $F'\mathcal{H}$ is an invariant subspace of $R(O)'$ and so $F' \in R(O)$. Thus $F' \in \mathcal{Z}(O)$, and as the third and sixth terms in (3.8) are equal, one has $F' = F$. Using this fact and the definition of F' one concludes that $P^{R(O)}_{E\Psi} = P^{R(O)}_{F\Psi}$, so that surely $P^{R(O)}_{E\Psi} \sim P^{R(O)}_{F\Psi} \bmod R(O)'$, which is equivalent to $P^{R(O)'}_{E\Psi} \sim P^{R(O)'}_{F\Psi} \bmod R(O)$, as proven in [7, p. 221]. Since $E, F \in R(O)$, and Ψ is cyclic for $R(O)'$, the left projection in the last formula is equal to E, and the right one to F. Thus we have obtained the desired $E \sim F \bmod R(O)$.

Now we reject the assumption that R has a cyclic vector, and consider the decomposition of $\mathcal{H}$ into a direct sum of cyclic subspaces of R. Let $\{G_\alpha\}_{\alpha\in\Delta}$ be the corresponding resolution of unity in R'. Applying the above result to the cyclic subspace $\mathcal{H}_\alpha = G_\alpha\mathcal{H}$, one gets $EG_\alpha \sim FG_\alpha \bmod R(O)\,G_\alpha$. Let F_α be the central support of G_α in $R(O)$; then one has also $F_\alpha E \sim F_\alpha F \bmod R(O)\,F$, due to [7, p. 18]. Taking the union over α, and in view of the fact that $\Sigma_{\alpha\in\Delta}\, G_\alpha = I$ implies $\cup_{\alpha\in\Delta} F_\alpha\mathcal{H} = \mathcal{H}$, one gets again $E \sim F \bmod R(O)$.

(2) $O_1 << O$. In this case the proof is rather cumbersome and it uses a number of auxiliary results. Therefore we do not present it here, referring the reader to Borchers' original work [79]. ❐

The results presented leave open for the algebra $R(O)$ the possibility of any infinite type, I_∞, II_∞, or III, or any combination of these types, if it is not a factor. There is a good collection of general arguments as well as model results (Section 3.3), which suggest that $R(O)$ for $O \in B(M)$ should have type III. Proposition 1.3.11 also supports this expectation because, for instance, many regions O can be covered by their subregions O_i such that $O_i < O$ or $O_i << O$, and in these cases $R(O)$ is generated by its subalgebras, in which all projections are infinite in $R(O)$. No example is known of a local algebra with the type different from type III, either in the axiomatic theory, or in concrete relativistic models.* (Under certain conditions, free fermion fields generate type-I algebras $R(O)$, but these fields do not satisfy the locality axiom [135].) Nevertheless, only in a few special cases has it been proved rigorously within the axiomatic formalism that $R(O)$ is indeed of type III. Such cases include, first of all, the above-mentioned classes of lightlike monotone regions (Proposition 1.3.7) and of infinite spacelike strips (Proposition 1.3.8), as well as a slight generalization of the latter class, obtained when we require that the region must be infinite along a ray in one direction only.

PROPOSITION 1.3.12 [164, 165, 387]. *Let Axioms* I–III, V *and* VI *hold, and suppose that there is a unique translation invariant vector* Ω. *Let* O *be a lightlike monotone region or an infinite spacelike strip. Then* $R(O)$ *is a type-*III *factor and in the former*

* On the contrary, in nonrelativistic theory, in many typical cases there is an isomorphism between the local algebra (now associated with a region in three-dimensional space or a finite sublattice of a lattice system) and the global algebra, and both algebras are of type I. Cf., e.g. [22, 252].

case the type is III_1. *Moreover, for the region of the form* $O = \cup_{\lambda \geq 0} (O_1 + \lambda e)$, *where* $O_1 \in B(M)$, *e is an arbitrary spacelike vector,* $R(O)$ *is a type-*III *algebra whose central decomposition includes only factors of type* III_1.

Proof. Elements of the Tomita–Takesaki theory are needed in the first case. As Ω is bicyclic for $R(O)$, it determines uniquely the achieved generalized Hilbert algebra $R(O)\Omega$ and the group Γ_t, $t \in \boldsymbol{R}$, of modular automorphisms. According to the Pedersen–Takesaki theorem [346], the algebra of fixed points of the transformations Γ_t, $R(O)^\Gamma \equiv R(O) \cap \{\Gamma_t, t \in \boldsymbol{R}\}'$, is given by

$$R(O)^\Gamma = \{A \in R(O) \mid (\Omega, AB\,\Omega) = (\Omega, BA\,\Omega), \forall\, B \in R(O)\}.$$

Hence we get, using the translation invariance of Ω,

$$A \in R(O)^\Gamma \Leftrightarrow A(a) \in R(O + a)^\Gamma, \forall\, a = \lambda_a n,$$

and for all $A \in R(O)^\Gamma$, it follows that

$$(\Omega, AT(-a)\,B\,\Omega) = (\Omega, BT(a)\,A\,\Omega), \quad \forall\, B \in R(O + a).$$

This formula being completely similar to (3.6), we can repeat the same arguments: using the analytic continuation in λ_a and the Liouville theorem, we conclude that $(\Omega, BT(a)\,A\,\Omega)$ does not depend on a; then the equality between the values of this function at $a = 0$ and $a = \lambda_a n$, $|\lambda_a| \to \infty$ in view of (3.7) implies

$$(\Omega, AB\,\Omega) = (\Omega, A\,\Omega)\,(\Omega, B\,\Omega).$$

Ω being cyclic for $R(O + a)$, this implies $A^*\Omega = (\Omega, A\,\Omega)\,\Omega$, and since Ω is also separating for $R(O)$, one concludes that $A^* = (\Omega, A\,\Omega)\,I$, and $R(O)^\Gamma = C(\mathcal{H})$.

Now suppose $R(O)$ is a semi-finite factor. Then all the automorphisms $\Gamma_t \cdot \Gamma_t^{-1}$ are inner, and the Connes invariant T for $R(O)$ is equal to $\boldsymbol{R}$. By the Connes theorem [122, Theorem 1.3.2], for all $t \in T$, one has $\Gamma_t \cdot \Gamma_t^{-1} = u_t \cdot u_t^{-1}$, where u_t are operators form the centre of $R(O)^\Gamma$. As proved above, all such operators are trivial, hence $R(O) = R(O)^\Gamma = C(\mathcal{H})$; this is a contradiction making us to conclude that $R(O)$ is a type-III factor. Further, applying Theorems 4.2.6 and 5.2.1 from [122] one can verify that $R(O)$ cannot have type III_λ, $\lambda \in (0, 1)$, or type III_0. Thus $R(O)$ is a type-III_1 factor.

For the other class of regions, $O = O + \lambda e$, $e^2 < 0$, $\lambda \in \boldsymbol{R}$, type III for the algebra $R(O)$ follows as a simple corollary from the theorem by Størmer [387, Theorem 2.2]: let N be a semi-finite von Neumann algebra, G its automorphism group and N^G the algebra of fixed points of N under the group G. Moreover, let $N^G = Z$, and N be G-finite in the sense of Kovacs and Szücs [286] (i.e. for any nontrivial positive operator $A \in N$, a normal G-invariant (see Section 1.3.5 below) state ρ on N can be found such that $\rho(A) \neq 0$). Then N is finite. Indeed, turning back to the proof of Proposition 1.3.8, we see that if $R(O)$ is semi-finite, then the Størmer theorem is valid for $R(O)$ with the automorphism group $G = \{T(\lambda e)\}_{\lambda \in R}$, because in this case the algebra of fixed points is $R(O)^G = Z(O) = C(\mathcal{H})$, and $R(O)$ is G-finite, since the state ω_Ω is a faithful conditional expectation* from $R(O)$ to $R(O)^G$ and so the main criterion of the G-finiteness [286, Theorem 2] is fulfilled. Hence, if $R(O)$ is semi-finite, then by the Størmer theorem it is also finite, in contradiction to Proposition 1.3.9.

Let us consider the last case [165]. Take the sequence of translations, $T(ne)$, $n = 1, 2, \ldots$, and note that the algebra $R(O)$ has the following properties:

(1) $T(ne)\,R(O)\,T(-ne) \subset R(O)$;

(2) $\underset{n\to\infty}{w\text{-lim}}\, A(ne) = \omega_\Omega(A)\,I \quad \forall A \in R(O)$;

(3) $\underset{n\to\infty}{s\text{-lim}}\, [A(ne), B] = 0 \quad \forall A, B \in R(O)$;

(in the last case the limit is in the strong operator topology). Indeed, the first property is obvious, as well as the second one, if (3.7) and the uniqueness of the vacuum are taken into account. The third property can be verified easily. Let A, $B \in R(O)$. By the Reeh–Schlieder theorem, for any $\varepsilon > 0$ an element $B_\varepsilon \in R(O)$ exists, such that $\|(B_\varepsilon - B)\,\Omega\| < \varepsilon$. Now, by the second property and Axiom III,

$$\lim_{n\to\infty} \|[A(ne), B]\,\Omega\| \le \lim_{n\to\infty} \|(A(ne)\,B_\varepsilon - B_\varepsilon A(ne))\,\Omega\| + \varepsilon\,\|A\|$$

$$= \lim_{n\to\infty} \|A(-ne)\,(B - B_\varepsilon)\,A\,\Omega\| + \varepsilon\,\|A\|$$

$$= \|(B - B_\varepsilon)\,\Omega\| \cdot \|A\,\Omega\| + \varepsilon\,\|A\| \le 2\,\varepsilon\,\|A\|.$$

Since Ω is separating for $R(O)$ and the sequence $\{[A(ne), B]\}_{n=1}^{\infty}$ is uniformly bounded, we get the desired result.

Now let us assume that there is a finite projection $E \in R(O)$. Then there is a normal finite trace τ on the finite von Neumann algebra $R(O)_E = ER(O)E\,|_{E\mathcal{H}}$ [7, p. 97]. As $R(O)_E$ has a separating vector $E\,\Omega$, τ is represented by a vector functional ω_Ψ, $\Psi \in E\,\mathcal{H}$ [7, p. 222]. It follows from properties 2 and 3 that for any $A, B \in R(O)$,

$$\lim_{n\to\infty} (\Psi, A(ne)\,EB(ne)\,\Psi) = \lim_{n\to\infty} (\Psi, (AB)(ne)\,\Psi) = \omega_\Omega(AB).$$

Hence the definition of a trace implies that $\omega_\Omega(AB) = \omega_\Omega(BA)$ for all $A, B \in R(O)$ and so ω_Ω is a nontrivial finite trace on $R(O)$. This is impossible, however, the algebra $R(O)$ being infinite (Proposition 1.3.9).

Finally, the assertion about the central decomposition of $R(O)$ is a particular case of

* The linear mapping $E : \mathcal{A} \to \mathcal{B}$ (where $\mathcal{B}$ is a C^*-subalgebra of the C^*-algebra $\mathcal{A}$) is called a faithful conditional expectation if $E(I) = I$; $E(A) > 0\ \forall A > 0$; $\|E(A)\| \le \|A\|$; $E(SAT) = SE(A)T$; $E(A)^*E(A) \le E(A^*A)\ \forall A \in \mathcal{A}, S, T \in \mathcal{B}$. See [23, pp. 100-101, 29, p. 332].

the results by Fredenhagen [193] concerning the structure of local algebras in theories with exact or asymptotic dilatation invariance. By explicit calculation of spectra of modular operators it was shown in [193, Section 4] that in such theories all factors in central decompositions of local algebras have the Connes invariant S equal to $\mathbf{R}_+$, and so they are of type III_1 (Section 1.1). This result is valid for our $R(O)$, since O is dilatation invariant. ❒

As for the most important class of regions, the double cones D, there is no axiomatic proof yet, but a number of results have been obtained under various assumptions suggesting that $R(D)$ should also be of type III, and more precisely, of type III_1 (see e.g. [245, 311, 193, 195]). The simplest result is as follows.

PROPOSITION 1.3.13 [410, 287]. *Let Axioms* Ia, IIa, III *and* IV *hold,* $R = \mathcal{B}(\mathcal{H})$ *and the strict locality condition (Section* 1.2) *be satisfied for a pair of regions* D *and* D'. *Then* $R(D)$ *cannot be a type-*I *factor. Moreover, if the duality condition is satisfied, the* $R(D)$ *is a factor of type* III.

Proof. As we have seen proving Proposition 1.3.6, Axioms Ia, IIa, IV imply

$$R(D \cup D') = \{R(D) \cup R(D')\}'' = R\,.$$

Hence, if $R(D)$ is a type-I factor, then $R = \mathcal{B}(\mathcal{H})$ together with Axiom III and the normalcy property of discrete factors [7, pp. 128, 287] imply the duality condition, $R(D) = R(D')'$. On the other hand, as we shall show analyzing strict locality (Proposition 1.3.32), strict locality for D and D' implies the following property: all projections from $R(D)$ are infinite in the algebra $R(D')' \supset R(D)$. Thus, if the factor $R(D)$ satisfies the duality condition, all its projections are infinite, i.e. $R(D)$ is a factor of type III. This proves both statements of the proposition: assuming that $R(D)$ is a type-I factor, one arrives at a contradiction; if one assumes instead the duality condition for $R(D)$ (together with strict locality for D and D') one concludes that $R(D)$ is a type-III factor. ❒

Finally, let us briefly discuss the hyperfiniteness property. Physical systems often enjoy this property, because in many cases they can be approximated by their finite-dimensional subsystems. This is the case, for example, in a large class of canonical systems described by representations of canonical commutation or anticommutation relations (see [10] and Section 3.3). In general Haag–Araki theory hyperfiniteness is closely connected with the split property. It is especially easy to see this connection, if the local net satisfies continuous isotony (Axiom Ia). In this case, choosing for a region O (which may be unbounded) a sequence of regions $O_k \in B(M)$ such that $O_k \ll O_{k+1}, \cup_{k=1}^{\infty} O_k = O$, and using the split property, we obtain a monotone increasing sequence of discrete factors M_k: $R(O_k) \subset M_k \subset R(O_{k+1}) \subset M_{k+1} \subset \dots$. Continuous isotony implies then

$$R(O) = \left\{ \bigcup_{k=1}^{\infty} R(O_k) \right\}'' = \left\{ \bigcup_{k=1}^{\infty} M_k \right\}''.$$

If the Hilbert space is separable, M_k are hyperfinite and so the local algebra $R(O)$ is hyperfinite too [123]. (This result can equally be obtained with the aid of a monotone decreasing sequence of regions, which means that any of the two continuity conditions in Axiom Ia is sufficient.)

If both hyperfiniteness and type are known for a local algebra, the description of its structure is nearly complete, considering that hyperfinite factor is unique for most of types (Section 1.1). In particular, if hyperfiniteness holds, then, whenever the local algebras are factors of type III_1 – and this is the most typical case, as we could see – they are isomorphic to one another and to the unique hyperfinite type-III_1 factor $\mathcal{R}$ (the Araki–Woods factor). If $R(O)$ is hyperfinite and is not a factor, but has central decomposition with the components of type III_1 (a.e.), then it is easy to prove [105] that

$$R(O) \simeq \mathcal{R} \bar{\otimes} Z(O),$$

where $\bar{\otimes}$ is the tensor product of W^*-algebras (see below, p. 59). This universal structure is gradually being proved for more and more theories and models. Besides the regions in Proposition 1.3.12, it is now established for double cones in the free field models (Section 3.3), light cones in free massless theories [245] and also for double cones in all theories having asymptotical dilatation invariance [193].

1.3.2. DEPENDENCE RELATIONS

Our next goal is to consider the relations between local algebras of different regions. There are basically two types of these relations.

(1) Relations stating that observables of a region O_2 are determined by the observables of some other region O_1, or, which is the same, observables of the region $O = O_1 \cup O_2$ are limited to the observables of its subregion O_1, $R(O) = R(O_1)$. (Note that the Wightman inequality is an example of the opposite kind of relation: if $O_1 << O$, the sets of observables of the region O and its subregion O_1 must be different.) Evidently, relations of this kind mean that no measurements performed in O_2 can produce any information on the system, besides that given by measurements in O_1, so that physical processes (actions, perturbations) may come to O_2 only from O_1. Or, in terms of states: if two states of a system cannot be distinguished by means of measurements performed in O_1, they cannot be distinguished by means of measurements in O_2 either. Therefore one can say that O_2 is a 'dependence region' for O_1, and it is natural to call relations like $R(O_1 \cup O_2) = R(O_1)$ *dependence relations*. These relations between spacetime regions, which express the propagation properties of relativistic quantum processes, bear a close resemblance to those familiar in the theory of hyperbolic partial differential equations (cf. e.g. [6, Vol.2]: if smooth solutions of wave equations coincide in some regions, they must also coincide in certain 'dependence regions'), which express propagation properties of classical wave processes. Similar relations take place in the theory of analytic functions of many complex variables [30, Sections 28, 29], where they are considered as quasianalyticity properties. In both cases, the analogy with quantum

field theory is by no means superficial, but quite essential, and the dependence relations for relativistic quantum processes can be derived on the basis of complex analysis as well as on the basis of theorems concerning the behaviour of solutions of hyperbolic equations (for this reason, one says sometimes that these relations demonstrate the hyperbolic propagation of relativistic quantum phenomena similar to that of classical wave processes).

(2) Relations stating that physical processes taking place in some regions O_1 and O_2 are mutually independent and do not affect each other. The physical basis of such relations is, of course, the Einstein causality principle, according to which events or processes are completely independent of each other, if they are localized in spacelike separated regions (hence such regions are often called *causally disjoint*). Mutual independence is, however, a very broad concept, which may be expressed in many concrete ways. Therefore relations of this kind may be not only derived from the axioms, but also proposed 'straight from physics' as possible formulations of the Einstein principle. The most important formulation is the locality axiom; some others were presented at the end of Section 1.2. Unlike the relations of the first kind, they all show how and where relativistic quantum processes cannot propagate, and so they can be called *independence relations* or *causality relations*, because they originate from the Einstein causality principle. It should be emphasized, however, that both kinds of relations originate from causality in a broad sense; the only difference is that relations of the second kind express mainly quantal causality, while those of the first kind express classical causality (even though they do not secure fully the classical causality principle, according to which the system state at any time moment $t = t_1 > 0$ is determined completely by its state at $t = 0$).

Of course, these two types do not exhaust all thinkable or even all known relations between local algebras. They do not include, for example, the Wightman inequality as well as the split property (which can be considered as an enhancement of the latter). As we shall see in Section 1.3.4, the split property is closely connected with independence relations, but its physical interpretation does not directly correspond to either type 1 or 2. Namely, it was shown in [107, 427] that the split property is equivalent to the so-called principle of local preparation of states. It states that any given state in a region O_1 can be prepared from any normal state in any region $O_2 >> O_1$ by means of the measurement of a single observable described by a certain projection from $R\,(O_2)$ [107]; according to [427], some more general preparation procedures would do as well.

In the problem of dependence relations one principal fact clearly stands out, around which most of the other results are centered. This is the 'spacelike hull theorem', found first by Borchers [72] in 1961 in its simplest form and later generalized and enhanced by Araki [38], Vladimirov [30], Wightman [433], and Kraus [287]. Here we present this theorem in the Wightman formulation because it is simple in its proof and still relatively general.

THEOREM 1.3.14. *Let Axioms* I, II, V, VI *hold, and let* $O_r \supset O$ *be the 'spacelike hull' of an arbitrary region* $O \in B\,(M)$, *defined as the union of all double cones* $D_{x,y}$, *for which the axes* $[x, y]$ *lie completely in* O :

$$O_r = \bigcup_{(x,y)\in\Sigma(O)} D_{x,y} \; ; \; \Sigma(O) = \{x, y \in M \mid x - y \in V^+ , \; [x, y] \subset O\} . \qquad (3.9)$$

Then $R(O_r) = R(O)$.

Proof. Let $[x, y]$ be a timelike segment, lying completely in O. As O is an open region, there is a neighbourhood $\mathcal{N}$ of the segment, which also lies in O. Let us choose a sub-neighbourhood $\mathcal{K} \subset \mathcal{N}$ such that $\mathcal{K} + a \subset \mathcal{N}$ for all a in a neighbourhood of the segment $1/2\,\lambda\,(x - y)$, $\lambda \in [-1, 1]$. Now, if Φ, Ψ are vectors analytic in the energy, and $A, B \in \mathcal{B}(\mathcal{H})$, the expectation values $(\Phi, A(a)\,B\,\Psi)$ and $(\Phi, BA(a)\,\Psi)$ are continuous functions of a. Due to spectrum condition, the former function is a boundary value of the function $(T(-(a + ib))\,\Phi\,,\; AT(-(a + ib))\,B\,\Psi)$, analytic in $a + ib$ for $b \in V^-$, while the latter is a boundary value of the function $(\Phi, BT(a + ib)\,AT(-(a + ib))\,\Psi)$, analytic in $a + ib$ for $b \in V^+$. If $A \in R(\mathcal{K})$, $B \in R(\mathcal{N})'$ here, then both functions coincide for all a from the indicated neighbourhood and $b = 0$. So, by the edge of the wedge theorem, they have a common analytic continuation into a neighbourhood of the double cone $D_{1/2(x-y),-1/2(x-y)}$ hence we get $R(\mathcal{N})' \subset R(D_{x,y})'$. Since $R(O)' \subset R(\mathcal{N})'$, taking the union over all timelike segments $[x, y] \subset O$, one gets $R(O)' \subset R(O_r)'$. □

As noted above, theorems of this type enable one to extend the class of regions, for which the local algebras are factors. Following Kraus [287], let us call a region $O \subset M$ *normal*, if the closure of $O_r \cup O'$ includes the complete spacelike slice $S^{(n)}$, or may be represented, in the spirit of Axiom Ia, as a projective limit of regions including such a slice.

COROLLARY 1. *Let Axioms* Ia, IIa, III–VI *hold and the region* O *be normal. Then* $\mathcal{Z}(O) \subset \mathcal{Z}$, $\mathcal{Z}(O') \subset \mathcal{Z}$. *In particular, if* R *is a factor, then* $R(O)$ *and* $R(O')$ *are factors too.*

Obviously, the class of normal regions includes cylinders (Figure 4). Next, we present the important

COROLLARY 2. (Borchers' tube theorem). *Let Axioms* I, II, V *and* VI *hold and suppose an unbounded region* O *is a 'timelike tube', in Borchers' sense, namely,* O *includes an infinite timelike curve*

$$\ell = \left\{x \in M \mid x = x(s),\, s \in \boldsymbol{R} \;\; \forall_{s_1 \neq s_2} \;\; (x(s_2) - x(s_1))^2 > 0,\; \lim_{s\to\pm\infty} x^0(s) = \pm\infty, \mid \boldsymbol{x} \mid < r < \infty\right\}.$$

Then $R(O) = R$.

Proof. It is sufficient to prove that for any $y \in M$ there is a region $\tilde{O}$, such that $y \in \tilde{O}$ and $R(\tilde{O}) \subset R(O)$. Let us associate with y the timelike slice

$$S^{(y)} = \{x \in M \mid y^0 - (\mid \boldsymbol{y} \mid + r + 1) < x^0 < y^0 + (\mid \boldsymbol{y} \mid + r + 1)\} .$$

For the region $O \cap S^{(y)}$, we construct the spacelike hull $(O \cap S^{(y)})_r$, as given in

(3.9), and applying Theorem 1.3.14, we get $R\,((O \cap S^{(y)})_r) \subset R\,(O)$. Repeating the procedure of constructing the spacelike hulls, after a finite number of steps we get a region $(O \cap S^{(y)})_{rr\,\dots\,r}$, which includes all the double cones D_{x_1,x_2}, where x_1 and x_2 are the intercepts of the curve ℓ with the upper and lower boundary hyperplanes of the layer $S^{(y)}$, respectively. Hence we get, by the construction of the slice $S^{(y)}$:

$$y \in (O \cap S^{(y)})_{rr\,\dots\,r}\,,\quad R\left((O \cap S^{(y)})_{rr\,\dots\,r}\right) \subset R\,(O)\,.$$ ❐

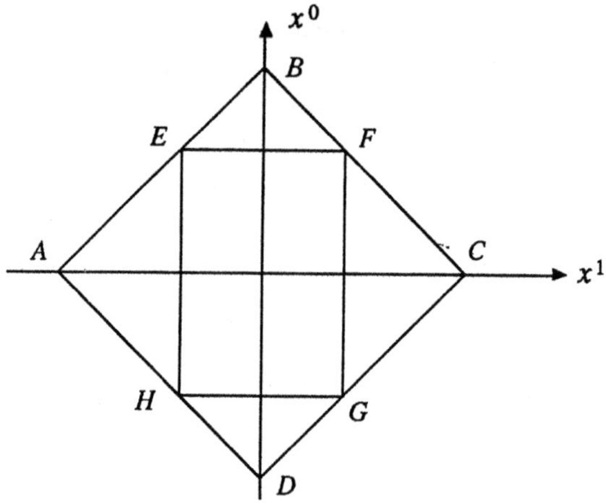

Fig. 4. A cylinder and its spacelike hull. $EFGH = Z$, $AEFCGH = Z_r$.

There are extensions of Theorem 1.3.14 based upon various modifications of the spacelike hull concept. In the construction by Araki [38], the hull $O_r^{(A)}$ was defined by means of introducing the five-dimensional space M_5, with an additional spacelike co-ordinate s,

$$O_r^{(A)} = (\varphi\, O)_r^{(A)} \cap \{s = 0\}\,, \tag{3.10}$$

where $\varphi\, O$ is the embedding of a region $O \subset M$ into M_5, and $(\varphi\, O)_r^{(A)}$ is the hull of $\varphi\, O$, defined as the union of double cones generated by all timelike segments belonging to $\varphi\, O$, as well as by all spheres in $\varphi\, O$, lying in spacelike hyperplanes in M_5 (the double cone associated with such a sphere has the latter as its base). Vladimirov [30, Section 28, 29] introduced a concept of the 'C-convex hull' of a region O; it is the union of all double cones, the apices of which can be connected by a timelike curve lying within O. Both the hull proposed by Araki and C-convex hull may be larger than that defined in (3.9); when this is the case, new extensions of the Borchers theorem can be obtained.

A special kind of dependence relations is given by 'total regions' (so called by analogy with total sets in vector spaces). By definition, $O \subset M$ is a total region, if $R\,(O) = R$ or, in other words, the whole Minkowski space is the dependence region for O. We have already dealt with examples of such regions:

(i) the complete spacelike slice $S_\varepsilon^{(n)}$, according to Axiom IV;

(ii) the region $D \cup D'$ for any double cone D, provided that Axioms Ia, IIa hold;

(iii) the disconnected region $Z \cup Z'$ for any cylinder Z;

(iv) the timelike tube, according to Borchers' theorem.

A sufficient condition of totality found by Sadowski and Woronowicz [373] enables one to obtain new classes of total regions and derive nontrivial enhancements of the tube theorem. This condition does not use locality, but relies on the existence of a unique and cyclic vacuum vector (which entails $R = \mathcal{B}(\mathcal{H})$).

Let us introduce the following class of 'regular commutator functions' $W(a)$ for the spectral measure $E(p)$ of the translation group $T(a)$:

$$W(a) = \int e^{-ipa}\, d\mu(p)\,, \quad \lim_{\lambda \to \infty} W(b + \lambda a) = 0 \tag{3.11}$$

for any $a, b \in M$, $a \neq 0$. Here $d\mu(p)$ is a finite, not necessarily positive measure with the support in the set $\Delta \cup (-\Delta)$, $\Delta = \operatorname{supp} E(p) \subset \overline{V^+}$. The name of the class is suggested by the fact that this class includes the vacuum expectation values of commutators $[A, B(a)]$ for any $A, B \in \mathcal{B}(\mathcal{H})$, $a \in M$. Indeed, the function $W(a) = (\Omega, [A, B(a)]\,\Omega)$ satisfies the first condition in (3.11) with the measure

$$d\mu(p) = (A^* \Omega, dE(p) B\,\Omega) - (B^* \Omega, dE(-p) A\,\Omega)\,, \tag{3.12}$$

whereas the second condition holds due to the cluster property which follows from the uniqueness of the vacuum (see Proposition 1.3.42).

PROPOSITION 1.3.15 [373]. *Let Axioms* I, II, V, VI *hold, and a unique and cyclic vacuum vector* Ω *exists in* $\mathcal{H}$. *Moreover, let a set* $Q \subset M$ *be such that*

$$W(a) = 0\,,\ a \in Q \Rightarrow W(a) \equiv 0 \tag{3.13}$$

for any regular commutator function W. *Then for any region* $O \subset M$ *the region*

$$O + Q = \bigcup_{a \in Q} (O + a) = \{x \in M \mid x = y + a\,,\ y \in O\,,\ a \in Q\} \tag{3.14}$$

is total.

Proof. Let us show that any operator $A \in R(O + Q)'$ is trivial. To this end we consider the regular commutator function corresponding to the commutator $[A, B(a)]$, $B \in R(O)$. Since $B(a) \in R(O + Q)$ for $a \in Q$, this function vanishes for any $a \in Q$; hence, it vanishes identically, according to (3.13). In view of (3.11), this function is the Fourier transform of the measure (3.12). Vanishing of this measure entails

$$(A^* \Omega\,, dE(p) B\,\Omega) = (B^* \Omega\,, dE(-p) A\,\Omega)\,.$$

Since the left-hand side here has its support in $\overline{V^+}$, and the right hand side in $\overline{V^-}$, they both have the support only in $\{p = 0\}$ which implies

$$(A^* \Omega, B\,\Omega) = \int (A^* \Omega, dE(p) B\,\Omega)$$

$$= \int_{\{0\}} (A^* \Omega, dE(p) B \Omega) = (A^* \Omega, \Omega)(\Omega, B \Omega)$$

$$= \alpha (\Omega, B \Omega), \ \alpha = (A^* \Omega, \Omega).$$

By the Reeh–Schlieder theorem, $\{B \Omega, B \in R(O)\}$ is a dense set, hence the relation obtained entails $A^* \Omega = \bar{\alpha} \Omega$. This equality will be valid also on any vector $C \Omega, C \in R(O + Q)$. The set of these vectors being dense, we get the desired result, $A^* = cI$. □

To use this criterion one must find examples of sets satisfying (3.13). Some of them are indicated in the following lemma.

LEMMA 1.3.16. *Let spectrum condition hold in the strong form* VIa. *Then the open backward light cone* V^- *satisfies* (3.13). *If spectrum condition holds in the standard form* VI, *condition* (3.13) *is satisfied by the sets*

$$Q_1 = \{x \in M \mid x^0 < -((x^2)^2 + (x^3)^2)^{1/2}\},$$

$$Q_2 = \{x \in M \mid |x^0 - x^1| < \tau\}$$

($\tau > 0$), as well as by any sets obtained from Q_1 and Q_2 by Poincaré transformations.

Proof. Representing an arbitrary function from the class (3.11) in the form

$$W(a) = \int_{-\infty}^{\infty} e^{-ip_0 a^0} d\mu_a(p_0)$$

(where $d\mu_a(p_0)$ is the finite projection of the finite measure $\exp(ipa)\, d\mu(p)$ on the p_0-axis), we see that in the case of strong spectrum condition the measure $d\mu_a(p_0)$ vanishes in the open interval $(0, E_0) \cup (-E_0, 0)$, $E_0 = \inf\{p_0 \in \mathbf{R} \mid p_0 \in \Delta, p_0 \neq 0\}$. If $W(a) = 0$ for $a \in V^-$, then $d\mu_a(p_0)$ is the boundary value of a function holomorphic in the upper half-plane, and so, by the analyticity principle,

$$d\mu_a(p_0) = 0, \ p_0 \in (0, E_0) \cup (-E_0, 0) \Rightarrow d\mu_a(p_0) \equiv 0,$$

whence $W(a) \equiv 0$.

Suppose now $W(a) = 0$ for $a \in Q_1$. For an arbitrary continuous bounded function φ on $\mathbf{R}$ we define a regular commutator function

$$W_\varphi(a) = \int e^{-ipa} d\mu_\varphi(p), \ d\mu_\varphi(p) = \varphi(p_1)\, d\mu(p).$$

Since Q_1 is invariant under translations along the x^1-axis, it is easy to verify that

$$W(a) = 0, \ a \in Q_1 \Rightarrow W_\varphi(a) = 0, a \in Q_1.$$

In view of $V^- \subset Q_1$, if φ is chosen to be vanishing in the open interval $(-E_0, E_0)$, $E_0 \geq \varepsilon > 0$, then $W_\varphi(a)$ enjoys all the properties of the function $W(a)$ considered in the preceding paragraph. Therefore $W_\varphi(\alpha) \equiv 0$, which implies $\varphi(p_1)\, d\mu(p) \equiv 0$ for any continuous bounded function $\varphi(p_1)$ vanishing in $(-E_0, E_0)$. Since E_0 is arbitrary, one has

$$\text{supp}\, d\mu(p) \subset \{p \in M \mid p_1 = 0\},$$

and therefore the Fourier transform $W(a)$ of the measure $d\mu(p)$ does not depend on a^1. Hence the second condition in (3.11) implies $W(a) \equiv 0$.

In the case when $W(a) = 0$ for $a \in Q_2$, we introduce the variable $m = (p_0^2 - p_1^2)^{1/2}$ and transform $W(a)$ to the form of the Jost–Lehmann–Dyson representation,

$$W(a) = \int_\Delta \exp(i(p_2 a^2 + p_3 a^3))\, W_{mp_2p_3}(a^0, a^1)\, d\sigma(m, p_2, p_3),$$

$$W_{mp_2p_3}(a^0, a^1) = \int \exp(-i(p_0 a^0 - p_1 a^1))\, d\mu_{mp_2p_3}(p_0, p_1),$$

where $d\sigma(\cdot)$ is a positive measure on the set $\Delta \subset \{(m, p_2, p_3) \mid m^2 \geq p_2^2 + p_3^2\}$, and $d\mu_{mp_2p_3}$ a finite measure on hyperbolas $m = \text{const}$ in the (p_0, p_1)-plane. The function $W_{mp_2p_3}$ satisfies the two-dimensional Klein–Gordon equation with mass m; moreover, if $W(a) = 0$ for $a \in Q_2$, then it follows from the Jost–Lehmann–Dyson representation for that $W_{mp_2p_3}(a^0, a^1) = 0$ in the region $\{(a^0, a^1) \in \mathbf{R}^2 \mid |a^0 - a^1| < \tau\}$. But any solution of the two-dimensional Klein–Gordon equation vanishing in this region is equal identically to zero. Hence it is solely the point $m = p_2 = p_3 = 0$ that contributes to the integral for $W(a)$. Using again the second condition in (3.11) we see that $W(a) \equiv 0$.

The last statement of the lemma results from the fact that if (3.13) holds for a set Q, then it also holds for any set $\Lambda Q + a$, $(a, \Lambda) \in \mathcal{P}_+^\uparrow$. ❐

As said above. the lemma makes it possible to construct new examples of total regions.

PROPOSITION 1.3.17 [373]. *Let Axioms* I, II, V, VI *hold, and a unique and cyclic vacuum vector* Ω *exist in* $\mathcal{H}$. *Then the following regions in Minkowski space are total* $(R(O) = \mathcal{B}(\mathcal{H}))$:

$O_1 = s + V^-$, *where* s *is a spacelike straight line in* M;

$O_2 = \mathcal{N}(s \cup \ell_x)$, *where* s *is a spacelike straight line,* ℓ_x *is a timelike curve going from* $x^0 = -\infty$ *to a certain point* $x \in s$ *and* $\mathcal{N}$ *is any open neighbourhood of* $s \cup \ell_x$;

$O_3 = x + Q_2$, $x \in M$;

$O_4 = \cup_{\lambda \in \mathbf{R}} (O_1 + \lambda n)$, $O_1 \subset M$, $n^2 = 0$ *(the tube obtained from an arbitrary region* O_1 *by means of translations along an isotropic direction* n*).*

If Axiom VIa *holds, the following regions are total as well:*

$O_5 = x + V_-$, $x \in M$;

$O_6 = \mathcal{N}(\ell_x)$, *where* ℓ_x *is a timelike curve going from* $x^0 = -\infty$ *to an arbitrary point* $x \in M$, $\mathcal{N}$ *is any open neighbourhood of* ℓ_x.

All regions obtained from the listed ones by Poincaré transformations are also total.

Proof. Using the lemma, one easily verifies that each region O_i, $i = 1, \ldots, 6$, can be represented in the form (3.14). In the case of O_1 one can assume without loss of generality that the line s is parallel to the x^1- axis; then $O_1 = O_1 + Q_1$.

The case of O_2 is reduced to that of O_1 : proceeding as in the proof of the tube theorem, we extend the region O_2 without changing the local algebra and obtain

$$R(O_2) \supset R(s + V^-) = \mathcal{B}(\mathcal{H}) .$$

In the case of O_3 , pointing out explicitly the τ dependence of Q_2 , we have

$$Q_2 = Q_2(\tau) = Q_2(\tau/2) + Q_2(\tau/2) ,$$

whence

$$O_3 = O + Q_2(\tau/2) , \quad O = x + Q_2(\tau/2) .$$

The case of O_4 is reduced to that of O_3 , like the case O_2 to that of O_1 : extending O_4 as in the proof of the tube theorem, we have

$$R(O_4) = R(O + Q_2(\tau = 0)) = \mathcal{B}(\mathcal{H}).$$

Finally, if Axiom VIa holds, then V^- satisfies (3.13), so that

$$O_5 = O + Q , \quad O = x + V^-, \quad Q = V^- .$$

The case of O_6 is similar to those of O_4 and O_2 , and we have

$$R(O_6) \supset R(x + V^-) = \mathcal{B}(\mathcal{H}) .$$

The last assertion being evident, the proof is over. ❒

In addition to the study of total regions, $O \subset M$, $R(O) = R$, one can ask the question: are there regions with the opposite property, $R(O) = C(\mathcal{H})$? The answer is positive. The first result in this direction was obtained by Wightman.

PROPOSITION 1.3.18 [433]. *Let a Haag–Araki theory satisfy Axioms* II, III, V, VI *and have a unique vacuum vector* Ω *cyclic for* Z' . *Then for any point* $\hat{x} \in M$,

$$R(\{\hat{x}\}) \equiv \bigcap_{O \ni \hat{x}} R(O) = C(\mathcal{H}) .$$

Proof. For an arbitrary operator $B \in R(\{\hat{x}\})$, let us put by definition,

$$B(y + \hat{x}) = T(y) B T(-y) .$$

Then the operator-valued function $B(y)$ has the property

$$y \in O \Rightarrow B(y) \in R(O).$$

Hence locality implies

$$[B(x), B(y)] = [B(x), B(y)^*] = 0 \text{ for } (x - y)^2 < 0 . \tag{3.15}$$

By the standard analytic continuation technique (cf. e.g. the proof of Proposition 1.3.15) one verifies that this function is trivial. Introducing continuous functions

$$F(x-y) = (\Omega, B(x)^* B(y) \Omega) = (\Omega, B^* T(y-x) B \Omega),$$

$$G(x-y) = (\Omega, B(y) B(x)^* \Omega) = (\Omega, BT(x-y) B^* \Omega),$$

we see that they are functions of positive type and, by the Bochner theorem, can be represented as

$$F(x-y) = \int e^{-ip(x-y)} \, d\mu(p), \quad G(x-y) = \int e^{-ip(x-y)} \, d\nu(p),$$

where μ and ν are positive measures with the supports in $\overline{V^+}$ and $\overline{V^-}$, respectively, owing to spectrum condition. Therefore F can be analytically continued as a holomorphic function into the backward tube $\mathcal{T}^- = \mathbf{R}^4 - iV^+$, and G into the forward tube $\mathcal{T}^+$. Moreover, $F(x) = G(x)$ for $x^2 < 0$, in view of (3.15). Therefore $F = G$ everywhere, and so both measures μ and ν are concentrated in $\{p = 0\}$ only. This implies $F(x-y) = \text{const} = F(0)$, whence it follows

$$T(y-x) B \Omega = B \Omega \quad \forall \, x, y \in M. \tag{3.16}$$

The vacuum being unique, this means that $B\Omega = \lambda \Omega$, $\lambda \in \mathbf{C}$, and since Ω is separating for all $R(O)$ with $O' \neq \emptyset$, we get $B = \lambda I$. □

Clearly, the crucial point in the proof is establishing that any 'bounded operator field', i.e. a function $B : M \to \mathcal{B}(\mathcal{H})$ which satisfies translation covariance, spectrum condition and locality (as given by (3.15)) is trivial. The result follows from this fact by means of a simple remark: translations $B(x)$ of observables of a 'point region' $B \in R(\{\hat{x}\})$ form a bounded operator field. A straightforward generalization of this result was obtained by Woronowicz [437] who noticed that observables of a certain class of two-dimensional bounded regions also form, in a sense, an analogue of the bounded operator field.

PROPOSITION 1.3.19. *Let a Haag–Araki theory satisfy Axioms* II, III, V, VI *and have a unique vacuum vector* Ω *cyclic for* Z'. *Let also a compact region* $K \subset M$ *satisfy the following condition: there exist two spacelike unit vectors* $n_1, n_2, n_1^2 = n_2^2 = -1$, $(n_1 - n_2)^2 \geq 0$, *such that for any* $a, b \in K, (a-b)^2 \leq 0$ *one has*

$$|(a-b)n_i| \leq \sqrt{-(a-b)^2}, \quad i = 1, 2.$$

Then

$$R_K = \bigcap_{O \supset K} R(O) = C(\mathcal{H}).$$

Proof is in its scheme identical to the preceding one. The scheme proves to be applicable due to the following property of the regions K: if $|x^0| < |\alpha x^1|$, $\alpha \in (0,1)$, then the sets K and $K + (x^0, x^1, 0, 0)$ have open neighbourhoods O and O_1, respectively, such that $O \times O_1$. Like (3.15), this property entails the coincidence of the functions $F(x)$ and $G(x)$ in a certain open neighbourhood (namely, for $x = (x^0, x^1, 0, 0)$, $|x^0| < |\alpha x^1|$, $\alpha \in (0, 1)$. Hence, as in Proposition 1.3.18, we derive that these functions coincide identically and an analogue of (3.16) is valid:

$$T(x) B \Omega = B \Omega, \ x = (x^0, x^1, 0, 0).$$

But, due to spectrum condition, any vector invariant under time translations is invariant under any translations, whence $B = \lambda I$ follows. ❐

As shown by Woronowicz [437], examples of the regions K are provided, for instance, by bounded subregions of two-dimensional spacelike hyperplanes. Within the formalism of field algebras (Section 3.2) and using a quite different technique, Driessler and Summers somewhat extended the class of regions with the trivial set of observables (as well as fields): as shown in [169], this class includes any region which is spacelike to some region of the form $(a, \Lambda)(W_R \cup W_L)$, where $(a, \Lambda) \in \mathcal{P}_+^\uparrow$ and W_R, W_L are the wedge regions (3.5). In particular, this class of regions includes arbitrary two-dimensional spacelike planes and any of their subregions.

The properties discussed above do not exhaust the list of all dependence relations considered in the Haag–Araki theory. Clearly, these relations include also the 'local primitive causality' (Axiom IVb): $R(O) = R(O_c)$ where O_c is the causal shadow of O. One more relation of this kind is the following 'diamond hypothesis' suggested by Araki: $R(O) = R(O'')$, i.e. O'' should be the dependence region for any $O \in B(M)$. This statement is an enhancement of Axiom IVb: due to the obvious inclusion $O \subset O_c \subset O''$, $R(O) = R(O'')$ implies $R(O) = R(O_c)$. Conversely, the 'diamond hypothesis' follows from Axiom IVb (and other axioms) for the class of regions $O \in B(M)$, for which the union of the causal shadow and the spacelike hull is a diamond, i.e. $O_r \cup O_c = D = D''$ (regions of this form were studied in [249]). Moreover, the diamond hypothesis (and, consequently, Axiom IVb) follows from the duality condition $R(O') = R(O)'$: since $O' = O'''$, duality implies $R(O) = R(O')' = R(O'')$. But none of these properties were derived thus far from canonical Axioms I–VI.

1.3.3. DUALITY

Now the only local properties that remain to be discussed are independence relations. Some were formulated at the end of Section 1.2: these are causal independence, strict locality, Schlieder property, extended locality and duality. Physically, they all can be treated as sufficiently reasonable properties expressing various aspects of Einstein's causality principle. We should find conditions securing them and their relations with one another and to the locality axiom, the central of all causal properties.

Of all the properties, duality is the most studied (though recently the emphasis appears to be shifting to causal independence). Later on we shall see it to play an important part in constructing fields from observables (Section 2.3) and in the structure of field algebras (Section 3.2.2). In Section 3.3 we shall show that duality for an arbitrary region $O \in B(M)$ does not follow from the Haag–Araki axioms and so by duality we shall usually mean the duality for double cones D. As for other regions, in some cases (including free and generalized free fields, see Section 3.3) the duality was proved for cylinders Z. Let us mention a few simple corollaries of this duality. Since Z'' is a double cone, duality for cylinders implies duality for double cones and the diamond hypothesis for cylinders; the latter in its turn entails primitive causality:

$$R(Z)' = R(Z') \begin{array}{l} \nearrow R(D)' = R(D') \\ \searrow R(Z) = R(Z'') \Rightarrow R(S_\varepsilon) = R\ . \end{array} \tag{3.17}$$

The last implication here becomes evident if we consider a sequence of cylinders of the constant height ε and increasing radii.

The meaning of duality can be clarified by simple facts showing its ties with the property of relative locality of local nets. As is well-known, for quantum fields this property is transitive [27, Theorem 4.21] and therefore it provides an equivalence relation splitting the set of all local fields (defined on the same space of states) into equivalence classes called *Borchers' classes.* In local quantum theory the definition of relative locality is straightforward: nets $\{R_1(O)\}$ and $\{R_2(O)\}$ can be called relatively local to each other if $R_1(O_1) \subset R_2(O_2)'$ for $O_1 \times O_2$. However, the transitivity of this property (and also of the usual locality which can be considered as its particular case) is not established. So, in a general case, we cannot speak here about equivalence classes but only about sets of relatively local theories.

PROPOSITION 1.3.20. *Let a set* $\{R_\gamma(O)\}_{O\in B(M),\gamma\in\Gamma}$ *of relatively local nets include at least one net* $\{R_\beta(O)\}, \beta \in \Gamma$, *satisfying duality. Then the net* $\{R_\beta(O)\}$ *is maximal in this set in the sense that*

$$\forall_{\gamma\in\Gamma} \ \forall_{D\in B(M)} \quad R_\gamma(D) \subset R_\beta(D)\,,$$

and unique in the sense that

$$\exists_{\gamma\in\Gamma} \ \forall_{D\in B(M)} \quad R_\gamma(D)' = R_\gamma(D') \Rightarrow R_\gamma(O) = R_\beta(O)\ \forall\ O \in B(M)\,.$$

Proof. The second statement follows from the first and the first from relative locality:

$$R_\gamma(D) \subset R_\beta(D')' = R_\beta(D)\,. \qquad \square$$

The problem of relative locality of local nets acquires more facets in theories which include various quantum fields $\varphi(f), \psi(f)$ as well as local nets $\{R_{\varphi,\psi}(O)\}$ associated with these fields. In such theories (studied in Section 3.2) fields are split into Borchers' classes and this raises the question, how the nets corresponding to the fields in the same class are related to each other. First of all, if the transition $\varphi \to R_\varphi(O)$ secures the locality of the net, then, clearly, the relative locality of the fields φ and ψ will secure the relative locality of $\{R_\varphi(O)\}$ and $\{R_\psi(O)\}$. In other words, the local nets corresponding to fields in the same Borchers class are relatively local. Next, according to Proposition 1.3.20, if in the same Borchers class the fields generate nets satisfying duality, then all these nets coincide with each other. Also, if some net $\{R_\chi(O)\}$ is known to satisfy duality, then all the nets $\{R_{\varphi_i}(O)\}$ corresponding to fields $\varphi_i(f)$

from the Borchers class of χ are subnets of $\{R_\chi(O)\}$, i.e. $R_{\varphi_i}(O) \subset R_\chi(O)$. In this case φ_i are said to be 'local functions' of χ [296]. The terminology fits the fact that for a free field (the local net of which satisfies duality, see Section 3.3) its Borchers class consists of its Wick polynomials only, i.e. local functions in the literal sense of the word. For other remarks on Borchers' classes in local quantum theory see [222, 296]. In [173, Section 6] Borchers' classes and local nets are analyzed for the fields satisfying 'generalized H-bound' (Section 3.1.2).

Despite considerable effort, only a few nontrivial results concerning duality were obtained in the purely axiomatic framework. This, obviously, suggests that duality touches deeper and more specific properties of quantum field systems than those reflected in the Haag–Araki axioms. This suggestion is well corroborated by the study of duality in various models, for nets of field algebras (Section 3.2) and for canonical systems. As for the pure Haag–Araki theory, there we have, first of all, an elementary necessary condition of duality showing its close ties with the global structure.

PROPOSITION 1.3.21. *Let Axioms* Ia, IIa, *and* IV *hold and duality be valid for at least one double cone* D. *Then* R' *is an abelian algebra.*

Proof. In proving Proposition 1.3.6 we have shown that our assumptions lead to $R' = R(D)' \cap R(D')'$. Together with the duality this gives $R' \subset R$. ❐

If R is a factor or $R(D)$ is a factor for at least one double cone D, the condition takes a stronger form, $R = \mathcal{B}(\mathcal{H})$. But it was never proved that this condition is already sufficient (though no counterexample showing the violation of duality for double cones in some irreducible Haag–Araki theory is known). Moreover, in general Haag–Araki theory no sufficient conditions of duality were found at all. The only exception is the following sufficient condition: $R = \mathcal{B}(\mathcal{H})$; $R(D)$ is a type I factor. This condition immediately follows from the above formula $R' = R(D)' \cap R(D')'$. When $R = \mathcal{B}(\mathcal{H})$ the formula means that $R(D)$ and $R(D')$ form a factorization [18, Section 3.4.4] and, if $R(D)$ is a type-I factor, then by virtue of the normalcy property of discrete factors [7, pp. 128, 287] this factorization is coupled, i.e. $R(D') = R(D)'$. However, all evidence of both axiomatic theory and models tells us that the type I is impossible for local algebras of quantum field theory, though the direct proof of this has thus far been absent. What was found is a certain condition under which the duality is 'representation-independent', i.e. its fulfillment in some irreducible locally normal* representation π of quasi-local C^*-algebra implies its fulfillment in all other representations of this class. The condition singles out the class of theories where duality is not linked to any representation dependent properties (such as the presence of vacuum). Its formulation, being somewhat overloaded by technicalities, fits, nevertheless, the situation in real models, and the condition proved usable. Relative intricacy of its proof is typical of all nontrivial results concerning duality.

* A representation π of a quasilocal algebra $\mathfrak{A}$ is called locally normal if its restriction to any local subalgebra $R(O)$ of $\mathfrak{A}$ is a normal representation of $R(O)$ [5, Section 2.6].

PROPOSITION 1.3.22 [167, 403]. *Let Axioms* I–III *and the split property hold and let us also assume that there exists an irreducible locally normal representation* π_0 *of* $\mathfrak{A}$ *with the following properties: for any double cones* D_1, D_2 with $D_1 << D_2$ *there is a double cone* D_3 *such that* $D_2 << D_3$ *and*

$$\pi_0 [R (D_1)' \cap R (D_2)] \subset \overline{\pi_0 [\mathfrak{A}(D_1{}') \cap R (D_3)]}^{w} \tag{3.18}$$

(closure in the weak operator topology). Then the duality in the representation π_0:

$$\overline{\pi_0 (\mathfrak{A}(D'))}^{w} = \pi_0 (R (D))'$$

implies the duality in any irreducible locally normal representation π *of* $\mathfrak{A}$.

Proof. Having fixed a double cone D, let us choose a decreasing sequence of double cones $\{D_n\}_{n=1}^{\infty}$ such that $D_{n+1} << D_n$, $\cap_{n=1}^{\infty} D_n = D$. Due to the split property there exists a sequence of discrete factors $\{M_n\}_{n=1}^{\infty}$ such that $R (D_{n+1}) \subset M_n \subset R (D_n) \subset M_{n-1} \subset \dots$. Then we choose an increasing sequence $\{\hat{D}_i\}_{i=1}^{\infty}$ of double cones covering in the limit $i \to \infty$ all the Minkowski space and such that $\hat{D}_i << \hat{D}_{i+1}$, $\hat{D}_i >> D_n$ for all $i, n = 1, 2, \dots$. For this sequence there is another sequence of discrete factors $\{N_i\}_{i=1}^{\infty}$ such that $R (\hat{D}_i) \subset N_i \subset R (\hat{D}_{i+1}) \subset N_{i+1} \subset \dots$. Now let us show that for any N_i there exists a double cone $\tilde{D}_i$ such that

$$\overline{\pi_0 [\mathfrak{A}(D') \cap R (\tilde{D}_i)]}^{w} \cap \pi_0 (N_i) = \pi_0 [R (D') \cap N_i] . \tag{3.19}$$

Indeed, we have

$$\pi_0 [R (D)' \cap N_i] \subset \pi_0 [R (D') \cap R (\hat{D}_{i+1})] \subset \overline{\pi_0 [\mathfrak{A}(D') \cap R (D_i)]}^{w}$$

where the former inclusion follows from $N_i \subset R (\hat{D}_{i+1})$ and the latter from $D << \hat{D}_{i+1}$ and (3.18); moreover, $\tilde{D}_i >> \hat{D}_{i+1}$. Taking here $\cap N_i$ we obtain the inclusion from the right-hand side in (3.19). The opposite inclusion is obvious by virtue of locality.

Due to the local normality of π_0, its restrictions to N_i are normal representations of N_i; in addition, they are faithful because N_i is a factor and its normal representation is an isomorphism. Owing to this, if we apply to (3.19) the mapping $\pi_0 \circ \pi_0^{-1}$ which is faithful and normal on every $\pi_0 (N_i)$ we find that (3.19) is valid in the representation π as well:

$$\overline{\pi_0 [\mathfrak{A}(D') \cap R (\tilde{D}_i)]}^{w} \cap \pi_0 (N_i) = \pi_0 [R (D)' \cap N_i] . \tag{3.20}$$

The desired duality in the representation π can be obtained by taking here the limit $i \to \infty$. To find the limit of the right-hand side let us show that

$$\pi [R (D)' \cap \mathfrak{A}] = \left\{ \overline{\bigcup_{i=1}^{\infty} \pi [R (D)' \cup N_i]}^{u} \right\} . \tag{3.21}$$

We have to prove that for any $\varepsilon > 0$ and $A \in \pi\,[R\,(D)' \cap \mathfrak{A}]$ there exist an index $i \in \mathbf{N}$ and an operator $C \in \pi\,[R\,(D)' \cap N_i]$ such that $\| A - C \| \leq \varepsilon$. Since $\mathfrak{A} = \overline{\cup_{i=1}^{\infty} N_i}^{u}$ and $\pi\,(\mathfrak{A}) = \overline{\cup_{i=1}^{\infty} \pi\,(N_i)}^{u}$, an index $i \in \mathbf{N}$ and an operator $C_1 \in \pi\,(N_i)$ exist such that $\| A - C_1 \| \leq \varepsilon/2$. Hence it is easy to verify that $C = E\,(A)$ can be used as the operator we need, where $E : \mathcal{B}\,(\mathcal{H}_\pi) \to \pi\,(N_i)$ is the conditional expectation from $\mathcal{B}\,(\mathcal{H}_\pi)$ onto the direct factor $\pi\,(N_i)$. Indeed, by the standard properties of the conditional expectation (see, e.g. [23, pp. 101–102])

$$\| E\,(A) - C_1 \| = \| E\,(A - C_1) \| \leq \| A - C_1 \| \leq \varepsilon/2\,;$$

next, if $B \in \pi\,[R\,(D)] \subset \pi\,(N_i)$ then $[E\,(A), B] = E\,([A, B]) = 0$, so that

$$E\,(A) \in \pi\,[R\,(D)]' \cap \pi\,(N_i) = \pi\,[R\,(D)' \cap N_i]\,.$$

Moreover, $\| A - E\,(A) \| = \| A - C_1 + C_1 - E\,(A) \| \leq \varepsilon$, and the formula (3.21) is proved. Since the sequence $\{N_i\}$ is monotone increasing and for any $r \in \mathbf{N}$,

$$\pi\,[R\,(D)' \cap N_r] = \bigcup_{i=1}^{r} \pi\,[R\,(D)' \cap N_i]\,,$$

we conclude that the sequence of C^*-algebras on the right-hand side of (3.20) has a C^*-inductive limit at $i \to \infty$ equal to $\pi\,[R\,(D)' \cap \mathfrak{A}]$.

On the left-hand side, due to $N_i \subset R\,(\tilde{D}_i)$, both sequences $\pi\,(N_i)$ and $\pi\,(R\,(\tilde{D}_i))$ have a C^*-inductive limit equal to $\pi\,(\mathfrak{A})$, whence for the whole left-hand side this limit equals $\overline{\pi\,[\mathfrak{A}(D')]}^{w} \cap \pi\,(\mathfrak{A})$. Thus in the limit $i \to \infty$ (3.20) becomes

$$\overline{\pi\,[\mathfrak{A}(D')]}^{w} \cap \pi\,(\mathfrak{A}) = \pi\,(R(D)' \cap \mathfrak{A})\,,$$

and by $\overline{\pi\,(\mathfrak{A})}^{w} = \mathcal{B}\,(\mathcal{H}_\pi)$ we get

$$\overline{\pi\,[\mathfrak{A}(D')]}^{w} = \overline{\pi\,(R\,(D)' \cap \mathfrak{A})}^{w}.$$

Now we have only to prove that

$$\overline{\pi\,(R\,(D)' \cap \mathfrak{A})}^{w} = \pi\,(R\,(D))'\,. \tag{3.22}$$

To this end we use the decreasing sequence of discrete factors $\{M_n\}_{n=1}^{\infty}$. Using duality in the representation π_0, we obtain

$$\bigcap_{n=1}^{\infty} M_n = R\,(D)\,,$$

whence

$$\pi\,[R\,(D)]' = \left\{ \bigcup_{n=1}^{\infty} \pi\,(M_n)' \right\}''. \tag{3.23}$$

To transform this, let us prove that

$$\pi(M_n)' = \overline{\pi(M'_n \cap \mathfrak{A})}^{w} . \tag{3.24}$$

Since $\hat{D}_i >> D_n$, $i, n = 1, 2, \dots$, we have $M_n \subset N_i$ and, due to the normalcy of the discrete factor M'_n with the von Neumann subalgebra N'_i,

$$N_i = \{M_n \cup (M'_n \cap N_i)\}'' \subset \{M_n \cup (M'_n \cap \mathfrak{A})\}'', \ i = 1, 2, \dots .$$

This relation holds also in the representation π, the latter being locally normal relative to $\{M_n\}$ and $\{N_i\}$. Hence we get

$$\left\{ \bigcup_{i=1}^{\infty} \pi(N_i) \right\}'' = \mathcal{B}(\mathcal{H}_\pi) = \{\pi(M_n) \cup \pi(M'_n \cap \mathfrak{A})\}''$$

$$= \left\{\pi(M_n) \cup \overline{\pi(M'_n \cap \mathfrak{A})}^{w} \right\}''$$

i.e. the abelian von Neumann algebras $\pi(M_n)$ and $\pi(M'_n \cap \mathfrak{A})$ form a factorization which is coupled because the factor is discrete. This means that these algebras satisfy (3.24). Inserting (3.24) into (3.23) and taking into account that for all $n \in \mathbf{N}$, $M'_n \cap \mathfrak{A} \subset R(D)' \cap \mathfrak{A}$, we get the inclusion from the right-hand side in (3.22). The inverse inclusion is trivial. ❐

The additional conditions in this proposition are both physically meaningful. As mentioned above, the split property is equivalent to a certain principle of local preparation of states. The other condition, (3.18), according to [167], means the absence of a definite class of observables, such as magnetic flux.

The criterion obtained can be used for proving duality in such models that have a nontrivial interaction but are constructed as representations of some canonical system satisfying duality (cf. Section 3.3; all locally Fock constructive models belong to this type, in particular). It is this way that was followed by the only existing proofs of duality in models with nontrivial interaction $(P(\varphi))_2$ and $(Y)_2$ [403]. The formulation of the criterion was also generalized to fit the formalism of field algebras for general Bose–Fermi systems.

Since the axiomatic Haag–Araki theory lacks both proofs and realistic sufficient conditions of duality, various weaker forms of the latter were repeatedly proposed. The first of them was the following condition called *weak duality*:

$$\overline{R(D)' \cap \mathfrak{A}}^{w} = R(D)' \cap R . \tag{3.25}$$

The usual duality readily implies the weak one because it follows from the definition of $R(D')$ that

$$R(D') = \overline{R(D') \cap \mathfrak{A}}^{w} \subset \overline{R(D)' \cap \mathfrak{A}}^{w} .$$

Condition (3.25) was introduced by Borchers [77] in 1965 in connection with the problem of constructing fields as intertwining operators of coherent superselection sectors (see Chapter 2). Later on it was analyzed by Doplicher, Haag and Roberts who found that it is "not sufficiently strong" to secure the existence of fields. Nevertheless, in recent studies by Buchholz and Fredenhagen [111] it was found to be useful if not for constructing the fields themselves then at least for introducing field-like properties in local quantum theory: it was proved that (3.25), like usual duality, excludes nonabelian coherent sectors (in which the global algebra is reducible) and, hence, all systems with nonabelian gauge groups and abnormal statistics (see Section 2.3).

Another useful weakening of duality is provided by so-called 'essential duality'.

DEFINITION 1.3.2 [296]. We say that a net of local algebras $\{R(O)\}$ satisfies essential duality if algebras $\hat{R}(D) \equiv R(D')'$ corresponding to double cones D, satisfy usual duality, $\hat{R}(D) = \hat{R}(D')'$.

It is obvious that usual duality entails essential duality. The converse is, in general, not true which is demonstrated by the example of generalized free fields. It will be shown in Section 3.3 that essential duality holds for all generalized free fields, whereas usual duality is violated for certain classes of them.

PROPOSITION 1.3.23. *Local net* $\{R(O)\}$ *satisfies essential duality if and only if the algebras* $\hat{R}(D)$ *satisfy locality. The net* $\{\hat{R}(O)\}$ *is the maximal local extension of* $\{R(O)\}$, *and on unbounded regions the both nets coincide,* $R(D') = \hat{R}(D')$.

Proof. In the first assertion the necessity is evident, whereas the sufficiency follows from the relations:

$$\hat{R}(D) \subset \hat{R}(D')' = R(D) \subset R(D')' = \hat{R}(D) .$$

Maximality of $\{\hat{R}(O)\}$ follows from Proposition 1.3.20 and the last assertion is evident. ❒

The proposition shows that essential duality is physically nothing other than transitivity of locality: we introduce a new net $\hat{R}(D)$ which is local relative to the initial local net: for all $O_1, O_2 \in B(M)$, $O_1 \times O_2$,

$$\hat{R}(O_1) \subset R(O_2)' ,$$

and fulfilling essential duality is tantamount to the fact that this new net is also local. Therefore, all statements giving sufficient conditions of essential duality can be interpreted as variants of the well known theorem [27, Theorem 4-21, Corollary] about the transitivity of locality in Wightman theory.

The good thing about essential duality is that for field algebras it can be derived from the Wightman axioms (Section 3.2), while in the Haag–Araki theory some realistic sufficient conditions of it are known. These conditions are obtained in the Bisognano–Wichmann formalism which is based on special properties of the wedge domains W_R, W_L. Development of this formalism has been one of the major advances in the technique of the algebraic approach. We shall treat it in full detail in Section 3.2 and present here only the initial elements relating to the essential duality problem.

It follows from definitions (3.5) that

$$W_R' = W_L\,,\ W_L' = W_R \tag{3.26}$$

(hence, these regions are another example of diamonds that are not double cones). Regions W_R and W_L are invariant under several subgroups of the Poincaré group $\mathcal{P}_+^\uparrow$:

(1) the subgroup of translations along the axes x^2 and x^3;

(2) the subgroup of rotations $R\,(e^1, \varphi), \varphi \in [0, 2\pi]$ in the (x^2, x^3) plane (e^1 is the unit vector along the x^1 axis);

(3) the subgroup $\Lambda\,(e^1, t)\,, t \in \mathbf{R}$, of Lorentz boosts along the x^1 axis.

The appropriate subgroups of the unitary group $U\,(a, \Lambda)$ implement spatial automorphisms of the algebras $R\,(W_R)$ and $R\,(W_L)$. Of all them, the subgroup $V\,(t) \equiv U\,(0, \Lambda\,(e^1, t))$ implementing the Lorentz boosts is of special importance since the complexification of the subgroup $\{\Lambda\,(e^1, t)\}_{t \in \mathbf{R}}$ includes the transformation

$$G : (x^0, x^1, x^2, x^3) \to (-x^0, -x^1, x^2, x^3)$$

(inversion in the (x^0, x^1) plane) mapping W_R onto W_L: indeed, we can identify G with $\Lambda\,(e^1, i\,\pi)$. Due to this property it proves possible to show under certain additional conditions that unitary groups $V\,(\pm\, t)$ coincide with modular groups of the algebras $R\,(W_R)$ and $R\,(W_L)$, respectively, and these algebras satisfy duality. The said result can be obtained as a corollary to the following theorem characterizing the modular group of two commuting von Neumann algebras with a common bicyclic vector.

RIGOTTI THEOREM [359]. *Let R_1 and R_2 be von Neumann algebras commuting with each other and possessing a common bicyclic vector χ. Let also a strongly continuous unitary group $W\,(t) = \exp\,(-itA)$ satisfy the following conditions*:

(1) $W\,(t)\, R_i\, W\,(t)^* = R_i\,,\ i = 1, 2,\ t \in \mathbf{R}$;

(2) $W\,(t)\, \chi = \chi\,,\ t \in \mathbf{R}$;

(3) $R_1\, \chi \subset \mathrm{dom}\, W\,(i)\,,\ W\,(i)\, R_1^h\, \chi \subset R_2^h\, \chi\,,$

where $R_i^h\,,\ i = 1, 2$ is the Hermitian part of R_i and $W\,(i) = \exp A$ is the closed unbounded operator from the complexification of $W\,(t)$.

Then $R_1 = R_2'$ and $W(t) = \Delta^{-it/2}$, where Δ is the modular operator of the pair (R_1, χ).

This theorem is directly applicable to the group $V\,(t)$ and algebras $R\,(W_R)$ and $R\,(W_L)$ of Haag–Araki theory with a cyclic vacuum: clearly, the Haag–Araki axioms ensure fulfillment of all the conditions of the theorem except, possibly, for the last one. Let $V\,(t) \to V\,(t + i\,\lambda)$ be the complexification of the unitary group $V\,(t)$ which can be represented, by the Stone theorem, as $V\,(t) = \exp\,(-itK)$ with a self-adjoint generator K; all $V\,(\tau), \tau \in \mathbf{C}$, are closed operators with the domains depending only on $\mathrm{Im}\,\tau$. ($V\,(\tau)$ form the representation of the corresponding subgroup of the complex Lorentz group, but they, in general, no longer define automorphisms of R or $R\,(W_{R,L})$, because the relativistic covariance axiom does not cover complex Poincaré transformations.) Then we can prove the following proposition.

PROPOSITION 1.3.24 [359]. *Let Axioms* II, III, V *and* VI *hold and the theory have a cyclic vacuum vector* Ω. *Let also the subgroup* $V(t)$, $t \in \mathbf{R}$, *and the operator* $V(i\pi)$ *from its complexification* $V(\tau)$, $\tau \in \mathbf{C}$, *satisfy the condition*

$$R(D)\,\Omega \subset \mathrm{dom}\, V(i\pi), \quad V(i\pi)\, R(D)^h\,\Omega \subset R(W_L)^h\,\Omega \tag{3.27}$$

for any double cone $D \subset W_R$. *Then the algebras* $R(W_R)$ *and* $R(W_L)$ *satisfy duality:* $R(W_R)' = R(W_L)$, $R(W_L)' = R(W_R)$, *the net* $R(O)$ *satisfies essential duality:*

$$\hat{R}(D) = \hat{R}(D')' \quad \forall D \in B(M),$$

and $V(\pm t)$ *coincides with the modular group of the pair* $(R(W_R), \Omega)$ *or* $(R(W_L), \Omega)$, *respectively:*

$$V(\pm t) = \Delta^{\mp it/2\pi},$$

$\Delta^{\pm 1} = \exp(\pm 2\pi K)$ *being the corresponding modular operator.*

Proof. Considering the Reeh–Schlieder theorem, the pair of algebras $R(W_R)$ and $R(W_L)$ satisfies the conditions of the Rigotti theorem, whereas the group $V(t)$ meets conditions 1 and 2 of the theorem. However (3.27) does not coincide with criterion 3 which singles the modular group out of all unitary automorphism groups leaving a given bicyclic vector invariant. But, as we readily see, (3.27) implies condition 3 for the pair of algebras $\vee_{D \subset W_R} R(D), R(W_L)$, which satisfy the other conditions of the theorem as well. According to [359], the theorem is also valid when R_1 is a weakly nonclosed *-algebra, giving in such a case the following result:

$$R(W_L)' = \left(\bigvee_{D \subset W_R} R(D) \right)'' = R(W_R),$$

which is, due to (3.26), duality for $R(W_R)$ and $R(W_L)$.

Essential duality, the only thing yet to prove, follows from the main bulk of the Bisognano–Wichmann formalism. As we show in Section 3.2, if a theory is defined initially by the net of local Op^*-algebras $\mathcal{P}(O)$ generated by Wightman fields and satisfies, besides the Wightman axioms, an additional condition similar to (3.27), then we can introduce the net of von Neumann algebras $F(O) = L(O')'_N$ and prove locality of this net (Proposition 3.2.10). (Here $L(O)$ is the set of all field operators $A(f)$ with $\mathrm{supp} f \subset O'$ and $L(O')'_N$ is the von Neumann commutant of this set, defined on p. 208).The result desired can be obtained by applying Proposition 3.2.10 instead of the net $\{F(O)\}$ to the net $\hat{R}(O) = R(O')'$, where $R(O)$ is the usual Haag–Araki net with the additional condition (3.27). Indeed, the nets $F(O)$ and $\mathcal{P}(O)$ are related to each other in the same way as the nets $\hat{R}(O)$ and $R(O)$. Hence, by repeating the proof of Proposition 3.2.10, we derive locality for $\hat{R}(O)$ from locality of $R(O)$ under the condition (3.27). By Proposition 1.3.23, this is essential duality for $R(O)$. ❐

Comparing the result obtained and Proposition 1.3.22 we see that both propositions relate to different classes of local quantum theories. Indeed, in a class described by

Proposition 1.3.22 duality is representation-independent, whereas in the case of Proposition 1.3.24 it is ensured by properties which are obviously representation-dependent (e.g., the existence of a cyclic vacuum) and, in general, is not transferable to other representations of quasilocal algebra.

As we shall see in Section 3.2 condition (3.27) is connected with the existence of *CPT* operator θ. In Wightman theory where θ is known to exist (due to the *CPT* theorem by Pauli–Lüders–Jost) with its help we shall prove (3.27) for the Op^*-algebra $\mathcal{P}(W_R)$ and the von Neumann algebra $F(W_R)$, i.e. in this theory (3.27) is a consequence of the existence of the *CPT* operator. It is not clear whether the conclusion holds for the Haag–Araki theory as well; in other words, whether it is sufficient to supplement this theory with the operator θ to obtain the condition (3.27) and, hence, all the statements of Proposition 1.3.24 together with other structural properties stemming from the existence of θ within the field theory framework (see Theorem 3.2.7). This is due to the fact that for deriving (3.27) and related properties specific technique of the Wightman theory is used which is based on the analytical continuation of Wightman functions and vector-valued functions of the form $A(f_1)\dots A(f_n)\,\Omega$ in parameters of both the translation group and the Lorentz group. In the Haag–Araki theory the similar technique has not been developed so far. It was found, however, that a certain enhancement of (3.27) is sufficient to prove the existence of *CPT* operator in the Haag–Araki theory. This sufficient condition is as follows [359]:

$$R(D)\,\Omega \subset \operatorname{dom} V(i\pi)\,,$$

$$V(i\pi)\,R(D)^h\,\Omega \subset R(D^j)^h\,\Omega\,,$$

$$D^j = \{x \in M \mid (-x^0, -x^1, x^2, x^3) \in D\} \tag{3.28}$$

for any double cone $D \subset W_R$. If it holds, the *CPT* operator will be $\theta = U(0, R(e^1, \pi))\,J$, where J is the modular involution for the pair $(R(W_R), \Omega)$. Mathematical meaning of such conditions will be clarified in Section 3.2, where the modular structure of wedge regions algebras will be fully described.

Coming back to duality, we can say now that the possibility of its deduction in Proposition 1.3.24 arises substantially due to the adding to the Haag–Araki theory some more assumptions about its symmetry properties. Mathematically, these assumptions mean the enhancement of relativistic covariance axiom by the requirement of definite analyticity properties in the Lorentz group parameters: so that the condition $U(0,\Lambda)\,R(O)\,U(0,\Lambda)^{-1} = R(\Lambda^{-1} O)$ could be analytically continued to Λ from the complex Lorentz group (see also Section 1.3.5). Another way of complementing the axiomatic formalism by new elements opens up for canonical systems and here it also proves possible to derive duality (Section 3.3).

Finally, we should mention that essential duality was used by Roberts [363–365] in his attempts to use the technique of Hermitian bimodules over a sheaf of von Neumann algebras for a more adequate description of superselection structure. In these studies another weak form of duality was also introduced, so-called *internal duality*:

$$\mathfrak{A}(O) = \mathfrak{A}(O)^{cc}, \tag{3.29}$$

where $\{\mathfrak{A}(O), O \in B(M)\}$ is a net of local C^*-algebras which can be treated as weakly dense subalgebras of $R(O)$, and $\mathfrak{A}(O)^c = \mathfrak{A}(O)' \cap \mathfrak{A}$ is the 'relative commutant' of $\mathfrak{A}(O)$ in $\mathfrak{A}$. It is easy to verify that (3.29) is, indeed, a weakening of duality: if for a region $O \in B(M)$ duality holds at least in one faithful locally normal representation π of a quasilocal algebra $\mathfrak{A}$, then internal duality holds for the algebra itself. Indeed, from $\mathfrak{A}(O)' \subset \mathfrak{A}(O)^c$ it follows that $\pi(\mathfrak{A}(O)^{cc}) \subset \pi(R(O'))' \cap \pi(\mathfrak{A})'' = \pi(R(O))$ (the last equality is duality for $\pi(R(O))$. Since π is faithful, the inclusion means that

$$\mathfrak{A}(O)^{cc} \subset R(O) \cap \mathfrak{A} = \mathfrak{A}(O).$$

The inverse inclusion being trivial, we have internal duality for $\mathfrak{A}(O)$. The result obtained means also that, unlike usual duality, internal duality is always representation independent.

1.3.4. CAUSAL INDEPENDENCE AND OTHER CAUSAL PROPERTIES

Now we proceed to the causal independence property which is sometimes called also local or statistical independence. It is one of the few properties for which the Haag–Araki and Haag–Kastler theories give markedly different description. The main source of the difference is the different treatment of the set of physical states. As we have seen in Section 1.2, for the Haag–Kastler theory it is natural to consider all states on an abstract C^*-observable algebra $\mathfrak{A}$ as physical states; while in the Haag–Araki theory, which can be treated as a concrete representation of the abstract theory, it is natural to identify physical states with normal states (density matrices) of this representation. As a result, there are two formulations of causal independence, respectively, for the abstract and concrete algebraic approaches:

$$\text{(CI)} \quad \forall_{O_1 \times O_2} \quad \forall_{\substack{\omega_i \in \mathfrak{A}(O_i)^{*+}_1 \\ i=1,2}} \quad \exists_{\omega \in \mathfrak{A}^{*+}_1} \quad \omega|_{\mathfrak{A}(O_i)} = \omega_i, \; i = 1, 2,$$

$$\text{(CI)}_N \quad \forall_{O_1 \times O_2} \quad \forall_{\substack{\omega_i \in N(R(O_i)) \\ i=1,2}} \quad \exists_{\omega \in N(R)} \quad \omega|_{R(O_i)} = \omega_i, \; i = 1, 2,$$

where $N(M) = M_* \cap M^{*+}_1$ for a von Neumann algebra M. Of course, within the Haag–Araki theory both conditions make sense and they are, in general, independent of each other.

Clearly, both conditions (CI) and (CI)_N can be treated as purely mathematical properties irrespective of any physics: to this end we only have to take in their formulations arbitrary commuting C^* or W^*-algebras instead of local algebras $\mathfrak{A}(O_i)$ or $R(O_i)$. The meaning of this generalized causal independence becomes evident due to its close link with three other structural properties of a pair of commuting C^* or W^*-algebras $\mathcal{A}_1$ and $\mathcal{A}_2$. The first of them is the Schlieder property, one more of the independence relations introduced in Section 1.2. This property we shall also provide with a purely mathematical formulation: a pair $\mathcal{A}_1$, $\mathcal{A}_2$ of commuting C^* or W^*-algebras is said to possess the Schlieder property if the following holds:

$$(S) \qquad \underset{\substack{A \in \mathcal{A}_1 \\ A \neq 0}}{\forall} \quad \underset{\substack{B \in \mathcal{A}_2 \\ B \neq 0}}{\forall} \quad A\,B \neq 0\,.$$

Next property in question is the existence on the algebra $\mathcal{A}_{12}$, the C^* or W^*-algebra generated by $\mathcal{A}_1$ and $\mathcal{A}_2$, of a so-called *product state* φ for which, by definition,

$$\varphi\,(AB) = \varphi\,(A)\,\varphi\,(B) \;\; \forall\, A \in \mathcal{A}_1\,,\; B \in \mathcal{A}_2$$

(if here φ is a state on some algebra $\mathcal{A} \supset \mathcal{A}_{12}$, then φ is called a product state for the pair $\mathcal{A}_1$, $\mathcal{A}_2$). And the last structural property in this group is the isomorphism of the algebra $\mathcal{A}_{12}$ and C^* or W^*-tensor product of the algebras $\mathcal{A}_1$ and $\mathcal{A}_2$.

Relations between the above structural properties we shall first analyze for abstract C^*-algebras (and condition (CI)) and then for von Neumann algebras (and condition $(CI)_N$). To start with, let us recall the basic concepts concerning tensor products of C^* and W^*-algebras (see [5, Section 2.7.2] and in greater detail [29, Chapter IV]). Let $\mathfrak{A}_1$ and $\mathfrak{A}_2$ be unital subalgebras of an abstract C^*-algebra $\mathfrak{A}$. We denote by $\mathfrak{A}_1 \odot \mathfrak{A}_2$ the direct algebraic product of $\mathfrak{A}_1$ and $\mathfrak{A}_2$, which is the set of all finite sums of the form $\Sigma_{i=1}^{n} A_i \otimes B_i\,, A_i \in \mathfrak{A}_1\,,\ B_i \in \mathfrak{A}_2$. On $\mathfrak{A}_1 \odot \mathfrak{A}_2$ one can introduce various topologies τ and then complete this set up to the topological tensor product $\mathfrak{A}_1 \otimes_\tau \mathfrak{A}_2$. In particular, topologies τ include C^*-norms, completions in which are C^*-algebras and cross-norms defined by the property $\| A \otimes B \| = \| A \| \cdot \| B \|$; a special role among them belongs to the so-called injective C^*-cross-norm $\| \cdot \|_\alpha$ (or, briefly, 'α-norm') which is the minimal of all C^*-norms. If $\mathfrak{A}_1$, $\mathfrak{A}_2$ are operator C^*-algebras in a Hilbert space $\mathcal{H}$, then $\| \cdot \|_\alpha$ coincides with the natural operator norm in a Hilbert space $\mathcal{H} \otimes \mathcal{H}$. In the case of von Neumann algebras R_1 and R_2 (acting, in general, in different Hilbert spaces $\mathcal{H}_1$ and $\mathcal{H}_2$) a W^*-tensor product $R_1 \bar{\otimes} R_2$ is defined that coincides with the von Neumann algebra in $\mathcal{H}_1 \otimes \mathcal{H}_2$ generated by all operators of the form $A \otimes I_2$, $A \in R_1$, and $I_1 \otimes B$, $B \in R_2$ (I_i is the identity operator in $\mathcal{H}_i$, $i = 1, 2$).

The following statement provides a comprehensive list of properties concerning causal independence in the abstract algebraic approach.

THEOREM 1.3.25 [370]. *Let $\mathfrak{A}_1$ and $\mathfrak{A}_2$ be commuting unital C^*-subalgebras of a C^*-algebra $\mathfrak{A}$. The following conditions are equivalent :*

(1) *$\mathfrak{A}_1$ and $\mathfrak{A}_2$ are causally independent in the sense of* (CI) *;*

(2) *$\mathfrak{A}_1$ and $\mathfrak{A}_2$ have the Schlieder property (S) ;*

(3) *for any pair of states φ_i on $\mathfrak{A}_i$, $i = 1, 2$, there exists a product state φ on $\mathfrak{A}$ such that $\varphi\,(AB) = \varphi_1\,(A)\,\varphi_2\,(B)\ \forall\, A \in \mathfrak{A}_1,\ B \in \mathfrak{A}_2$;*

(4) *there is an algebraic isomorphism $\Phi:\ \mathfrak{A}_1 \vee \mathfrak{A}_2 \to \mathfrak{A}_1 \odot \mathfrak{A}_2$* such that for all *$A_i \in \mathfrak{A}_1$, $B_i \in \mathfrak{A}_2$,*

$$\Phi\left(\sum_{i=1}^{n} A_i B_i\right) = \sum_{i=1}^{n} A_i \otimes B_i\,; \tag{3.30}$$

and imply

(5) *there is a continuous *-homomorphism*

$$\bar{\Phi};\ \overline{\mathfrak{A}_1 \vee \mathfrak{A}_2}^{\,u} \to \mathfrak{A}_1 \otimes_\alpha \mathfrak{A}_2\ .$$

Proof. We shall prove $1 \to 2 \to 4 \to 3 \to 1$ and $4 \to 5$. Of these, the most nontrivial are the implications connected with tensor products: $2 \to 4$ and $4 \to 5$. Their proofs will be given schematically; for the complete exposition see the original work by Roos [370].

$\underline{1 \to 2}$. Let us suppose that there are non-zero $A \in \mathfrak{A}_1$, $B \in \mathfrak{A}_2$ such that $AB = 0$. Then for elements $A' = \beta^{-1} A^*A \in \mathfrak{A}_1$ (where $\beta \in \sigma(A^*A)$, $\beta \neq 0$, $\sigma(\cdot)$ is the spectrum of a C^*-algebra element, see [8, Section 1]) and $B' = BB^* \in \mathfrak{A}_2$ we have

$$A'B' = 0\,,\ A' \neq 0\,,\ B' \neq 0\,,\ A'^* = A'\,,\ 1 \in \sigma(A')\,,$$

whence

$$C \equiv (I - A')^2 \geq 0,\ 0 \in \sigma(C)\,,\ \beta \in \sigma(C + \beta)\,. \tag{3.31}$$

On the subspace of $\mathfrak{A}$ generated by operators I and C we define a functional $\tilde{\varphi}_1$ by the conditions

$$\tilde{\varphi}_1(I) = 1\,,\quad \tilde{\varphi}_1(C) = 0.$$

This functional is positive, since by virtue of (3.31), $\lambda I + \mu C \geq 0$, $\lambda, \mu \in \mathbf{R}_+$, implies $\tilde{\varphi}_1(\lambda I + \mu C) = \lambda \geq 0$. Hence it can be extended to a state φ_1 on $\mathfrak{A}_1$, for which $\varphi_1((I - A')^2) = 0$. Moreover, due to the inequality $|\varphi_1(A)|^2 \leq \|\varphi_1\|\, \varphi_1(A^*A)$ for all $A \in \mathfrak{A}_1$, we also have $\varphi_1(I - A') = 0$.

Now let us choose a state φ_2 on $\mathfrak{A}_2$ such that $\varphi_2(B') \neq 0$ and show that φ_1 and φ_2 cannot have a common extension onto $\mathfrak{A}$. Indeed, if there is such an extension φ then, by the Schwartz inequality

$$|\varphi((I - A')(I + B'))|^2 \leq \varphi((I - A')^2)\, \varphi((I + B')^2) = \varphi_1((I - A')^2)\, \varphi_2((I + B')^2) = 0.$$

On the other hand, however,

$$\varphi((I - A')(I + B')) = \varphi(I - A' + B') = \varphi_1(I - A') + \varphi_2(B') = \varphi_2(B') \neq 0\,,$$

and we obtain that violation of (S) implies violation of (CI).

$\underline{2 \to 4}$. As we can check easily, the mapping Φ defined by (3.30) is the required isomorphism of *-algebras $\mathfrak{A}_1 \vee \mathfrak{A}_2$ and $\mathfrak{A}_1 \odot \mathfrak{A}_2$, if the following property holds which is called 'algebraic independence' of $\mathfrak{A}_1$ and $\mathfrak{A}_2$: let $\{A_i\}_{i=1,\ldots,n}$, $\{B_k\}_{k=1,\ldots,m}$ be some sets of linearly independent elements from $\mathfrak{A}_1$ and $\mathfrak{A}_2$, respectively; then $\{A_iB_k\}_{i=1,\ldots,n,\,k=1,\ldots,m}$ is a set of linearly independent elements from $\mathfrak{A}$. Hence, it suffices to prove that the Schlieder property implies the algebraic independence of $\mathfrak{A}_1$ and $\mathfrak{A}_2$. The proof of this fact [370, Corollary] is rather cumbersome, though simple by its idea, and consists of several assertions about linear independence properties for elements of the form $A_i B_i, A_i \in \mathfrak{A}_1, B_i \in \mathfrak{A}_2$. According to the main one, if the Schlieder property holds, then the property $\Sigma_{i=1}^{n} A_i B_i = 0$ implies linear independence of both sets $\{A_i\}$ and $\{B_i\}$, $i = 1, \dots, n$.

4 → 3. For arbitrary states φ_1 on $\mathfrak{A}_1$ and φ_2 on $\mathfrak{A}_2$ we introduce a functional $\tilde{\varphi} = \varphi_1 \otimes \varphi_2$ on the *-algebra $\mathfrak{A}_1 \odot \mathfrak{A}_2$ and note that it is also a state [29, Chapter IV, Section 4]. By virtue of 4, $\tilde{\varphi}$ determines on $\mathfrak{A}_1 \vee \mathfrak{A}_2$ the functional φ_{12} defined by the formula

$$\varphi_{12}(\cdot) = \varphi(\Phi(\cdot)) = \varphi_1 \otimes \varphi_2 \circ \Phi$$

and this functional is also a state, because its positivity follows from the fact that properties $A \in \mathfrak{A}_1 \vee \mathfrak{A}_2$, $A \geq 0$ imply $\Phi(A) \geq 0$. Clearly, (3.30) guarantees that φ_{12} is a product state:

$$\varphi_{12}(AB) = (\varphi_1 \otimes \varphi_2)\, \Phi(AB) = \varphi_1(A)\, \varphi_2(B) \quad \forall A \in \mathfrak{A}_1, B \in \mathfrak{A}_2 .$$

Finally, φ_{12} can be extended to a state φ on $\mathfrak{A}$, and condition 3 is obtained.

The implication 3 → 1 is evident.

4 → 5. It is easy to show that an isomorphism Φ satisfying (3.30) is continuous relative to the α-norm. To this end let us introduce on $\mathfrak{A}_1 \odot \mathfrak{A}_2$ the 'β-norm'

$$\left\| \sum_{i=1}^{n} A_i \otimes B_i \right\|_{\beta} = \left\| \sum_{i=1}^{n} A_i B_i \right\| .$$

It is a norm due to the condition 4 and we can check directly that it satisfies the C^*-condition, i.e.

$$\left\| \left(\sum_{i=1}^{n} A_i \otimes B_i \right)^* \left(\sum_{i=1}^{n} A_i \otimes B_i \right) \right\|_{\beta} = \left\| \left(\sum_{i=1}^{n} A_i \otimes B_i \right) \right\|_{\beta}^{2} .$$

Since the α-norm is the minimal one of all C^*-norms, we have

$$\left\| \Phi \left(\sum_{i=1}^{n} A_i B_i \right) \right\|_{\alpha} = \left\| \sum_{i=1}^{n} A_i \otimes B_i \right\|_{\alpha} \leq \left\| \sum_{i=1}^{n} A_i \otimes B_i \right\|_{\beta} = \left\| \sum_{i=1}^{n} A_i B_i \right\| .$$

This means that Φ is continuous relative to $\| \cdot \|_{\alpha}$, and so it can be extended to the homomorphism

$$\bar{\Phi} : \overline{\mathfrak{A}_1 \vee \mathfrak{A}_2}^{\,u} \to \mathfrak{A}_1 \otimes_{\alpha} \mathfrak{A}_2 . \qquad \square$$

It is obvious why condition 5 of this theorem is, in general, weaker than 1–4. For an arbitrary homomorphism $\bar{\Phi}$ we cannot prove that its restriction to $\mathfrak{A}_1 \vee \mathfrak{A}_2$ would have the property (3.30) (in the presence of which the mapping $\bar{\Phi}$ might be called 'natural'). Because of this we cannot deduce either 4 or 3 from 5. In the case of W^*-algebras the situation, however, proves different. Here the analogue of condition 5 will be not a homomorphism but an isomorphism of corresponding algebras and one can find a sufficient condition, under which the isomorphism is natural (see Proposition 1.3.27

The following can be considered as a direct complement to Theorem 1.3.25.

PROPOSITION 1.3.26 [370]. *Let* $\mathfrak{A}_1$, $\mathfrak{A}_2$ *be unital commuting* C^**-subalgebras of a* C^**-algebra* $\mathfrak{A}$ *satisfying* (CI). *Then for any pair of pure states* φ_1 *on* $\mathfrak{A}_1$ *and* φ_2 *on* $\mathfrak{A}_2$ *there exists a pure state* φ *on* $\mathfrak{A}$ *which is a common extension of* φ_1 *and* φ_2 .

Proof. With the notations $\mathfrak{A}_{12} = \overline{\mathfrak{A}_1 \vee \mathfrak{A}_2}$ and $\ker \bar{\Phi}$ for the kernel of the homomorphism $\bar{\Phi}$ we have that algebras $\mathfrak{A}_{12}/\ker\bar{\Phi}$ and $\mathfrak{A}_1 \otimes \mathfrak{A}_2$ are isomorphic and, hence, the dual mapping $\Phi^*\colon \tilde{\varphi} \to \tilde{\varphi} \circ \bar{\Phi}$ implements the isomorphism of sets of states

$$\Phi^* : (\mathfrak{A}_1 \otimes_\alpha \mathfrak{A}_2)^{*+}_1 \to S_{\bar{\Phi}} \doteq \left\{\varphi \in (\mathfrak{A}_{12})^{*+}_1 \mid \varphi|_{\ker\bar{\Phi}} = 0\right\} \simeq (\mathfrak{A}_{12}/\ker\bar{\Phi})^{*+}_1 ,$$

which maps pure states into pure states on $\mathfrak{A}_{12}/\ker\bar{\Phi}$. Therefore, since for pure states φ_1 on $\mathfrak{A}_1$ and φ_2 on $\mathfrak{A}_2$ the state $\varphi_1 \otimes \varphi_2 \in (\mathfrak{A}_1 \otimes_\alpha \mathfrak{A}_2)^{*+}_1$ is also pure, the functional $\varphi_{12} = \varphi_1 \otimes \varphi_2 \circ \bar{\Phi}$ is a pure state on $\mathfrak{A}_{12}/\ker\bar{\Phi}$. Now we consider φ_{12} as a state on $\mathfrak{A}_{12}$ (i.e. as an element of $S_{\bar{\Phi}} \subset (\mathfrak{A}_{12})^{*+}_1$) and suppose that in this case φ_{12} is not pure, i.e. there is a majorized pure state $\varphi' \dot{<} \varphi_{12}$, $\varphi' \in (\mathfrak{A}_{12})^{*+}_1$. Then it follows from $\varphi_{12}(A) = 0 \ \forall A \in \ker\bar{\Phi}$ that $\varphi'(A) = 0 \ \forall A \in \ker\bar{\Phi}$ as well, i.e. $\varphi' \in S_{\bar{\Phi}}$. So it defines a pure state on the algebra $\mathfrak{A}_{12}/\ker\bar{\Phi}$ which is majorized by φ_{12} in the sense of this algebra. But φ_{12} is pure on $\mathfrak{A}_{12}/\ker\bar{\Phi}$; hence $\varphi' = \varphi_{12}$, and φ_{12} is pure as a state on $\mathfrak{A}_{12}$. Extending φ_{12} to a pure state on $\mathfrak{A}$ we obtain the pure state which is a common extension of φ_1 and φ_2 . ❐

Returning to the physical situation ($\mathfrak{A}_{1,2} = \mathfrak{A}(O_{1,2})$, $O_1 \times O_2$) we see that the results obtained only offer some necessary and sufficient conditions of (CI) but do not prove its actual fulfillment in local quantum theory. In the abstract Haag–Kastler theory (and only there (CI) is an adequate formulation of causal independence, as we noted above) neither of the conditions has been deduced so far from the axioms. Therefore causal independence still remains here only as a possibility, just as one of physically plausible causal properties whose relation to the system of Haag–Kastler axioms is still to be elucidated. In Haag–Araki theory, however, the property S, as we shall see later (Propositon 1.3.34 and its Corollary), can be deduced from the axioms (under some restrictions on a pair of regions O_1, O_2) . Hence the condition (CI) in Haag–Araki theory also follows from the axioms though this fact does not solve the problem of causal independence because the adequate formulation of the latter is here $(\mathrm{CI})_N$ rather than (CI).

The set of results connected with $(\mathrm{CI})_N$ looks somewhat different from that connected with (CI). In general, it shows closer connections with the structure of local quantum theory, but even here we can and should single out a group of results which depend only on locality (out of all physical properties) and, hence, can be formulated without reference to Haag–Araki theory.

PROPOSITION 1.3.27. *Let* R_1 *and* R_2 *be commuting von Neumann algebras and* $R_{12} = \{R_1 \vee R_2\}''$. *Then each of the conditions below implies all which follow it:*

(1) R_1 *and* R_2 *have the split property, i.e. there are discrete factors* M_1 *and* M_2 *such*

that $R_1 \subset M_1 \subset M'_2 \subset R'_2$;

(2) *there is an isomorphism of von Neumann algebras*

$$\Phi : R_{12} \to R_1 \bar{\otimes} R_2$$

which is natural in the sense of (3.30);

(3) *for any pair of normal states* ω_i *on* R_i, $i = 1, 2$, *there is a normal state* ω *on* R_{12} *such that*

$$\omega(AB) = \omega_1(A)\,\omega_2(B) \;\; \forall\, A \in R_1 ,\; B \in R_2 ;$$

(4) *there is a normal state* ω *on* R_{12} *which is a product state for* R_1 *and* R_2 :

$$\omega(AB) = \omega(A)\,\omega(B) \;\; \forall\, A \in R_1 ,\; B \in R_2 .$$

If, in addition, R_1 *and* R_2 *are factors, then the implication* $4 \to 3$ *holds and if* R_{12} *is a factor, then all conditions* $2 \to 4$ *are equivalent.*

Proof. To prove $1 \to 2$ we shall show that the property 2 always holds for the pair M_1, M_2 .

LEMMA 1.3.28. *Let* M_1, M_2 *be commuting von Neumann algebras and at least one of them be a type-*I *factor. Then there exists a natural (in the sense of* (3.30)) *isomorphism*

$$\psi : M_{12} \to M_1 \bar{\otimes} M_2 ,\; M_{12} = \{M_1 \vee M_2\}'' .$$

Proof of the Lemma. Let M_1 be a discrete factor. Then there are Hilbert spaces $\tilde{\mathcal{H}}$, $\tilde{\mathcal{H}}'$ and an isometry $U : \mathcal{H} \to \tilde{\mathcal{H}} \otimes \tilde{\mathcal{H}}'$ such that

$$U M_1 U^* = \mathcal{B}(\tilde{\mathcal{H}}) \bar{\otimes} C(\tilde{\mathcal{H}}') ,\; U M'_1 U^* = C(\tilde{\mathcal{H}}) \bar{\otimes} \mathcal{B}(\tilde{\mathcal{H}}') , \tag{3.32}$$

whence we also have

$$U M_2 U^* = C(\tilde{\mathcal{H}}) \bar{\otimes} \tilde{M}_2 , \tag{3.33}$$

where $\tilde{M}_2$ is a von Neumann algebra in $\tilde{\mathcal{H}}'$. A simple calculation gives also

$$U M_{12} U^* = \mathcal{B}(\tilde{\mathcal{H}}) \bar{\otimes} \tilde{M}_2 . \tag{3.34}$$

Now we introduce mappings of $\mathcal{B}(\tilde{\mathcal{H}})$ and $\tilde{M}_2$ into $\mathcal{B}(\tilde{\mathcal{H}} \otimes \tilde{\mathcal{H}}')$:

$$\pi_1 : \tilde{A} \to \tilde{A} \otimes I_{\tilde{\mathcal{H}}'} \;\; \forall\, \tilde{A} \in \mathcal{B}(\tilde{\mathcal{H}})$$

$$\pi_2 : \tilde{B} \to I_{\tilde{\mathcal{H}}} \otimes \tilde{B} \;\; \forall\, \tilde{B} \in \tilde{M}_2 .$$

Since π_1 and π_2 are isomorphisms, due to (3.32) and (3.33) their composition with the spatial isomorphism $U \cdot U^*$ produces isomorphisms

$$\chi_1 : \mathcal{B}(\tilde{\mathcal{H}}) \to M_1 ,\; \chi_1(\cdot) = U^* \pi_1(\cdot)\, U ;$$

$$\chi_2 : \tilde{M}_2 \to M_2 ,\; \chi_2(\cdot) = U^* \pi_2(\cdot)\, U .$$

As we know (see e.g. [29, Corollary IV, Section 5.3]) in such a case there exists also an isomorphism of corresponding W^*-tensor products:

$$\chi : \mathcal{B}(\tilde{\mathcal{H}}) \bar{\otimes} \tilde{M}_2 \to M_1 \bar{\otimes} M_2 ,\; \chi(\tilde{A} \otimes \tilde{B}) = \chi_1(\tilde{A}) \otimes \chi_2(\tilde{B}) \;\; \forall\, \tilde{A} \in \mathcal{B}(\tilde{\mathcal{H}}), \tilde{B} \in \tilde{M}_2$$

which generates, due to (3.34), the required isomorphism ψ. Indeed, putting $\psi = \chi \circ (U \cdot U^*)$, we check trivially that ψ is natural, since for any $A = U^* \tilde{A} \otimes I_{\tilde{\mathcal{H}}'} U \in M_1$, $B = U^* I_{\tilde{\mathcal{H}}} \otimes BU \in M_2$, $\psi(AB) = \chi(\tilde{A} \otimes \tilde{B}) = A \otimes B$. ❐

Returning to the proof of $1 \to 2$, let us consider the restriction Φ of the isomorphism ψ to the subalgebra $R_{12} \subset M_{12}$:

$$\Phi = \psi|_{R_{12}} : R_{12} \to M_1 \bar{\otimes} M_2 .$$

With ψ being natural, Φ is clearly a mapping onto $R_1 \bar{\otimes} R_2$, and since a restriction of a normal and isomorphic map inherits these properties we get condition 2.

2 → 3. As we know (see, e.g. [7, p. 67]), for any pair of normal states ω_i on R_i there exists a unique normal state $\tilde{\omega}$ on $R_1 \bar{\otimes} R_2$ such that

$$\tilde{\omega}(A \otimes B) = \omega_1(A)\, \omega_2(B) \ \forall A \in R_1 ,\ B \in R_2 .$$

Obviously, the functional $\omega = \tilde{\omega} \circ \Phi$ is a normal state on R_{12} satisfying condition 3.

Since $3 \to 4$ is evident, the first part of our statement is proved. Let now R_1 and R_2 be factors and let a normal product state ω exist on R_{12}. We notice that any pair of states ω_i on R_i determines a state $\omega_1 \tilde{\times} \omega_2$ on *-algebra $A_{12} = R_1 \vee R_2$ defined as follows:

$$(\omega_1 \tilde{\times} \omega_2)\left(\sum_{i=1}^{n} A_i B_i\right) = \sum_{i=1}^{n} \omega_1(A_i)\, \omega_2(B_i) \ \forall A_i \in R_1,\ B_i \in R_2,\ i = 1, \dots, n . \qquad (3.35)$$

If ω_i are normal, this state is ultraweakly continuous and has a unique extension to a state $\omega_1 \times \omega_2$ on R_{12} which is, by construction, a product state satisfying the equality in the condition 3. Therefore it is sufficient to show that under the assumption made the state $\omega_1 \times \omega_2$ is normal.

Denoting $\varphi_i = \omega_i|_{R_i}$, $i = 1, 2$ (evidently, $\varphi_i \in R_{i*}$), we consider the sets of functionals

$$V_1 = L\{\varphi_1(a_1 \cdot a_2),\ a_1, a_2 \in R_1\} \subset R_{1*} ,$$

$$V_2 = L\{\varphi_2(b_1 \cdot b_2),\ b_1, b_2 \in R_2\} \subset R_{2*} .$$

By construction, V_i is a linear set in R_{i*}, $i = 1, 2$, invariant under the action of $R_i \subset R_i^{**}$ on R_i^* [29, p. 123]; if R_i is a factor, then any such linear set is dense in R_{i*} in its Banach norm [29, Theorem III.2.7]. But, as we can easily verify, for any $\chi_i \in V_i$, $i = 1, 2$, the state $\chi_1 \times \chi_2$ on R_{12} is normal. Indeed, from (3.35) and $\varphi_1 \times \varphi_2 = \varphi \in R_{12}$ we have

$$\chi_1 \times \chi_2 = \varphi_1(a_1 \cdot a_2) \times \varphi_2(b_1 \cdot b_2) = \varphi(a_1 b_1 \cdot a_2 b_2) \in R_{12*} .$$

Therefore the density of V_1 in R_{1*} and V_2 in R_{2*} gives us the desired result: the state $\omega_1 \times \omega_2$ is normal for any $\omega_1 \in R_{1*}$ and $\omega_2 \in R_{2*}$.

Now we have only to prove $3 \to 2$ assuming that R_{12} is a factor (in this case commutativity of R_1 and R_2 readily implies that R_1 and R_2 are factors as well). We define a mapping $\psi : R_1 \odot R_2 \to R_1 \vee R_2$ in a natural way:

$$\psi\left(\sum_{i=1}^{n} A_i \otimes B_i\right)=\sum_{i=1}^{n} A_i B_i \quad \forall A_i \in R_1, B_i \in R_2, i = 1, \dots , n; n = 1, 2, \dots .$$

Since the pair of commuting factors R_1 and R_2 has the Schlieder property (Lemma 1.3.33), by virtue of Theorem 1.3.25 ψ is an algebraic isomorphism. Now let us prove that ψ is isometric. Denoting the set of all functionals $\omega_1 \times \omega_2$ by $R_{1*} \times R_{2*}$ and taking into account that this set is invariant under the algebra A_{12}, which is ultra-weakly dense in R_{12}, we obtain from Theorem III.2.7 in [29] that if R_{12} is a factor, then $R_{1*} \times R_{2*}$ is dense in R_{12*}. Using this we obtain the following chain of equalities (where $\mathfrak{A}_{12} = \overline{A_{12}}^{u}$, Rep $\mathfrak{A}_{12}$ is the set of all representations of $\mathfrak{A}_{12}$ and Π, Q are the arguments of the numerator and denominator, respectively, in the fourth term in the chain):

$$\left\| \sum_{i=1}^{n} A_i B_i \right\| = \sup \left\{ \left\| \pi \left(\sum_{i=1}^{n} A_i B_i \right) \right\|, \pi \in \text{Rep. } \mathfrak{A}_{12} \right\}$$

$$= \sup \left\{ \left\| \pi_\omega \left(\sum_{i=1}^{n} A_i B_i \right) \right\|, \omega \in \mathfrak{A}_{12}^{*+} \right\}$$

$$= \sup \left\{ \frac{\omega \left[\left(\sum_{j=1}^{m} a_j^* b_j^* \right) \left(\sum_{i=1}^{n} A_i^* B_i^* \right) \left(\sum_{i=1}^{n} A_i B_i \right) \left(\sum_{j=1}^{m} a_j b_j \right) \right]^{1/2}}{\omega \left[\left(\sum_{j=1}^{m} a_j^* b_j^* \right) \left(\sum_{j=1}^{m} a_j b_j \right) \right]}, \omega \in \mathfrak{A}_{12}^{*+}, a_j \in R_1, b_j \in R_2 \right\}$$

$$= \sup \left\{ \frac{\omega (\Pi)^{1/2}}{\omega (Q)^{1/2}}, \omega \in R_{12}^{*+} \right\} = \sup \left\{ \frac{(\omega_1 \times \omega_2)(\Pi)^{1/2}}{(\omega_1 \times \omega_2)(Q)^{1/2}}, \omega_1 \in R_1^{*+}, \omega_2 \in R_2^{*+} \right\}$$

$$= \sup \left\{ \frac{(\omega_1 \otimes \omega_2) \left[\left(\sum_{j=1}^{m} a_j \otimes b_j \right)^* \left(\sum_{i=1}^{n} A_i \otimes B_i \right)^* \left(\sum_{i=1}^{n} A_i \otimes B_i \right) \left(\sum_{j=1}^{m} a_j \otimes b_j \right) \right]^{1/2}}{(\omega_1 \otimes \omega_2) \left[\left(\sum_{j=1}^{m} a_j \otimes b_j \right)^* \left(\sum_{j=1}^{m} a_j \otimes b_j \right) \right]^{1/2}}, \right.$$

$$\left. \omega_1 \in R_1^{*+}, \omega_2 \in R_2^{*+} \right\} = \left\| \sum_{i=1}^{n} A_i \otimes B_i \right\| ,$$

where the explicit form of the α-norm on the tensor product of C^*-algebras is used (see, e.g., [29, Theorem IV.9]).

The mapping ψ being isometric, it can be extended to an isometric isomorphism of C^*-algebras $\bar{\psi} : R_1 \otimes_\alpha R_2 \to \mathfrak{A}_{12}$. However, to prove that the weak closures are isomorphic as well, we have to use dual spaces. The dual mapping

$$^t\bar{\psi} : \mathfrak{A}_{12}^* \to (R_1 \otimes_\alpha R_2)^*$$

is also isometric. We shall show that this mapping generates the isometric mapping

$$\rho : R_{12*} \to (R_1 \bar{\otimes} R_2)_* \ .$$

To prove the existence of ρ, it is sufficient to find two sets of functionals V and $V_\otimes$ such that

(1) V is dense in R_{12*} and $V_\otimes$ is dense in $(R_1 \bar{\otimes} R_2)_*$,

(2) restrictions to $\mathfrak{A}_{12}$ of the functionals from V and restrictions to $R_1 \otimes_\alpha R_2$ of the functionals from $V_\otimes$ are mapped into each other by the mapping

$$^t\bar{\psi}\,[V|_{\mathfrak{A}_{12}}] = V_\otimes|_{R_1 \otimes_\alpha R_2} \,. \tag{3.36}$$

(Indeed, in this case (3.36) determines naturally the mapping $\tilde{\rho} : V \to V_\otimes$ which can be extended by continuity to the mapping we look for.) It is easy to see that

$$V = R_{1*} \times R_{2*} \,, \ V_\otimes = R_{1*} \odot R_{2*}$$

can be used as such sets.

Indeed, we have already pointed out the density of V in R_{12*}. The density of $V_\otimes$ (algebraic tensor product of the preduals of R_1 and R_2) in $(R_1 \bar{\otimes} R_2)_*$ is obvious from the definition [29, p. 221]: $(R_1 \bar{\otimes} R_2)_* = R_{1*} \bar{\otimes} R_{2*}$. The property (3.36) can be checked explicitly: for all $\varphi_1 \in R_{i*}$, $i = 1, 2$, we have

$$\left(^t\bar{\psi}\,[\varphi_1 \times \varphi_2]\right)\left(\sum_{i=1}^n A_i \otimes B_i\right) = (\varphi_1 \times \varphi_2)\left[\bar{\psi}\left(\sum_{i=1}^n A_i \otimes B_i\right)\right]$$

$$= (\varphi_1 \times \varphi_2)\left(\sum_{i=1}^n A_i\, B_i\right) = (\varphi_1 \otimes \varphi_2)\left(\sum_{i=1}^n A_i \otimes B_i\right),$$

and the existence of ρ is proved. Next, it is easy to verify that ρ is isometric on V: this follows from the fact that the norms of functionals from V and $V_\otimes$ coincide with those of their restrictions to the corresponding C^*-algebras. Indeed, due to the Kaplansky density theorem, for all $\varphi \in V$,

$$\|\varphi\| = \sup\,\{|\varphi(A)|\,,\ A \in R_{12}\,,\ \|A\| = 1\}$$

$$= \sup\,\{|\varphi(A)|\,,\ A \in \mathfrak{A}_{12}\,,\ \|A\| = 1\} = \|\varphi|_{\mathfrak{A}_{12}}\|\,,$$

and similarly for $\varphi \in V_\otimes$. Due to the density of V in R_{12*}, this implies that ρ is isometric.

Thus, there exists for ρ a dual mapping

$$^t\rho : R_1 \bar{\otimes} R_2 \to R_{12}\,,$$

which is an isometric isomorphism and inherits the property (3.34) due to its link with ψ. The inverse mapping $(^t\rho)^{-1} : R_{12} \to R_1 \bar{\otimes} R_2$ is the natural isomorphism required by the condition 2. ❐

Obviously, the property 3 in this proposition is a direct enhancement of the causal independence $(\mathrm{CI})_N$. Therefore we have the following trivial corollary:

COROLLARY. *Each of the structural properties* 1–3 *in Proposition* 1.3.27 *is a sufficient condition of* $(\mathrm{CI})_N$; *if local algebras* $R(O_1)$ *and* $R(O_2)$, $O_1 \times O_2$, *are factors, then the property* 4 *is also such a sufficient condition of* $(\mathrm{CI})_N$.

This corollary is only a part of the conclusions which can be drawn from analyzing the properties 1–4 in the Haag–Araki framework. Specific features of local quantum theory produce new relations between these properties, so that eventually they all prove to be mutually equivalent, though, as before, only sufficient for $(\mathrm{CI})_N$. In particular, we shall obtain necessary and sufficient conditions for the split property.

Before we turn to obtaining such results, it is necessary to notice that the existence of a normal product state $\omega^{(p)}$ for local algebras $R(O_1)$ and $R(O_2)\,(O_1 \times O_2)$ under the condition $(O_1 \cup O_2)' \neq \emptyset$ (excluding, in particular, the pairs of regions $O_1 = D$, $O_2 = D'$), and in the presence of a cyclic vector for Z', is tantamount to the existence of a vector product state $\omega_\Phi^{(p)}$. Indeed, $\omega^{(p)}$ has a normal state on $R(O_1 \cup O_2)$ as its restriction, which by virtue of $\overline{R(O_1) \vee R(O_2)}^{\,w} \subset R(O_1 \cup O_2)$ is also a product state for $R(O_1)$ and $R(O_2)$. But this state is, in addition, a vector state because if $(O_1 \cup O_2)' \neq \emptyset$ and there is a cyclic vector for Z', then W^*-algebra $R(O_1 \cup O_2)$ has a separating vector (Proposition 1.3.4) and hence all its normal states are vector ones [7, p. 222]. Extension of the vector state from $R(O_1 \cup O_2)$ onto R produces the vector state $\omega_\Phi^{(p)}$ which is a product state for $R(O_1)$ and $R(O_2)$. For the same reasons, if there is a cyclic vector for Z' and $(O_1 \cup O_2)' \neq \emptyset$, the condition $(\mathrm{CI})_N$, i.e. causal independence in the class of normal states, proves to be equivalent to causal independence in the class of vector states, i.e. to the condition

$$(\mathrm{CI})_V \qquad \underset{O_1 \times O_2}{\forall} \quad \underset{\Phi_1, \Phi_2 \in S_1}{\forall} \quad \underset{\Phi \in S_1}{\exists} \quad \omega_\Phi \big|_{R(O_i)} = \omega_{\Phi_i}^{(O_i)}, \quad i = 1, 2,$$

where $\omega_\Phi^{(O)}$ stands for a vector state on $R(O)$. Indeed, under the assumptions made all normal states on all $R(O)$, $O' \neq \emptyset$, are vector ones, which implies $(\mathrm{CI})_V \Rightarrow (\mathrm{CI})_N$. Conversely, if $(\mathrm{CI})_N$ holds, then for any pair of vector states $\omega_{\Phi_i}^{(O_i)}$, $i = 1, 2$, there exists a common extension $\omega \in N(R)$. Its restriction to $R(O_1 \cup O_2)$ is a vector state $\omega_\Phi^{(O_1 \cup O_2)}$, the extension of which to a vector state ω_Φ on R is also a common extension of 'local states' $\omega_{\Phi_1}^{(O_1)}$ and $\omega_{\Phi_2}^{(O_2)}$. Eventually we have

$$(\mathrm{CI})_N \Leftrightarrow (\mathrm{CI})_V.$$

We turn now to deriving the above mentioned relations between the structural properties from Proposition 1.3.27 and the condition $(\mathrm{CI})_N$ in the Haag–Araki theory. Such relations are relatively easy to obtain using the following lemma by Buchholz (which is, by contrast, not so easy to prove).

LEMMA 1.3.29 [99]. *Let the Haag–Araki theory be given satisfying Axioms* I–III, V *and* VI *and such that* $R = \mathcal{B}(\mathcal{H})$. *Let regions* O_1 *and* O_2 *be such that* $O_1 \times O_2$ *and there exists a neighbourhood* $\mathcal{N}$ *of zero in* M, *for which* $(O_1 + \mathcal{N}) \times (O_2 + \mathcal{N})$ *(this property we shall denote as* $O_1 \times\!\!\times O_2$*). Then the existence of a normal product state* ω *for the algebras* $R(O_1 + \mathcal{N})$, $R(O_2 + \mathcal{N})$ *implies the existence of a vector product state* ω_χ *for the algebras* $R(O_1)$ *and* $R(O_2)$ *such that the vector* χ *is bicyclic for* $\overline{R(O_1) \vee R(O_2)}^{w}$ *and, in addition,*

$$\omega_\chi \big|_{R(O_i)} = \omega_\xi \big|_{R(O_i)}, \quad i = 1, 2, \tag{3.37}$$

where ξ *is any vector given beforehand from the set of common bicyclic vectors of all local algebras.**

Proof. Denoting $O_i + \mathcal{N} \equiv \hat{O}_i$, $i = 1, 2$, we notice that by contracting the neighbourhood $\mathcal{N}$ it is always possible to obtain $(\hat{O}_1 \cup \hat{O}_2)' \neq \varnothing$, so that we can hold without loss of generality that $R(\hat{O}_1 \cup \hat{O}_2)$ has a separating vector and the state ω is a vector one, i.e. $\omega = \omega_\Psi$. Denoting $P_i = P_\Psi^{R(\hat{O}_i)}$, $i = 1, 2$, and using the property of product states:

$$(\Psi, AB\,\Psi) = (\Psi, A\,\Psi)\,(\Psi, B\,\Psi) \quad \forall\, A \in R(\hat{O}_1),\ B \in R(\hat{O}_2),$$

we obtain the relations

$$P_1 B P_1 = \omega_\Psi(B)\, P_1 \quad \forall\, B \in R(\hat{O}_2),$$

$$P_2 A P_2 = \omega_\Psi(A)\, P_2 \quad \forall\, A \in R(\hat{O}_1). \tag{3.38}$$

From here it is easy to prove that $\omega_{P_1\xi/\|P_1\xi\|}$ is also a product state for $R(\hat{O}_1)$ and $R(\hat{O}_2)$. Indeed, for $A \in R(\hat{O}_1)$ and $B \in R(O_2)$ we have from (3.38)

$$(P_1\xi,\ ABP_1\xi) = (P_1\xi,\ AP_1\xi)\,(\Psi, B\,\Psi)$$

and taking here $A = I$ we get $\omega_\Psi(B) = \omega_{P_1\xi/\|P_1\xi\|}(B)$. Next, the state $\omega^{(O_1)}_{P_1\xi/\|P_1\xi\|}$ is faithful since for any $A \in R(O)$ the condition

$$\omega^{(O_1)}_{P_1\xi/\|P_1\xi\|}(A^*A) = \|AP_1\,\xi\| = 0$$

implies $AP_1 = 0$ because ξ is cyclic for all local algebras. Since the assumptions made imply the Schlieder property (see Proposition 1.3.34) it follows that $A = 0$ as well. A faithful normal state can always be represented by a cyclic vector [23, p. 104] and hence there exists a vector $\Phi \in S_1$ such that

$$\mathcal{H}_\Phi^{R(O_1)} = \mathcal{H}; \quad \omega_\Phi^{(O_1)} = \omega^{(O_1)}_{P_1\xi/\|P_1\xi\|}.$$

* In particular, if there are vacuum vectors, then for any such vector Ω_j there is ω_χ satisfying (3.37) with $\xi = \Omega_j$.

The second equality here means that the vector $P_1 \xi / \| P_1 \xi \|$ belongs to the $\mathcal{H}$-image of the local state $\omega_{\Phi}^{(O_1)}$; therefore, according to 3, we can find a partially isometric operator $U_1 \in R(O_1)'$ such that

$$U_1^* U_1 = P_{\Phi}^{R(O_1)} = I \; ; \; U_1 U_1^* = P_{P_1\xi/\|P_1\xi\|}^{R(O_1)} = P_1 \, ,$$

if $R(O_1) \subset R(\hat{O}_1)$. Substituting the second formula into (3.38) and taking into account the first we obtain

$$U_1^* B U_1 = \omega_{\Psi}(B) I \, , \quad B \in R(\hat{O}_2) \, .$$

Thus for $A \in R(O_1)$, $B \in R(\hat{O}_2)$ we have

$$(U_1 \xi, AB U_1 \xi) = (\xi, A\xi)(\Psi, B\Psi), \, (\xi, A\xi) = (U_1 \xi, A U_1 \xi), \, (\Psi, B\Psi) = (U_1 \xi, B U_1 \xi) \, ,$$

i.e. $\omega_{U_1\xi}$ is the product state for algebras $R(O_1)$ and $R(\hat{O}_2)$, and $\omega_{U_1\xi}^{(O_1)} = \omega_{\xi}^{(O_1)}$. Thus, proceeding from the product state ω_{Ψ} for $R(\hat{O}_1), R(\hat{O}_2)$, we have constructed a product state $\omega_{U_1\xi}$ for $R(O_1)$, $R(\hat{O}_2)$ which coincides with ω_{ξ} on $R(O_1)$. Now proceeding from $\omega_{U_1\xi}$ we can, by analogy, construct the product state ω_{Φ} for $R(O_1)$ and $R(O_2)$ coinciding with ω_{ξ} on $R(O_1)$ and $R(O_2)$.

It remains to construct a state ω_{χ} which has all the properties of ω_{Φ} and, in addition, is such that χ is bicyclic for $\overline{R(O_1) \vee R(O_2)}^{w}$. To this end consider the regions obtained from O_i by small translations along the x^0-axis:

$$O_{i,\varepsilon} = \bigcup_{|t|<\varepsilon} (O_i + t) \, , \; i = 1, 2 \, .$$

For a reasonably small ε they are related to $\hat{O}_i$ exactly in the same way as O_i and therefore, as proved above, there exists a normal product state ω for $R(O_{1,\varepsilon})$ and $R(O_{2,\varepsilon})$ such that

$$\omega|_{R(O_{i,\varepsilon})} = \omega_{\xi}^{(O_{i,\varepsilon})} \, , \quad i = 1, 2 \, .$$

Picking a positive function h on $\mathbf{R}$ such that $\int_{-\varepsilon}^{\varepsilon} h(t)\, dt = 1$, define the state

$$\omega_h(A) = \int_{-\varepsilon}^{\varepsilon} dt h(t)\, \omega(T(t) A T(t)^{-1})$$

on R. Clearly, ω_h is a normal product state for $R(O_1)$ and $R(O_2)$ that coincides with ω_{ξ} on these algebras. We shall prove that ω_h is faithful as a state on $\overline{R(O_1) \vee R(O_2)}^{w}$. If $\omega_h(Q^*Q) = 0$ for some $Q \in \overline{R(O_1) \vee R(O_2)}^{w}$, by the definition of ω_h one has

$$\omega(T(t)\, Q^* Q\, T(t)^{-1}) = 0, \; |t| < \varepsilon \, .$$

For a vector state $\omega_{\eta} \prec \omega$ (which always exists) one has

$$QT(t)^{-1} \eta = Q(t)\, \eta = 0, \; |t| < \varepsilon \, . \tag{3.39}$$

The following result holds, the proof of which we omit.

SCHLIEDER'S THEOREM [376]. *Let Axioms* I–III, V, VI *and the condition* $R = \mathcal{B}(\mathcal{H})$ *hold in Haag–Araki theory. Then for any* $Q \in R(O)$, $O' \neq \emptyset$, *and any vector* $\eta \in \mathcal{H}$ *the set of all points* t *on the* x^0*-axis, for which* $Q(t)\eta \neq 0$, *is dense in this axis.*

Proof. By this theorem, one has $Q = 0$ in (3.39) and so ω_h is a faithful state on $\overline{R(O_1) \vee R(O_2)}^w$. Therefore, this algebra has a bicyclic vector χ such that ω_χ coincides with ω_h. Hence ω_χ has all the required properties. ❐

Using this lemma, we now readily obtain

PROPOSITION 1.3.30. *Let Haag–Araki theory satisfy Axioms* I–III, V, VI *and the condition* $R = \mathcal{B}(\mathcal{H})$. *Then for any pair of regions* O_1, O_2 *with* $O_1 \Join O_2$ *the following properties of local algebras are equivalent:*

(1) $R(O_1)$ *and* $R(O_2)$ *have the split property, i.e. there exist type-*I *factors* M_1 *and* M_2 *such that*

$$R(O_1) \subset M_1 \subset M_2' \subset R(O_2)'$$

(which is tantamount to the formulation of the split property given above);

(2) *there exists a normal and natural isomorphism of von Neumann algebras*

$$\Phi : \overline{R(O_1) \vee R(O_2)}^w \to R(O_1) \bar{\otimes} R(O_2);$$

(3) *for any pair of normal states* ω_i *on* $R(O_i)$, $i = 1, 2$, *there is a normal state* ω *on* R *such that*

$$\omega(AB) = \omega_1(A)\, \omega_2(B), \ A \in R(O_1), \ B \in R(O_2);$$

(4) *there is a normal state on* R *which is a product state for* $R(O_1)$ *and* $R(O_2)$; *and imply the property*

(5) $R(O_1)$ *and* $R(O_2)$ *are causally independent in the sense of* $(\mathrm{CI})_N$ *or, equivalently,* $(\mathrm{CI})_V$.

Proof. By Proposition 1.3.27, for the equivalence of all conditions 1–4 it is sufficient to derive $4 \to 1$. To prove the split property for $R(O_1)$ and $R(O_2)$ we shall construct explicitly the discrete factors M_1 and M_2. Namely, if ω_χ is the special product state constructed in Lemma 1.3.29 and $P_i = P_\chi^{R(O_i)}$, $i = 1, 2$, we shall check that

$$M_1 = \{R(O_1), P_2\}'', \ M_2 = \{R(O_2), P_1\}''$$

will do. (Condition $O_1 \Join O_2$ is required only here to apply Lemma 1.3.29.) In complete analogy with the deduction of (3.38) one gets for ω_χ:

$$P_1 B P_1 = \omega_\chi(B) P_1 \quad \forall B \in R(O_2),$$

$$P_2 A P_2 = \omega_\chi (A) P_2 \quad \forall A \in R(O_1),$$

and these equalities are readily extendable to all $B \in M_2$ and $A \in M_1$, respectively. Hence it follows that P_1 is an abelian projection in M_2 and so is P_2 in M_1.

Show that M_1 and M_2 are factors. For a projection $E \in M_1 \cap M'_1$ the equalities $\omega_\chi (E) P_2 = P_2 E P_2 = E P_2$ entail $\omega_\chi (E) = \| E \chi \|^2 = 0$ or I, since $E P_2$ is a projection. In the first case $E\chi = 0$ implies $E = 0$, because $E \in R(O_1)' \cap R(O_2)'$ and χ is separating for this algebra. In the second case one has $(\chi, (I - E)\chi) = 0$ which implies $E = I$ for the same reason. Thus, there are only trivial projections in the M_1 centre. The factoriality of M_2 is proved identically. Due to the presence of abelian projections, the factors M_1 and M_2 are discrete [7, p. 123]. Inclusions $R(O_1) \subset M_1$ and $R(O_2)' \supset M_2'$ being evident, it remains only to prove $M_1 \subset M_2'$, for which the equality $P_1 P_2 = P_2 P_1$ is sufficient. For any $A \in R(O_1)$, $B \in R(O_2)$ one has the string of equalities

$$\begin{aligned} P_1 P_2 AB \chi &= P_1 P_2 A P_2 B \chi = (\chi, A\chi) P_1 B P_1 \chi \\ &= (\chi, A\chi)(\chi, B\chi)\chi = P_2 P_1 B P_1 A \chi = P_2 P_1 AB \chi, \end{aligned}$$

from which due to the cyclicity of χ for $\overline{R(O_1) \vee R(O_2)}^{w}$ we obtain the required. $3 \to 5$ being evident, the proposition is proved. ❐

Remark. There are regions $O_1 \times O_2$, for which product states exist, but are certainly not normal. The example is given, e.g., by a pair O_1, O_2 of mutually spacelike lightlike monotone regions (see Definition 1.3.1) such that $O_1 + a \subset O_1, O_2 + a \subset O_2$ for some translations $a \in M$ (in particular, one can take the wedges W_R, W_L). Local algebras of such regions are factors (Proposition 1.3.7) and the Schlieder property holds for the said pair (Proposition 1.3.34), hence, the assumptions of Theorem 1.3.25 are fulfilled. On the other hand, using the property $R(O_i + na) \subset R(O_i)$, $i = 1, 2$, and the formula $\lim_{n\to\infty} A(na) = \omega_\Omega (A) I$, one can easily prove [99] that if there is a normal product state ω for the pair O_1, O_2, the vacuum ω_Ω is also a product state for these regions. Indeed, for any $A \in R(O_1)$, $B \in R(O_2)$ one has

$$\begin{aligned} \omega_\Omega (AB) &= \lim_{n\to\infty} \omega (A(na) B(na)) \\ &= \lim_{n\to\infty} \omega (A(na))\, \omega (B(na)) = \omega_\Omega (A)\, \omega_\Omega (B). \end{aligned}$$

But the vacuum vector being cyclic for $R(O_1), R(O_2)$, ω_Ω cannot be a product state for these algebras. Hence, there is no normal product state for $R(O_1)$ and $R(O_2)$.

Furthermore, we should examine the relation between the condition $(\mathrm{CI})_N$ and the Schlieder property as well as Haag–Araki axioms. There are many ways to verify easily that $(\mathrm{CI})_N$ implies the Schlieder property. We shall prove it by means of restating the both properties together with strict locality in terms of $\mathcal{H}$-images

$\mathfrak{M}_\Phi(O) \equiv \mathfrak{M}_\Phi^{R(O)}$ of states $\omega_\Phi^{(O)}$. Such a restatement immediately reveals the ordering among these properties.

PROPOSITION 1.3.31 [410]. In the Haag–Araki theory, causal independence $(\mathrm{CI})_N$, strict locality (SL), and the Schlieder property (S) can be formulated as follows: for all O_1, O_2 with $O_1 \times O_2$ and all $\Phi_1, \Phi_2, \in S_1$,

$$(\mathrm{CI})_V \quad \mathfrak{M}_{\Phi_1}(O_1) \cap \mathfrak{M}_{\Phi_2}(O_2) \neq \varnothing\,; \tag{3.40}$$

$$(\mathrm{SL}) \quad \mathcal{H}_{\Phi_1}^{R(O_1)'} \cap \mathfrak{M}_{\Phi_2}(O_2) \neq \varnothing\,; \tag{3.41}$$

$$(\mathrm{S}) \quad \mathcal{H}_{\Phi_1}^{R(O_1)'} \cap \mathcal{H}_{\Phi_2}^{R(O_2)'} \neq \varnothing\,. \tag{3.42}$$

The condition $(\mathrm{CI})_V$ *can also be represented as* $\Phi_1 = T_1^* T_2 \Phi_2$, *where* $T_i \in R(O_i)'$, $T_i^* T_i = P_{\Phi_i}^{R(O_i)}$, $T_i T_i^* = P_{T_i\Phi_i}^{R(O_i)}$, $i = 1, 2$.

Proof. To prove (3.40), rewrite the usual form of $(\mathrm{CI})_V$ in terms of $\mathcal{H}$-images:

$$\underset{O_1 \times O_1}{\forall} \quad \underset{\Phi_1, \Phi_2 \in S_1}{\forall} \quad \underset{\Phi \in S_1}{\exists} \quad \mathfrak{M}_\Phi^R \cap \mathfrak{M}_{\Phi_i}(O_i) \neq \varnothing\,, i = 1, 2\,.$$

Hence it follows that there exists $\chi \in \mathfrak{M}_\Phi^R \cap \mathfrak{M}_{\Phi_1}(O_1)$, and for this vector one has

$$\mathfrak{M}_\chi^R = \mathfrak{M}_\Phi^R\,,\ \mathfrak{M}_\chi(O_1) = \mathfrak{M}_{\Phi_1}(O_1)\,,\ \mathfrak{M}_\chi^R \subset \mathfrak{M}_\chi(O_1)\,.$$

These formulae imply the equivalence

$$\mathfrak{M}_\Phi^R \cap \mathfrak{M}_{\Phi_1}(O_1) \neq \varnothing \Leftrightarrow \mathfrak{M}_\Phi^R \subset \mathfrak{M}_{\Phi_1}(O_1)\,.$$

Obviously, the similar relation holds for Φ_2, O_2 as well. This means that $(\mathrm{CI})_N$ is equivalent to the condition

$$\underset{\Phi \in S_1}{\exists} \quad \mathfrak{M}_\Phi^R \subset \mathfrak{M}_{\Phi_1}(O_1) \cap \mathfrak{M}_{\Phi_2}(O_2)\,,$$

which is, in its turn, equivalent to (3.40). The other formulation can be obtained from (3.40) by applying the theorem on the general form of $\mathcal{H}$-image [408]: for arbitrary vector $\eta \in S_1$ and von Neumann algebra M,

$$\mathfrak{M}_\eta^M = \{\xi \in S_1 \,|\, \xi = T\eta,\ \ T \in M'\,,\ T^*T = P_\eta^M\,,\ TT^* = P_{T\eta}^M\}\,, \tag{3.43}$$

To obtain (3.41) note that the definition of strict locality in Section 1.2 means nontriviality of the intersection of $\mathfrak{M}_{\Phi_2}(O_2)$ with any invariant subspace of $R(O_1)'$, while (3.41) means that the intersection of $\mathfrak{M}_{\Phi_2}(O_2)$ with any cyclic subspace of $R(O_1)'$ is nontrivial. Clearly, the former property entails the latter and the converse is also true since any invariant subspace of a von Neumann algebra includes a cyclic subspace.

To obtain (3.42) show that the original formulation (S) is equivalent to that in terms of projections:

$$\underset{O_1 \times O_2}{\forall} \quad \underset{E_1, E_2 \neq 0}{\forall} \quad E_1 E_2 \neq 0\,,\ E_1 \in R\,(O_1)^P\,,\ E_2 \in R\,(O_2)^P, \tag{3.44}$$

where M^P is the set of all projections in a *-algebra M. Assume that there are non-zero $A_i \in R\,(O_i)$, $i = 1, 2,$ such that $A_1 A_2 = 0$. Then neither A_1 nor A_2 can be unitary and hence in every pair of operators $(A_1^* A_1\,; A_1 A_1^*)$, $(A_2^* A_2\,; A_2 A_2^*)$ at least one is not equal to unity. Assume, for example, that $A_1^* A_1\,, A_2 A_2^* \neq I$; then they have nontrivial spectral projections $E_1 \in R\,(O_1), E_2 \in R\,(O_2)$ and $A_1^* A_1 A_2^* A_2 = 0$ entails $E_1 E_2 = 0$. The formulation (3.44) is proved. It means that any two invariant subspaces of $R\,(O_1)'$ and $R\,(O_2)'$, respectively, have nontrivial intersection, and the end of the proof of (3.42) repeats that of of (3.41). ❐

COROLLARY 1. *Let* $O_1, O_2 \subset M$ *be such that they can be embedded into mutually spacelike regions* $\hat{O}_1, \hat{O}_2 : O_i \subset \hat{O}_i, i = 1, 2,$ $\hat{O}_1 \times \hat{O}_2$. *Then the fulfillment of any of properties* $(\mathrm{CI})_V$, (SL) *or* (S) *for the pair* $\hat{O}_1, \hat{O}_2$ *implies the corresponding property for the pair* O_1, O_2.

COROLLARY 2. *In any Haag–Araki theory there are the following implications between causal independence, strict locality, and the Schlieder property*

$$(\mathrm{CI})_V \Rightarrow (\mathrm{SL}) \Rightarrow (\mathrm{S})\,. \tag{3.45}$$

(3.45) is evident since $\mathfrak{M}_\Phi(O) \subset \mathcal{H}_\Phi^{R(O)'}$, due to (3.43).

We shall see soon that the string of implications can be continued because the Schlieder property trivially entails extended locality. As for the relations of $(\mathrm{CI})_V$ (or its equivalent $(\mathrm{CI})_N$) and other causal properties to Haag–Araki axioms, only the Schlieder property and extended locality, the weakest requirements, according to (3.45), have been derived from the axioms. Strict locality has been derived only for regions O_1 and O_2, $O_1 \times O_2$, which are bounded and subject to some other conditions (see Corollary to Proposition 1.3.32). But $(\mathrm{CI})_V$ and $(\mathrm{CI})_N$ have not been obtained either from the axioms or from the (S) property (in contrast to the implication (S) ⇒ (CI) proved in Theorem 1.3.25). On the other hand, there are no counterexamples proving that $(\mathrm{CI})_V$ and $(\mathrm{CI})_N$ are independent of the Haag–Araki axioms and can be violated when they all are fulfilled. (Note that for the split property such counterexample is provided by certain generalized free fields, see Section 3.3.) As a result, like in the case of (CI), relation of $(\mathrm{CI})_V$ and $(\mathrm{CI})_N$ to the axioms is not yet sufficiently elucidated. Special attention was paid to the relation between causal independence and the locality axiom taken separately, i.e. commutativity of algebras $R\,(O_1)$ and $R\,(O_2)$. The conclusion obtained was that neither of these properties entails the other. Commutativity of arbitrary C^*- or W^*-algebras $\mathcal{A}_1$ and $\mathcal{A}_2$ cannot imply their causal independence: indeed, the latter implies extended locality, i.e. triviality of $\mathcal{A}_1 \cap \mathcal{A}_2$, which can be violated for commuting $\mathcal{A}_1$ and $\mathcal{A}_2$. Conversely, the commutativity of $\mathcal{A}_1$ and $\mathcal{A}_2$

cannot follow from their causal independence, as proved by a number of examples [333, 132, 133]. Our concluding remark about causal independence is that this property is expected to be broken in gauge theories, as Ekstein and Avishai have pointed out [56]. This is suggested by the following loose argument. Let O_1, O_2 be the two halves of a sphere S in $\mathbf{R}^3$, and Q_1, Q_2 observables corresponding to the flux of electromagnetic field through O_1 and O_2, respectively. Then one has $O_1 \times O_2$ and there exist local states $\omega_1^{(O_1)}$ and $\omega_2^{(O_2)}$ such that $\omega_1^{(O_1)}(Q_1) + \omega_2^{(O_2)}(Q_2) \neq 0$. These states do not have the common extension to a global state ω since, by the Gauss theorem, the flux of electromagnetic field through S in the absence of internal sources is equal to zero and, therefore, for any physical state one has $\omega(Q_1 + Q_2) = 0$.

Let us now turn to strict locality (SL). Physically it is close to causal independence. Projections $P \in R(O_1)$ correspond to properties measurable in a region O_1, and ω_χ with $\chi \in P\mathcal{H}$ are physical states having the property P. And strict locality requires that a state ω_Φ could have the property P irrespective of the results of measurements in this state in a region O_2 spacelike to O_1. It is easy to reformulate this condition for normal states. Mathematically, (SL) is closely connected with types of local algebras and equivalence properties of certain classes of projections in these algebras. Such connections are the most visible in the case of regions D, D'. According to Corollary 1 to Proposition 1.3.31, strict locality for pairs of bounded regions O_1, O_2 is, in general, a weaker requirement.

PROPOSITION 1.3.32. *Let Axioms* Ia–III, V *and* VI *hold, and the algebra* Z' *have a cyclic vector. Then the following conditions are equivalent:*

(1) $R(D)$ *is a type-III factor:*

(2) $\forall_{P \in R(D)} \quad P \sim I \bmod R(D)$;

(3) $\forall_{P \in R(D)} \; \exists_{W \in R(D)} \; \forall_{\Phi \in S_1} \quad W\Phi \in P\mathcal{H} \cap \mathfrak{M}_\Phi(D')$,

where W *is a partially isometric operator and imply the properties:*

(1′) *all the projections in* $R(D)$ *are infinite relative to* $R(D')'$;

(2′) $\forall_{P \in R(D)} \quad P \sim I \bmod R(D')'$;

(3′) *strict locality for the pair of regions* D, D'.

Furthermore, relations $2' \to 3' \to 1'$ *hold, and if* R *has a cyclic vector,* $3' \to 2'$ *holds as well; if, moreover,* $R(D')$ *is a factor, conditions* $1', 2', 3'$ *are equivalent.*

Proof. Let us verify the equivalence of 1–3. $1 \to 2$ follows from the fact that $R(D)$ is a σ-finite factor (since it has a separating vector, by Proposition 1.3.4, and in such a factor all infinite projections are mutually equivalent [7. p. 320]).

$\underline{2 \to 3}$. One has for all $P \in R(D)$: $P = WW^*$, $W^*W = I$, where $W \in R(D)$. Hence it is clear that $W\Phi \in \mathfrak{M}_\Phi(D')$ for all $\Phi \in S_1$; on the other hand, it is evident that $W\Phi \in P\mathcal{H}$, and 3 is obtained.

3 → 1. Since the partial isometry W differs from zero only on the subspace $W^*W\mathcal{H}$, it follows from $W\Phi \in \mathfrak{M}_\Phi(D') \subset S_1 \ \forall \Phi \in S_1$, that $W^*W = I$. On the other hand, $W\Phi \in P\mathcal{H}$ implies $WW^* = P' \leq P$. It means that for any $P \in R(D)$, $P < I \bmod R(D)$, whence both 2 and 1 follow.

Next, let us deduce $2' \to 3' \to 1'$; then, due to the obvious $2 \to 2'$ it will be also proved that any of conditions 1–3 entails any of 1′–3′. $2' \to 3'$ is deduced identically to $2 \to 3$.

3′ → 1′. Let $\Psi \in P\mathcal{H} \cap \mathfrak{M}_\Phi(D')$. By the theorem on the general form of $\mathcal{H}$-image [408], $P_\Phi^{R(D')} \sim P_\Psi^{R(D')} \bmod R(D')'$; on the other hand, $P_\Phi^{R(D')} \leq P_\Psi^{R(D)'} \leq P$, since $\mathcal{H}_\Psi^{R(D)'}$ is the smallest invariant subspace of $R(D')$ including Ψ. We obtain

$$\forall_{\Phi \in \mathcal{H}} \quad P \succ P_\Phi^{R(D')} \bmod R(D')' . \tag{3.46}$$

Since the algebra $R(D')'$ is infinite, it includes infinite cyclic projections. Hence (3.46) entails condition 1.

If we now assume that R has a cyclic vector, then, by Proposition 1.3.4, there is a cyclic vector χ for $R(D')$ and one has from 3:

$$\forall_{P \in R(D)} \quad P\mathcal{H} \cap \mathfrak{M}_\chi(D') \neq \emptyset.$$

If $\Psi \in P\mathcal{H} \cap \mathfrak{M}_\chi(D')$, then, by (3.43),

$$\Psi = W\chi, \ W \in R(D')', \ W^*W = P_\chi^{R(D')} = I, \ WW^* = P_\Psi^{R(D')} \leq P,$$

whence 2′ follows. The last assertion is obtained as follows. If there is a cyclic vector for $R(D')$, the algebra $R(D')'$ is σ-finite [7, p. 302]; if, in addition, $R(D')$ is a factor, then all infinite projections in $R(D')'$ are equivalent and we obtain $1' \to 2'$. Thus conditions 1′–3′ are equivalent. □

COROLLARY. *Let Axioms* I–III, V *and* VI *hold and* R *be a factor with a cyclic vector. Let also regions* O_1, O_2 $(O_1 \times\!\!\times O_2)$ *be double cones or such that they may be included into mutually spacelike double cones* $(O_i \subset D_i,\ i = 1, 2;\ D_1 \times\!\!\times D_2)$. *Then strict locality holds for the pair* O_1, O_2.

Proof. If D_1 and D_2 are double cones and $D_1 \times\!\!\times D_2$, then there is a double cone D_3 such that $D_1 \ll D_3 \subset D'_2$. By Proposition 1.3.11, all projections in $R(D_1)$ are infinite relative to $R(D_3)$ and, hence, relative to $R(D_2)'$ as well. It means that in Proposition 1.3.32 taken for regions D_1, D_2 condition I holds. Since $R(D_2)$ is a factor (Proposition 1.3.6), conditions 1′–3′ are equivalent and so one has strict locality for the regions D_1 and D_2. The general case follows from Corollary 1 to Proposition 1.3.31. □

Another corollary to Proposition 1.3.32 is Proposition 1.3.13 above which concerns the type of local algebras.

From Proposition 1.3.31 strict locality and the Schlieder property might seem very close to each other: indeed, the only difference between (SL) and (S) is that the $\mathcal{H}$-image of the vector state $\omega_{\Phi_2}^{(O_2)}$ is replaced by the support $P_{\Phi_2}^{R(O)'}$ of the latter in (3.41). But this replacement amounts to a considerable weakening of the condition imposed. The reason for this lies in the fact that the $\mathcal{H}$-image of a vector state determines the state uniquely whereas for the support it is not the case at all. The weakening manifests itself in the fact that while (SL) is closely connected with the type of $R(O)$, (S) is connected only with its factoriality, as shown by the following elementary lemma.

LEMMA 1.3.33. *Let R_1 and R_2 be commuting von Neumann algebras and R_1 be a factor. Then R_1 and R_2 have the Schlieder property:*

$$A_i \in R_i\,,\ A_i \neq 0\,,\ i = 1, 2 \Rightarrow A_1 A_2 \neq 0\,.$$

Proof. Let $A_1 A_2 = 0$ and the vector $\Phi \in \mathcal{H}$ be such that $A_1\,\Phi \neq 0$. Taking account of $Z_1 = C(\mathcal{H})$, one has

$$\mathcal{H} = \mathcal{H}^{Z'_1}_{A_1\Phi} = \mathcal{H}^{R_1}_{A_1\mathfrak{M}}\,,\quad \mathfrak{M} \equiv \mathcal{H}^{R'_1}_{\Phi}\,.$$

On the other hand, since $A_2 \in R_2 \subset R'_1$, one has

$$A_2\,\mathcal{H} = A_2\,\mathcal{H}^{R_1}_{A_1\mathfrak{M}} = \mathcal{H}^{R_1}_{A_2 A_1\mathfrak{M}} = 0\,,$$

which contradicts $A_2 \neq 0$. ❐

Hence we immediately obtain the conditions for the fulfillment of the Schlieder property in the Haag–Araki theory.

PROPOSITION 1.3.34. *Let Axioms* I *and* III *hold. Then the Schlieder property is valid for any pair of regions O_1 and O_2 such that $O_1 \times O_2$ and, in addition, there is a region O_3 for which $O_1 \subset O_3 \subset O'_2$ and $R(O_3)$ is a factor.*

COROLLARY. *Let Axioms* Ia, IIa, III *and* IV *hold and R be a factor. Then the Schlieder property is valid for any pair of regions O_1 and O_2 such that $O_1 \times O_2$ and, in addition, there is a double cone D for which either $O_1 \subset D \subset O'_2$ or $O_1 \subset D' \subset O'_2$.*

The corollary immediately follows from Propositions 1.3.34 and 1.3.6. It implies, in particular, that the Schlieder property holds for any pair D, D'.

Another proof of (S) suggested by Schlieder [376] uses different assumptions (namely, Axioms I–III, V and VI and existence of a unique and cyclic vacuum) and is valid for pairs O_1, O_2 with $O_1 \times\!\!\times\, O_2$. The proof is based on the following result.

Borchers' theorem [79]. Let a continuous unitary representation $U(t)$ of a one-parameter group with semibounded spectrum (e.g., a representation of the group of translations in time in a Haag–Araki theory satisfying Axioms V and VI) act in Hilbert space $\mathcal{H}$ and let P_1 and P_2 be projections in $\mathcal{H}$ such that

$$P_1 P_2 = 0\,,\ [U(t) P_1 U(t)^{-1},\ P_2] = 0,\ t \in (-\varepsilon, \varepsilon)\,,\ \varepsilon < 0\,.$$

Then for all $t \in \mathbf{R}$,

$$U(t) P_1 U(t)^{-1} P_2 = 0\,.$$

Now we shall show that the Schlieder property implies extended locality, i.e. the above-mentioned implication holds

$$(\mathrm{S}) \Rightarrow (\mathrm{EL})\,. \tag{3.47}$$

The proof is trivial: if $R(O_1) \cap R(O_2) = C(\mathcal{H})$, there is a projection $P \in R(O_1) \cap R(O_2)$, $P \neq 0, I$, and then the equality $P(I - P) = 0$ means that the Schlieder property is violated.

Due to (3.45) and (3.47), extended locality is the weakest of all causal properties we have considered. Obviously it holds in any Haag–Araki theory satisfying the assumptions of Proposition 1.3.34 or its corollary. Its physical meaning and reasons for its introduction are quite clear: if extended locality is violated then, doing measurements in a region O_1 we can exactly determine values of certain observables in a region O_2 spacelike to O_1. Using Lorentz-covariance, it is possible to obtain a nontrivial enhancement of extended locality: Landau has shown that it holds for pairs of double cones D_1, D_2 even if the latter are not mutually spacelike.

PROPOSITION 1.3.35 [294]. *Let Axioms* I–III, V *and* VI *hold*, R *be a factor and regions* D_1 *and* D_2 *be double cones such that* $(D_1 + \mathcal{N}) \cap D_2 = \emptyset$ *for a certain neighbourhood* $\mathcal{N}$ *of the origin. Then for* D_1 *and* D_2 *one has extended locality,* $R(D_1) \cap R(D_2) = C(\mathcal{H})$.

Proof will be given briefly. Due to Axiom III and factoriality of R, one sufficient condition for (EL) is

$$R(D'_1)' \cap R(D'_2)' = R'\,. \tag{3.48}$$

When D_1 and D_2 are mutually spacelike, (3.48) is derived trivially by using Axioms Ia and IIa (see the proof of Proposition 1.3.6). But in the case in question the proof is not so simple. It relies on the technique which is based on the uniqueness theorem for five-dimensional wave equation and was first developed in the theory of the Jost–Lehmann–Dyson representation [30, Section 32]; in the algebraic approach it was used by Araki [38] in his proof of the spacelike hull theorem (see Proposition 1.3.14 and its discussion). Let $\mathcal{N}_1$ and $\mathcal{N}_2$ be neighbourhoods of the origin in M such that $\tilde{D}_1 \cap \tilde{D}_2 = \emptyset$, where $\tilde{D}_i = D_i + \mathcal{N}_i$, $i = 1, 2$, are double cones. There exists a spacelike hyperplane S_{12} such that $\tilde{D}_1$ and $\tilde{D}_2$ lie on different sides of it. Consider the function $f(x) = (\Phi, [A, B(x)]\,\Psi)$, where Φ, Ψ are vectors analytic in the energy, $A \in R(D'_1)' \cap R(D'_\varepsilon)'$, $B \in R(D_\varepsilon)$, D_ε is a double cone with the centre in the origin and such that $(D_\varepsilon + x_i) \subset \tilde{D}_i$ for all $x_i \in D_i$, $i = 1, 2$. One easily sees that $f(x) = 0$ for $x \in \tilde{D}'_1 \cap \tilde{D}'_2$, and, due to weak additivity, it is sufficient for obtaining (3.48) to prove that $f(x) \equiv 0$. To this end the Lorentz transformation is used which maps S_{12} into the hyperplane $S_0 = \{x \in M \mid x^0 = 0\}$. Then $f(x)$ is identified with the boundary value at $s = 0$ of a smooth solution $F(x, s)$ of five-dimensional wave equation. The required extension of the region where $f(x)$ vanishes is obtained with the help of the following property (see, e.g. [38, Lemma 6]): if a smooth solution of five-dimensional wave equation vanishes with all its derivatives on a certain timelike segment, it also vanishes in the five-dimensional double cone generated by the segment (the intersection of this double cone with $\{s = 0\}$ is, in general, larger than the original region of vanishing of $f(x)$). Analyzing the support properties of $f(x)$, one verifies that the region of vanishing obtained in this way includes the whole hyperplane S_0 whence, the wave equation being hyperbolic, one has $F(x, s) \equiv 0$ and therefore $f(x) \equiv 0$ as well. ❐

Apart from the discussed properties, extended locality turns out to be closely connected with the global structure. It will be shown in Section 2.1 that it is equivalent to several important global properties, including the simplicity of quasilocal C^*-algebra.

1.3.5. SYMMETRY PROPERTIES

We shall conclude our exposition of axiomatic formalism with a discussion of the symmetry problem. Broadly speaking, this problem confronts the algebraic approach as it does any other physical theory: a theory describing a certain class of physical phenomena must have ways and means of describing, inter alia, the symmetry properties of these phenomena. Speaking concretely, the task of this chapter, study of rigorous consequences of the Haag–Araki–Kastler axioms, includes the study of consequences of Axioms V and VI, according to which all relativistic quantum phenomena possess the symmetry under the Poincaré transformations.

Algebraic formalism for the description of symmetries is highly universal (in particular, for the cases of quantum field theory and statistical mechanics all its basic elements are completely identical). The treatment of symmetries as automorphisms of some or other 'physical' algebras proved to be both convenient and fruitful. It provided the basis for mathematical theory of many phenomena and problems in many areas: phenomena of spontaneous symmetry breaking and vacuum degeneration, studies of asymptotically abelian systems, various problems connected with invariant and ergodic states, to give just a few examples. Due to widespread interest in these applications, the algebraic treatment of symmetries, unlike the algebraic approach as a whole, has been expounded several times in detail: only over the relatively recent period it was done in the monographs by Ruelle [22], Emch [10], Bratteli and Robinson [5]. These books, especially the last two, expound in detail the general algebraic theory of symmetries based on the modern concept of a C^*-dynamical system; however, physical applications are chosen only from the area of statistical mechanics.

Such a situation clearly dictated to us the manner of our exposition. Obviously, it would be unreasonable to reproduce here again all the foundations of the general theory: the reader can easily find the rigorous results and proofs in the book by Bratteli and Robinson and the extensive physical discussion in the book by Emch. On the other hand, in the general study of local nets $\{R(O)\}$ the analysis of their symmetry properties is a matter of necessity, both per se and because of its ties with the other properties: as we saw above, properties of the translation automorphisms of the local net are actually the main tool in proving structural theorems in the Haag–Araki theory. Keeping this in mind, we present below all the necessary information on symmetry properties in local quantum theory, but only in the concise form, omitting the proofs and discussion of results when it can be found in the said books. On the other hand, we try to single out and emphasize the few facts which are not parts of the general formalism and belong specifically to relativistic quantum theory.

Let us begin with the group of translations. Analyzing properties of the translation automorphisms it is natural to start with those implied solely by the theory of group representations and independent of the structures of local quantum theory. This category includes, first of all, the formulas for the projection P_0 on the subspace $\mathcal{H}_0$ of all translationally invariant vectors. Hilbert space $\mathcal{H} \supset \mathcal{H}_0$ in this section can be assumed to be an arbitrary space where a strongly continuous unitary representation $T(a)$ of the group $\hat{M}$ is acting. First, let us recall some simple facts. For any unitary operator U the projection E_0 on the subspace of all its invariant vectors (coinciding with the kernel of $U-I$) has the representation by means of the von Neumann ergodic theorem:

$$E_0 = s\text{-}\lim_{n \to \infty} \frac{1}{n} \sum_{k=1}^{n} U^k .$$

Any strongly continuous one-parameter unitary group U_t, $t \in \mathbf{R}$, by the Stone theorem, has a spectral decomposition, and the projection E_0 on the set of all invariant vectors of the group is one of the projections in this decomposition, $E = E(\{0\})$. Whence it follows that E_0 lies in the von Neumann algebra generated by U_t:

$$E_0 \in \{U_t\,,\ t \in \mathbf{R}\}'' .$$

These elementary facts lead us to the basic classical result enhancing and generalizing them. This is the Alaoglu–Birkhoff ergodic theorem [5, Theorem 4.3.4] which states that for any subgroup $\mathcal{U}$ of contracting mappings of a Hilbert space (in particular, any unitary group), the projection E_0 on the set of all invariant vectors of $\mathcal{U}$ belongs to the strong closure of the convex hull of $\mathcal{U}$:

$$E_0 \in \overline{\mathrm{Co}\ \mathcal{U}}^{\,s} ,$$

where, by definition, the convex hull of a set is the set of all convex combinations of elements of the latter,

$$\mathrm{Co}\ \mathcal{U} = \left\{ A = \sum_{i=1}^{n} \lambda_i U_i\,,\ U_i \in \mathcal{U}\,,\ \lambda_i \geq 0\,,\ \sum_{i=1}^{n} \lambda_i = 1\,,\ n = 1\,,2\,,\ldots \right\} .$$

Thus in the case $\mathcal{U} = T(\hat{M})$ one has

$$P_0 = s\text{-}\lim_{\gamma} \sum_{i=1}^{n_\gamma} \lambda_{i,\gamma} T(a_i)\,,\ \lambda_{i,\gamma} \geq 0\,,\quad \sum_{i=1}^{n_\gamma} \lambda_{i,\gamma} = 1\,, \tag{3.49}$$

the limit being the strong operator limit of a directed set [20, Vol. 1, Section IV.2] of elements of $\mathrm{Co}T(\hat{M})$. If $\mathcal{U}$ is a strongly continuous representation of a group G with the invariant mean (see, e.g. [16, Section 9]; such groups are called *amenable*), besides the Alaoglu–Birkhoff theorem there exists another representation for P_0 belonging to the same class of theorems which are sometimes called mean ergodic theorems [22, Proposition 6.2.15]. In this representation a directed set, strongly approximating the projection P_0, instead of convex combinations of elements of $\mathcal{U}$, is formed by integrals over the group G. In this case instead of normed families of positive coefficients $\lambda_{i,\gamma}$ in (3.49), one uses normed positive functions $\chi_\gamma(g)$ on G forming so-called M-net [22, Chapter 6 and Appendix D6]. But we shall not really need the general formalism of invariant means over topological groups, only a few of its elements. Specifically, in this case it suffices to note that for any Lebesque measurable region $O \subset M$ the integral $\int_O d^4a T(a)$ is well defined (which is tantamount to saying that the translation group $\hat{M}$ is amenable). An arbitrary M-net on $\hat{M}$ is obtained if a system of Borel sets $\{O_\gamma\}$ is chosen in the Minkowski space such that for any $a \in M$

$$\lim_{\gamma} \frac{|O_\gamma \wedge (O_\gamma + a)|}{|O_\gamma|} = 0$$

(here $|O_\gamma|$ is the Lebesque measure of the set O_γ, and for arbitrary sets M and N, $\mu \wedge \mathcal{N} \equiv (\mu \cup \mathcal{N}) \setminus (\mu \cap \mathcal{N})$). Elements of the net are then characteristic functions of the sets O_γ, and the analogue of (3.49) is the formula [5, Example 4.3.5]

$$P_0 = s\text{-}\lim_{\gamma} \frac{1}{|O_\gamma|} \int_{O_\gamma} d^4a\, T(a)\,. \tag{3.50}$$

The standard choice of the system $\{O_\gamma\}$ is the set of increasing 4-cubes:

$$O_\gamma = O_L = \{a \in M \mid |a^\mu| \le L\}\,,\ |O_L| = L^4\,.$$

General theory assuming only that the group $T(\hat{M})$ is unitary and amenable cannot produce anything stronger than (3.50). This property, however, is still too weak compared to those required in relativistic quantum theory. The representation for the projection P_0 which is generally used here (see, e.g., Propositions 1.3.7 and 3.1.13) is the enhancement of (3.50) that can be derived only taking into account that $T(\hat{M})$ is a subgroup of a representation $U(\mathcal{P}_+^\uparrow)$ of the Poincaré group. This enhancement was first pointed out by Borchers as far back as 1962.

PROPOSITION 1.3.36 [314]. *Let a strongly continuous unitary representation $U(a, \Lambda)$ of the Poincaré group $\mathcal{P}_+^\uparrow$ act in a Hilbert space $\mathcal{H}$. Then for any 4-vector $e \in M$,*

$$P_0 = w\text{-}\lim_{\lambda \to \infty} U(\lambda e, I)\ . \tag{3.7}$$

Proof. Let $E(p_0, \mathbf{p})$ be the spectral function of the 4-vector generator $\hat{p}_\mu$ of the representation $U(a, I)$ and $E_e(p)$ be the spectral function of the generator $e^\mu \hat{p}_\mu$ of the one-parameter subgroup $U(\lambda e, I)$. For any $\Phi, \Psi \in \mathcal{H}$ one has

$$(\Phi, U(\lambda e, I)\Psi) = (\Phi, \exp(i\lambda e^{\mu}\hat{p}_{\mu})\Psi) = \int_{-\infty}^{\infty} e^{i\lambda p}\,[\frac{d}{dp}\,(\Phi, E_e(p)\Psi)]\,dp$$

$$= \int_{-\infty}^{\infty} e^{i\lambda p}\,[\frac{d}{dp}\,(\Phi, P_0 E_e(p)\Psi)]\,dp + \int_{-\infty}^{\infty} e^{i\lambda p}\,[\frac{d}{dp}\,(\Phi, (I - P_0) E_e(p)\Psi)]\,dp\,. \tag{3.51}$$

From the obvious $E_e(\{0\}) \geq P_0$ one has

$$P_0 E_e(p) = E_e(p) P_0 = \theta(p) P_0\,,$$

and since $d\theta(p)/dp = \delta(p)$, the first term in (3.51) is equal to $(\Phi, P_0\Psi)$. Thus it suffices to prove that the second term vanishes in the limit $\lambda \to \infty$. And for this it is sufficient to show that the function $(\Phi, (I - P_0) E_e(p)\Psi)$ is absolutely continuous on $\mathbf{R}$: then its derivative belongs to $L^1(\mathbf{R})$ and, by the Riemann–Lebesque Lemma (e.g. [20, Vol. 2, Theorem IX.7]) the said limit equals zero.

By definition, the absolute continuity of $(\Phi, (I - P_0) E_e(p)\Psi)$ means that

$$(\Phi, (I - P_0) E_e(\Delta)\Psi) = 0 \quad \forall\, \Delta \subset \mathbf{R}, |\Delta| = 0\,. \tag{3.52}$$

Taking account for the Schwartz inequality, it is sufficient to check this property for $\Phi = \Psi$. To do this, exploit the presence of the representation $U(a, \Lambda)$ of the Poincaré group in $\mathcal{H}$. $U(a, \Lambda)$ has the decomposition into irreducible representations $U_{(m,s)}(a, \Lambda)$; the corresponding decomposition of $\mathcal{H}$ is as follows (see e.g. [4, Chapter 2, Section 3]):

$$\mathcal{H} = \Big(\int_s \oplus\; \mathcal{H}_{(m,s)}\, d\rho(m^2)\Big) \oplus \mathcal{H}_0\,, \quad \mathcal{H}_{(m,s)} = L^2\big(V^+_{(m)}, d\mu; E^{(2s+1)}\big)\,,$$

$$d\mu = d\mu(p) = \delta(p^2 - m^2)\, d^4p\,,$$

where $E^{(2s+1)}$ is the $(2s+1)$-dimensional space of the spin degrees of freedom. Denoting elements of $\mathcal{H}_{(m,s)}$ by

$$\Phi_{(m,s)}(p) = \Big\{\Phi^{\ell}_{(m,s)}(p),\ \ell = -s,\ -s+1, \ldots, s\,, p_0 = (\mathbf{p}^2 + m^2)^{1/2}\Big\}$$

and taking into account that the momentum operators have in each $\mathcal{H}_{(m,s)}$ a dense domain $D_{(m,s)}$ of their eigenvectors:

$$\big(\hat{p}_{\mu}\Phi_{(m,s)}\big)^{\ell}(p) = p_{\mu}\Phi^{\ell}_{(m,s)}(p), \quad (e^{\mu}\hat{p}_{\mu})\,\Phi_{(m,s)}(p) = (e^{\mu}p_{\mu})\,\Phi_{(m,s)}(p)\,,$$

one can verify the fulfillment of (3.52) for $\Phi = \Psi \in \mathcal{H}_{(m,s)}$. Using the expression for the spectral function of an operator A in terms of its resolvent $R_z(A)$:

$$E_A(\Delta) = \frac{1}{2\pi i} \oint_{\Gamma(\Delta)} R_z(A)\, dz\,,$$

one finds by the explicit calculation that for an arbitrary measurable set $\Delta \subset \mathbf{R}$,

$$\left(\Phi_{(m,s)}, E_e(\Delta)\, \Phi_{(m,s)}\right) = \sum_{\ell=-s}^{s} \int \left|(\Phi^{\ell}_{(m,s)}(p)\right|^2 \chi_\Delta (e^\mu p_\mu)\, \delta(p^2 - m^2)\, d^4 p\,,$$

where $\chi_\Delta(\cdot)$ is the characteristic function of Δ. Hence it is already evident that if $|\Delta| = 0$, the integral in the right-hand side is also equal to zero. (Indeed, it suffices to consider the cases $e = (1, 0, 0, 0)$, $(0, 1, 0, 0)$, $1/\sqrt{2}\ (1, 0, 0, 1)$.) ❐

The group theoretical meaning of the property (3.7) was better elucidated in [347] where it was shown that one-parameter subgroups with this asymptotic behaviour are typical of unitary representations of any simple noncompact Lie group.

The central place among the properties of translation automorphisms belongs to numerous implications of the cardinal fact that a quasilocal algebra $\mathfrak{A}$ and the translation group $T(\hat{M})$ acting on it form an asymptotically abelian C^*-dynamical system. It is time to define these notions.

DEFINITION 1.3.3. The C^*-dynamical system* is a triplet consisting of C^*-algebra $\mathfrak{A}$, locally compact group G, and a strongly continuous homomorphism α of G into the group of *-automorphisms of $\mathfrak{A}$. Covariant representation of a dynamical system $(\mathfrak{A}, G, \alpha)$ is a triplet $(\mathcal{H}, \pi, U)$ where $\mathcal{H}$ is a Hilbert space, π is a non-degenerate representation of the algebra $\mathfrak{A}$ in $\mathcal{H}$, U is a strongly continuous representation of the group G in $\mathcal{H}$ such that for all $A \in \mathfrak{A}$, $g \in G$,

$$\pi(\alpha_g(A)) = U_g\, \pi(A)\, U_g^* \,. \tag{3.53}$$

This definition means that for each $g \in G$, α_g is a *-automorphism of $\mathfrak{A}$, $\alpha_e = \mathrm{Id}$, and the mapping $g \to \alpha_g(A)$ is norm-continuous at any $A \in \mathfrak{A}$ (e is the identity of G, Id is the identical mapping of $\mathfrak{A}$). Functional or state ω on the algebra $\mathfrak{A}$ of a dynamical system is called *G-covariant* if π_ω is a covariant representation; it is called *G-invariant* if

$$\omega(\alpha_g(A)) = \omega(A)$$

for all $g \in G$, $A \in \mathfrak{A}$. We shall denote by S_{inv} or, if necessary, by $S^{(G)}_{\mathrm{inv}}$, the set of all G-invariant states. Extending the Gelfand–Naimark–Segal construction to the triplet $(\mathfrak{A}, G, \alpha)$, one readily verifies that each G-invariant state ω canonically generates a cyclic covariant representation $(\mathcal{H}_\omega, \pi_\omega, U_\omega)$ and the cyclic vector Ω_ω of this representation is invariant under the unitary group U_ω:

$$U_\omega(g)\, \Omega_\omega = \Omega_\omega \quad \forall\, g \in G\,. \tag{3.54}$$

The representation U_ω of G is determined uniquely by conditions (3.53) and (3.54).

For a covariant representation π we shall denote by $E_{0,\pi}$ the projection on the

* Borchers once rightly remarked that in relativistic quantum theory this commonly used term is hardly a good choice. Indeed, the structures $(\mathfrak{A}, G, \alpha)$ are used here mainly for describing kinematic and not dynamical properties, unlike nonrelativistic theories where α is frequently taken to be the group of dynamic evolution $\exp(iHt) \cdot \exp(-iHt)$.

subspace $\mathcal{H}_{0,\pi}$ of all G-invariant vectors in the space $\mathcal{H}_\pi$; if $\pi = \pi_\omega$, these objects will be denoted by $E_{0,\omega}$ and $\mathcal{H}_{0,\omega}$, respectively; and if $G = \hat{M}$, the notations $P_{0,\pi}$ and $P_{0,\omega}$ will be preserved. For the projections $E_{0,\pi}$ the representation (3.49) given by the Alaoglu–Birkhoff theorem holds, and also the representation of (3.50) type if the group G is amenable. Clearly, in any Haag–Araki or Haag–Kastler theory its quasilocal C^*-algebra $\mathfrak{A}$, the Poincaré group $\mathcal{P}_+^\uparrow$, and the automorphisms $\alpha_{(a,\Lambda)}$, $(a, \Lambda) \in \mathcal{P}_+^\uparrow$ introduced by the Axiom V, form a C^*-dynamical system. Similarly, such a system is formed by any triplet $(\mathfrak{A}, \Pi, \alpha_{(a,\Lambda)})$ where Π is a certain subgroup of $\mathcal{P}_+^\uparrow$ and $(a, \Lambda) \in \Pi$. Vacuum states of relativistic quantum theory are, by definition, invariant states of the system $(\mathfrak{A}, \mathcal{P}_+^\uparrow, \alpha_{(a,\Lambda)})$. Nevertheless the term is routinely applied to all translationally invariant states which, strictly speaking, should be treated only as 'candidates' for true vacuum states.

DEFINITION 1.3.4. The C^*-dynamical system $(\mathfrak{A}, G, \alpha)$ with a noncompact group G will be called:

(1) G-abelian, if for any G-invariant state ω the set of operators $E_{0,\omega}\, \pi_\omega(A)\, E_{0,\omega}$, $A \in \mathfrak{A}$, is abelian;

(2) having a large symmetry group G, if for every $\omega \in S_{\text{inv}}^{(G)}$ and for all $A = A^* \in \mathfrak{A}$,

$$\overline{\text{Co}\left\{\pi_\omega(\alpha_g(A)), g \in G\right\}}^{\,w} \cap \pi_\omega(\mathfrak{A})' \neq \emptyset ;$$

(3) weakly asymptotically abelian, if for any state ω on $\mathfrak{A}$, any $A, B \in \mathfrak{A}$ and any way of going to infinity on the group G,

$$\lim_{g\to\infty} \omega([\alpha_g(A), B]) = 0 ;$$

(4) asymptotically abelian in the sense of Størmer, if for every element $A = A^* \in \mathfrak{A}$ there exists a sequence $\{g_n(A)\}_{n=1}^{\infty} \subset G$ such that $g_n \to \infty$ as $n \to \infty$, and for all $B \in \mathfrak{A}$,

$$\lim_{g\to\infty} \| [\alpha_{g_n}(A), B] \| = 0 ; \tag{3.55}$$

(5) uniformly asymptotically abelian, if for any $A, B \in \mathfrak{A}$,

$$\lim_{g\to\infty} \| [\alpha_g(A), B] \| = 0 .$$

The following implications hold between the above properties:

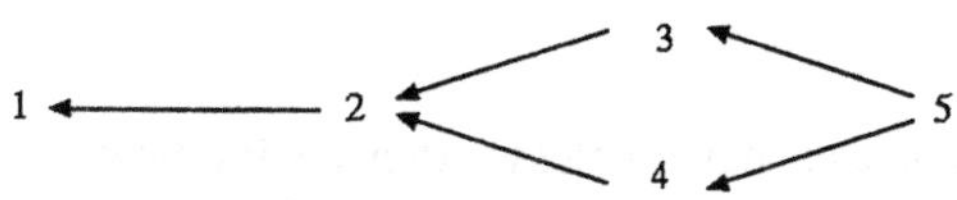

(3.56)

These implications can be easily derived if the following equivalent definitions of properties 1 and 2 are used:

(1′) the system $(\mathfrak{A}, G, \alpha)$ is G-abelian, if for every $\omega \in S_{\text{inv}}$ and for all $A, B \in \mathfrak{A}$,

$$\inf_{A' \in \mathrm{Co\,Orb}_G A} |\omega([A', B])| = 0,$$

where the symbol $\mathrm{Co\,Orb}_G A$ stands for the convex hull of a G-orbit [16, Section 2.2] of an element A in the algebra $\mathfrak{A}$, i.e. the set

$$\mathrm{Orb}_G A = \{\alpha_g(A), g \in G\}$$

(for the equivalence of I and I' see [22, Theorem 6.2.8].

(2′) the system $(\mathfrak{A}, G, \alpha)$ has a large symmetry group, if for every $\omega \in S_{\mathrm{inv}}$ and every finite set of elements $A, B_1, \ldots, B_n, C$ of $\mathfrak{A}$,

$$\inf_{A' \in \mathrm{Co\,Orb}_G A} |\omega(C^* [A', B_k] C)| = 0, \quad k = 1, \ldots, n$$

(for the equivalence of 2 and 2′ see [386, Theorem 3.5]).

These reformulations make evident the implications $1 \leftarrow 2$, $2 \leftarrow 3$ and $2 \leftarrow 4$; the implications $4 \leftarrow 5$ and $3 \leftarrow 5$ are trivial.

Let us note that in formulation 1 the set of operators $E_{0,\omega} \pi_\omega(\mathfrak{A}) E_{0,\omega}$, in general, may not be an algebra. As for formulation 1′, taking into account that $S^{(H)}_{\mathrm{inv}} \supset S^{(G)}_{\mathrm{inv}}$ for any subgroup $H \subset G$, it readily implies the following: if H is a subgroup of G and the system $(\mathfrak{A}, H, \alpha)$ is H-abelian, then the system $(\mathfrak{A}, G, \alpha)$ is G-abelian. It is evident from the definition that the same property is shared by the systems with a large symmetry group: in full compliance with the name, if any subgroup of G is a large symmetry group, then G is a large symmetry group as well.

In the context of local quantum theory, the properties of asymptotic abelianness (relative to spacelike translations) appear most naturally, being automatic implications of local commutativity.

PROPOSITION 1.3.37. *Let local quantum theory satisfy Axioms* III *and* V. *Then for any* $A, B \in \mathfrak{A}$,

$$\lim_{\substack{|a| \to \infty \\ a^2 < 0}} \| [A(a), B] \| = 0, \quad |a| \equiv (a^0)^2 + \boldsymbol{a}^2. \tag{3.57}$$

Proof [10, p. 284] is a trivial verification, since A and B can be approximated in the norm topology with arbitrary precision by elements A_0 and B_0 from the algebra of all local observables $\mathcal{A}$, and at sufficiently large spacelike $a \in \hat{M}$, one has $[A_0(a), B_0] = 0$. □

This elementary statement produces, nevertheless, extensive implications. Let us first point out its immediate corollaries.

COROLLARY 1. *In local quantum theory satisfying Axioms* III *and* V *operators of the Poincaré transformations cannot be quasilocal observables:*

$$U(a, \Lambda) \notin \mathfrak{A} \;\; \forall (a, \Lambda) \in \mathcal{P}_+^\uparrow.$$

Proof [10, p. 285]. For the operators $T(a) = U(a, I)$ we check directly that the assumption $T(a) \in \mathfrak{A}$ for some $a_0 \in \hat{M}$, $a_0 \neq 0$, leads to a contradiction: $A(a_0) = A$ for all $A \in \mathfrak{A}$. In the case of the operators $U(0, \Lambda)$ the group law $U(0, \Lambda)\, T(a)\, U(0, \Lambda)^{-1} = T(\Lambda a)$ should also be taken into account. ❐

This corollary does not exclude, however, that Poincaré transformation can be locally implemented by operators from $\mathfrak{A}$: for some $O \in B(M)$ and $(a, \Lambda) \in \mathcal{P}_+^{\uparrow}$ there may be, in general, unitary operators $U(O; a, \Lambda)$ such that

$$U(O; a, \Lambda)\, R(O)\, U(O; a, \Lambda)^{-1} = R(\Lambda^{-1}(O - a)).$$

COROLLARY 2. *In the Haag–Araki theory for any* $A \in \mathfrak{A}$ *and* $B \in Z'$,

$$\underset{\substack{|a| \to \infty \\ a^2 < 0}}{w\text{-lim}}\ [A(a), B] = 0\,.$$

Proof [10, p. 286–287] is a simple generalization of the proof of Proposition 1.3.37. ❐

COROLLARY 3. *Let in a Haag–Araki or Haag–Kastler theory the states* ω_1, $\omega_2 \in \mathfrak{A}_1^{*+}$ *be vector ones in a certain factorial representation of* $\mathfrak{A}$. *Then*

$$\lim_{\substack{|a| \to \infty \\ a^2 < 0}} [\omega_1(A(a)) - \omega_2(A(a))] = 0\,.$$

Proof [10, p. 287] is just the application of Corollary 2. ❐

As a simple exercise, the reader can reconstruct the details of the proof of Proposition 1.3.37 and its Corollaries.

Since Proposition 1.3.37 is restricted to spacelike directions, it does not mean that a C^*-dynamical system $(\mathfrak{A}, \hat{M}, \alpha_a)$ is uniformly asymptotically abelian: this property would require the fulfillment of (3.57) for any directions in M. But all systems of the form $(\mathfrak{A}, \hat{M}_s, \alpha_a)$, where $\hat{M}_s$ is some subgroup of spacelike translations, are uniformly asymptotically abelian; in particular, so is the system with the subgroup $\hat{\mathbf{R}}^3$ of purely spatial translations. Therefore all such systems are also G-abelian (for $G = \hat{M}_s$). Hence it follows due to $\hat{M}_s \subset \hat{M} \subset \mathcal{P}_+^{\uparrow}$ that the G-abelianness holds also for the group of all translations and the Poincaré group, i.e. the system $(\mathfrak{A}, \hat{M}, \alpha_a)$ is $\hat{M}$-abelian and $(\mathfrak{A}, \mathcal{P}_+^{\uparrow}, \alpha_{(a,\Lambda)})$ is $\mathcal{P}_+^{\uparrow}$-abelian. Similarly, all $\hat{M}_s$ and, in particular, $\hat{\mathbf{R}}^3$, are large symmetry groups (by virtue of (3.56)) and, hence $\hat{M}$ and $\mathcal{P}_+^{\uparrow}$ are also large symmetry groups. Thus the whole complex of properties of G-abelian systems as well as systems with large symmetry groups is valid for local quantum theory.

Turning to these properties, we shall first mention direct implications of G-abelianness. Since the representation of a C^*-dynamical system $(\mathfrak{A}, G, \alpha)$ corresponding to a G-covariant state ω includes both the representation π_ω of $\mathfrak{A}$ and the representation U_ω of G, the von Neumann algebra generated by the union of the two families of operators is an important object. So we introduce the special symbols:

$$R_\omega = \pi_\omega(\mathfrak{A})'' ,\ Z_\omega = R_\omega \cap R'_\omega ,\ R^{(U)}_\omega = \{\pi_\omega(\mathfrak{A}) \cup U_\omega(G)\}'' ,\ Z^{(U)}_\omega = R^{(U)}_\omega \cap R^{(U)'}_\omega$$

($R^{(T)}_\omega$ and $Z^{(T)}_\omega$, respectively, in the case of a quantum field C^*-system $(\mathfrak{A}, \hat{M}, \alpha_a)$) and prove straightforward that the algebra $R^{(U)}_\omega$ is closely related to structural properties of the G-abelian system.

PROPOSITION 1.3.38. *Let a Haag–Kastler theory satisfy Axioms* III *and* V, *and* ω *be a translationally invariant state on* $\mathfrak{A}$. *Then*

(1) $P_{0,\omega}\, \pi_\omega(\mathfrak{A})\, P_{0,\omega}$ *is a set of commuting operators;*

(2) $R_{0,\omega} \equiv P_{0,\omega} R_\omega P_{0,\omega}$ *is a maximal abelian von Neumann algebra in the space* $\mathcal{H}_{0,\omega}$;

(3) $R^{(T)'}_\omega$ *is an abelian von Neumann algebra in the space* $\mathcal{H}_\omega$.

Similarly, in a Haag–Araki theory with an arbitrary nontrivial subspace $\mathcal{H}_0$ *of* $\hat{M}$*-invariant vectors one has that:*

(1) $P_0\, \mathfrak{A}\, P_0$ *is a set of commuting operators;*

(2) $R_0 = P_0 R P_0$ *is an abelian von Neumann algebra in* $\mathcal{H}_0$;

(3) *for any vector* $\Omega \in \mathcal{H}_0$, $(\mathfrak{A}_{P^R_\omega} \cup T(\hat{M})_{P^R_\omega})'$ *is an abelian von Neumann algebra in the subspace* $\mathcal{H}^R_\Omega$.

Proof. Property I follows from the $\hat{M}$-abelianness of the system $(\mathfrak{A}, \hat{M}, \alpha_a)$; property 3 is always implied by 1 [5, Proposition 4.3.7]. But property 2 is a considerable enhancement of 1 and it relies not only on the $\hat{M}$-abelianness of $(\mathfrak{A}, \hat{M}, \alpha_a)$. We shall consider its assertions one by one.

(a) The fact that $R_{0,\omega}$ is a von Neumann algebra follows from the Kovacs–Szücs ergodic theorem (see, e.g., [5, Theorem 4.3.8]) which has already been used in Proposition 1.3.12 and asserts the existence and uniqueness of a normal G-invariant mapping (projection) $\Phi : R_\omega \to R_\omega \cap T(\hat{M})'$ in the case when $F_0 = P^{R'_\omega}_{\mathcal{H}_{0,\omega}} = I_\omega$. By the Kovacs–Szücs theorem, for any $A \in R_\omega$ one has

$$\Phi(A)\, P_{0,\omega} = P_{0,\omega}\, A\, P_{0,\omega} .$$

Hence at $F_0 = I_\omega$,

$$R_{0,\omega} = \Phi(R_\omega) P_{0,\omega} = (R_\omega \cap T(\hat{M})')\, P_{0,\omega} .$$

The last term here is a von Neumann algebra since

$$P_{0,\omega} \in T(\hat{M})'' \subset (R_\omega \cap T(\hat{M})')' .$$

In the general case $(F_0 \neq I_\omega)$ the result holds true for the algebra $F_0 R_\omega F_0$ and hence for $R_{0,\omega}$ as well, owing to $P_{0,\omega} \leq F_0$.

(b) The abelianness of $R_{0,\omega}$ is the direct enhancement of property I, and as such it is established [154, Lemma 4] with the help of stronger assumptions: namely, it is assumed that $\mathcal{H}_{0,\omega}$ is a cyclic subspace of $\pi_\omega(\mathfrak{A})$ and $T(\hat{M})$ is a large symmetry group. Under these assumptions the desired result follows by the straightforward application of the Kovacz–Szücs theorem to the algebras R'_ω and Z_ω.

(c) The maximal abelianness of $R_{0,\omega}$ follows from the fact that the vector $\Omega_\omega \in \mathcal{H}_{0,\omega}$ corresponding to the functional ω and cyclic for R_ω is also cyclic for $R_{0,\omega}$.

The Kovacs–Szücs theorem is a general statement about a von Neumann algebra N with a group U_g of spatial automorphisms and a nontrivial U_g-invariant subspace $\mathcal{H}_{\rm inv}$; its corollary [154, Corollary 2] which asserts that $P_{\rm inv}\, N\, P_{\rm inv}$ is a von Neumann algebra is a statement of the same kind. Therefore by applying this corollary to the triplet $(R, T(\hat{M}), \mathcal{H}_0)$ in the general Haag–Araki theory we immediately obtain 2′. The property 1′ follows from 2′. Finally, 3′ is the exact analogue (or a particular case, for that matter) of 3 since the induction $\mathfrak{A} \to \mathfrak{A}_{P^R_\Omega}$ coincides with the representation π_{ω_Ω} corresponding to the invariant state ω_Ω and acting in the subspace $\mathcal{H}^R_\Omega$; due to this, the von Neumann algebra in the condition 3′ is exactly $R^{(T)'}_\omega$.

Remark. Clearly, unlike 3′, the properties 1′ and 2′ are stronger than the direct Haag–Araki analogues of the Haag–Kastler assertions 1 and 2 (such analogues, just as 3′, are formulated for $\mathfrak{A}_{P^R_\Omega}$ and $\mathcal{H}^R_\Omega$). But we cannot make 3′ stronger in the same way. Such a stronger form which is, obviously,

$$(3'')\ \{\mathfrak{A} \cup T(\hat{M})\}' \subset \{\mathfrak{A} \cup T(\hat{M})\}''$$

does not follow from the Kovacs–Szücs theorem. It should be derived from 1′ in the manner in which 3 is derived from 1; but the implication $1 \to 3$ is specific to representations corresponding to invariant states [5, Proof of Proposition 4.3.7]. As a result, if a Haag–Araki theory has no cyclic vacuum vector, we obtain for it, in general, only property 3′ rather than the stronger property 3″. In the case of systems satisfying the spectrum condition, when the translations turn out to be inner automorphisms of R (see Theorem 1.3.43) this means that in the absence of a cyclic vacuum the Haag–Araki axioms imply only that R is discrete (see Proposition 1.3.46) but not that R' is abelian. The last property is markedly stronger and it excludes, for example, physical theories with parastatistics (see Section 2.3).

The results of the general theory of asymptotically abelian systems fall roughly into two groups dealing, respectively, with ergodic properties and cluster properties. For both these groups of results their main physical applications belong to statistical mechanics. In relativistic quantum theory they are of lesser importance and so we shall discuss only those with the direct application there. First we shall consider properties of ergodic states.

DEFINITION 1.3.5. The invariant state ω on a C^*-dynamical system $(\mathfrak{A}, G, \alpha)$ is called ergodic or extremal invariant if it cannot be represented as a mixture of invariant states:

$$\omega \neq \sum_i \lambda_i \omega_i, \; \lambda_i \geq 0, \; \sum_i \lambda_i = 1, \; \omega_i \in S_{\text{inv}}.$$

The set of all ergodic states will be denoted by S_{erg} or $S^{(G)}_{\text{erg}}$; obviously, S_{erg} is the set of all extremal points of the set S_{inv}.

The condition of ergodicity of a state (introduced by Segal in 1951) is a certain weakening of the condition of the purity (indecomposability) which naturally appears in the description of physical systems with definite symmetry properties. Any pure invariant state is obviously ergodic, whereas the converse is, in general, not true since an extremal point of the set of invariant states may not be extremal in the set of all states. The difference between the ergodicity and purity properties is very noticeable: if pure states correspond to irreducible representations, ergodic states in typical cases correspond to factorial representations. Necessity of ergodic states is clearly visible, above all, in the problem of decomposition of states or representations. Obviously, in the absence of any additional conditions the elementary components into which we decompose a physical system should be characterized by pure (indecomposable) states. However, if a system possesses a symmetry under some group and we want this symmetry to be preserved in the components of the decomposition we should decompose not over pure states but over ergodic states. Such states still can be decomposed, in general, but if we do this, we leave the class of invariant states and lose the symmetry of our description.

General properties of ergodic states for various groups of symmetry are thoroughly analyzed in the above-mentioned books. What is more, in local quantum theory there is practically no need for such a general analysis (save for the problem of internal symmetries and their breaking which we shall not touch upon). Thus we give here only the most important of these properties, for the case of the translation group.

PROPOSITION 1.3.39. *Let a Haag–Araki or Haag–Kastler theory satisfy Axioms* III *and* V. *Then for any invariant state of the C*-dynamical system* $(\mathfrak{A}, \hat{M}, \alpha_a)$ *the following conditions are equivalent* *:

(1) *the state* ω *is ergodic;*

(2) ω *is a unique normal* $\hat{M}$*-invariant state on* R_ω *;*

(3) ω *is a unique vector* $\hat{M}$*-invariant state on* R_ω *;*

(4) $\dim \mathcal{H}_{0,\omega} = 1$;

(5) $R^{(T)}_\omega$ *is a factor ;*

(6) $R^{(T)}_\omega = \mathcal{B}(\mathcal{H}_\omega)$;

(7) $R_\omega \cap R^{(T)'}_\omega = C(\mathcal{H}_\omega)$.

* We formulate the conditions for the Haag–Kastler theory only; the modifications required in the case of the Haag–Araki theory are obvious.

Proof. Conditions 1–7 are among the conditions of Theorem II.2.8 in [10, Chapter 2]. According to [10, p. 185] for any C^*-dynamical system with a large symmetry group all the conditions of Theorem 8 are equivalent. ❒

It should be noted that for an arbitrary G-abelian system, conditions 1–7 are not completely equivalent, and so the proof relies on the fact that the C^*-system $(\mathfrak{A}, \hat{M}, \alpha_a)$ in local quantum theory is not only $\hat{M}$-abelian but also has a large symmetry group. A similar assertion obtained from Proposition 1.3.39 by replacing $T_\omega(\hat{M})$ with $U_\omega(\mathcal{P}_+^\uparrow)$ is valid for the C^*-system $(\mathfrak{A}, \mathcal{P}_+^\uparrow, \alpha_{(a,\Lambda)})$.

We emphasize next that in the Haag–Araki theory, the ergodicity of a vacuum state ω_Ω, as proved above, is tantamount to its uniqueness only in the subspace $\mathcal{H}_\Omega^R$ but not in $\mathcal{H}$; only in the case of a cyclic vacuum vector Ω', the ergodicity of $\omega_{\Omega'}$ entails the uniqueness of both the vacuum state and the vacuum vector.

Finally, we notice that the proposition obtained immediately makes evident the above-mentioned connection between the ergodicity of state and its factoriality.

COROLLARY. *Let* ω *be a factorial* $\hat{M}$*-invariant state. Then* ω *is ergodic and the representation* π_ω *satisfies all the conditions of Proposition* 1.3.39. *If* $Z_\omega \subset R_\omega^{(T)'}$, *the converse is also true, i.e. ergodicity of a state implies its factoriality.*

Proof. For a factorial state, by definition, $Z_\omega = C(\mathcal{H}_\omega)$, which implies condition 7. Conversely, if $Z_\omega \subset R_\omega^{(T)'}$ and ω is ergodic, the implication $1 \to 5$ in Proposition 1.3.39 entails $Z_\omega = C(\mathcal{H}_\omega)$. ❒

Later on, when spectrum condition will be taken into account, we shall return to the properties of ergodic states to see that in this case they are simplified considerably. Now we turn to the discussion of cluster properties. In Section 1.3.4. we have analyzed product states on $\mathfrak{A}$ which have the factorizability property $\omega(AB) = \omega(A)\,\omega(B)$, when A and B are taken from commuting subalgebras of $\mathfrak{A}$. Now, cluster properties (which play an important part in studying asymptotic behaviour of physical systems in both quantum field theory and statistical mechanics) can be considered as certain approximate or asymptotic properties of factorizability. Indeed, they are typically relationships of the form:

$$\lim_{\substack{|a|\to\infty \\ a^2<0}} [\omega(A(a)B) - \omega(A(a))\,\omega(B)] = 0, \quad \forall A, B \in \mathfrak{A}, \tag{3.58}$$

and since the elements $A(a)$ and B at sufficiently large $a^2 < 0$ belong to commuting subalgebras of $\mathfrak{A}$, (3.58) is related to the factorizability property exactly as the asymptotic abelianness is related to the usual one. This parallel goes still further. In the case of an abelian C^*-algebra $\mathfrak{A}$, all its pure states have the factorizability property [5, Corollary 2.3.21]; in the case of an asymptotically abelian algebra all its ergodic states, i.e. the analogues of pure states for systems with symmetry groups, have, as we shall see, cluster properties, i.e. the asymptotic factorizability. The same parallel can be revealed in the physical meaning of the both groups of properties. Mutual commutativity of two subalgebras of $\mathfrak{A}$ expresses locality or causality of the theory, that is mutual independence of 'effects' or 'excitations' or 'particles' described by observables from these subalgebras. Hence asymptotic abelianness should be interpreted as 'asymptotic causality', that is a weaker form of local commutativity expressing the same mutual independence of 'excitations', etc. but only for such groups (clusters, beams) of 'excitations', which are not only mutually spacelike, but also asymptotically detached from one another. And finally, cluster properties, as is already clear, are related to asymptotic abelianness exactly as causal independence is related to usual locality: they express similar physical contents but in terms of states. In short, if asymptotic abelianness is an asymptotic (and hence, weaker) analogue of locality, then cluster properties are an asymptotic (and hence, weaker) analogue of causal independence and also of factorizability of states. Two reservations should be made, however. Firstly, the links pointed out in this loose reasoning refer only to the physical meaning, not to the exact formulations of the properties in question. We saw repeatedly that the proximity of physical meaning of certain properties (for example, causal properties in Section 1.3.4) does not guarantee at all the equivalence of their mathematical formulations. Secondly, these considerations deal, of course, with relativistic quantum theory only; in statistical mechanics cluster properties are given a different physical interpretation related to the concepts of correlations and mixing.

Cluster properties are numerous. We shall point out just basic ones, proceeding from weaker to stronger (though there is no strict ordering among the known cluster properties, unlike the various forms of asymptotic abelianness).

In the first place it is natural to put the property of uniform clustering which does not need any symmetry group. Like all cluster properties, it is also a certain property of asymptotic factorizability of states, but now the asymptotic separation of clusters is defined with respect to the net structure of a quasilocal algebra $\mathfrak{A}$, rather than in terms of symmetry group parameters. Accordingly, the class of systems for which this property is formulated includes not C^*-dynamical systems $(\mathfrak{A}, G, \alpha)$ but generalized quasilocal systems [5, Section 2.6], for which the algebra is equipped not with a symmetry group but with a net of C^*-subalgebras $\{\mathfrak{A}_\gamma\}_{\gamma\in\sigma}$, satisfying the generalized locality condition $[\mathfrak{A}_\gamma, \mathfrak{A}_\beta] = 0$ for all $\gamma \perp \beta$. Here we denote by $\gamma \perp \beta$ a relation of orthogonality on the set σ, i.e. an abstract analogue of the relation $O_1 \times O_2$ on the set $B\,(M)$, having the same properties. Evidently, generalized quasilocal systems provide the basis for unified treatment of many classes of theories, both relativistic and nonrelativistic. Such a level of generality is not needed here, however, and we shall consider the uniform clustering only within the framework of local quantum theory.

DEFINITION 1.3.6. State ω on a quasilocal algebra $\mathfrak{A}$ is called uniformly clustering if for any $A, B \in \mathfrak{A}$ and $\varepsilon > 0$ there is a region $O \in B\,(M)$ such that for all $B \in \mathfrak{A}(O')$,

$$|\,\omega\,(AB) - \omega\,(A)\,\omega\,(B)\,| \le \varepsilon\,\|\,B\,\|\,.$$

PROPOSITION 1.3.40. *Let a Haag–Araki or Haag–Kastler theory satisfy Axioms* I *and* III. *Then any factorial state on the quasilocal algebra* $\mathfrak{A}$ *is uniformly clustering.*

Proof consists of the following two assertions:

(1) The state ω on a quasilocal algebra $\mathfrak{A}$ is uniformly clustering if and only if the 'algebra of asymptotic observables' $Z_{\omega,as}$ of the representation π_ω defined as

$$Z_{\omega,as} = \bigcap_{O\in B(M)} R_\omega(O'), R_\omega(O') = \left\{ \bigcup_{\substack{\tilde{O}\subset O' \\ \tilde{O}\in B(M)}} \pi_\omega(\tilde{O}) \right\}''$$

is trivial;

(2) $Z_{\omega,as} \subset Z_\omega$.

The first assertion is the Lanford–Ruelle theorem, the proof of which can be found in [10, p. 294]. The second assertion is elementary: since $Z_{\omega,as} \subset R_\omega$ and $R_\omega(O') \subset \pi_\omega(O)'$ for any $O \in B(M)$, one has

$$Z_{\omega,as} \subset R_\omega \cap \left(\bigcap_{O\in B(M)} \pi_\omega(O)' \right) = Z_\omega .$$ ❑

We see from this proof that the properties of factoriality and uniform clustering are equivalent for all states ω satisfying the conditions $Z_{\omega,as} = Z_\omega$. It was proved [5, Theorem 2.6.10] that if a net of local subalgebras of $\mathfrak{A}$ has the split property, the said condition is met by all locally normal states on $\mathfrak{A}$. It is also worth noting that the algebra of asymptotic observables proves to be useful in studying many properties of local nets. In particular, in Section 2.2 it will be applied to the analysis of quasi-equivalence properties of coherent superselection sectors.

The main kind of cluster properties are the cluster properties found in ergodic states. In particular, the well known cluster property in the Wightman theory which is the necessary and sufficient condition for the uniqueness of vacuum (see, e.g. [4, Section 3.2.3]) belongs to this kind. The usual, called sometimes weak, cluster property is the property of asymptotic factorizability of ergodic states of a C^*-dynamical system with an arbitrary noncompact group G, having, in general, no properties of asymptotic abelianness. The standard two-element weak cluster property for a G-invariant state ω is as follows:

$$\inf_{A' \in \mathrm{Co\,Orb}_G A} | \omega(A'B) - \omega(A)\,\omega(B) | = 0 . \tag{3.59}$$

The asymptotic nature of the property is not very clear in this form. For this reason the similar two-element property is often formulated in terms of directed sets $\{A_\gamma\} \subset \mathrm{Co\,Orb}_G A$:

$$\lim_{\gamma} | \omega(\alpha_g(A_\gamma) B) - \omega(A)\,\omega(B) | = 0, \tag{3.60}$$

where the limit is approached uniformly with respect to g. Finally, for amenable groups, of which we only consider the translation group $\hat{M}$, the similar property can obviously be written down as

$$\lim_{L\to\infty} \frac{1}{L^4} \int_{|a^\mu|\le L} d^4a\, [\omega(A(a)B) - \omega(A)\,\omega(B)] = 0 . \tag{3.61}$$

Let us also point out a three-element weak cluster property

$$\inf_{A' \in \operatorname{Co}\operatorname{Orb}_G A} \mid \omega\,(CA'B) - \omega\,(A)\,\omega\,(CB) \mid = 0 \qquad (3.62)$$

for all $A, B, C \in \mathfrak{A}$. Obviously, this property also admits formulations similar to (3.60) and (3.61), which we omit for the sake of brevity.

The two-element properties (3.59) and (3.60) may hold for any C^*-dynamical systems without any restrictions on the group G (in the class of locally compact noncompact groups) or on the asymptotic behaviour of automorphisms α_g. They mean restrictions on the state ω only. Specifically, according to [5, Theorem 4.3.22], for an arbitrary group G and any G-invariant state ω the properties (3.59) and (3.60) (and in the case of an amenable group, (3.61) as well) are equivalent to each other and to the property $\dim \mathcal{H}_{0,\omega} = 1$. (The proof is not difficult and follows by the application of the Alaoglu–Birkhoff theorem.) If the system $(\mathfrak{A}, G, \alpha)$ is not G-abelian, the latter property is, in general, only sufficient but not necessary for the ergodicity of ω. Hence, we conclude: the weak cluster property of a state is equivalent to the uniqueness of an invariant vector in the representation space and is generally stronger than the ergodicity of the state. For G-abelian systems this property becomes equivalent to the ergodicity, whereas for systems with large symmetry groups it is also equivalent to all the conditions of Proposition 1.3.39. As for the three-element property (3.62), for an arbitrary system it is a stronger requirement as compared with the two-element one, but for systems with asymptotic abelianness as well as for faithful states these properties are equivalent [5, Section 4.3]; see also [151, Theorem 4].

As we can see from (3.59)–(3.62), typical of all the weak cluster properties is that the product state $\omega\,(A)\,\omega\,(B)$ or $\omega\,(\alpha_g\,(A))\,\omega\,(B)$ is approached not by the states $\omega\,(\alpha_g\,(A)\,B)$ as such but only by some of their convex combinations with varying values of the group parameters or by their integrals over the group if the latter is amenable. Unlike this situation, the strong cluster properties, to which we turn now, ensure the convergence (in g) of $\omega\,(\alpha_g\,(A)\,B)$, but under some additional conditions on the group G or the asymptotics of automorphisms α_g. We have already met something similar in studying the formulae for the projection $E_{0,\omega}$ on the subspace of all G-invariant vectors of a representation π_ω: 'weak' formulas (3.49) and (3.50) were derived for arbitrary groups whereas the 'strong' formula (3.7) has been obtained under rather special conditions.

The following formulation of the strong cluster property [276, p. 159] is generally accepted: the state ω of a C^*-dynamical system (in general, not G-invariant) is called *strongly clustering*, if for a certain directed set $\{g_\gamma\} \subset G$ and any $A, B \in \mathfrak{A}$,

$$\lim_{\gamma} \mid \omega\left(\alpha_{g_\gamma}(A)B\right) - \omega\left(\alpha_{g_\gamma}(A)\right)\omega\,(B) \mid = 0\,. \qquad (3.63)$$

We see from this definition that any G-invariant strongly clustering state is also weakly clustering in the sense of (3.59) or (3.60). (It is also evident that any noninvariant but strongly clustering state enjoys a generalized form of the weak cluster property obtained from (3.59) of (3.60) by replacing $\omega\,(A)$ with $\omega\,(A')$ or

$\omega(\alpha_g(A_\gamma))$, respectively). The class of states and systems having property (3.63) is described by the following result [22, Exercise 6E; 5, example 4.3.24]. Let $(\mathfrak{A}, G, \alpha)$ be a C^*-dynamical system which is weakly asymptotically abelian or asymptotically abelian in Størmer's sense. Then every factorial state ω on $\mathfrak{A}$ is strongly clustering. (However, in the case of systems asymptotically abelian in Størmer's sense, the directed set $\{g_\gamma\}$ in (3.63) may depend on A; in particular, one can take the sequence $\{g_\gamma\} = \{g_n(A)\}$ from (3.55).) It was also proved [276, Section 6] that the converse is, in general, not true, that is the strong cluster property of a state is only the necessary but not sufficient condition for its factoriality.

For G-invariant states the strong cluster property takes a particularly simple form

$$\lim_\gamma \ \omega\left(\alpha_{g_\gamma}(A)B\right) = \omega(A)\,\omega(B) \tag{3.64}$$

for some directed set $\{g_\gamma\} \subset G$. If the system is asymptotically abelian in Størmer's sense and the state ω satisfies the following additional condition:

$$Z_\omega = Z_\omega \cap U_\omega(G)',$$

the strong cluster property (3.64) proves to be the necessary and sufficient condition of the factoriality of ω [386, Corollary 5.5]. Since $Z_\omega \cap U_\omega(G)' = R_\omega \cap R_\omega^{(T)'}$, we see from Proposition 1.3.39 that in this case the both properties are also equivalent to the ergodicity of ω, to all the other conditions of Proposition 1.3.39, and, taking account of the above results, to the weak cluster property (3.60) as well.

Properties such as (3.64) can also arise in the absence of the asymptotic abelianness, owing to some special properties of the group G and the chosen class of its representations. Assume, for instance, that the group G and the G-invariant state ω are such that the 'ergodic' formulae of the form (3.49) and (3.50) are enhanced in the same way as in the case of the translation group (Proposition 1.3.36):

$$E_{0,\omega} = w\text{-}\lim_\gamma \ U_\omega(g_\gamma)\,. \tag{3.65}$$

Then we immediately obtain that for all $A, B \in \mathfrak{A}$ and for any $\omega \in S_{\text{inv}}$,

$$\lim_\gamma \ \omega\left(\alpha_{g_\gamma}(A)\,B\right) = \omega(A\,E_{0,\omega}B)\,, \tag{3.66}$$

and if, moreover, $\dim \mathcal{H}_{0,\omega} = 1$, then, taking into account the G-invariance of ω, one has

$$\lim_\gamma \ \omega\left(\alpha_{g_\gamma}(A)\,B\right) = \omega(A)\,\omega(B)\,.$$

Thus, if the enhanced form (3.65) of the formula for the projection $E_{0,\omega}$ holds, the uniqueness of a G-invariant vector ($\dim \mathcal{H}_{0,\omega} = 1$) entails the strong cluster property (3.64). This property, however, is the direct enhancement of the weak cluster property (3.60) and so, by the above-mentioned theorem [5, Theorem 4.3.22] it, in its turn, implies $\dim \mathcal{H}_{0,\omega} = 1$. As a result, in this case, due to the special features of the group structure, the strong and the weak cluster properties are again equivalent to each other and to the uniqueness of a G-invariant vector (and imply the ergodicity of the state in question).

Two concrete cases are known when such a situation takes place. In both cases the key condition (3.65) is fulfilled due to the fact that the representation π_ω turns out to be covariant with respect to a larger group $\tilde{G} \supset G$ which is the semi-direct product of G and some group G_1, so that $U_\omega(g) = \tilde{U}_\omega(g, e_{(1)})$ (where $\tilde{U}_\omega$ is a representation of $\tilde{G}$ in $\mathcal{H}_\omega$ and $e_{(1)}$ is the identity of the group G_1); and this extended covariance ensures additional properties of the unitary group U_ω necessary for (3.65). What are these properties, we have seen in proving the Proposition 1.3.36, where (3.65) was established for the case when $G = \hat{M}$, $\tilde{G} = \mathcal{P}_+^\uparrow$, $G_1 = L_+^\uparrow$. It is the first of the two cases. In the second case, which is very similar, $G = \hat{\mathbf{R}}^n$ (the group of translations of the Euclidean space $\mathbf{R}^n$), $\tilde{G} = E^n$ (the group of motions of $\mathbf{R}^n$), $G_1 = O(n)$ (the group of rotations in $\mathbf{R}^n$). The proof of (3.65) for this case given by Robinson and Ruelle [369] is repeated in [5, Example 4.3.25].

We see that for the system $(\mathfrak{A}, \hat{M}, \alpha_a)$ both ways of obtaining the strong cluster property (3.64) are possible. Their results are, however, not completely identical. If we proceed from asymptotic abelianness, then by comparing (3.57) with (3.55) we conclude immediately that the system $(\mathfrak{A}, \hat{M}, \alpha_a)$ is asymptotically abelian in Størmer's sense and the sequence $\{g_n(A)\}$ in (3.55) can be chosen as an arbitrary sequence $\{a_n\} \subset \hat{M}$ tending to infinity in the spacelike region $(|a| \to \infty, a^2 < 0)$. Hence for any such sequence and any factorial state ω the cluster property (3.63) holds true; if moreover, ω is $\hat{M}$-invariant, one has (3.64). (In this case, as noticed above, the property $\dim \mathcal{H}_{0,\omega} = 1$ is fulfilled too.) On the other hand, if we proceed from the formula (3.65) for $P_{0,\omega}$, we obtain (3.64) for a different class of directed sets $\{a_\gamma\} \subset \hat{M}$, namely, $a_\gamma = \gamma e$, $\gamma \in \mathbf{R}$ and e is an arbitrary 4-vector. (The conditions on the state are: $\omega \in S_{\text{inv}}^{(\hat{M})}$, $\dim \mathcal{H}_{0,\omega} = 1$.) Hence, as Maison [314] first noticed, in this case we establish the strong cluster property not only for spacelike, but also for arbitrary directions, with a less general way, however, of tending to infinity (the direction e should stay fixed: this was used in the proof of Proposition 1.3.36). One can expect that the extension of the cluster property to timelike directions might have some implications for the structure of local quantum theory, but such implications have not been studied so far.

Using the fact that for any G-invariant state

$$\omega\left(\alpha_{g_1}(A_1)\,\alpha_{g_2}(A_2)\right) = \omega\left(\alpha_{g_1 g_2^{-1}}(A_1)\,(A_2)\right) \quad \forall\, A_1, A_2 \in \mathfrak{A} \tag{3.67}$$

and choosing directed sets $\{g_\gamma^{(i)}\} \subset G$, $i = 1, 2, \ldots$ such that for $i \neq k$ the property (3.64) holds for any directed set $g_\gamma^{(i)}\left(g_\gamma^{(k)}\right)^{-1}$, it is easy to derive by induction the following multi-element generalization of this property:

$$\lim_\gamma \, \omega\left(\alpha_{g_\gamma^{(1)}}(A_1) \ldots \alpha_{g_\gamma^{(n)}}(A_n)\right) = \omega\left(A_1\, E_{0,\omega}\, A_2 \ldots E_{0,\omega}\, A_n\right) \tag{3.68}$$

[5, Example 4.3.24]. One can also obtain another generalization, similar to the three-element weak property (3.62): the 'strong analogue' of this property is clearly

$$\lim_\gamma \, \omega\left(C\, \alpha_{g_\gamma}(A)\, B\right) = \omega(A)\, \omega(CB)\,. \tag{3.69}$$

Later on we shall derive this property in the case of the translation group.

Cluster properties of a state ω are very closely connected with asymptotic (with respect to the group parameters) properties of operators $\alpha_g(A)$ of the representation π_ω. Such properties are actively exploited in local quantum theory and so it is convenient to provide them with special names.

DEFINITION 1.3.7. We say that the representation π_ω of a C^*-dynamical system $(\mathfrak{A}, G, \alpha)$ possesses the weak asymptotic property, if for all $A \in \mathfrak{A}$,

$$\underset{\gamma}{w\text{-lim}} \left\{\pi_\omega(\alpha_g(A_\gamma)) - \omega(A_\gamma) I_\omega\right\} = 0 \tag{3.70}$$

(where $\{A_\gamma\}$ is some directed set chosen from the set $\mathrm{Co\,Orb}_G A$ and I_ω is the identity operator in $\mathcal{H}_\omega$); it possesses the strong asymptotic property, if for all $A \in \mathfrak{A}$ and some directed set $\{g_\gamma\} \subset G$,

$$\underset{\gamma}{w\text{-lim}} \left[\pi_\omega(\alpha_{g_\gamma}(A)) - \omega(\alpha_{g_\gamma}(A)) I_\omega\right] = 0 . \tag{3.71}$$

As we have already seen both for cluster properties and representations for the projection $E_{0,\omega}$, if the symmetry group G is amenable, the averaging over the convex hull of a G-orbit is generally performed by means of integrals over the group. Hence, in particular, for $\hat{M}$ the weak asymptotic property can be formulated in the form

$$\underset{L\to\infty}{w\text{-lim}} \frac{1}{L^4} \int_{|a^\mu| \le L} d^4a \left(\pi_\omega(A(a)) - \omega(A(a)) I_\omega\right) = 0 .$$

It is evident from the above definitions that the 'asymptotic property' implies the corresponding cluster property. Indeed, the product in $\mathcal{B}(\mathcal{H}_\omega)$ being continuous in the weak operator topology, if we multiply (3.70) or (3.71) from the right by an arbitrary $\pi_\omega(B), B \in \mathfrak{A}$, and remember that ω is a vector state in its own representation, we obtain the weak (resp., strong) cluster property. As for the converse, i.e the possibility of deriving asymptotic properties from cluster ones, it is evident that an asymptotic property is equivalent to the corresponding three-element cluster property. Take, for example, the weak property: let the equality

$$\lim_{\gamma} \left[\omega\left(C\, \alpha_g(A_\gamma) B\right) - \omega\left(\alpha_g(A_\gamma)\right) \omega(CB)\right] = 0 \tag{3.72}$$

hold for all $A, B, C \in \mathfrak{A}$. Then, since in the representation π_ω $\omega = \omega_\Omega$, where the vector Ω is cyclic for $\pi_\omega(\mathfrak{A})$, (3.72) readily implies (3.70). The converse is also true: the three-element cluster property follows from the asymptotic property (3.70) in the same way as the two-element one: we only need, in addition, to multiply (3.70) from the left by $\pi_\omega(C), C \in \mathfrak{A}$. It is equally trivial to verify that the three-element strong cluster property (cf. (3.69))

$$\lim_{\gamma} \left[\omega\left(C\, \alpha_{g_\gamma}(A) B\right) - \omega\left(\alpha_{g_\gamma}(A)\right) \omega(CB)\right] = 0$$

is equivalent to the strong asymptotic property (3.71). Thus the first elementary conclusions are as follows: an asymptotic property always entails the corresponding cluster property; it is always equivalent to the corresponding three-element cluster property, it is equivalent to the cluster property if and only if this two-element property is equivalent to the three-element one. The last equivalence is also valid for most typical cases, but here we have to consider strong and weak properties separately.

In the case of weak properties we notice first that the weak asymptotic property, like the similar cluster one, is used, as a rule, in the representations π_ω, $\omega \in S_{\text{inv}}$, when it assumes the form

$$w\text{-}\lim_{\gamma} \pi_\omega(\alpha_g(A_\gamma)) = \omega(A) I_\omega . \tag{3.73}$$

Relations of this property to cluster and other properties of a C^*-dynamical system are fully described by the following result [10, Chapter 2, Theorem 8], cf. also [154, Theorem 3]; [151, Theorem 4a]:

> For any G-abelian C^*-dynamical system the weak asymptotic property (3.73) of a G-invariant state ω is equivalent to the weak cluster property (3.60) and, hence, to the ergodicity of ω, to one-dimensionality of the G-invariant subspace $\mathcal{H}_{0,\omega}$ and to all the other necessary and sufficient conditions of ergodicity (see Proposition 1.3.39).

Hence, it becomes clear that general theory of asymptotically abelian systems is primarily concerned with weak asymptotic properties since they are inherent in all ergodic states, whereas strong asymptotic properties need some additional conditions. On the contrary, in local quantum theory weak properties are not used at all because, as we shall now see, the required additional conditions (for the case $G = \hat{M}$) are fulfilled here automatically, so that strong asymptotic properties are always at hand. They are actively used in local quantum theory and we shall now consider them in more detail.

PROPOSITION 1.3.41. *Each of the following conditions implies the strong asymptotic property* (3.71)*:*

(1) *a C^*-system is asymptotically abelian in Størmer's sense and the state ω is strongly clustering;*

(2) *a C^*-system is asymptotically abelian in Størmer's sense and the state ω is factorial;*

(3) *the state ω is faithful and strongly clustering.*

Proof. Let a system $(\mathfrak{A}, G, \alpha)$ be asymptotically abelian in Størmer's sense, and $\{g_n\} = \{g_n(A)\}$ be the corresponding sequence in (3.55). For an arbitrary $A \in \mathfrak{A}$ consider the following set of bounded operators in $\mathcal{H}_\omega$:

$$\left\{\pi_\omega\left(\alpha_{g_n}(A)\right) - \omega\left(\alpha_{g_n}(A)\right) I_\omega\right\}_{n=1}^{\infty} . \tag{3.74}$$

Since this set is norm-bounded:

$$\| \pi_\omega(\alpha_g(A)) - \omega(\alpha_g(A)) I_\omega \| \leq \| \pi_\omega(\alpha_g(A)) \| + | \omega(\alpha_g(A)) | \leq 2 \| A \|$$

(because any homomorphism of a C^*-algebra does not increase the norm and each of its automorphisms is isometric) it is weakly compact, and there exist subsequences of indices n_k, $k = 1, 2, \ldots$ for which the subsequences of operators are weakly convergent, i.e.

$$\underset{k \to \infty}{w\text{-lim}} \left[\pi_\omega\left(\alpha_{g_{n_k}}(A)\right) - \omega\left(\alpha_{g_{n_k}}(A)\right) I_\omega\right] = C\,, \tag{3.75}$$

where C is a bounded operator in $\mathcal{H}_\omega$, depending, in general, on A and on the sequence $\{n_k\}$. To obtain the strong asymptotic property it suffices to show that $C = 0$: indeed, in this case C is independent of the subsequence n_k, and 0 is the only weak limit point of the sequence (3.74); hence, this sequence converges weakly to zero.

First we shall show that $C \in Z_\omega$. Since the representation π_ω is continuous in the norm topology of $\mathfrak{A}$, due to the asymptotic abelianness in Størmer' sense, one has for all $B \in \mathfrak{A}$:

$$\lim_{n \to \infty} \left\| \left[\left(\pi_\omega\left(\alpha_{g_n}(A)\right) - \omega\left(\alpha_{g_n}(A)\right) I_\omega\right),\, \pi_\omega(B)\right] \right\| = 0\,.$$

Hence for any $\varepsilon > 0$ and any family $\Phi_1, \ldots, \Phi_k, \Psi_1, \ldots, \Psi_k$ of unit vectors in $\mathcal{H}_\omega$ one can find a sufficiently large N such that for all $n > N$ and all $i = 1, \ldots, k$,

$$\left|\left(\Phi_i, \left(\left[C, \pi_\omega(B)\right] - \left[\left(\pi_\omega\left(\alpha_{g_n}(A)\right) - \omega\left(\alpha_{g_n}(A)\right) I_\omega\right),\, \pi_\omega(B)\right]\right) \Psi_i\right)\right| < \varepsilon/2$$

and, moreover,

$$\left\| \left[\left(\pi_\omega\left(\alpha_{g_n}(A)\right) - \omega\left(\alpha_{g_n}(A)\right) I_\omega\right),\, \pi_\omega(B)\right] \right\| < \varepsilon/2\,.$$

Taken together these inequalities entail

$$\left|\left(\Phi_i, \left[C, \pi_\omega(B)\right] \Psi_i\right)\right| < \varepsilon \quad \forall\, i = 1, \ldots, k\,,$$

and since $\varepsilon, k, \Phi_i, \Psi_i$ are arbitrary, one obtains $[C, \pi_\omega(B)] = 0$ and $C \in R'_\omega$. $C \in R_\omega$ is clear from the definition.

Assume now that ω is strongly clustering and one can take in (3.63) $\{g_\gamma\} = \{g_n(A)\}$. Then for the subsequence $\{n_k\}$ in (3.75) one has the following: if Ω is a cyclic vector in $\mathcal{H}_\omega$ corresponding to the state ω, for all $B \in \mathfrak{A}$

$$\lim_{k \to \infty} \left(\Omega, \left[\pi_\omega\left(\alpha_{g_{n_k}}(A)\right) - \omega\left(\alpha_{g_{n_k}}(A)\right) I_\omega\right] \pi_\omega(B)\, \Omega\right) = 0\,. \tag{3.76}$$

Comparing this formula with (3.75) we get

$$\left(\Omega, C\, \pi_\omega(B)\, \Omega\right) = 0\,.$$

Choosing here $B = B_1^* B_2$ with arbitrary $B_1, B_2 \in \mathfrak{A}$ and taking into account that $C \in R'_\omega$ and $\pi_\omega(B_1)\,\Omega$ as well as $\pi_\omega(B_2)\,\Omega$ are dense sets in $\mathcal{H}_\omega$, we obtain $C = 0$.

Similarly, assuming the factoriality of ω, one has $C = \lambda I_\omega$, $\lambda \in \mathbf{C}$, and (3.75) entails

$$0 = \lim_{k \to \infty} \left(\Omega, \left[\pi_\omega\left(\alpha_{g_{n_k}}(A)\right) - \omega\left(\alpha_{g_{n_k}}(A)\right) I_\omega\right] \Omega\right) = (\Omega, C\,\Omega) = \lambda\,,$$

which gives again $C = 0$.

Finally, if ω is faithful, the corresponding cyclic vector $\Omega \in \mathcal{H}_\omega$ is also cyclic for R'_ω. If, in addition, ω is strongly clustering, the formula (3.76) holds, in which, due to the cyclicity of Ω for $\pi_\omega(\mathfrak{A})$, one can replace $\pi_\omega(B)\Omega$ with any $\Phi \in \mathcal{H}_\omega$. Choosing $\Phi = T^*\Psi, T \in R'_\omega$, and noting that one can now take in (3.76) the whole sequence g_n instead of the subsequence g_{n_k}, one has

$$\lim_{n\to\infty} \left|\left(T\Omega, \left[\pi_\omega\left(\alpha_{g_n}(A)\right) - \omega\left(\alpha_{g_n}(A)\right) I_\omega\right]\Psi\right)\right| = 0,$$

whence, $R'_\omega \Omega$ being a dense set, the strong asymptotic property follows. (At the same time, one remark made above is proved: if ω is faithful, the two-element cluster property is equivalent to the three-element one.) ❐

Obviously, all the sufficient conditions of Proposition 1.3.41 are also valid in the case of a G-invariant state when the strong asymptotic property has the form

$$w\text{-}\lim_{\gamma} \pi_\omega\left(\alpha_{g_\gamma}(A)\right) = \omega(A) I_\omega. \tag{3.77}$$

A minor addition to these conditions can be obtained if we recall that the strong cluster property also follows from the formula (3.65) for the projection $E_{0,\omega}$. In this case it holds for all the directed sets $\{g_\gamma\}$ for which (3.65) is true. Indeed, it is evident from the foregoing that the strong asymptotic property can be derived for any directed set satisfying (3.65) if the derivation does not rely on asymptotic abelianness. The possibility of such derivation is ensured by the condition 3 in Proposition 1.3.41, and hence we obtain:

COROLLARY. *Let the group G and the G-invariant state ω be such that* (3.65) *holds. Let also the state ω be faithful and the G-invariant subspace $\mathcal{H}_{0,\omega}$ one-dimensional. Then for any directed set $\{g_\gamma\}$ satisfying* (3.65) *the strong asymptotic property* (3.77) *holds.*

As already noticed, in the case $G = \hat{M}$ the condition (3.65) makes it possible to obtain the strong cluster property for any direction in Minkowski space. Now we see that this is also true of the strong asymptotic property. Unfortunately, the class of states possessing the strong asymptotic property for all directions is made much narrower by the additional condition that the state is faithful.

Concluding the discussion of cluster and asymptotic properties it is worthwhile to summarize its results as applied to the translation group in local quantum theory.

THEOREM 1.3.42. *Let the Haag–Araki or Haag–Kastler theory satisfy Axioms* III *and* V. *Then,*

A: *States of the theory possess the following cluster properties:*

(1) *any factorial state ω is strongly clustering in spacelike directions*

$$\lim_{\substack{|a|\to\infty \\ a^2<0}} \left|(\omega(A(a)B) - \omega(A(a))\,\omega(B)\right| = 0 \tag{3.78}$$

for all $A, B \in \mathfrak{A}$ *;*

(2) *any translationally invariant state* ω *such that* $\dim \mathcal{H}_{0,\omega} = 1$, *is strongly clustering in any fixed direction:*

$$\lim_{\lambda \to \infty} \omega (A (\lambda e) B) = \omega (A) \omega (B) \tag{3.79}$$

for any $e \in M$ *;*

(3) *the three-element strong cluster property is valid:*

$$\lim_{\substack{|a| \to \infty \\ a^2 < 0}} \left| (\omega (CA (a) B) - \omega (A (a)) \omega (CB) \right| = 0$$

for all $A, B, C \in \mathfrak{A}$ *and if the state* ω *is factorial;*

(4) *the multi-element cluster properties are valid: for all* $A_1, \ldots, A_n \in \mathfrak{A}$

$$\lim_{\substack{|a_i - a_j| \to \infty \\ (a_i - a_j)^2 < 0}} \omega (A_1 (a_1) \ldots A_n (a_n)) = \omega (A_1) \ldots \omega (A_n) ,$$

if ω *is translationally invariant and factorial, and*

$$\lim_{\substack{\lambda_1, \ldots, \lambda_{n-1} \to \infty \\ \lambda_i / \lambda_{n-1} = \text{const}}} \omega \big(A_1 (\lambda_1 e_1) \ldots A_n (\lambda_n e_n)\big) = \omega (A_1) \ldots \omega (A_n)$$

(vectors $e_1, \ldots, e_n$ *are arbitrary), if* ω *is translationally invariant and such that* $\dim \mathcal{H}_{0,\omega} = 1$.

B: *States of the theory possess the following asymptotic properties:*

(1) *in any factorial representation* π_ω *the strong asymptotic property holds for spacelike directions:*

$$w\text{-}\lim_{\substack{|a| \to \infty \\ a^2 < 0}} \left[\pi_\omega (A (a)) - \omega_1 (A (a)) I_\omega\right] = 0 ,$$

where ω_1 *is any vector state of the representation* π_ω *;*

(2) *in any factorial representation* π_ω *corresponding to the* $\hat{M}$*-invariant state one has a strong asymptotic property of the form*

$$w\text{-}\lim_{\substack{|a| \to \infty \\ a^2 < 0}} \pi_\omega (A (a)) = \omega (A) I_\omega .$$

Proof. The asymptotic properties in B follow from Proposition 1.3.41 and Corollary 3 to Proposition 1.3.37. The three-element cluster property A(3), as shown above, is equivalent to the asymptotic property B(1) and, obviously, entails the two-element cluster property (3.78). The property (3.79) was proved previously. The multi-element cluster properties A(4) follow from (3.78) and (3.79), respectively; the restriction on the way in which λ_i tends to infinity is clear from (3.65) and

$$\lambda_i e_i - \lambda_{i-1} e_{i-1} = \lambda_{i-1} ((\lambda_i / \lambda_{i-1}) e_i - e_{i-1}) .$$ □

It should be noted that the asymptotic property B(1), the strongest of the assertions of the theorem, was derived by Araki [41] as far back as 1964 when the theory of asymptotically abelian systems did not yet exist. In the early period of local quantum theory studies in cluster and asymptotic properties were known as the 'method of large translations'. In a more restricted sense, this name was taken to mean a way, or rather a supposed possibility, of constructing from an arbitrary Haag–Araki theory one with a nontrivial vacuum subspace $\mathcal{H}_0$ and using to this end the limit of infinitely large spacelike translations. The idea of such a construction was suggested in 1963 by Borchers, Haag and Schroer [90]. Summing up the earlier studies by Haag, Araki, Ruelle et al. (see, e.g. [227, 48, 371]), they pointed out the following: if in a Haag–Araki theory there exists a limit of the kind $w\text{-lim}_{|a| \to \infty, a^2<0} A(a)$, then the limit operator $C(A)$ should belong to the centre of the global algebra R; in addition, it should be linear, positive when A is positive and commuting with the translations. (All these properties are readily checkable by the same arguments as in the proof of Proposition 1.3.41.) Accordingly, if R is a factor, then by assuming the existence of the said limit for all quasilocal operators $A \in \mathfrak{A}$ we obtain

$$\underset{\substack{|a| \to \infty \\ a^2 < 0}}{w\text{-lim}} A(a) = \omega_0(A) I , \qquad (3.80)$$

where ω_0 is a translationally invariant state on $\mathfrak{A}$. Thus, if the existence of the limit in (3.80) is proved, we draw the important conclusion: the asymptotic construction (3.80) enables us to associate with any Haag–Araki theory where R is a factor (and then, by means of the central decomposition, with an arbitrary theory) another theory which has a vacuum state. And if, in addition, we succeed in proving that the fulfillment of the Haag–Araki axioms in the original theory implies their fulfillment in the new theory defined by the vacuum ω_0, then we could claim the complete solution of the problem concerning the status of 'vacuum postulate' in local quantum theory (cf. discussion on p. 19).

Looking at the above results we see, however, that the desired program is not fulfilled so far. Its most problematic and still missing point is the very initial one: the proof of the existence of the limit in (3.80). Clearly, all the results of the theory of asymptotically abelian systems collected in Theorem 3.42 are still much weaker than the condition (3.80). The strong asymptotic property B(2) coincides outwardly with (3.80), but the translationally invariant state is assumed there to be given, not constructed, and the existence of the large-translations limit is only proved in the space $\mathcal{H}_\omega$ with $\mathcal{H}_{0,\omega} \neq \{0\}$. The asymptotic property B(1) is valid for arbitrary theories, including those without a vacuum subspace; however, in contrast to (3.80), the existence of the weak limit is proved there only for the differences $\pi_\omega(A(a)) - \omega_1(A(a)) I$, which does not yet open up the possibility to define a vacuum state. Hence, the lesson is clear: unlike the above properties, the relationship (3.80) is not one of general facts of the theory of asymptotically abelian systems. Its proof, if it is possible at all, should rely more strongly on concrete structures of local quantum theory. This is confirmed by more recent results: it was found that properties similar to (3.80) can be obtained, assuming the existence of a local conserved current [110] or localized charged states [111], cf. Section 2.3, or demanding that all positive representations of $\mathfrak{A}$ were of type 1 [95] (cf. Proposition 1.3,46 and p. 139).

Thus far our discussion of translation automorphisms did not take into account spectrum condition. Implications of this condition are not so numerous, but important. They are all centred around the well known Borchers theorem which states that the translation operators $T(a)$ in the Haag–Araki theory without loss of generality can be assumed to be global observables, $T(a) \in R$. In the particular case of a theory with a cyclic vacuum subspace the theorem has a very simple proof pointed out in 1964 by Araki ([41], see Proposition 1.3.45). The general case, however, needs more sophisticated mathematical tools than those used before in this book. Several versions of the proof are known; the earlier [78, 134] were more cumbersome and used the technique of derivations on von Neumann algebras, while the recent rely on the Arveson spectral theory for the representations of groups in Banach spaces. This theory is a generalization of the spectral theory of unitary representations, familiar to all physicists and based on the Stone theorem (or its generalized form, the SNAG theorem [394]), to the representations of groups by automorphisms of C^*- or W^*-algebras. From the viewpoint of local quantum theory such a generalization is perfectly natural. Indeed, it is the automorphisms of the algebra $\mathfrak{A}$ or R that actually describe here symmetry properties of relativistic quantum systems, including spacetime symmetries. Even in the Haag–Araki theory where Axioms V and VI define a unitary representation of the Poincaré group in a Hilbert space, one assumes that this representation is not unique and may, if desired, be replaced with some other unitary representation implementing the same automorphisms of the algebra $\mathfrak{A}$. The basic object in the algebraic framework for physical symmetries is undoubtedly the group of automorphisms of a quasilocal algebra $\mathfrak{A}$, or, in other words, the automorphic representation of a certain symmetry group G in the Banach space $\mathfrak{A}$. When corresponding automorphisms are spatial, the symmetry in question is also described by a unitary representation of G, but this representation is usually far from being unique: for example, if a unitary group V_t implements automorphisms of a von Neumann algebra N, the same automorphisms are implemented by any unitary group U_t such that $V_t U_t^* \in N'$ for all t. For the majority of problems this nonuniqueness is of no importance, and we work, having chosen any of the unitary representations corresponding to a given automorphic representation. In the case of the Borchers theorem the situation is different, since the theorem deals with the set of all unitary representations implementing the same group of automorphisms of a von Neumann algebra N, and states that this set always includes a representation lying in N. The theory of automorphic representations and, in particular, the Arveson spectral theory are clearly the adequate formalism in this case. It might be added that the Arveson theory provides an efficient formalism for many other group-theoretical problems in the algebraic approach, such as the search for conditions of unitary implementability of various external and internal symmetries, treatment of the phenomena of spontaneous symmetry breaking, etc. Its first applications to such problems have been already reported (see, e.g. [64, 284, 87]) and one can expect that their number will grow.

The generalization of the spectral theory is based, naturally, on a generalization of the concept of spectrum of a representation. Let us describe this generalization. Let G be a locally compact abelian group, $\hat{G}$ the dual group whose elements are characters $\chi(t)$ of the group G, and $L^1(G)$ the group algebra of G, i.e. the set of all functions f on G, integrable with respect to the Haar measure ds, equipped with an involution

$$f^*(t) \doteq f(-t) \quad \forall\, t \in G$$

and a multiplication defined as the convolution operation:

$$(f \times g)(t) \doteq \int_G ds\, f(t-s)\, g(s)\,.$$

(For group-theoretical concepts used in this paragraphs see, e.g. [16, Sections 9–12]). The group $\hat{G}$ is endowed with the topology of uniform convergence on compact subsets; $L^1(G)$ is considered as a Banach *-algebra with respect to the L^1 norm. Generalized Fourier transformation is mapping $L^1(G)$ into the set $C_1(\hat{G})$ of all continuous functions on $\hat{G}$ vanishing at infinity:

$$f \in L^1(G) \to \hat{f}(\chi) = \int_G dt\, f(t)\, \chi(t) \in C_1(\hat{G})\,.$$

Furthermore, let X be a Banach space, F be a norm closed subspace of the dual space X^*, and $\sigma(X, F)$ the topology on X generated by functionals of F (see e.g., [20, Vol. 1, Section IV.5]); let also $B_F(X)$ be set of all linear bounded operators on X continuous with respect to the $\sigma(X, F)$ topology. A $\sigma(X, F)$-continuous homomorphism $U: t \to U_t$ of the group G into the group of invertible isometric operators from $B_F(X)$ is called a representation of the group G in the space X. (It is also possible to develop the theory under weaker conditions on the operators U_t, demanding only $\| U_t \| \leq M$ for all $t \in G$; see [5]). Every such representation induces a representation of the group algebra $L^1(G)$ into $B_F(X)$ denoted by the same symbol U and defined by the formula

$$U: f \in L^1(G) \to U(f) = \int_G dt\, f(t)\, U_t \in B_F(X)\,,$$

whence $U(f \times h) = U(f)\, U(h)$. Associating with an arbitrary subset $Z \subset X$ the set

$$I_Z^U = \{ f \in L^1(G) \mid U(f) A = 0 \quad \forall\, A \in Z \}$$

(which is a closed *-ideal in $L^1(G)$, as one easily verifies), we are ready to formulate the following definition.

DEFINITION 1.3.8. The spectrum (or the Arveson spectrum) of a representation U of a locally compact abelian group G in a Banach space X is the set

$$\sigma(U) = \left\{ \chi \in \hat{G} \mid \hat{f}(\chi) = 0 \quad \forall f \in I_X^U \right\}\,.$$

The spectrum (or the Arveson spectrum) of an arbitrary subset $Z \subset X$ is the set

$$\sigma_U(Z) = \left\{ \chi \in \hat{G} \,|\, \hat{f}(\chi) = 0 \quad \forall f \in I_Z^U \right\}.$$

The spectral subspace (or the spectral M-subspace) corresponding to a subset $K \subset \hat{G}$ is the set

$$M^U(K) = \overline{\{A \in X \,|\, \sigma_U(A) \subset K\}}^{\sigma(X,F)}$$

The associated subspace (or the spectral R-subspace) corresponding to a subset $K \subset \hat{G}$ is the set

$$M^U(K) = \overline{L\ \{\ A \in X \,|\, A = U(f)B,\ B \in X,\ \operatorname{supp} \hat{f} \in K\}}^{\sigma(X,F)}$$

In the particular case of the usual spectral theory for unitary representations of compact abelian groups the new definitions coincide with the standard ones. Specifically, if U_t, $t \in \mathbf{R}$, is a continuous one-parameter unitary group in a Hilbert space $\mathcal{H}$, then, by the Stone theorem, it possesses the spectral decomposition

$$U_t = \int_{-\infty}^{\infty} e^{-ipt}\, dE(p)$$

in terms of which one defines the spectrum of the group:

$$\operatorname{sp} U = \operatorname{supp} E(p) \subset \mathbf{R},$$

and spectral subspaces $\mathcal{H}_\Delta \subset \mathcal{H}$ corresponding to subsets $\Delta \subset \mathbf{R}$:

$$\mathcal{H}_\Delta = E(\Delta)\, \mathcal{H}.$$

In this case $X = F = \mathcal{H}$, $G = \hat{G} = \mathbf{R}$, $\chi(t) = \exp(ipt)$, and it is easy to verify that the Arveson spectrum coincides with the usual one, $\sigma(U) = \operatorname{sp} U$, and spectral subspaces in the new and the old definition coincide, too:

$$M^U(\Delta) = E(\Delta)\, \mathcal{H}.$$

Further information on the Arveson theory see in [5, Chapter 3.2; 19, Chapter 8].

Returning to the Borchers theorem, we notice that the case of a σ-weakly continuous representation of a group G by automorphisms of a von Neumann algebra N corresponds to $X = N$, $F = N_*$ in Definition 1.3.8. A very superficial use of the Arveson theory is sufficient to prove the Borchers theorem in the case of one-parameter group. Spectrum condition means in this case that the spectrum of the unitary representation of this group is non-negative.

The Borchers–Arveson theorem. Let $t \to \alpha_t$, $t \in \mathbf{R}$, be a σ-weakly continuous one-parameter group of *-automorphisms of a von Neumann algebra N in a Hilbert space $\mathcal{H}$. Then the following conditions are equivalent:

(1) there is in $\mathcal{H}$ a strongly continuous unitary group U_t, $t \in \mathbf{R}$, with the non-negative spectrum and such that for all $A \in N$ and $t \in \mathbf{R}$,

$$\alpha_t(A) = U_t A U_t^*;$$

(2) there is in $\mathcal{H}$ a strongly continuous unitary group V_t, $t \in \mathbf{R}$, with the non-negative spectrum and such that $V_t \in N$ for all $t \in \mathbf{R}$, and for all $A \in N$ and $t \in \mathbf{R}$,

$$\alpha_t(A) = V_t A V_t^* ;$$

(3) $$\bigcap_{t \in \mathbf{R}} M^{\alpha}([t, \infty])\,\mathcal{H} = \{0\} ,$$

where $M^{\alpha}(\Delta)$ are spectral subspaces of the representation α.

Proof given in [5, Theorem 3.2.46] needs only the most general properties of the spectrum and spectral subspaces and exploits the relations between the spectral subspaces $M^{\alpha}(\Delta) \subset N$ of the automorphic representation α and the usual spectral subspaces $E_U(\Delta)\,\mathcal{H}, E_V(\Delta)\,\mathcal{H}$ of the unitary representations U and V.

The largest class of groups for which the Borchers theorem is established is that of connected abelian groups. Subgroups of such groups may be groups $\hat{\mathbf{R}}^n$ as well as compact connected abelian groups. In the case of a compact group G spectrum condition demands that the spectrum $\sigma(U)$ of the representation U of G was contained in a certain cone in the discrete space $\hat{G}$; in the case of $\hat{\mathbf{R}}^n$ spectrum condition is imposed in a very weak form: $\sigma(U) \subset \Lambda$, where Λ is an arbitrary convex set in $\mathbf{R}^n$ which does not include any whole straight line [19, Chapter 8.4]. Clearly, the said class of groups and spectrum conditions includes the translation group and spectrum condition of local quantum theory, whence we obtain the following theorem.

THEOREM 1.3.43 (the Borchers theorem). *Let a Haag–Araki theory with a global algebra R satisfy Axioms* V *and* VI. *Then there exists a strongly continuous unitary representation $\tilde{T}(a)$ of the translation group which also satisfies these axioms and is such that $\tilde{T}(\hat{M}) \subset R$, and for all $A \in R, a \in \hat{M}$*

$$\tilde{T}(a)\,A\,\tilde{T}(-a) = T(a)\,A T(-a) ,$$

where $T(a)$ is the original representation of the translation group specified by Axiom V.

Proof. Statement of the theorem is the particular case of Corollary 8.4.12 [19, p. 332] ❐

COROLLARY 1. *For any vacuum vector $\Omega \in \mathcal{H}_0$,*

$$\mathcal{H}_{\Omega}^{R'} \subset \mathcal{H}_0 , \ \mathfrak{M}_{\Omega/\|\Omega\|}^{R} \subset \mathcal{H}_0 .$$

Proof. If the representation $T(a)$ is chosen in such a way that $T(\hat{M}) \subset R$, then the first inclusion follows from

$$\mathcal{H}_{\Omega}^{R'} \subset \mathcal{H}_{\Omega}^{T(\hat{M})'} \subset \mathcal{H}_0 ,$$

and the second from the formula $\mathfrak{M}_{\Omega/\|\Omega\|}^{R} \subset \mathcal{H}_{\Omega}^{R'}$ which is evident from (3.43). ❐

COROLLARY 2. *Energy and momentum operators $\hat{p}_{\mu}$ in the Haag–Araki theory are purely global observables.*

Proof. Properties of the representation $\widetilde{T}(a)$ imply that it possesses self-adjoint generators $\hat{p}_\mu$ forming the 4-vector of energy-momentum and affiliated to the global algebra R. As we argued in Section 1.2, such operators can also be considered as observables. Furthermore, due to Corollary 1 to Proposition 1.3.37, the operators $\hat{p}_\mu$ cannot be affiliated to any algebra $R(O), O \in B(M)$; i.e. they do not belong to local observables. ❐

It is convenient to mention here one more simple but important property of operators $\hat{p}_\mu$. It was pointed out by Kadison [270] and the elementary proof which we give is due to Borchers [80].

PROPOSITION 1.3.44. *Let a Haag–Araki theory satisfy Axioms* II, V *and* VI. *Then the energy-momentum operators* $\hat{p}_\mu$ *cannot be bounded operators, and the group* $T(a)$ *cannot be uniformly continuous.*

Proof. If $\hat{p}_\mu$ is a bounded operator, then any vector $\Phi \in \mathcal{H}$ is analytic for it, and due to the Reeh–Schlieder theorem, one has $\mathcal{H}_\Phi^{R(O)} = \mathcal{H}_\Phi^{R}$ for any region $O \in B(M)$. If it is true for all $\Phi \in \mathcal{H}$, then the sets of all cyclic projections of the algebras $R(O)$ and R coincide and hence the algebras coincide too, which contradicts Lemma 1.3.10. Due to relativistic covariance, if $\hat{p}_0$ is unbounded, then so are the other components of the 4-vector $\hat{p}_\mu$. By [5, Proposition 3.1.1], the last assertion of the Proposition is equivalent to the first one. ❐

For theories with a cyclic vacuum subspace Theorem 1.3.43 is enhanced: any representation $T(a)$ satisfying Axioms V and VI should now belong to the algebra R.

PROPOSITION 1.3.45 [41]. *Any Haag–Araki theory satisfying Axioms* V *and* VI *and the condition* $\mathcal{H}_{\mathcal{H}_0}^{R} = \mathcal{H}$ *is such that* $T(\hat{M}) \subset R$.

Proof. For arbitrary operators $A \in R$ and $B \in R'$, taking account of $T(a)P_0 = P_0$ and

$$[A, B(a)] = T(a)[A(-a), B]\,T(-a) = 0,$$

one has for any $\Phi \in \mathcal{H}$

$$(\Phi, P_0 AT(a) B P_0 \Phi) = (\Phi, P_0 BT(-a) A P_0 \Phi). \tag{3.81}$$

In view of spectrum condition, the left- (right-) hand side of this equality is the Fourier transform of a complex measure with the support in $\overline{V^+}$ (respectively, $\overline{V^-}$). Since the equality holds for all $a \in \hat{M}$, each of the two measures can have the support only in the common part $\overline{V^+} \cap \overline{V^-} = \{0\}$. The Fourier transform of such a measure is a constant and due to this, taking $A = A_1^* A_2, A_1, A_2 \in R$, one has from (3.81):

$$(A_1 P_0 \Phi, [B(a) - B] A_2 P_0 \Phi) = 0.$$

Since $\mathcal{H}_{\mathcal{H}_0}^{R} = \mathcal{H}$, it implies $B(a) = B$, whence $T(a) \in R$. ❐

Of course, the Borchers theorem entails that all elements of the centre Z commute with the 'internal' representation $\tilde{T}(a)$, i.e. $Z \subset \tilde{T}(\hat{M})'$. One can prove, however, that the centre commutes also with the initial representation $T(a)$ which may not lie in R; in other words, one has for any representation $T(a)$ satisfying Axioms V and VI:

$$T(\hat{M}) \subset Z'.$$

This result is independent of the Borchers theorem and is the subject of the well known Araki theorem [41, Proposition 1]; see also [264]. We shall omit its proof because the Borchers theorem in its practical applications is always understood as the possibility of assuming from the outset that in the Haag–Araki theory $T(\hat{M}) = \tilde{T}(\hat{M}) \subset R$, and with this assumption the Araki theorem turns to be trivial.

The Borchers theorem (understood in the above sense) makes it possible to enhance or simplify many of the above-mentioned results. Some new conclusions can also be drawn. First of all, we discover that the existence of a nontrivial vacuum subspace puts strong restrictions on the global algebra, making its structure very close to that of discrete von Neumann algebras.

PROPOSITION 1.3.46. *Let a Haag–Araki theory satisfy Axioms* III, V *and* VI *and be such that* $\mathcal{H}_0 \neq \{0\}$. *Then the von Neumann algebras* R_{P_0} *and* $R'_{P^R_\Omega}$ $(\Omega \in \mathcal{H}_0)$ *are abelian. If the vacuum subspace* $\mathcal{H}_0$ *is cyclic for* Z', *then* R *and* R' *are of type* I, *and the algebra* R_{P_0} *is isomorphic to a subalgebra of* R'. *If there exists a cyclic vacuum vector, then* R' *is an abelian von Neumann algebra.*

Proof. If $T(\hat{M}) \subset R$, the first assertion follows directly from the conditions 2 and 3 of Proposition 1.3.38. The second assertion follows from the first: if $P^{Z'}_{\mathcal{H}_0} = I$, then the induction $R' \to R'_{P_0}$ is an isomorphism so that R' is isomorphic to the algebra R_{P_0} with the abelian commutant; the concluding part of this assertion is evident from $R_{P_0} \subset R'_{P_0}$. The last assertion trivially follows from the first one. ❒

COROLLARY [41]. *Let a Haag–Araki theory satisfy Axioms* III, V, VI *and its global algebra* R *have the central decomposition*

$$R = \int^{\oplus} R_\xi \, d\nu(\xi), \quad \mathcal{H} = \int^{\oplus} \mathcal{H}_\xi \, d\nu(\xi).$$

Then the representation $T(a)$ *and the subspace* $\mathcal{H}_0$ *of this theory are also decomposable with respect to the measure* ν:

$$T(a) = \int^{\oplus} (a)_\xi \, d\nu(\xi), \quad \mathcal{H}_0 = \int^{\oplus} \mathcal{H}_{0,\xi} \, d\nu(\xi), \tag{3.82}$$

and in each space $\mathcal{H}_\xi$ *the Haag–Araki theory with the global algebra* R_ξ *is defined, satisfying Axioms* III, V *and* VI. *For each* ξ *such that* $\mathcal{H}_{0,\xi} \neq \{0\}, R_\xi$ *is a factor of type* I.

Proof. By the definition of central decomposition (Section 1.1), the set of all decomposable operators is Z', so that (3.82) follows from $T(\hat{M}) \subset Z'$, $P_0 \in Z'$. The spectral function $E(p)$ of the representation $T(\hat{M})$ is also decomposable, and it is evident that $\operatorname{supp} E(p)_\xi \subset \operatorname{supp} E(p)$. The last assertion is obvious from Proposition 1.3.46. ❐

Furthermore, the property $T(\hat{M}) \subset R$ brings some changes into the picture of ergodic properties or, in other words, the picture of relations between the uniqueness of vacuum, its cyclicity and the irreducibility of the global algebra. For an arbitrary representation corresponding to an invariant state this picture is described by Proposition 1.3.39. In the case in question one should put in its conditions

$$R_\omega^{(T)} = \left(\pi_\omega(\mathfrak{A}) \cup T_\omega(\hat{M})\right)'' = R_\omega .$$

As a result, we obtain that for any translationally invariant state ω on $\mathfrak{A}$ the following conditions are equivalent:

(1) ω is an ergodic state;

(2) ω is the only normal invariant state on R_ω;

(3) ω is the only vector invariant state on R_ω;

(4) $\dim \mathcal{H}_{0,\omega} = 1$;

(5) R_ω is a factor;

(6) $R_\omega = \mathcal{B}(\mathcal{H}_\omega)$.

Hence, it is evident, first of all, that in the case $T_\omega(\hat{M}) \subset R_\omega$ the difference is lost between the ergodicity of an invariant state and its purity (the equivalence $1 \leftrightarrow 6$) and, equally, between the factoriality and purity (equivalence $5 \leftrightarrow 6$). As for the uniqueness of vacuum (vacuum vector), in a general Haag–Araki theory with $\mathcal{H}_0 \neq \{0\}$ one easily obtains the following chain of relations:

$$\begin{pmatrix} R = \mathcal{B}(\mathcal{H}) \\ \mathcal{H}_0 \neq \{0\} \end{pmatrix} \Leftrightarrow \begin{pmatrix} \dim \mathcal{H}_0 = 1 \\ \mathcal{H}_\Omega^R = \mathcal{H} \end{pmatrix} \Rightarrow \dim \mathcal{H}_0 = 1 \Rightarrow R p_\Omega^R$$
$$= \mathcal{B}\left(\mathcal{H}_\Omega^R\right) \Leftrightarrow \dim\left(\mathcal{H}_0 \cap \mathcal{H}_\Omega^R\right) = 1 , \tag{3.83}$$

which includes, in particular, the well known property: uniqueness and cyclicity of vacuum taken together are equivalent to irreducibility of the global algebra.

Based on this, it is easy to obtain the complete description of vacuum structure in a certain class of local quantum theories which is sufficiently large for the majority of physical problems. We mean the so-called '*P*-class theories' which have global algebras of the form $R = \oplus_{\gamma \in \sigma} R_\gamma$, where R_γ are discrete factors, and correspond to quantum systems with arbitrary discrete superselection rules (see Section 2.1). In such theories the subspaces $\mathcal{H}_\gamma$ of the decomposition $\mathcal{H} = \oplus_{\gamma \in \sigma} \mathcal{H}_\gamma$ are interpreted as coherent superselections sectors, whereas the subspaces $\mathcal{H}_\gamma$ such that $\mathcal{H}_{0,\gamma} \equiv \mathcal{H}_0 \cap \mathcal{H}_\gamma \neq \{0\}$ are known as *vacuum superselection sectors*. The structure of vacuum sectors is thoroughly described by

PROPOSITION 1.3.47 [409]. *Let a Haag–Araki theory satisfy Axioms* III, V *and* VI *and have the P-class global algebra*

$$R = \bigoplus_{\gamma \in \sigma} R_\gamma, \quad \mathcal{H} = \bigoplus_{\gamma \in \sigma} \mathcal{H}_\gamma$$

(R_γ are factors of type I_∞*) and the vacuum subspace* $\mathcal{H}_0 \neq \{0\}$. *Let us assume* that* $\mathcal{H}_{0,\gamma} \neq \{0\}$ *for all* $\gamma \in \sigma$. *Then*

(1) *in every factorial subspace* $\mathcal{H}_\gamma$ *of* R *there is one and only one vacuum state, and this state is pure:*

$$\mathcal{H}_{0,\gamma} \cap S_1 = \mathfrak{M}^{R}_{\Omega_\gamma}, \quad \omega_{\Omega_\gamma} \in PV(R),$$

where Ω_γ *is any unit vector from* $\mathcal{H}_{0,\gamma}$ *and* $PV(R)$ *is the set of all pure vector states on* R *;*

(2) *in every irreducible subspace of* R *there is one and only one unit vacuum vector;*

(3) $P_{0,\gamma} = P^{R'}_{\Omega_\gamma} = P^{T(\hat{M})'}_{\Omega_\gamma}$,

where $P_{0,\gamma}$ *is a projection on* $\mathcal{H}_{0,\gamma}$ *;*

(4) $\mathcal{H}_0 = \bigoplus_{\gamma \in \sigma} \mathcal{H}_{0,\gamma}$.

Proof. Let us derive the first assertion. For an arbitrary vector $\Omega_\gamma \in \mathcal{H}_{0,\gamma} \cap S_1$, the vector state ω_{Ω_γ} is invariant and such that $R_{\omega_{\Omega_\gamma}} = R P^{R}_{\Omega_\gamma}$ is a factor (as an induction of the factor R_γ). Therefore the equivalence of conditions (5) and (6) above implies $R_{\omega_{\Omega_\gamma}} = \mathcal{B}(\mathcal{H}^{R}_{\Omega_\gamma})$, and ω_{Ω_γ} is a pure state. Next, if Ω_1 is any other unit vector from $\mathcal{H}_{0,\gamma}$, then the state ω_{Ω_γ} is also pure, and the subspaces $\mathcal{H}^{R}_{\Omega_\gamma}$ and $\mathcal{H}^{R'}_{\Omega_\gamma}$ have a non-trivial intersection (this property of a Hilbert space of a discrete factor readily follows from the unitary equivalence of all its irreducible subrepresentations, see Section 2.1). Hence, taking into account (3.83) and Corollary 2 to Theorem 1.3.43 one has

$$\{0\} \neq \mathcal{H}^{R}_{\Omega_\gamma} \cap \mathcal{H}^{R'}_{\Omega_1} \subset \mathcal{H}^{R}_{\Omega_\gamma} \cap \mathcal{H}_0 = \{c\,\Omega_\gamma\}_{c \in \mathbf{C}},$$

and therefore $\Omega_\gamma \in \mathcal{H}^{R'}_{\Omega_1} \cap S_1$. Due to the formula for the $\mathcal{H}$-image of a pure state,

$$\mathfrak{M}^{R}_{\Omega_1} = \mathcal{H}^{R'}_{\Omega_1} \cap S_1$$

(which follows from (3.43) and will be derived in Section 2.1) it means that $\omega_{\Omega_\gamma} = \omega_{\Omega_1}$.

* Without loss of generality, since otherwise all the assertions can be applied to the algebra $\tilde{R} = \oplus_{\gamma \in \sigma' \subset \sigma} R_\gamma$ such that $\mathcal{H}_{0,\gamma} \neq \{0\}$ for all $\gamma \in \sigma'$.

Next, any irreducible subspace $\mathcal{H}_1$ of R lies in one of the factorial subspaces $\mathcal{H}_\gamma$ and has, as noted above, a nontrivial intersection with the subspace $\mathcal{H}^{R'}_{\Omega_\gamma}$, $\Omega_\gamma \in \mathcal{H}_{0,\gamma}$. Since $\mathcal{H}^{R'}_{\Omega_\gamma} \subset \mathcal{H}_0$, then $\mathcal{H}_1 \cap \mathcal{H}_0 \neq \{0\}$, and, due to (3.83), this intersection is one-dimensional.

Assertion (3) is evident from the inclusions

$$\mathcal{H}^{R'}_{\Omega_\gamma} \subset \mathcal{H}^{T(\hat{M})'}_{\Omega_\gamma} \subset \mathcal{H}_{0,\gamma} \subset \overline{L\{\mathfrak{M}^{R}_{\Omega_\gamma}\}} \subset \mathcal{H}^{R'}_{\Omega_\gamma},$$

where the first follows from Theorem 1.3.43, the third from assertion (1), and the fourth from (3.43).

Finally, since $[P_0, P_\gamma] = 0$, one has

$$P_0 = \Big(\sum_{\gamma\in\sigma} P_\gamma\Big) P_0 = \sum_{\gamma\in\sigma} P_{0,\gamma}$$

and the Proposition is proved. ❒

Symmetries related to the Lorentz group still remain insufficiently studied. In this respect local quantum theory differs unfavourably from Wightman theory where Lorentz covariance (combined with locality and spectrum condition) provided the basis for extensive studies devoted mainly to analyticity properties of Wightman functions W_n. These studies produced an impressive complex of results: it was found that the analyticity domain of W_n extended with the help of Lorentz covariance, includes real points; a sound generalization of locality (weak local commutativity) was discovered and the rigorous proof of the CPT theorem was obtained. Only a small fraction of such results is reproduced within the framework of local quantum theory, since its dissimilarity to Wightman theory strongly prevents this. First of all, local quantum theory lacks such an object as Wightman function or Wightman functional, and we have to start with looking for some analogue of it. An obvious (though not unique, perhaps) option is to endow local observables $A \in R(O)$ with the argument $x \in M : A \to A(x) = T(x)AT(-x)$, treating the resultant function from M into $\mathcal{B}(\mathcal{H})$ as an analogue of Wightman field, and vacuum expectations

$$\Phi_n(x_1, \dots, x_n) \equiv (\Omega, A(x_1) \dots A(x_n)\,\Omega)$$

as analogues of Wightman functions. Unfortunately, the analogy proves to be very incomplete. Locality for the functions $\Phi_n(x_1, \dots, x_n)$ is fulfilled in a weaker form only:

$$\Phi_n(x_1, \dots, x_i, x_{i+1}, \dots, x_n) = \Phi_n(x_1, \dots, x_{i+1}, x_i, \dots, x_n), \text{ if } (O + x_i) \times (O + x_{i+1}).$$

The Lorentz covariance condition is weaker too. In Wightman theory fields $A(f) = \Sigma_{j=1}^{r} A_j(f_j)$ possess definite spins, which implies for them a simple linear transformation law under Lorentz transformations [4, Chapter 3, Section 1.1]: in the state space $\mathcal{H}$ there is a unitary representation $U(a, K)$ of the universal covering group $\bar{\mathcal{P}}^{\uparrow}_{+}$ of the Poincaré group such that

$$U(a,K)\,A(f)\,U(a,K)^{-1} = A\big(f_{(a,K)}\big) = \sum_{j=1}^{r} V_{ij}\left(K^{-1}\right) A_j\left(f_j\left(\Lambda^{-1}(x-a)\right)\right), \tag{3.84}$$

where $V(K)$ is a r-dimensional representation of the universal covering group $\bar{L}_+^{\uparrow} = SL(2, \mathbf{C})$, and the connection of Lorentz transformations $\Lambda \in L_+^{\uparrow}$ with elements of $\bar{L}_+^{\uparrow}$ is given by the formula

$$\Lambda_\mu^{\ \nu} = \Lambda_\mu^{\ \nu}(K) = \frac{1}{2} \operatorname{Tr} (\sigma_\mu K \sigma_\nu K^*) \tag{3.85}$$

(σ_0 is the unit 2×2-matrix, σ_i are Pauli matrices). Of course, in local quantum theory (3.84) does not hold, and virtually no information is available on the representation $U(a, \Lambda)$ defined by Axiom V (except for its decomposability into irreducible representations that was used in proving Proposition 1.3.36). Moreover, the assumption of the existence of a vacuum vector $\Omega \in \mathcal{H}$ here is no longer a part of the axioms and restricts the class of theories under consideration. As a result, the functions $\Phi_n(x_1, \dots, x_n)$ by no means have all the necessary properties of Wightman functions; in particular, their analytic and asymptotic properties are much weaker and much more difficult to prove (see e.g., [47, 184, 185]). Moreover, physically and structurally they are by far not as important as Wightman functions, and the same is true of operator-valued functions $A(x)$ (usually called Haag, or Haag–Araki, fields) whose physical significance certainly cannot equal that of true quantum fields $A(f)$. Considering all this, the problem of reproducing in local quantum theory all the results of the Wightman approach based on Lorentz covariance seems to be neither practical nor rewarding enough.

On the other hand, there are a number of questions related to Lorentz covariance which local quantum theory cannot avoid. These obviously include:

(1) conditions ensuring that generators of the representation $U(a, \Lambda)$ of $\mathcal{P}_+^{\uparrow}$, like generators of translations, are physical observables, i.e. are affiliated to a global algebra R;

(2) properties of Lorentz-invariant vectors and Lorentz-invariant states; and, since discrete symmetries are closely related to the Lorentz group and its complexification, we should list here also:

(3) description of discrete symmetries.

It should be stressed here that the formalism of observables and states, unlike field theory, has no obvious way of defining operations P, T and C, so that not only the properties of these objects but their existence as well is not known from the outset. The most important question here concerns the possibility of introducing a CPT-operator and proving the CPT-theorem.

All these questions pertain directly to physics. It is self-evident for question (1). Question (2) is an indispensable part of studying the vacuum structure of relativistic quantum systems; our study of this structure was so far markedly incomplete, since no account was taken of the fact that vacuum states must have the Lorentz invariance in addition to the translation invariance. As for question (3), it is well known that discrete symmetries and the CPT-theorem belong to experimentally verifiable properties of systems of elementary particles. Nevertheless, the results in all these questions are still incomplete. Probably, the best situation is with question (1), for which a possibility of a positive solution, i.e. of proving that $U(a, \Lambda) \in R$, is suggested already by the following elementary facts.

(1) All spatial automorphisms of a factor of type I are inner. Therefore, if in the Haag–Araki theory the algebra R is a factor and $\mathcal{H}_0 \neq \{0\}$, then, by Proposition 1.3.46, Poincaré automorphisms $\alpha_{(a,\Lambda)}$ are inner, and for each $(a, \Lambda) \in \mathcal{P}_+^\uparrow$ there is an operator $\tilde{V}(a, \Lambda) \in R^U$ such that for all $A \in R$,

$$\alpha_{(a,\Lambda)}(A) = U(a, \Lambda)\, A U(a, \Lambda)^{-1} = \tilde{V}(a, \Lambda)\, A\, \tilde{V}(a, \Lambda)^{-1}\,, \tag{3.86}$$

where $U(a, \Lambda)$ is the initial strongly continuous representation defined by Axiom V.

(2) If R is a direct sum of factors, and α_g is a strongly continuous connected group of automorphisms, then the centre Z is point-wise invariant under the automorphisms α_g.

(Indeed, in this case Z is generated by its minimal projections, and α_g, as we can easily check, must map these projections onto each other. R being a discrete sum of algebras, the set of all minimal projections in Z is discrete with respect to the strong operator topology. Hence, due to the strong continuity and connectedness of the group α_g, the minimal projections are left invariant by α_g, and Z is point-wise invariant.) The well known criterion [7, p. 241] implies then that if R is an algebra of type I, the automorphisms α_g are inner. As a result, if in the Haag–Araki theory R is an algebra of P-class (i.e. a direct sum of factors of type I), the Poincaré automorphisms are again inner, and there exists a family of operators $\tilde{V}(a, \Lambda)$ satisfying (3.86). At the same time, for the original $U(a, \Lambda)$ defined by Axiom V, we obtain that $U(a, \Lambda) \in Z'$. For global algebras of the form $R = \oplus_{\gamma \in \sigma} \mathcal{B}(\mathcal{H}_\gamma)$ which correspond to systems with the normal Bose–Fermi statistics (see Section 2.3) this implies $U(a, \Lambda) \in R$.

In both cases we have only concluded that each automorphism $\alpha_{(a,\Lambda)}$ is inner (or, which is the same, in the Haag–Araki theory there is an automorphic representation $\alpha : \mathcal{P}_+^\uparrow \to \operatorname{Int} R$ where $\operatorname{Int} R$ is the group of all inner automorphisms of R). No information has been obtained about the set of operators $\tilde{V}(a, \Lambda)$. But this set may, in general, not implement a representation of $\mathcal{P}_+^\uparrow$, since the fulfillment of the group law for some group of automorphisms $\tau_g = V_g \cdot V_g^{-1}$ does not imply the fulfillment of this law for the unitary operators V_g. Hence the facts found still do not imply the desired properties of the Poincaré generators. To obtain these properties we have to prove that the automorphisms $\alpha_{(a,\Lambda)}$ can be implemented by operators which, in addition to lying in R, form a (weakly or strongly) continuous representation of $\mathcal{P}_+^\uparrow$. It is sufficient for this representation to be not unitary but just projective, i.e. to be a unitary representation of the universal covering group $\tilde{\mathcal{P}}_+^\uparrow$. The existence of such a projective representation can be verified again by elementary arguments. Indeed, if $\alpha_{(a,\Lambda)}$ form the group of automorphisms of a W^*-algebra, then the operators $\tilde{V}(a, \Lambda)$ implementing these automorphisms may not satisfy the group law (as said above), but they always enjoy the following easily checkable property:

$$\tilde{V}(L_1 L_2)\, \tilde{V}(L_2)^{-1}\, \tilde{V}(L_1)^{-1} \in R\,,$$

where we used the notation $L = (a_i, \Lambda_i), i = 1, 2$; $L_1 L_2 = (a_1 + \Lambda_1 a_2, \Lambda_1 \Lambda_2)$. With

$\tilde{V}(L) \in R$ it gives

$$\tilde{V}(L_1 L_2) = Z(L_1, L_2)\, \tilde{V}(L_1)\, \tilde{V}(L_2), Z(L_1, L_2) \in \mathcal{Z}.$$

Hence, if R is a factor, then $\tilde{V}(a, \Lambda)$ form a projective representation of $\mathcal{P}_+^\uparrow$. Furthermore, if $R = \oplus_{\gamma \in \sigma} R_\gamma$ is a direct sum of factors, it is obvious that we can also obtain a projective representation of $\mathcal{P}_+^\uparrow$, using the projective representations in subspaces where the algebras R_γ are acting. Thus only the continuity of the resultant representation remains to be proved. This is a far more complex task.

Simple arguments of this kind were applied to the study of relativistic covariance in local quantum theory already in the pioneer work by Araki [41] who obtained the following proposition.

PROPOSITION 1.3.48. *Let in a Haag–Araki theory Axioms* II, V *and* VI *hold, the subspace* $\mathcal{H}_0$ *be cyclic for* $\mathcal{Z}'$ *and each vector in* $\mathcal{H}_0$ *be Lorentz invariant. Then there exists a unitary representation* $V(a, K)$ *of the universal covering group* $\bar{\mathcal{P}}_+^\uparrow$ *such that* $V(a, K) \in R^U$ *and for all* $A \in R$,

$$V(a, K)\, AV(a, K)^{-1} = U(a, \Lambda(K))\, AU(a, \Lambda(K))^{-1}, \qquad (3.87)$$

where the elements K *and* $\Lambda(K)$ *are connected according to* (3.85).

Proof. By Theorem 1.3.43, $\mathcal{Z}\mathcal{H}_0 = \mathcal{H}_0$, whence due to $U(a, \Lambda)\, \mathcal{H}_0 = \mathcal{H}_0$ one has for all $Z \in \mathcal{Z}$:

$$U(a, \Lambda)\, ZU(a, \Lambda)^{-1} \big|_{\mathcal{H}_0} = Z \big|_{\mathcal{H}_0}. \qquad (3.88)$$

Since the transformation $U(a, \Lambda) \cdot U(a, \Lambda)^{-1}$ is an automorphism of R, it leaves $\mathcal{Z}$ invariant, that is $U(a, \Lambda)\, ZU(a, \Lambda)^{-1} \in \mathcal{Z}$. Hence, multiplying (3.88) from the left by an arbitrary operator from $\mathcal{Z}'$ and taking into account that $\mathcal{Z}'\mathcal{H}_0$ is dense in $\mathcal{H}$ one has

$$U(a, \Lambda) \in \mathcal{Z}' \quad \forall (a, \Lambda) \in \mathcal{P}_+^\uparrow.$$

This means that the operators $U(a, \Lambda)$ are decomposable in the central decomposition of R (see Section 1.1):

$$R = \int^\oplus R_\xi \, d\nu(\xi), \qquad U(a, \Lambda) = \int^\oplus U(a, \Lambda)_\xi \, d\nu(\xi),$$

and since every component R_ξ is a factor of type I (Proposition 1.3.46 which uses locality), all the automorphisms of R_ξ are inner. Hence, for each $(a, \Lambda) \in \mathcal{P}_+^\uparrow$ there is an operator $\tilde{V}(a, \Lambda)_\xi \in R_\xi^U$ such that

$$U(a, \Lambda)_\xi\, A_\xi\, U(a, \Lambda)_\xi^{-1} = \tilde{V}(a, \Lambda)_\xi\, A_\xi\, \tilde{V}(a, \Lambda)_\xi^{-1}$$

for all $A_\xi \in R_\xi$. As noted above, since R_ξ is a factor, we can by choosing appropriate scalar multipliers, transform the family of operators $\{\tilde{V}(a, \Lambda)_\xi\}$ into a unitary representation $V(a, K)_\xi$ of the universal covering group $\bar{\mathcal{P}}_+^\uparrow$. The direct integral of the resultant field of representations

$$V(a,K) = \int^{\oplus} V(a,K)_{\xi}\, d\nu(\xi)$$

defines the required representation of the universal covering group satisfying (3.87). ❐

This result does not tell anything about the continuity of the representation $V(a,K)$. Its proof needs more careful analysis which was carried out (for a more general case) in a much later work by Zinoviev and Sushko.

THEOREM 1.3.49 [447]. *Let us take a Haag–Araki theory with a strongly continuous unitary representation $U(a,\Lambda)$ of the Poincaré group and a global algebra of P-class. Then there exists a strongly continuous unitary representation $V(a,K)$ of the universal covering group $\bar{\mathcal{P}}_+^{\uparrow}$ such that for all $(a,K) \in \bar{\mathcal{P}}_+^{\uparrow}$ the following properties hold: $V(a,K) \in R^U$ and*

$$V(a,K)\, A V(a,K)^{-1} = U(a,\Lambda(K))\, A U(a,\Lambda(K))^{-1} \quad \forall A \in R\,.$$

Proof. As shown above, we have only to establish continuity of the projective representation implemented by operators $\tilde{V}(a,\Lambda)$ in (3.86) basing on the continuity of the automorphic representation $\alpha_{(a,\Lambda)} : \mathcal{P}_+^{\uparrow} \to \operatorname{Int} R$. To solve this problem we shall reduce it to the analysis of inner automorphisms of the algebra $\mathcal{B}(\mathcal{H})$. This is possible due to the assumption that $R = \oplus_{\gamma\in\sigma} R_\gamma$, where all R_γ are discrete factors.

Since $\tilde{V}(a,\Lambda) \in R$, the representation α is direct sum $\alpha = \oplus_{\gamma\in\sigma}\, \alpha^{\gamma}$ of representations $\alpha^{\gamma} : \mathcal{P}_+^{\uparrow} \to \operatorname{Int} R_\gamma$. Indeed, for all $A \in R$ one has $A = \oplus_{\gamma\in\sigma} A_\gamma$, $A_\gamma \in R_\gamma$, whence

$$\begin{aligned}\alpha_{(a,\Lambda)}(A) &= \tilde{V}(a,\Lambda)\, A\, \tilde{V}(a,\Lambda)^{-1} = \tilde{V}(a,\Lambda) \bigoplus_{\gamma\in\sigma} A_\gamma \tilde{V}(a,\Lambda)^{-1} \\ &= \bigoplus_{\gamma\in\sigma} \tilde{V}(a,\Lambda)\, A_\gamma\, \tilde{V}(a,\Lambda)^{-1} = \bigoplus_{\gamma\in\sigma} \alpha^{\gamma}_{(a,\Lambda)}(A_\gamma)\,. \end{aligned} \tag{3.89}$$

For any discrete factor R_γ there is a Hilbert space $\hat{\mathcal{H}}_\gamma$ such that an ultraweakly continuous isomorphism $\Phi_\gamma : R_\gamma \to \mathcal{B}(\hat{\mathcal{H}}_\gamma)$ holds. This isomorphism induces the representation $\hat{\alpha}^{\gamma} : \mathcal{P}_+^{\uparrow} \to \operatorname{Int} \mathcal{B}(\hat{\mathcal{H}}_\gamma)$ defined as $\hat{\alpha}^{\gamma} = \Phi_\gamma \circ \alpha^{\gamma} \circ \Phi_\gamma^{-1}$. As explained above, we want to define a projective representation of the Poincaré group in terms of the representation $\hat{\alpha}^{\gamma}$. To this end let us consider an algebraic isomorphism

$$\mu_\gamma : \operatorname{Int} \mathcal{B}(\hat{\mathcal{H}}_\gamma) \to B_\gamma^U / C_1(\hat{\mathcal{H}}_\gamma)$$

(where B_γ^U is the group of all unitary operators in $\hat{\mathcal{H}}_\gamma$ and $C_1(\hat{\mathcal{H}}_\gamma)$ is the set of all scalar operators of unit norm in $\hat{\mathcal{H}}_\gamma$) defined explicitly as follows:

$$\mu_\gamma(V \cdot V^{-1}) = \{V e^{i\varphi}\}_{\varphi\in\mathbf{R}}$$

for each $V \in B_\gamma^U$. Since the isomorphism Φ_γ induces faithful mappings:

$$\Phi_\gamma|_{R_\gamma^U} : R_\gamma^U \to B_\gamma^U, \qquad \tilde{\Phi}_\gamma : R_\gamma^U / C_1(\mathcal{H}_\gamma) \to B_\gamma^U / C_1(\hat{\mathcal{H}}_\gamma)\,,$$

we can define a projective representation in $\mathcal{H}_\gamma$ by the formula

$$\tilde{\alpha}^{\gamma} = \tilde{\Phi}_{\gamma}^{-1} \circ \mu_{\gamma} \circ \hat{\alpha}^{\gamma} : \mathcal{P}_{+}^{\uparrow} \to R_{\gamma}^{U} / C_1 (\mathcal{H}_{\gamma}) .$$

The definitions of the mappings in this formula imply that

$$\tilde{\alpha}^{\gamma}_{(a,\Lambda)} (A_{\gamma}) = \alpha^{\gamma}_{(a,\Lambda)} (A_{\gamma}) \quad \forall\, A_{\gamma} \in R_{\gamma} .$$

Hence, we get using (3.89), that if $V_{\gamma} (a, \Lambda) \in R_{\gamma}^{U}$ is an arbitrary member of the equivalence class

$$[V_{\gamma} (a, \Lambda)] = \{e^{i\varphi} V_{\gamma} (a, \Lambda)\}_{\varphi \in \mathbf{R}} \in R_{\gamma}^{U} / C_1 (\mathcal{H}_{\gamma}) ,$$

then we have for all $A_{\gamma} \in R_{\gamma}$, and all $(a, \Lambda) \in \mathcal{P}_{+}^{\uparrow}$

$$\tilde{\alpha}^{\gamma}_{(a,\Lambda)} (A_{\gamma}) = V_{\gamma} (a, \Lambda) A_{\gamma} V_{\gamma} (a, \Lambda)^{-1} = \tilde{V} (a, \Lambda) A_{\gamma} \tilde{V} (a, \Lambda)^{-1} .$$

Taking here the direct sum $\oplus_{\gamma \in \sigma}$, for all $A = \oplus_{\gamma \in \sigma} A_{\gamma} \in R$, due to (3.86), one has

$$V (a, \Lambda) A V (a, \Lambda)^{-1} = U (a, \Lambda) A U (a, \Lambda)^{-1} , \tag{3.90}$$

where $V (a, \Lambda) \equiv \oplus_{\gamma \in \sigma} V_{\gamma} (a, \Lambda) \in R^{U}$. By construction, the operators $V (a, \Lambda)$ implement a projective representation of $\mathcal{P}_{+}^{\uparrow}$ into the group $R^{U} / C_1 (\mathcal{H})$.

To prove the strong continuity of the projective representation $\tilde{\alpha} = \oplus_{\gamma \in \sigma} \tilde{\alpha}^{\gamma}$, it suffices to establish the continuity of the isomorphism μ_{γ} since $\tilde{\Phi}_{\gamma}^{-1}$ and $\hat{\alpha}^{\gamma}$ obviously enjoy the necessary continuity properties. In this proof the groups $B_{\gamma}^{U} / C_1 (\hat{\mathcal{H}}_{\gamma})$ and $\mathrm{Int}\, \mathcal{B} (\hat{\mathcal{H}}_{\gamma})$ must be endowed, respectively, with the topology induced by the strong operator topology on $\hat{\mathcal{H}}_{\gamma}$ and the topology in which the representation $\hat{\alpha}^{\gamma}$ is continuous. The last topology can be defined by the family of neighbourhoods

$$\mathcal{N} (\tau_2 ; \varphi, \psi ; \varepsilon) = \left\{ \tau_1 \in \mathrm{Int}\, \mathcal{B} (\hat{\mathcal{H}}_{\gamma}) \,\middle|\, \sup_{S_{\psi}} \left| \varphi \left(\tau_1 (A_1) - \tau_2 (A_2)\right) \right| \le \varepsilon \right\} ,$$

where φ, ψ are normal positive functionals on $\mathcal{B} (\hat{\mathcal{H}}_{\gamma})$ and

$$S_{\psi} = \left\{ A_1 , A_2 \in \mathcal{B} (\hat{\mathcal{H}}_{\gamma}) \,\middle|\, \left| \psi (A_1 - A_2) \right| \le \varepsilon \right\} .$$

Using the weak compactness of the unit ball in $\mathcal{B} (\hat{\mathcal{H}}_{\gamma})$, we can verify that this topology is stronger than the bounded-weak topology defined by the neighbourhoods

$$\mathcal{N}_{bw} (\tau_2 ; \chi ; \varepsilon) = \left\{ \tau_1 \in \mathrm{Int}\, \mathcal{B} (\hat{\mathcal{H}}_{\gamma}) \,\middle|\, \sup_{\|A\| \le 1} \left| \left(\chi, (\tau_1 (A) - \tau_2 (A)\, \chi)\right) \right| \le \varepsilon \right\} ,$$

where $\chi \in \hat{\mathcal{H}}_{\gamma}$. Therefore it is sufficient to prove that μ_{γ} is continuous when the group $\mathrm{Int}\, \mathcal{B} (\hat{\mathcal{H}}_{\gamma})$ is endowed with the bounded-weak topology. To this end we have to verify that

$$\tau_1 \in \mathcal{N}_{bw} (\tau_2 ; \chi ; \varepsilon) \Rightarrow \mu_{\gamma} (\tau_2) \in \mathcal{N}_s (\mu_{\gamma} (\tau_1) ; \chi ; \varepsilon') , \tag{3.91}$$

where $\varepsilon' \to 0$ at $\varepsilon \to 0$ and

$$\mathcal{N}_S(0\,;\chi\,;\varepsilon) = \left\{ [V] \in \overset{U}{B_\gamma}/C_1(\hat{\mathcal{H}}_\gamma) \;\middle|\; \inf_{V' \in [V]} \| V'\chi \|^2 \le \varepsilon \right\}$$

are the neighbourhoods defining the topology on $\overset{U}{B_\gamma}/C_1(\hat{\mathcal{H}}_\gamma)$ induced by the strong operator topology on $\hat{\mathcal{H}}_\gamma$. Taking into account that $\tau_i = V_i \cdot V_i^{-1}$, $V_i \in \overset{U}{B_\gamma}$, $i = 1, 2$, we note that

$$\inf_{\substack{V_i' \in [V_i] \\ i=1,2}} \| (V_1' - V_2'\chi \|^2 = \inf_{\varphi \in \mathbf{R}} \| (V_1 - e^{i\varphi} V_2)\chi \|^2 = 2\left(1 - |(V_1\chi\,,V_2\chi)|\right)$$

Hence, denoting by P_1 the projection on the ray of the vector $V_1^{-1}\chi \in \hat{\mathcal{H}}_\gamma$ we get

$$\sup_{\|A\| \le 1} |(\chi\,,(\tau_1(A) - \tau_2(A))\,\chi)| \ge |(\chi\,,(\tau_1(P_1) - \tau_2(P_1))\,\chi)|$$

$$= \left(1 - |(V_1\chi, V_2\chi)|^2\right) \ge \left(1 - |(V_1\chi, V_2\chi)|\right).$$

From here (3.91) is evident. Thus the operators $V(a, \Lambda) \in R^U$ in (3.90) form a strongly continuous projective representation of $\mathcal{P}_+^\uparrow$. By the classical results of Wigner and Bargmann, any such representation is generated by a strongly continuous unitary representation of the universal covering group $\bar{\mathcal{P}}_+^\uparrow$. In our case this representation will lie in R^U, and we may keep the notation $V(a, K)$ for it. ❐

COROLLARY 1. *In the Haag–Araki theory with a global algebra of P-class one can define in a neighbourhood* $\mathcal{N}$ *of identity of the Poincaré group a strongly continuous unitary representation* $\tilde{U} : \mathcal{N} \to R^U$ *such that for any* $(a, \Lambda) \in \mathcal{N}$ *and* $A \in R$

$$U(a, \Lambda)\,A\,U(a, \Lambda)^{-1} = \tilde{U}(a, \Lambda)\,A\,\tilde{U}(a, \Lambda)^{-1}.$$

Proof. It is evident, since the projective unitary representation $V(a, \Lambda)$ in (3.90) is locally isomorphic to a unitary representation. ❐

COROLLARY 2. *In local quantum theory the generators* M_k^i *of the Lorentz group are purely global observables.*

Proof. Operators M_k^i can be considered as generators of the representation introduced in Corollary 1. They are then affiliated to R. By Corollary 1 to Proposition 1.3.37, they cannot be affiliated to any of the algebras $R(O)$ with $O \in B(M)$. ❐

As said above, it is the existence of an automorphic representation of the Poincaré group that has the direct physical meaning as a condition of relativistic covariance. The proof of Theorem 1.3.49 shows that this condition is equivalent, under some continuity assumptions, to the existence of a projective representation of $\mathcal{P}_+^\uparrow$ or a unitary representation of its universal covering; and the projective representation without loss of generality can be chosen from the global algebra. But the existence of a unitary representation of $\mathcal{P}_+^\uparrow$, i.e. the standard formulation of relativistic covariance in the

Haag–Araki theory, does not, in general, follow from the existence of the automorphic representation and is an additional restriction justified by convenience more than by physical arguments.This suggests that, in principle, Axiom V in the Haag–Araki theory could be weakened down to the requirement of the existence of a unitary representation of $\tilde{\mathcal{P}}_+^\uparrow$ instead of $\mathcal{P}_+^\uparrow$. In this case the result of Theorem 1.3.49 for the Poincaré group is the exact parallel to the result of the Borchers theorem (Theorem 1.3.43) for the group of translations: both theorems state that a continuous unitary representation corresponding to the automorphic representation given by Axiom V, can be chosen from the global algebra. It is precisely in this way, i.e. for $\tilde{\mathcal{P}}_+^\uparrow$, that one formulates relativistic covariance in field theory (cf. (3.84)) and in the formalism of localized morphisms by Doplicher–Haag–Roberts (Section 2.3) where it is also proved that a strongly continuous unitary representation of $\tilde{\mathcal{P}}_+^\uparrow$ lies in R [150, Lemma 2.2].

Let us proceed to the problem of Lorentz-invariant states and vectors. Information on their properties has long remained scarce and, in some degree, conflicting, being limited essentially to the following two facts: as far back as 1962 Borchers proved [73, Theorem 3] that in the Wightman theory with a cyclic Poincaré invariant vector, any translationally invariant vector is Lorentz invariant; in 1964 Araki [41] proved by an explicit example that in local quantum theory certainly not any translationally invariant vector is Lorentz invariant. In fact, in the Araki example none of translationally invariant vectors is Lorentz invariant. Let us describe this example. Let $d\mu(\Lambda)$ be the invariant measure on the restricted Lorentz group, and let us have for every $\Lambda \in L_+^\uparrow$ a certain Haag–Araki theory satisfying Axioms III, V and VI, with Poincaré covariance replaced by translational covariance. Let us also assume all these theories $\{\mathcal{H}_\Lambda, R(O)_\Lambda, T(a)_\Lambda\}$ to be copies of the same theory $\{\mathcal{H}, R(O), T(a)\}$ obtained by the identical mapping $S(\Lambda): \mathcal{H} \to \mathcal{H}_\Lambda$. Let us construct the direct integral of these theories (see Section 1.1 and [7, Chapter II]):

$$\tilde{\mathcal{H}} = \int^{\oplus} \mathcal{H}_\Lambda \, d\mu(\Lambda), \quad \tilde{R}(O) = \int^{\oplus} R(\Lambda O)_\Lambda \, d\mu(\Lambda), \quad \tilde{T}(a) = \int^{\oplus} T(\Lambda a)_\Lambda \, d\mu(\Lambda).$$

The resultant Haag–Araki theory in the space $\tilde{\mathcal{H}}$ can be turned into a Poincaré-covariant theory by means of introducing in $\tilde{\mathcal{H}}$ a unitary representation $\tilde{U}(\Lambda)$ of the Lorentz group:

$$(\tilde{U}(\Lambda)\Phi)_{\Lambda'} = S(\Lambda')\, S(\Lambda'\Lambda)^{-1}\, \Phi_{\Lambda'\Lambda}$$

(where $\Phi = \int^{\oplus} \Phi_\Lambda d\mu(\Lambda)$) and a unitary representation of the Poincaré group: $\tilde{U}(a,\Lambda) = \tilde{T}(a)\,\tilde{U}(\Lambda)$. Then the objects $\{\tilde{R}(O), \tilde{U}(a,\Lambda), \tilde{\mathcal{H}}\}$ define a Haag–Araki theory satisfying Axioms III, V and VI. In this theory, translationally invariant vectors cannot be Lorentz invariant. Indeed, such vectors in $\tilde{\mathcal{H}}$ are represented by vector fields $\Phi = \int^{\oplus} \Phi_\Lambda \, d\mu(\Lambda)$, since the translation operators are decomposable in the direct integral. On the other hand, the operators $\tilde{U}(\Lambda)$ are, by construction, not decomposable and do not leave any vector field invariant.

Further elucidations followed only recently, although the main result turned out to be simple and easy.

PROPOSITION 1.3.50 [172]. *Let Axioms* III, V *and* VI *be valid in a Haag–Araki theory and the subspace* $\mathcal{H}_0$ *include a cyclic vector* Ω *for* R. *Then the following conditions are equivalent:*

(1) *all translationally invariant vectors are Lorentz invariant;*

(2) $U(\mathcal{P}_+^\uparrow) \subset R$;

(3) *operators of Poincaré transformations are decomposable in the central decomposition of* R .

Proof. 1 → 2. By Proposition 1.3.46, in our theory $R' = Z$. Taking account of $Z\mathcal{H}_0 = \mathcal{H}_0$, one has for any $T \in Z$ and $(a, \Lambda) \in \mathcal{P}_+^\uparrow$:

$$T\Omega = U(a, \Lambda)\, T\Omega = U(a, \Lambda)\, TU\ (a, \Lambda)^{-1}\,\Omega\,. \tag{3.92}$$

Since the unitary transformation $U(a, \Lambda) \cdot U\ (a, \Lambda)^{-1}$ leaves Z invariant, one has in (3.92) $U(a, \Lambda)\, TU\ (a, \Lambda)^{-1} \in Z$. But Ω is a separating vector for Z and, therefore, (3.92) yields

$$T = U(a, \Lambda)\, TU\ (a, \Lambda)^{-1} \quad \forall\, (a, \Lambda) \in \mathcal{P}_+^\uparrow\ ,$$

whence 2 follows.

2 → 1. Since the translations form the invariant subgroup in $\mathcal{P}_+^\uparrow$ and, moreover, $U(a, \Lambda)\, T(\hat{M})\, U(a, \Lambda)^{-1} = T(\hat{M})$, the subspace $\mathcal{H}_0$ is invariant under all $U(a, \Lambda)$, which is equivalent to $P_0 \in U(\mathcal{P}_+^\uparrow)'$. This, in its turn, implies that the operators $V_0(\Lambda) \equiv P_0\, U(0, \Lambda)\, P_0$, $\Lambda \in L_+^\uparrow$, form a strongly continuous unitary representation of $L_+^\uparrow$ in $\mathcal{H}_0$. On the other hand, by Proposition 1.3.46, the set of operators $P_0 RP_0$ is commutative so that 2 entails the commutativity of $V_0(\Lambda)$. For the Lorentz group in Minkowski space only the trivial representation enjoys this property which entails the desired: $U(a, \Lambda)\ \Phi = V_0(\Lambda)\,\Phi = \Phi$ for any $\Phi \in \mathcal{H}_0$ and $\Lambda \in L_+^\uparrow$.

2 → 3 is obvious since 3 means $U(a, \Lambda) \in Z'$. □

COROLLARY. *Let the global algebra* R *of a Haag–Araki theory be a direct sum of discrete factors* R_γ. *Then the spaces* $\mathcal{H}_\gamma$ *of* R_γ *reduce the operators of Poincaré transformations, and one has in each* $\mathcal{H}_\gamma$ *a Haag–Araki theory satisfying Axioms* I–VI. *If the original theory possesses a cyclic vacuum, it has the properties* 1–3 *of Proposition* 1.3.50.

Proof. As noted above,(p. 111),for the the global structure in question $U(a, \Lambda) \in Z' = \oplus_\gamma \mathcal{B}(\mathcal{H}_\gamma)$, which makes it easy to verify that Axioms I–VI hold for the local nets $\{R(O)|_{\mathcal{H}_\gamma}\}$. In the presence of a cyclic vacuum one has $R_\gamma = \mathcal{B}(\mathcal{H}_\gamma)$ and $U(a, \Lambda) \in R$. □

Thus the discreteness of the central decomposition of R is the sufficient condition for the fulfillment of properties 1–3 in theories with a cyclic vacuum. (Hence the breaking of these properties, which means the spontaneous breaking of Lorentz invariance, according to Definition 1.3.10, is only possible in theories with a continuous central decomposition, such as one in the Araki example. We notice, however, that in the absence of a cyclic vacuum the continuity of the central decomposition may not be necessary.) Apart from this global condition, Driessler and Summers derived also a local sufficient condition for properties 1–3 in terms of the algebras of wedge regions (Section 1.3.2, 3.2.2). Namely, if we denote $W_j = \{x \in M \mid x^j > |x^0|\}, j = 1, 2, 3$ (obviously, $W_1 = W_R$) and $V_j(t)$ is the corresponding unitary group of Lorentz boosts implementing automorphisms of $R(W_j)$ (in the notations of Section 1.3.2, $V_1(t) = V(t)$), then the following condition is sufficient for the fulfillment of properties 1–3 [172, Proposition 5.3]: the groups $V_j(t)$ coincide with the modular groups Δ_j^{it} of the pairs $(R(W_j), \Omega)$, $j = 1, 2, 3$. We shall omit the proof of this result, noting only that it uses the relation $Z(W_j) = Z$, which generalizes Proposition 1.3.7 to the case of a theory with vacuum degeneration. By its meaning, this condition, like the Borchers theorem [73, Theorem 3] mentioned above, proves the absence of the spontaneous breaking of Lorentz invariance in field theories: as explained in Section 1.3.2 and proved in Section 3.2.2 (Theorem 3.2.7) it is connected with the existence of a *CPT*-operator and always holds in the Wightman theory if it has a vacuum cyclic for the von Neumann field algebra $F(W_j)$.

Finally, we shall discuss very briefly discrete symmetries. When we begin to consider other symmetries, besides those in the restricted Poincaré group, the character of problems changes markedly. The axioms do not guarantee anymore that a given symmetry or a group of symmetries is unitarily implemented in the state space. Hence, the first problem is to find out how to describe this symmetry. A wide range of different ways of the description are possible. By an intuitive physical idea, the presence of a symmetry means that characteristics of a physical system in question (i.e. observables and states in the case of algebraic formalism) can be changed in a definite way without changing either kinematics or dynamics of the system (i.e. the expectations of all observables in all states will not change). Unitary operators preserve the scalar product in a Hilbert space and so offer a self-evident way of describing symmetries. But this way is not unique and not the most general (on the contrary, it proves to be the most restrictive). Clearly, the most general definition of symmetry is one which does not put any conditions at all on the form of the transformation of observables and states, apart from the basic physical condition of the invariance of expectations. In particular, the said form can be different for different regions in Minkowski space. Summing up, we can adopt as a starting definition of a symmetry in local quantum theory the following

DEFINITION 1.3.9*. A quantum system described by a net of local algebras will be

* We do not wish to enter the vast area of the general symmetry theory for quantum systems. Its principal ideas were developed mainly by Weyl and Wigner whereas the systematic formalism of the algebraic description of symmetries is largely based on results by Kadison [267, 269]. The problem of the most general definition of a symmetry transformation was investigated by Roberts and Roepstorff [366]. Hunziker [256] provided an elementary derivation of several basic properties of symmetry transformations exploiting only simple properties of mappings in a Hilbert space rather than automorphisms of C^*-algebras. Rigorous treatment of symmetries based on various algebraic structures was developed in several schemes of axiomatic quantum mechanics (see, e.g. [10, 14]). In quantum field theory and especially in the algebraic approach rigorous expositions of the symmetry theory are scarce, and the old but very well written review by Reeh [353] still can be useful. See also [12, 339, 438, 329, 330].

said possessing a local presymmetry $\{\langle \nu^O, \tau^O \rangle\}_{O\in B(M)}$, if for every region $O \in B(M)$ there exists a pair of mappings

$$\nu^O : S(O) \to S \subset \mathfrak{A}_1^{*+}, \quad \tau^O : R(O)^h \to \mathcal{A}^h \subset \mathcal{A}$$

(where $S(O)$ is a full family* of states over the algebra $R(O)$ and $R(O)^h$, $\mathcal{A}^h$ are the sets or Hermitian elements in $R(O)$ and $\mathcal{A}$, respectively) such that

$$(\nu^O[\omega^{(O)}])\,(\tau^O(A)) = \omega^{(O)}(A)$$

for any $\omega^{(O)} \in S(O)$ and $A \in R(O)^h$.

As shown in [447], this definition implies that ν^O and τ^O are linear one-to-one mappings, τ^O is isometric and extendable to a mapping $\tilde{\tau}^O : R(O) \to \mathcal{A}$ preserving the involution and the identity. If, moreover, we demand that a symmetry transformation does not break isotony and local commutativity of local algebras, then all ν^O prove to be restrictions (to appropriate regions) of the same affine isometric mapping $\nu : \mathfrak{A}_1^{*+} \to \mathfrak{A}_1^{*+}$, i.e. for any $A \in R(O)$ and $\omega^{(O)} \in S(O)$,

$$(\nu^O[\omega^{(O)}])\,(A) = (\nu[\omega])\,(A),$$

where $\omega^{(O)} = \omega|_{R(O)}$, $\omega \in \mathfrak{A}_1^{*+}$. Similarly, all $\tilde{\tau}^O$ are restrictions (to appropriate $R(O)$) of the same mapping $\tau : \mathfrak{A} \to \mathfrak{A}$ which is a Jordan *-automorphism. Transformations determined by pairs $\langle \nu, \tau \rangle$ form a sufficiently large and general class of physical symmetries.

The Jordan or J^*-automorphism is defined as a linear bijective and involution-preserving mapping φ of a *-algebra $\mathfrak{M}$ such that $\varphi(A)^2 = \varphi(A^2)$ for all $A \in \mathfrak{M}$, or equivalently, $\varphi(AB + BA) = \varphi(A)\,\varphi(B) + \varphi(B)\,\varphi(A)$. These conditions do not imply $\varphi(AB) = \varphi(A)\,\varphi(B)$, and so the class of J^*-automorphisms is larger than that of automorphisms, including, besides the latter, also *-antiautomorphisms defined by the property $\varphi(AB) = \varphi(B)\,\varphi(A)$. It means that the class of symmetry transformations compatible with the axioms of local quantum theory is larger than the class of automorphisms of a quasilocal algebra and particularly the class of spatial automorphisms implementable by unitary operators. This going beyond the class of automorphisms of a quasilocal algebra is directly related to the problem of describing discrete symmetries: in field theory the operations T, PT, CPT are implemented by anti-unitary operators and this property originating in general Wigner theorems must be preserved in the algebraic approach as well (if wc succeed in introducing there the said operations).

Finally, one more physically natural condition (condition 1): symmetry transformations transform any local algebra into a local algebra (in a different region, in general) so that for any $O \in B(M)$ there is $\tilde{O} \in B(M)$ such that $\tau[R(O)] \subset R(\tilde{O})$, leads to the division of all symmetry properties into external and internal. Formulating another condition (condition 2): symmetry transformations commute with translations,

* The convex subset $S \subset \mathfrak{A}_1^{*+}$ is called full, if $\omega(A) \geq 0$ for all $\omega \in S$ entails $A \geq 0$.

$$\tau\,[A\,(a)] = (\tau\,[A\,])\,(a) \quad \forall\, A \in \mathfrak{A}\,, \quad a \in \hat{M}\,,$$

we obtain another important division, namely, into exact and broken symmetries.

DEFINITION 1.3.10. A physical symmetry $\langle \nu, \tau \rangle$ satisfying condition 1 is called internal, if for all $O \in \mathcal{B}(M)$,

$$\tau\,[R\,(O)] = R\,(O)\,.$$

Otherwise the symmetry is called external. A physical symmetry $\langle \nu, \tau \rangle$ satisfying conditions 1 and 2 is called exact (or conserved, or a Wigner symmetry), if it leaves a vacuum state (or states) ω_0 of a system invariant, $\nu\,[\omega_0] = \omega_0$. Otherwise the symmetry is called spontaneously broken (or nonconserved, or a Goldstone symmetry). Local quantum theory includes a number of results concerning the rigorous study of broken and exact symmetries. Here the Goldstone and Coleman theorems are analyzed, various methods of defining and studying charges and currents are considered, etc. We do not discuss these results, just point out a few principal papers [277, 390, 353, 209, 339, 147, 155, 438, 329, 330]. On the other hand, for analyzing discrete symmetries no new methods were suggested, whereas attempts of borrowing techniques and/or results of field theory are not very efficient, as explained above. The crucial fact is that the transformation law (3.84) being invalid for 'Haag fields' $A\,(x)$, one was not able to prove for vacuum expectations $\Phi_n\,(x_1, \dots, x_n)$ of these fields the Bargmann–Hall–Wightman theorem which is the cornerstone of the entire complex of results concerning *CPT*-invariance [15, Chapters 4,5]. Functions Φ_n admit an analytic continuation into the extended tube $\mathcal{T}_n = \cup_{\Lambda \in L_+(\mathrm{C})}\, \Lambda\, \mathcal{T}_n^+$ (where $L_+\,(\mathrm{C})$ is the connected component of the complex Lorentz group) and possess the same analyticity domain as Wightman functions. However, although this domain is invariant under $L_+\,(\mathrm{C})$, no transformation law under this group was obtained for the functions Φ_n (see [37, 185]). For these reasons conditions for the existence of a *CPT*-operator and validity of the *CPT*-theorem in local quantum theory are still not sufficiently elucidated though this problem was dealt with in several papers [184, 359, 446].

CHAPTER 2

From the Theory of Observables to the Theory of Quantum Fields

2.1. Global Theory of Superselection Rules

The results of Chapter 1 provide a consistent and developed picture of local quantum theory conceived as the 'axiomatic algebraic approach' in quantum field theory. The picture, however, does not look finished since the information available on many properties of local nets (such as types of the algebras $R(O)$, their modular structure, duality, etc.) is only fragmentary. Another unsatisfactory point concerns applications of the formalism. The whole complex of the results of Chapter 1 is still too poor for the needs of describing concrete models and processes of interaction of elementary particles. It lacks a wide range of notions and properties inherent in quantum field systems but not reflected in Axioms I–VI (cf. Introduction). As a result, the Haag–Araki or Haag–Kastler theory, like other axiomatic approaches, provides an essentially incomplete formulation of quantum field theory and should not be considered as a self-contained 'axiomatic theory' in the strict sense of axiomatic theories in mathematics. It is rather a starting ground, a base set of firmly established facts, which still needs to be expanded and complemented (obviously, on some other principles, not purely axiomatic any more).

The most cardinal gap in the axiomatic formalism described in Chapter 1 is the absence of a notion of quantum field. All relativistic quantum systems are effectively systems of quantum fields and the notions 'relativistic quantum theory' and 'quantum field theory' are always considered as identical. Many of the basic properties of relativistic quantum systems are defined in terms of fields and considered as 'field-like properties': these are internal quantum numbers, the presence of antiparticle for every elementary particle, connection between the spin and statistics, etc. At the same time, fields may not belong to observables: for example, all charged fields are unobservable because they are not gauge invariant while all Fermi or para fields are unobservable since they do not satisfy the locality axiom. For an arbitrary field the character of its relation to observables is far from evident. Thus local quantum theory is confronted with the task of clarifying this relation and complementing its formalism with the concept of quantum field. Or, at least, if fields are not introduced explicitly, we should know how to express all the related properties in terms of observables and states. In this chapter we shall discuss the solution of this problem.

The starting phase in the process of developing the axiomatic algebraic approach into a realistic field theory is the analysis of superselection rules. The phenomenon of superselection plays a crucial role (not fully assessed until long after its discovery) in the relations between fields and observables. In the context of quantum theory, superselection rules appear as certain conditions on the set of all observables, both local and global; in other words, as conditions specifying the global structure of the theory. Such conditions which were absent in the axiomatic formalism of Chapter 1 serve as a logical compliment to Axioms I–VI specifying the local structure of the theory.

The so-called superselection rules were first discovered by Wick, Wightman and Wigner [429] as far back as 1952. According to their description of this phenomenon*, its essence consists in the existence of orthogonal decompositions $\mathcal{H} = \oplus \mathcal{H}_k$ of a state space $\mathcal{H}$ such that the probabilities of all transitions between different $\mathcal{H}_k$ caused by any observable are equal to 0. Apart from this, it is required that all superpositions of vectors from different $\mathcal{H}_k$ correspond to mixtures and not pure states. Subspaces $\mathcal{H}_k$ are called *superselection sectors* and are eigenspaces of the operator of a certain conserved (in the standard quantum-mechanical sense of the commutativity of the corresponding operator with the Hamiltonian $H = \hat{p}_0$) physical quantity. Any such quantity determines a certain superselection rule, and [429] pointed out two such rules associated, respectively, with the operator of the total electric charge and with the so-called operator of univalence $(-)^{2s}$ (s is the spin of a state). Obviously, any such operator, called a *superselection operator*, must have the form $T = \Sigma\lambda_k P_k$, where P_k are the projections on $\mathcal{H}_k$. Since all observable operators cannot bring vectors from one $\mathcal{H}_k$ into another, all $\mathcal{H}_k$ are invariant subspaces of the global algebra R, and the operator T as well as all projections P_k commute with R.

In the name given to the new phenomenon the authors wanted to stress the analogy with selection rules well known from quantum mechanics and atomic physics. Selection rules also mean the vanishing of all transition probabilities between different eigenspaces of the operator of a conserved quantity; but in this case the vanishing holds only for an isolated system and can be broken when external perturbations appear. In other words, a selection rule forbids spontaneous transitions between eigenstates of a conserved quantity whereas a superselection rule forbids any transitions. Discovery of such rules implied a certain modification in the conceptual framework of quantum mechanics: it turned out that, contrary to the usual postulates, in quantum theory not all Hermitian operators correspond to observables ($R \neq \mathcal{B}(\mathcal{H})$) and not all rays in a Hilbert space correspond to pure physical states. (Clearly, these two properties are equivalent to each other.) It is exactly here that the physical essence of superselection rules lies: according to the concise formula by Strocchi and Wightman [402], "in its most general form the

* The theory of superselection rules is one of the subjects where the very fact of switching to the algebraic formalism was of great help in clarifying the situation. Earlier works, which were trying to use the traditional language of quantum mechanics, based the description on rather vague and mathematically ill-defined notions, such as physically realizable and unrealizable vectors, unmeasurable phases, etc. This made the picture of the phenomenon rather confused. Transition to the general formalism of observables and states has easily eliminated this confusion. Specifically, the central concept of the 'physically realizable state' [27, Section 1–1] was given the precise definition as the pure vector state on a global algebra. (This concept is not to be mistaken for the notion of physical state in the algebraic approach, not including, of course, the requirement of purity.)

superselection rule in the quantum theory can be defined as any restriction on a set of observables in a given theory". (Later on we shall have to refine this definition; see Definition 2.1.3 and discussion on p. 138.)

Study of superselection rules involves problems of two kinds. First, we have a physical problem: it is necessary to find out which superselection rules are present in real quantum systems. Second, we have a mathematical problem: it is necessary to describe the structure of the sets of observables and states in systems with superselection rules. One readily agrees that in relativistic quantum theory where we never know exactly the set of all observables (see Section 1.1) the presence of a superselection rule can hardly be proved formally and rigorously. Any such rule relates to a definite observable whose measurement is compatible with those of all other observables. But as long as we have no explicit description of all possible (in principle) measuring procedures, any statement about the universal compatibility of some measurement remains to a certain degree problematic and relying on physical arguments only. The only properties of observables which do not need any analysis of measuring procedures for their substantiation are the axioms of local quantum theory. But superselection rules cannot be considered as axioms if only for the reason that they are global properties whereas local quantum theory can admit only local conditions as its axioms (see Sections 1.1 and 1.2). As a result, statements about the presence of superselection rules in real physical systems are neither axioms nor theorems, nor unconditional inferences from experiments (since no experiment can demonstrate the compatibility of a given observable with all others) and so their substantiation is never absolutely rigorous. This explains, at least, partially, why even now the viewpoint is sometimes expounded which claims that superselection rules do not exist at all. Leaving this exotic position aside, however (see [322] including the history of the problem and references) we shall now list the main superselection rules that are considered as established.

We start with a general remark. All superselection rules are connected with symmetry properties and conservation laws of systems of elementary particles. As we have seen, every superselection rule is related to a certain family of operators in R': all bounded operators of the form $\Sigma\lambda_k P_k$, $\lambda_k \in \mathbf{C}$, lie in R'. In particular, this family includes unitary operators $V = V^{*-1}$, for which $|\lambda_k| = 1$ for all k. It means that every superselection rule corresponds to a certain symmetry $V \cdot V^*$ (see Section 1.3.5). The converse is, of course, not true, and so one can say that superselection rules correspond to a special subclass of symmetries (note that the term 'supersymmetry' was first introduced by Jauch and Misra in 1961 [265] precisely for these 'superselecting' symmetries leaving all observables invariant). Due to this connection between superselection rules and symmetries, the division of symmetries into external (geometrical or kinematic, related to transformations of space-time) and internal (dynamical, related to the character of interaction) entails a similar division for superselection rules. It turns out to be very important here: it is the dynamical or, synonymously, internal superselection rules that provide the key link between fields and observables. (Our terminology does not mean, however, that dynamical superselection rules are only inherent to systems with nontrivial dynamics. They are connected with charges, and, of course, free fields can be charged: e.g., the free Dirac field has nontrivial electric charge and the corresponding dynamical superselection rule.)

The list of examples we open with the simplest superselection rule that was also the first discovered.

(1) The superselection rule for univalence. This rule acts in quantum systems including fermions (as well as bosons, possibly) and separates the sets of states with integer and half-integer spins, respectively. Thus it induces the decomposition of the state space into two sectors collecting state vectors with the Bose and Fermi statistics, respectively:

$$\mathcal{H} = \mathcal{H}_{\mathrm{B}} \oplus \mathcal{H}_{\mathrm{F}}$$

(such decompositions as well as corresponding superselection rules are called dichotomic). A Hermitian and unitary 'univalence operator' is usually chosen as a superselection operator:

$$U = P_{\mathrm{B}} - P_{\mathrm{F}} = (-)^{2S},$$

where S is the total spin operator. U can be identified with the operator implementing in $\mathcal{H}$ the rotation by 2π [241]; it can also be connected with the operator U_T implementing the time inversion [429]. Obviously, we deal here with a kinematic superselection rule.

(2) The superselection rule for the 'type' is another kinematic rule first pointed out by Wightman in 1959 [432]. Symmetries generating this rule are connected with the inversion: superselection sectors are here eigenspaces of the operators $(-)^{2S} U_T^2$ and $(-)^{2S} (U_P U_T)^2$ (P is the space inversion). Due to antiunitarity of U_P and U_T, each of these operators has two eigenvalues, ± 1, so that $\mathcal{H}$ is decomposed into the orthogonal sum of four superselection sectors. In Wightman's terminology, each of these sectors collects states 'of a definite type'. All known elementary particles belong to the subspace (+, +) which was called the subspace of states of type I or normal type.

(3) It is natural to expect that a superselection rule should separate states of particles from states of antiparticles. Such a superselection rule was, indeed, found, but only for systems, satisfying a certain condition relating to the behaviour under the time inversion. Let us give the precise formulation of the result [54].

Let $\mathcal{H}_+$ and $\mathcal{H}_-$ be mutually orthogonal subspaces of states containing, respectively, only particles and only antiparticles of a definite kind. Let also a representation of the full Poincaré group including inversions, act in $\mathcal{H}$, and the antiunitary operator U_T be defined in such a way that it leaves $\mathcal{H}_+$ and $\mathcal{H}_-$ invariant. Then $\mathcal{H}_+$ and $\mathcal{H}_-$ are separated by a superselection rule if and only if all $\mathcal{P}_+^\uparrow$-invariant observables are also T-invariant, i.e.,

$$[A, U(\mathcal{P}_+^\uparrow)] = 0 \Rightarrow [A, U_T] = 0$$

for any $A \in R$. This condition is satisfied not in all systems of elementary particles: as shown in [54], it is violated, in particular, for the doublet of K^0 and

$\bar{K}^0$ mesons. (Since the existence of noninvariant observables implies existence of noninvariant states, we see that there are particles whose states are not invariant under the time inversion T).

(4) One more kinematic superselection rule pointed out in [117] is connected with the conformal group $SO(4,2)/C^2$. It was shown that systems invariant under this group (like massless particles at high energies) and, possibly, systems with the broken conformal symmetry (like massless particles at intermediate energies) display the superselection rule connected with the conservation law for total helicity (i.e. the projection of total spin on the momentum direction). A possible choice for the superselection operator is here $R = \exp[i\pi(I - \lambda)]$, where λ is the helicity operator. Evidently, it has four eigenvalues $\pm 1, \pm i$, and gives rise to the decomposition of the state space into four superselection sectors.

(5) In 1954 Bargmann [59] found that quantum systems with the symmetry under the Galilei group have the superselection rule for mass. Obviously, this rule (which is now referred to as the Bargmann superselection rule) is a kinematic superselection rule in nonrelativistic quantum theory. The superselection operator for it is the mass operator. It has a purely point spectrum filling all the real positive semi-axis. Accordingly, the set of superselection sectors is now infinite and even uncountable, and the Hilbert space of states is decomposed into a continuous direct sum of pairwise orthogonal subspaces.

(6) Superselection rules are very often present in various schemes of the theory of quantum measurements. This is due to the fact that superselection observables are interpreted as classical, macroscopic characteristics of a quantum system (see below). Typical examples are the Machida–Namiki theory [46] where superselection observables are related to the measuring device or Piron formalism [348] where superselection rules characterize the measuring act. In some other schemes superselection describes effects due to the external (macroscopic) surrounding.

Now we turn to dynamical superselection rules which are generated by internal symmetries of elementary particles. It is universally accepted that the class of exact internal symmetries includes only four continuous one-parameter groups $U(1)$, the generators of which are electric Q, baryon B and two leptonic L_e and L_μ charges. These charges are considered as absolutely conserved, and each of them generates a superselection rule. Therefore, we have:

(7–10) Dynamical superselection rules for 'superconserved' charges Q, B, L_e, L_μ. The corresponding symmetry groups are unitarily implemented in the state space by gauge transformations of the first kind (acting on field operators). The charges are generators of these gauge groups and, at the same time, superselection operators (hence, observables, in contrast to charged fields, are gauge invariant). They have an integer point spectrum and generate countable families of superselection sectors.

This list of examples is not meant to be exhaustive. Quite a number of other superselection rules are known in various branches of quantum theory: say, the superselection rule for the particle number in Fock formalism or for temperature and chemical potential in statistical mechanics. Moreover, for the time being we do not mention any systems outside conventional quantum theory, such as gauge theories or models with topological charges. As we shall see below (Section 3.3), such systems give rise to new kinds of dynamical superselection rules which may need a different physical interpretation. The important fact, however, is that despite all physical differences the algebraic structure of any quantum field system with superselection rules, not only global but local as well (see Section 2.2), is nearly the same. We turn now to describing this universal structure.

Let us list again formal properties characterizing a system with superselection rules. Take first the simplest, dichotomic case: $\mathcal{H} = \mathcal{H}_1 \oplus \mathcal{H}_2$.

As noted above, the superselection sectors $\mathcal{H}_1$, $\mathcal{H}_2$ must be invariant subspaces of the global algebras:

$$R\,\mathcal{H}_k \subset \mathcal{H}_k\,, \qquad k = 1, 2\,. \tag{1.1}$$

We can easily check that the following property is equivalent to (1.1): if Φ_λ is an arbitrary normed superposition of vectors from $\mathcal{H}_1$ and $\mathcal{H}_2$, i.e.

$$\Phi_\lambda = \frac{1}{\sqrt{2}}\left(\Phi_1 + e^{i\lambda}\,\Phi_2\right), \qquad \Phi_k \in \mathcal{H}_k\,, \qquad k = 1, 2,$$

then

$$\left(\Phi_{\lambda_1}\,,\, A\,\Phi_{\lambda_1}\right) = \left(\Phi_{\lambda_2}\,,\, A\,\Phi_{\lambda_2}\right) \tag{1.2}$$

for any $A \in R$ and $\lambda_1, \lambda_2 \in \mathbf{R}$. Thus the vectors Φ_{λ_1} and Φ_{λ_2} correspond to the same physical state and hence the phase factor $\exp[i(\lambda_1 - \lambda_2)]$ differentiating them is unobservable. This is the exact formulation of the statement about 'unobservability of the relative phase between superselection sectors' which we can often meet in the physical literature.

The condition (1.1) (or its equivalent, (1.2)) expresses the separation between superselection sectors. It is easy to see that this separation entails one more condition. Indeed, in the formalism of observables and states, the physical condition of belonging to a definite superselection sector must obviously refer not to individual vectors $\Phi \in \mathcal{H}$, but to physical states ω_Φ on R, and a sector is meant to be not merely a set of vectors but a set of physical states. The mathematical formulation of this condition is evident: if $\mathcal{H}_k$ is a superselection sector, then $\Phi \in \mathcal{H}_k$ implies that the whole $\mathcal{H}$-image of the vector state ω_Φ is also in $\mathcal{H}_k$:

$$\forall_{\Phi \in \mathcal{H}_k} \quad \mathfrak{M}^{R}_{\Phi} \subset \mathcal{H}_k\,, \qquad k = 1, 2\,. \tag{1.3}$$

It is easy to show that this property is equivalent to the invariance of $\mathcal{H}_k$ under R':

$$R'\,\mathcal{H}_k \subset \mathcal{H}_k\,, \qquad k = 1, 2\,. \tag{1.3'}$$

To this end we obtain the following general property of $\mathcal{H}$-images:

$$\forall_{\Phi \in \mathcal{H}} \quad \overline{L\left\{\mathfrak{M}^{R}_{\Phi}\right\}} = \mathcal{H}^{R'}_{\Phi}\,. \tag{1.4}$$

For any $\Psi \in \mathfrak{M}_{\Phi}^{R}$ one has

$$\| P_{\Phi}^{R'} \Psi \|^2 = \omega_{\Psi}(P_{\Phi}^{R'}) = \omega_{\Phi}(P_{\Phi}^{R'}) = 1$$

whence $\Psi \in \mathcal{H}_{\Phi}^{R'}$. By applying the operation $\overline{L\{\cdot\}}$ to the formula

$$R'^{U}\Phi \subset \mathfrak{M}_{\Phi}^{R} \subset \mathcal{H}_{\Phi}^{R'} \tag{1.5}$$

(where the left inclusion is obvious) we have (1.4). Equivalence of (1.3) and (1.3') follows immediately from (1.4).

The next formal property is the existence of a superselection operator for a given superselection rule. As said above, such an operator T commutes with R and superselection sectors are its eigenspaces; moreover, it is always connected with a certain symmetry and corresponds to a certain physical quantity, i.e. it is an observable. Hence, if the number of superselection sectors is finite and so the superselection operator $T = \Sigma\, \lambda_k P_k$ is bounded, it must belong to R. Thus we obtain that T belongs to the centre Z.

Finally, one more necessary property was noted above: superpositions of vectors from different superselection sectors always correspond to mixed states on R.

As a result, a system with a dichotomic superselection rule is characterized by the decomposition of the state space $\mathcal{H} = \mathcal{H}_1 \oplus \mathcal{H}_2$, which satisfies the following set of properties.

I. Superpositions of vectors from $\mathcal{H}_1$ and $\mathcal{H}_2$ correspond to mixed states.

II. (A) Any observables cannot induce transitions between $\mathcal{H}_1$ and $\mathcal{H}_2$,

$$(\Phi_1, A\,\Phi_2) = 0 \;\; \forall A \in R, \qquad \Phi_k \in \mathcal{H}_k, \qquad k = 1, 2,$$

or, equivalently, in any superposition Φ_λ of vectors from $\mathcal{H}_1$ and $\mathcal{H}_2$ the relative phase is unobservable:

$$(\Phi_{\lambda_1}, A\,\Phi_{\lambda_1}) = (\Phi_{\lambda_2}, A\,\Phi_{\lambda_2}) \quad \forall \lambda_1, \lambda_2 \in \mathbf{R},$$

or equivalently, the sectors $\mathcal{H}_k$ are invariant subspaces of R.

(B) For any state ω_Φ, $\Phi \in \mathcal{H}_k$, its $\mathcal{H}$-image belongs to $\mathcal{H}_k$:

$$\forall_{\Phi \in \mathcal{H}_k} \quad \mathfrak{M}_{\Phi}^{R} \subset \mathcal{H}_k, \qquad k = 1, 2.$$

or, equivalently, the sectors $\mathcal{H}_k$ are invariant subspaces of R'.

III. There exists an operator $T \in Z$ such that $\mathcal{H}_1$ and $\mathcal{H}_2$ are its eigenspaces.

Clearly, these properties are not independent. First of all, conditions II and III are both equivalent to the fact that the projections P_k lie in Z, which is, in its turn, equivalent to the decomposition of R into the direct sum of von Neumann algebras acting respectively in $\mathcal{H}_1$ and $\mathcal{H}_2$:

$$R = R_1 \oplus R_2, \qquad R_k \subset \mathcal{B}(\mathcal{H}_k), \qquad k = 1, 2.$$

The complete set of relations between properties I–III is described by the following simple

PROPOSITION 2.1.1. *Let a quantum system with the global algebra* R *be given, and* $\mathcal{H} = \mathcal{H}_1 \oplus \mathcal{H}_2$ *be an arbitrary dichotomy of its state space. Then the following relations are valid for this dichotomy:*

$$I \Leftarrow II \Leftrightarrow III\ . \tag{1.6}$$

Proof is elementary and is omitted.

It is evident that the missing implication in (1.6) is not valid for an arbitrary von Neumann algebra: the set of 'pure vectors' (i.e., vectors $\Phi \in \mathcal{H}$ such that $\omega_{\Phi/\|\Phi\|}$ is pure) of such an algebra can be empty, and then condition I is fulfilled trivially for any dichotomy $\mathcal{H}_1 = \mathcal{H}_1 \oplus \mathcal{H}_2$ and does not imply anything. This situation in which condition I is meaningless, does not, however, conform with the generally accepted idea of a superselection rule. Model physical descriptions of the phenomenon, starting with [429], invariably assume that condition I is completely sound and is an unmistakable indication of a superselection rule. Therefore, the situation in question must be excluded, and condition I, as well as II and III, must be sufficient condition of a superselection rule. It means that some new condition must be introduced so as to ensure the last implication in (1.6). Clearly, this condition, unlike I–III, will restrict not only the admissible partition $\mathcal{H} = \mathcal{H}_1 \oplus \mathcal{H}_2$, but also the admissible global algebras: condition I is not meaningless only if $\mathcal{H}$ includes pure vectors, and so the equivalence of I to II and III is expected to require a certain 'sufficient amount' of such vectors. Let us derive the condition.

DEFINITION 2.1.1. A von Neumann algebra R is said to be an algebra of P-class (where P stands for 'pure') if the set $\mathcal{P}(R)$ of all its pure vectors is total in $\mathcal{H}$:

$$\mathcal{H} = \overline{L\ \{\mathcal{P}(R)\}} \equiv \mathcal{H}_p\ , \qquad \mathcal{P}(R) = \left\{ \Phi \in \mathcal{H} \,|\, \omega_{\Phi/\|\Phi\|} \in PV(R) \right\}.$$

PROPOSITION 2.1.2. *Let* R *be a von Neumann algebra. The following conditions are equivalent:*

(1) *for any decomposition* $\mathcal{H} = \mathcal{H}_1 \oplus \mathcal{H}_2$ *each of the properties* I–III *implies the remaining two;*

(2) R *is an algebra of* P*-class.*

Proof. To derive $1 \to 2$ we need a simple lemma:

LEMMA 2.1.3. *Any von Neumann algebra* R *can be represented as* $R = R_p \oplus R_m$, *where* R_p *is an algebra of P-class and* R_m *is a von Neumann algebra without pure vector states.*

Proof of the lemma. Presenting $\mathcal{H}$ as $\mathcal{H} = \mathcal{H}_p \oplus \mathcal{H}_p^{\perp}$, we shall show that the projection P_p on $\mathcal{H}_p$ lies in Z. It suffices to show that for any $\Phi \in \mathcal{P}(R)$,

$$Z'\,\Phi \subset \mathcal{H}_p\ . \tag{1.7}$$

One has $Z'\,\Phi = \overline{L\ \{R'\,R\,\Phi\}}$. Since Φ is pure only if $\mathcal{H}^{R}_{\Phi}$ is an irreducible subspace of R, all vectors from $R\,\Phi$ are pure. Next, for any $\Psi \in \mathcal{P}(R)$ vectors from $R'\Psi$ are also pure because, as we can easily see, the functional $\omega_{T\Psi}$, $T \in R'$, is majorized by the pure functional ω_Ψ and, hence, coincides with it. Hence (1.7) follows and we have $P_p \in \mathcal{Z}$. Due to this, $R = R_p \oplus R_m$, where R_p and R_m are the inductions of R by the projections P_p and $I - P_p$, respectively. Clearly,

$$\mathcal{P}(R) = \mathcal{P}(R) \cap \mathcal{H}_p = \mathcal{P}(R_p)\,,\quad \mathcal{P}(R_m) = \emptyset\,,$$

and the lemma is proved. ❒

Returning to Proposition 2.1.2, assume that R is not of P-class. Then, by the lemma, $R = R_p \oplus R_m$, where $P_m \neq 0$. By the definition of $\mathcal{H}_m$, there is a subspace $\mathcal{H}_3 \subset \mathcal{H}_m$ such that the projection P_3 is not in the centre of R_m. Consider the decomposition $\mathcal{H} = \mathcal{H}_1 \oplus \mathcal{H}_2$ with $\mathcal{H}_1 = \mathcal{H}_p \oplus \mathcal{H}_3$; clearly, $P_1, P_2 \notin \mathcal{Z}$. This decomposition enjoys property I (all pure vectors being in $\mathcal{H}_p$) and does not enjoy properties II and III (since they are tantamount to $P_1, P_2 \in \mathcal{Z}$).

2 → 1. Let R be an algebra of P-class and $\mathcal{H} = \mathcal{H}_1 \oplus \mathcal{H}_2$ be an arbitrary dichotomy of $\mathcal{H}$. Taking account of Proposition 2.1.1, it suffices to prove that if property I is fulfilled for this dichotomy, then II and III are fulfilled too. To this end we only need to derive $P_1, P_2 \in \mathcal{Z}$. First of all, I entails

$$\mathcal{P}(R) = (\mathcal{P}(R) \cap \mathcal{H}_1) \cup (\mathcal{P}(R) \cap \mathcal{H}_2)\,, \tag{1.8}$$

whence due to $R\,\mathcal{P}(R) \subset \mathcal{P}(R)$, $R'\,\mathcal{P}(R) \subset \mathcal{P}(R)$ one has

$$R\,(\mathcal{P}(R) \cap \mathcal{H}_k) \subset \mathcal{P}(R) \cap \mathcal{H}_k\,,\quad R'\,(\mathcal{P}(R) \cap \mathcal{H}_k) \subset \mathcal{P}(R) \cap \mathcal{H}_k\,,\quad k = 1, 2\,. \tag{1.9}$$

Moreover, (1.8) implies

$$L\,\{\mathcal{P}(R)\} \cap \mathcal{H}_k = L\,\{\mathcal{P}(R) \cap \mathcal{H}_k\}\,,\quad k = 1, 2\,. \tag{1.10}$$

If R belongs to P-class, then $\overline{L\,\{\mathcal{P}(R)\}} = \mathcal{H}_1 \oplus \mathcal{H}_2$, and this, together with (1.10) implies $\overline{L\,\{\mathcal{P}(R) \cap \mathcal{H}_k\}} = \mathcal{H}_k$, $k = 1, 2$, whence by virtue of (1.9) one has $R\,\mathcal{H}_k \subset \mathcal{H}_k$, $R'\,\mathcal{H}_k \subset \mathcal{H}_k$, $k = 1, 2$. ❒

COROLLARY 2.1.4. *Properties* I–III *characterizing a superselection structure are mutually equivalent if and only if* R *is an algebra of* P*-class.*

Thus, in view of the foregoing discussion, the property that a global algebra belongs to P-class should be accepted as the last of necessary conditions of the presence of a superselection rule. The analysis of the dichotomic superselection rule is now completed, and we can summarize it by formulating the following definition.

DEFINITION 2.1.2. Let a quantum system with the global algebra R of P-class be given as well as a decomposition of the state space $\mathcal{H}$ in the form $\mathcal{H} = \mathcal{H}_1 \oplus \mathcal{H}_2$ satisfying any of conditions I–III (and hence all these conditions). In this case we say that the system has a dichotomic superselection rule described by superselection sectors $\mathcal{H}_1$ and $\mathcal{H}_2$ and superselection operators $T \in \mathcal{Z}$ of the form $T = \lambda_1 P_1 + \lambda_2 P_2$,

$\lambda_{1,2} \in \mathbf{R}^{*}$.

Furthermore, we have to analyze in a similar way the structure of an arbitrary superselection rule. Generally speaking, superselection sectors can now form the set of any power: finite, countable, continual, etc. On the other hand, structural conditions I–III, as well as the condition that the global algebra belongs to P-class, characterize physical contents of not just dichotomic but any superselection rule as such. Therefore they must be preserved, though their formulations may need to be modified. Appropriate modifications are easy to find.

Let us assume that

$$\mathcal{H} = \bigoplus_{\gamma \in \sigma} \mathcal{H}_\gamma , \tag{1.11}$$

σ being an arbitrary set. If the subspaces $\mathcal{H}_\gamma$ are to be interpreted as superselection sectors, they must satisfy some generalizations of properties I–III. For properties I and II the generalization is straightforward: any superpositions of vectors taken from different $\mathcal{H}_\gamma$ correspond to mixed states (the generalization of I); for every pair of subspaces $\mathcal{H}_\gamma$, $\mathcal{H}_\beta$, $\gamma, \beta \in \sigma$, conditions (1.1) and (1.3) are valid, and hence, also their equivalents (1.2) and (1.3′) (the generalization of II). Next, a superselection operator T also retains its physical meaning as an operator corresponding to a superselection quantum number. Accordingly, it must commute with R and also be an observable whereas all the subspaces $\mathcal{H}_\gamma$, $\gamma \in \sigma$, must be its eigenspaces. So it must be of the form $T = \Sigma_{\gamma \in \sigma} \lambda_\gamma P_\gamma$, $\lambda_\gamma \in \mathbf{R}$, $P_\gamma \in \mathcal{Z}$ (where P_γ are the projections on $\mathcal{H}_\gamma$). When σ is infinite, depending on the set of coefficients (or, more exactly, on the choice of a bounded function $\lambda : \sigma \to \mathbf{R}$), T can be unbounded. As a result, we have the following generalization of III: there exists an operator $T \eta\, \mathcal{Z}$ (i.e. affiliated to $\mathcal{Z}$ and, in general, unbounded) such that all $\mathcal{H}_\gamma$, $\gamma \in \sigma$, are its eigenspaces. (Thus P_γ are spectral projections of T, and the spectrum of T is purely point one, without a continuous part). Like in the dichotomic case, it is immediately obvious that the generalizations of II and III are both equivalent to the property $P_\gamma \in \mathcal{Z}$, $\forall \gamma \in \sigma$, and also to the decomposition of the global algebra into a direct sum of global algebras of superselection sectors:

$$R = \bigoplus_{\gamma \in \sigma} R_\gamma , \; R_\gamma \subset \mathcal{B}(\mathcal{H}_\gamma) . \tag{1.12}$$

Denoting the generalizations obtained by II'–III', we readily verify that they satisfy the same relationships as in the dichotomic case, i.e. the following implications are valid:

$$\mathrm{I}' \Leftarrow \mathrm{II}' \Leftrightarrow \mathrm{III}',$$

and if and only if R is an algebra of P-class, the last implication holds as well, that is the following conditions are equivalent:

(1) for any decomposition $\mathcal{H} = \oplus_{\gamma \in \sigma} \mathcal{H}_\gamma$ each of the properties I'–III' implies the remaining two;

* If $\mathrm{Im}\, \lambda_i \neq 0$, the operator T is not Hermitian and, therefore, is not an observable.

(2) R is an algebra of P-class.

Proofs of these facts repeat those of Proposition 2.1.1 and 2.1.2. Hence we can adopt the following definition.

DEFINITION 2.1.3. Let a quantum system with the global algebra R of P-class be given as well as the decomposition of the state space $\mathcal{H}$ in the form $\mathcal{H} = \oplus_{\gamma\in\sigma} \mathcal{H}_\gamma$ (σ is, in general, an arbitrary set) satisfying any of conditions I'–III' (and hence, all these conditions). In this case we say that the system has a superselection rule (of a general type) described by superselection sectors $\mathcal{H}_\gamma$ and superselection operators T (generally, unbounded) of the form $T_\eta\ Z, T = \Sigma_{\gamma\in\sigma} \lambda_\gamma P_\gamma$.

Mathematically, it is worthwhile to look at this definition from the viewpoint of the decomposition theory of von Neumann algebras; see [7, Chapter II, or 5, Chapter 4]. In terms of this theory, the structure associated with an arbitrary superselection rule is nothing but the decomposition (1.11), (1.12) of the algebra R and its space $\mathcal{H}$ for which the algebra of diagonalizable operators is the subalgebra $Z_T \subset Z$ generated by the corresponding superselection operator T. By one of our physical requirements, T (and hence all the operators from Z_T) must have a purely point spectrum; due to this, the decompositions (1.11) and (1.12) generated by the algebra Z_T take the form of direct sums rather than direct integrals. Conversely, if we demand that these decompositions should be direct sums, then we obtain that T must have a purely point spectrum. This connection with the decomposition theory will soon prove useful for us.

Further questions concern the completeness and generality of our scheme. According to the standard views, two types of superselection rules, discrete and continuous, are possible for quantum systems and the difference between them is that the state space is in the first case the direct sum and in the second case the direct integral of superselection sectors. Our scheme admits only the first type of the decomposition. So one can ask whether Definition 2.1.3 really provided the most general form of superselection structure and embraces all known and all possible superselection rules.

Since our scheme is not postulated but follows rigorously from definite physical assumptions (namely, conditions I'–III' and the condition $\mathcal{H} = \mathcal{H}_p$), any structures deviating from Definition 2.1.3 are only possible if some of these assumptions are rejected. In view of this, the question must be asked as follows: are there any algebraic structures which are different from those in our definition but still preserve the 'most essential' part of their physical properties (e.g., make it possible to define superselection sectors and superselection operators related to the centre of R)? These structures could indeed be considered as superselection rules of some new type. It would be the case if, for example, Z would include operators with a continuous spectrum: they could be treated as superselection operators, though the decomposition of $\mathcal{H}$ generated by them would be a direct integral of Hilbert spaces. To answer the question we need additional information of the structure of P-class algebras.

PROPOSITION 2.1.5. *For an arbitrary von Neumann algebra R the following conditions are equivalent:*

(1) *R is an algebra of P-class;*

(2) *$R = \oplus_{\gamma \in \sigma} R_\gamma$, where R_γ are factors of type I;*

(3) *R is an algebra of type* I, *and $\mathcal{Z}$ includes only operators with a purely point spectrum.*

Proof. The simplest way is to prove the equivalence of 2 to 1 and 3. $2 \to 3$ is obvious. $3 \to 2$ follows from the fact that abelian algebras of operators with purely point spectra are the nearest generalization of finite-dimensional abelian algebras: if any $T \in \mathcal{Z}$ is of the form $T = \Sigma_{\gamma \in \sigma(T)} \lambda_\gamma P_\gamma^{(T)}$, then there exists some family $\{P_\gamma\}_{\gamma \in \sigma}$ of mutually orthogonal minimal projections which generates $\mathcal{Z}$ (cf. [212]):

$$\sum_{\gamma \in \sigma} P_\gamma = I, \quad P_\gamma P_\beta = 0 \quad \forall \gamma, \beta \in \sigma, \left\{ \bigcup_{\gamma \in \sigma} P_\gamma \right\}'' = \mathcal{Z}.$$

Obviously, this entails discrete decompositions for $\mathcal{Z}$ and $\mathcal{Z}'$ and the discrete central decomposition for R:

$$R = \bigoplus_{\gamma \in \sigma} R_\gamma, \quad \mathcal{Z} = \bigoplus_{\gamma \in \sigma} \mathcal{Z}_\gamma, \quad \mathcal{Z}_\gamma = R_\gamma \cap R'_\gamma = C(\mathcal{H}_\gamma). \tag{1.13}$$

If R is an algebra of type I, then so are the factors R_γ.

$\underline{2 \to 1}$. Representing a discrete factor R_γ as $R_\gamma = \mathcal{B}(\mathcal{H}_1) \otimes C(\mathcal{H}_2)$, one easily sees that R_γ is an algebra of P-class. Hence so is $R = \oplus_{\gamma \in \sigma} R_\gamma$.

$\underline{1 \to 2}$. Let us consider the set $\mathcal{P}_1(R)$ of all pure unit vectors of R. Any pair of vectors $\Phi_1, \Phi_2 \in \mathcal{P}_1(R)$ corresponds to the pair of irreducible representations of R, which can be either disjoint or unitarily equivalent so that their central supports $P_{\Phi_1}^{\mathcal{Z}'}$ and $P_{\Phi_2}^{\mathcal{Z}'}$ either coincide with one another or are mutually orthogonal. Let $\mathcal{P}_1^0$ be a maximal set of vectors from $\mathcal{P}_1^0$ for which all $\mathcal{H}_\Phi^{\mathcal{Z}'}$ are pairwise orthogonal. Owing to $\mathcal{P}(R) \subset \cup_{\Phi \in \mathcal{P}_1^0} \mathcal{H}_\Phi^{\mathcal{Z}'}$ one has from 1

$$\overline{L\{\mathcal{P}(R)\}} = \mathcal{H} = \bigoplus_{\Phi \in \mathcal{P}_1^0} \mathcal{H}_\Phi^{\mathcal{Z}'}. \tag{1.14}$$

Since the projections corresponding to (1.14) lie in $\mathcal{Z}$, this decomposition of $\mathcal{H}$ generates the decomposition of R as well:

$$R = \bigoplus_{\Phi \in \mathcal{P}_1^0} R_{P_\Phi^{\mathcal{Z}'}}. \tag{1.15}$$

Next, the induction $R_{P_\Phi^{\mathcal{Z}'}} \to R_{P_\Phi^R}$ is an isomorphism, because the central support of $R_{P_\Phi^R}$ in $R_{P_\Phi^{\mathcal{Z}'}}$ is the identity on $\mathcal{H}_\Phi^{\mathcal{Z}'}$. Hence the centres of these algebras are isomorphic, and $R_{P_\Phi^R} = \mathcal{B}(\mathcal{H}_\Phi^R)$ implies that $R_{P_\Phi^{\mathcal{Z}'}}$ is a factor and is of type I (being a W^*-algebra isomorphic to $\mathcal{B}(\mathcal{H}_\Phi^R)$ in a certain $\tilde{\mathcal{H}}$). Thus (1.15) is the required decomposition of R. ❐

Let us now return to the problem of continuous superselection rules. A priori, three situations are thinkable when one can speak about the presence of such rules:

(1) the set σ of all superselection sectors is continuous rather than discrete, i.e. card $\sigma > \aleph_0$;

(2) there is a superselection operator with the continuous rather than discrete spectrum;

(3) the state space is decomposed into a direct integral of superselection sectors rather than a direct sum:

$$\mathcal{H} \simeq \int^{\oplus} \mathcal{H}(\xi)\, d\mu(\xi)\,, \qquad R \simeq \int^{\oplus} R(\xi)\, d\mu(\xi)\,. \tag{1.16}$$

One easily notices that the last two cases are equivalent. Indeed, the proof of the implication $3 \to 2$ shows that the central decomposition of R represents a direct integral rather than a direct sum if and only if Z includes operators with a continuous spectrum. Similarly, since the algebraic structure for a separate superselection rule with the superselection operator T is the decomposition of R with the diagonal algebra $Z_T \subset Z$, this structure represents a direct integral rather than a direct sum if and only if T has a continuous spectrum. But, by Proposition 2.1.5, if R is an algebra of P-class, then Z and, therefore, the set of its affiliated operators as well, do not include operators with a continuous spectrum. Thus continuous superselection rules in the sense of the above type 2 or, equivalently, 3 described by direct integrals of superselection sectors, are only possible if R is not an algebra of P-class. Hence they are excluded in our scheme. A possibility of using global algebras which do not belong to the P-class will be discussed later on; but it is immediately clear that structures of the form (1.16) lack the major part of physical properties included in our definition of a superselection rule. Components of the direct integral cannot be subspaces of $\mathcal{H}$, and the structure (1.16), in addition to breaking the condition $\mathcal{H} = \mathcal{H}_p$, cannot satisfy any of the properties II' and I'.

By contrast, our scheme readily admits of continuous superselection rules in the sense of the above type 1 for which superselection sectors, still are subspaces of $\mathcal{H}$ and satisfy all conditions I'–III', but form an uncountable set (card $\sigma > \aleph_0$). Such rules are characterized by a simple mathematical condition: obviously, an algebra of P-class induces superselection structures with card $\sigma > \aleph_0$ if and only if its centre Z is not a σ-finite algebra. It is easy to see that in such a case the algebras R and Z can have neither cyclic nor separating vectors. A superselection structure of this type corresponds, in particular, to the above-mentioned Bargmann superselection rule in the Galilei-covariant quantum theory.

Furthermore, we have to describe the algebraic structure generated by the set of all superselection rules present in a given quantum system; we need also to characterize this set in terms of the algebra R. These problems are easily solved with the help of Proposition 2.1.5. It says that for an algebra of P-class any operator lying in or affiliated to Z has a purely point spectrum and, by Definition 2.1.3, it corresponds to a definite superselection rule. For a given superselection rule, the superselection operator $T_\eta\, Z$ is not unique, but the set $\{P_\gamma\}_{\gamma \in \sigma}$ of orthogonal projections onto superselection sectors or, equivalently, the partition of identity lying in Z is, obviously, unique. Thus we obtain the following characterization of the set of all superselection rules:

> There is a one-to-one correspondence between the set of all superselection rules of given quantum system and the set of all partitions of identity in the centre $\mathcal{Z}$ of its global algebra (supposed to be of P-class).

The algebraic structure generated by this set of superselection rules is clearly visible. Since any pair of superselection rules is associated with two mutually commuting partitions of identity $\{P_\gamma\}_{\gamma\in\sigma_1}$ and $\{P_\gamma\}_{\gamma\in\sigma_2}$ (the 'commutativity of superselection rules' holding due to $P_\gamma \in \mathcal{Z}$), any two superselection structures generate the third one which is associated with the partition of identity $\{P_\gamma P_\beta\}_{\gamma\in\sigma_1,\, \beta\in\sigma_2}$ and is 'finer' than the first two, in the sense that every sector is the intersection of two sectors taken from the first and the second structure, respectively. Similarly, condition 2 of Proposition 2.1.5 implies that the set of all superselection structures also generates a unified superselection structure which is the finest of all and coincides with the central decomposition of R. The sectors $\mathcal{H}_\gamma$ of this structure or, more exactly, the pairs $\langle R_\gamma, \mathcal{H}_\gamma\rangle$, $R_\gamma \equiv R\,|\,_{\mathcal{H}_\gamma}$, are called the *coherent superselection sectors*. They have quite a number of important properties, such as:

(1) any vector in a coherent sector is the common eigenvector of all superselection operators. In other words, any coherent sector is the intersection of a family of superselection sectors including one and only one sector from each superselection structure;

(2) coherent sectors are 'minimal' in the sense that no superselection sector can be a proper subspace of a coherent sector;

(3) global algebras R_γ of coherent sectors are factors, whereas for usual sectors corresponding to individual superselection rules they can be direct sums of factors.

Thus the central decomposition of a global algebra of P-class:

$$R = \bigoplus_{\gamma\in\sigma} R_\gamma, \qquad R_\gamma \text{ is a factor of type I,}$$

is interpreted as the decomposition of R and $\mathcal{H}$ into minimal or coherent superselection sectors $\langle R_\gamma, \mathcal{H}_\gamma\rangle$. Obviously, for a given quantum system it is unique. The existence of this decomposition or, what is the same, the existence of a minimal, finest superselection structure is closely connected with the commutativity of superselection rules. Like the commutativity, it is ensured automatically in our scheme. (Other treatments of the superselection problem usually assume to this end a special postulate, the 'Wightman postulate of the commutativity of superselection rules'.)

The fact that the decomposition into coherent sectors coincides with the central decomposition of the global algebra is physically very meaningful. Let us recall first that in algebraic statistical mechanics the central decomposition of a global algebra $\pi_\omega(\mathfrak{A})''$ (where $\mathfrak{A}$ is a quasilocal algebra of a statistical system and π_ω its representation for a physical state ω) is typically the decomposition of the state ω into pure thermodynamic phases. Thus the notion of the coherent sector in quantum field theory represents a parallel to the notion of the pure phase in statistical mechanics. This is a far reaching parallel, both mathematical and physical. For example, a specific feature of states corresponding to pure phases is that macroscopic observables take fixed values in them [22, Section 6.5]; the same property, as we can easily see, is specific of states in coherent sectors. What are, indeed, macroscopic observables? Macroscopic systems exhibit classical behaviour and all their observables are mutually compatible (i.e., observable algebras of such systems must be abelian, which is confirmed by the existing examples (see, e.g. [25, 120, 213])) . As for a quantum system, its classical or macroscopic observables are, obviously, those which are compatible with all other observables. In the W^*-description, such observables lie in or are affiliated to the centre $\mathcal{Z}$ of a global algebra, and hence they take fixed values in every coherent sector. Discussion of the connection between coherent sectors and pure phases (which will be continued in Section 3.1.2), leads us to the well known problem of the 'global nature of superselection rules'. We see that the decomposition into coherent sectors is one in which the complete separation between the classical (macroscopic) and quantum (microscopic) properties of the system is achieved. The centre $\mathcal{Z}$ is the algebra of classical observables and the components of the central decomposition are purely quantum systems in which this algebra is trivial and which differ from each other by values of superselection observables lying in $\mathcal{Z}$ and hence classical. In other words, physical characteristics for which there is a superselection are always classical (macroscopic) by their nature. In Section 2.2, where we study the spacetime structure of systems with superselection rules, this conclusion will be given the mathematical formulation.

As for the global theory of superselection rules, its construction is basically completed. We need only to note some physical implications of the fact that global algebras of coherent sectors are not, in general, irreducible, and may be arbitrary discrete factors.

The first group of implications concerns the superposition principle. Sectors with $R_\gamma \neq \mathcal{B}(\mathcal{H})$ (which are called nonabelian since the commutant R'_γ is for them a non-abelian algebra) include not only pure vectors and, as a result, the fulfillment of this principle in such sectors is not obvious at all. Indeed, the usual formulation of the principle states that the superposition of vectors corresponding to pure physical states also correspond to a pure state, i.e.

$$\omega_\Phi\,,\ \omega_\Psi \in PV(R) \Rightarrow \omega_{N^{-1}(\Phi + e^{i\lambda}\Psi)} \in PV(R)\,, \tag{1.17}$$

$$\omega_{N^{-1}\chi} \equiv \omega_{\|\chi\|^{-1}\chi}\,.$$

This formulation is, however, too restrictive since the physical contents required of the superposition principle does not inevitably lead to (1.17). As explained by Dirac (cf. also [293]), this contents is as follows: we say that the superposition principle is fulfilled for a certain set of states (or a certain subspace $\mathcal{M} \subset \mathcal{H}$) if the states obtained by means of superposing (combining) pure states from this set (or states of the form ω_Φ with pure vectors $\Phi \in \mathcal{M}$) are again pure. But the superposition operation can be defined in different ways. Clearly, in our case superposing means taking linear combinations not of the states ω_{Φ_i} (these combinations $\Sigma\,\lambda_i\,\omega_{\Phi_i}$ are mixtures by definition) but of the vectors Φ_i. But the set of vectors corresponding to each state ω_{Φ_i} is all the (many-dimensional, in general) $\mathcal{H}$-image $\overset{R}{\mathfrak{M}}_{\Phi_i}$. This means that we must consider as superpositions of vector states ω_Φ and ω_Ψ not only the states $\omega_{N^{-1}(\Phi + e^{i\lambda}\Psi)}$ but also the states $\omega_{N^{-1}(\Phi_1 + e^{i\lambda}\Psi_1)}$ with arbitrary $\Phi_1 \in \overset{R}{\mathfrak{M}}_\Phi$, $\Psi_1 \in \overset{R}{\mathfrak{M}}_\Phi$. (We shall call such states *generalized superpositions* of ω_Φ and ω_Ψ). The usual superposition principle demands that all the superpositions should be pure states. It is easy to verify that in the Hilbert space $\mathcal{H}_\gamma$ of a discrete factor R_γ this requirement is violated: the set of all pure vectors in $\mathcal{H}_\gamma$ is not linear ($\Phi, \Psi \in \mathcal{P}(R_\gamma)$ give $\Phi + \Psi \notin \mathcal{P}(R_\gamma)$ if and only if Φ and Ψ belong to different irreducible subspaces of R_γ and also to different irreducible subspaces of R'_γ [251, Lemma 3]). A natural generalization of this requirement is the condition that at least one of the generalized superpositions is a pure state. And this condition holds for R_γ, as we shall see now.

DEFINITION 2.1.4. Let $\mathcal{H}$ be the Hilbert space of a von Neumann algebra R. We say that the generalized superposition principle holds in a subspace $\mathcal{M} \subset \mathcal{H}$ if for any two pure vector states ω_Φ and ω_Ψ with $\Phi, \Psi \in \mathcal{M}$ there are vectors $\Phi_1 \in \overset{R}{\mathfrak{M}}_\Phi \cap \mathcal{M}$ and $\Psi_1 \in \overset{R}{\mathfrak{M}}_\Phi \cap \mathcal{M}$, whose normed superposition also corresponds to a pure state:

$$\underset{\substack{\Phi,\Psi \in \mathcal{M}\cap S_1 \\ \Phi,\Psi \in \mathcal{P}(R)}}{\forall} \quad \underset{\Phi_1,\Psi_1 \in \mathcal{M}\cap S_1}{\exists} \quad \omega_{\Phi_1} = \omega_\Phi\,,\ \omega_{\Psi_1} = \omega_\Psi\,,\ \omega_{N^{-1}(\Phi_1 + \Psi_1)} \in PV(R)\,.$$

The state $\omega_{N^{-1}(\Phi_1 + \Psi_1)}$ will be called the *coherent superposition* of states ω_Φ and ω_Ψ.

PROPOSITION 2.1.6. *The generalized superposition principle holds in any coherent sector* $\langle R_\gamma, \mathcal{H}_\gamma \rangle$.

Proof. First of all, we need the following property of pure vector states.

LEMMA 2.1.7. *For an arbitrary von Neumann algebra R a vector state ω_Φ is pure if and only if its $\mathcal{H}$-image coincides with the unit sphere in the subspace* $\overset{R'}{\mathcal{H}}_\Phi$:

$$\omega_\Phi \in PV(R) \Leftrightarrow \overset{R}{\mathfrak{M}}_\Phi = \overset{R'}{\mathcal{H}}_\Phi \cap S_1\,.$$

Proof of the Lemma. Necessity. As noted above, it is easy to check that $\omega_{T\Phi/\|T\Phi\|} \prec \omega_\Phi$ for all $T \in R'$. Hence $\omega_\Phi \in PV(R)$ entails $(R'\Phi) \cap S_1 \subset \overset{R}{\mathfrak{M}}_\Phi$. Taking account of (1.5) we obtain

$$(R'\Phi) \cap S_1 \subset \overset{R}{\mathfrak{M}}_\Phi \subset \overset{R'}{\mathcal{H}}_\Phi \cap S_1\,. \tag{1.18}$$

It is easy to show [407] that any $\mathcal{H}$-image $\overset{R}{\mathfrak{M}}_\Phi$ is a closed set. Hence (1.18) implies the required property.

Sufficiency. Prove that if $\overset{R}{\mathfrak{M}}_\Phi = \overset{R'}{\mathcal{H}}_\Phi \cap S_1$, then any positive functional ω majorized by ω_Φ coincides with it. When Φ is a cyclic vector of R, it is well known [18, Chapter IV, Section 19] that

$$\omega \prec \omega_\Phi \Rightarrow \omega = \omega_{T\Phi}, \; T \in R'.$$

One can easily verify [407, Theorem 1] that this result is also valid for an arbitrary $\Phi \in S_1$. Hence the required property follows immediately: since $T\Phi / \|T\Phi\| \in \mathcal{H}^{R'}_{\Phi} \cap S_1 = \mathfrak{M}^R_\Phi$, one has $\omega_{T\Phi/\|T\Phi\|} = \omega = \omega_\Phi$. ❐

Now, to construct a coherent superposition of pure states ω_Φ and ω_Ψ, let us use the following property of discrete factors:

$$\forall_{\Phi_1, \Phi_2 \in \mathcal{P}(R)} \quad \mathcal{H}^R_{\Phi_1} \cap \mathcal{H}^{R'}_{\Phi_2} \neq \{0\}. \tag{1.19}$$

(The reader can easily derive (1.19) from (1.4) and Lemma 2.1.7.) Taking any vector $\chi \in \mathcal{P}(R_\gamma)$, we note that due to (1.19) and Lemma 2.1.7

$$\exists_{\Phi_1, \Psi_1} \quad \mathcal{H}^R_\chi \cap \mathfrak{M}^R_\Phi \ni \Phi_1 \neq 0, \; \mathcal{H}^R_\chi \cap \mathfrak{M}^R_\Psi \ni \Psi_1 \neq 0.$$

By construction, $\omega_{\Phi_1} = \omega_\Phi$, $\omega_{\Psi_1} = \omega_\Psi$, $\Phi_1 + \Psi_1 \in \mathcal{H}^R_\chi \subset \mathcal{P}(R_\gamma)$. Hence the state $\omega_{N^{-1}(\Phi_1 + \Psi_1)} \in PV(R)$ is a coherent superposition of ω_Φ and ω_Ψ. ❐

Use of the term 'coherent superposition' for the superposition of pure states in Definition 2.1.4 is in accordance with the physical meaning of the coherence property. Indeed, pure states are called mutually coherent if, superimposing, they produce again pure states. (Hence, it is clear that coherence and the superposition principle are closely related. In fact, the coherence of a certain set of pure states and the fulfillment of the superposition principle for this set are merely different formulations of the same physical property.) Thus Proposition 2.1.6 shows that the name 'coherent sector' for the Hilbert space of a discrete factor is justified physically because all pure states in such a sector are mutually coherent. The converse is also true: any two mutually coherent states always lie in the same coherent sector. To show this, check the equivalent assertion: pure states from different coherent sectors cannot form coherent superpositions. Indeed, if $\Phi \in \mathcal{H}_\gamma$, $\Psi \in \mathcal{H}_\beta$, $\gamma \neq \beta$, then also $\mathfrak{M}^R_\Phi \subset \mathcal{H}_\gamma$, $\mathfrak{M}^R_\Psi \subset \mathcal{H}_\beta$, and for any pair $\Phi_1 \in \mathfrak{M}^R_\Phi$, $\Psi_1 \in \mathfrak{M}^R_\Psi$ the vector $\Phi_1 + \Psi_1$ is not pure. Hence it follows another property of our notion of coherence: two pure states are mutually coherent if and only if they are unitarily equivalent. It is easy to show [251] that the coherent superposition of pure states is unitarily equivalent to these states and is determined uniquely up to the relative phase factor.

Other treatments of the superposition principle were developed within the framework of the Jauch–Piron quantum mechanical axiomatics [262, 263], in quantum logics [118], and quantum theory of measurements [242]. Comparison with these treatments as well as more detailed study of coherent superpositions and the generalized superposition principle can be found in [251, 181].

The generalized superposition principle differs from the usual (1.17) only in non-abelian coherent sectors. Hence the question arises whether such sectors are really necessary. Which quantum systems do they correspond to? So far we have no answer to this question: we have studied nonabelian sectors for the reason that they are, in principle, allowed by our scheme. And the complete answer is, indeed, unobtainable in the purely global formalism. Physical properties distinguishing nonabelian sectors are properly elucidated only when the local structure is taken into account. The main properties will be pointed out in Section 2.3. In particular, the 'nonabelianness parameter' κ'_γ (multiplicity of the discrete factor R_γ equal to unity for an abelian sector and some natural $n > 1$ or ∞ for a nonabelian sector) turns out to be the parameter which determines the type of (para) statistics of the corresponding sector. At the global level, systems with nonabelian sectors exhibit the following property: these (and only these) systems must have other restrictions on their sets of observables, apart from superselection rules. Indeed, if superselection rules are the only restrictions on a global algebra R, then all Hermitian operators commuting with superselection operators are observables, i.e. $R = Z'$. But this means $R' = Z$, and all coherent sectors are abelian. There is nothing strange of exotic about the presence of such 'nonsuperselection' restrictions on a set of observables. As Wigner *et al.* [255] pointed out in 1965, "it is highly improbable that all restrictions on measurability of selfadjoint operators could be presented in the form of superselection rules". Here is, for instance, a large class of theories with nonabelian sectors. In models of quantum field theory observables are typically determined as physical quantities invariant under internal symmetries. In these cases there is a unitary representation $U(G)$ of the internal symmetry group G in the state space, and the global algebra R is defined as the commutant of this representation. $R = U(G)'$. Using the Peter–Weyl theorem one easily verifies that if G is a compact nonabelian group, R is an algebra of P-class and its central decomposition is a direct sum of nonabelian coherent sectors. Each of these sectors is the unitary equivalence class of some irreducible representation of G and its 'nonabelianness parameter' κ'_γ is equal to the dimension of this representation.

In conclusion let us summarize what can be said about the status of the global structure of P-class. While discussing continuous superselection rules we have already noted that abandoning this structure involves the loss of many properties present in the canonical picture of the superselection phenomenon: the notions of the relative phase between sectors and the superposition of vectors from different sectors disappear; structural conditions I'–II' cannot be formulated, etc. But still more important is that P-class is the only global structure for which the state space includes pure (vector or normal) states. As Lemma 2.1.3 shows, if there are any pure vector states, they always form a structure of P-class; the set of all pure vectors of a von Neumann algebra is either empty or total (or is a trivial combination of these cases, $\mathcal{H} = \mathcal{H}_p \oplus \mathcal{H}_m$). Hence the only alternative to the structure of P-class is offered by global structures for which the state space has no pure states at all. They are quite common, for example, in statistical mechanics, where global algebras of pure phases with nonzero temperature are factors of type III. They appear also in the description of infinite nonrelativistic systems by means of C^* or W^* dynamical systems (see [32] and references therein; continuous superselection rules with the direct integrals of superselection sectors are typical in this case). However, both in conventional quantum mechanics and quantum field theory their traditional physical interpretation seems to make unacceptable the idea of a state space without a single pure state. As (relatively) recently as in the book by Streater and Wightman [314], the term 'state' was used as synonymous to 'pure state'. Moreover, by Proposition 1.3.46 in local quantum theory with a cyclic vacuum subspace the global algebra is always of type I (though still it can have no pure vector states if its central decomposition is purely continuous). Thus, so far as the traditional physical interpretation of quantum theory is respected, algebras of P-class must be considered as the universal class of global algebras for relativistic quantum systems. But is must be added that the said interpretation is quite frequently extended or even rejected in modern theories. The study of new classes of quantum systems such as gauge fields, chiral and soliton models, models with phase transitions, often produces formalisms where the standard interpretation needs change and generalization. In all these cases adequacy (or non-adequacy) of the global structure of P-class is not immediately evident; analysis of vacuum structure in Section 3.1 will give us an example of this situation. Another example is provided by massless theories: their global algebras can have positive (but vacuumless, of course) representations of types II and III, which was proved both for free [162, 163, 148, 96] and interacting [106] systems. Physical implications of these results are not unexpected: in the presence of massless particles we have to revise the treatment of the superselection phenomenon, superposition principle and many elements of interpretation of quantum theory. This revision is yet to be developed, however.

2.2. Local Theory of Superselection Rules: Equivalence Properties of Coherent Sectors

As stressed above, the theory of superselection rules should be a starting point for introducing field objects and properties into the formalism of local observables*. Simple physical arguments display clearly the close connection between fields and superselection structures. Intuitively, we imagine a quantum field as a stream of quanta, which are elementary carriers of quantum numbers inherent to a relativistic quantum system, such as charge, spin, strangeness, baryon and lepton number, etc. If a state of the system is characterized by definite values of quantum numbers, then field quanta, acting on the

* Obviously, if we want to associate with a local net scalar fields carrying no nontrivial quantum numbers, then superselection rules are irrelevant. In this case, however, we endow local quantum theory not with new physical properties, but only with a new mathematical object, an operator-valued distribution $A(f)$. A certain method of introducing such an object was first suggested in [197] and then studied and generalized in [356, 435, 436]. The idea is to construct fields from observables by means of formulas of the type

$$A(f)\,\Phi = \operatorname*{s-lim}_{n\to\infty} \int T(-a) f(a)\,\Phi\, d^4 a\,, \quad f \in S(\mathbf{R}^4)\,,$$

where Φ are vectors from a certain linear dense domain in $\mathcal{H}$, $\{A_n\}$ a sequence of local observables independent of f and such that $A_n \in R(O_n)$, $O_1 \supset O_2 \supset \ldots, \bigcap_{n=1}^{\infty} O_n = \{0\}$. In such a way one can obtain fields $A(f)$, whose closures are affiliated to $R(O)$ [197] or local relative to $R(O)$ [435] when $\operatorname{supp} f \subset O$ and satisfy some regularity conditions. The form of the construction, with a sequence of regions O_n shrinking to a point, supports the modern view (see, e.g. [233, 336, 196]) that fields, being highly non-unique for a given local net, can be regarded as a *sui generis* coordinatization of the latter.

state, change these values. Quantum numbers are related to conservation laws, symmetries and also to superselection rules. In the last case the operator corresponding to the quantum number in question is a superselection operator and the set of all states with a given value of this quantum number forms a superselection sector. (In particular, superselecting quantum numbers include various charges, because the latter generate groups G of gradient transformations of fields, and, as shown above, representations $U(\hat{G})$ produce the superselection structure for the observable algebra defined as $R = U(G)'$). Clearly, in such a case the field will transfer states from one superselection sector into another (whereas observables leave the sectors invariant!) Hence, we obtain the first elementary conclusion: local quantum theory with dynamical superselection rules includes not only observables but also (though implicitly) field objects which must correspond to some operators in the state space mapping one superselection sector onto the other.

Certainly, not any such operator can be a field operator. Like observables, fields are local quantities related to spacetime regions and possessing both global and local (spacetime) properties. The behaviour of fields with respect to the superselection, i.e. global, structure of a quantum system belongs, obviously, to the category of global properties. As for the local properties of fields, they must include, above all, the main property of localization in spacetime regions (due to which the set of fields, like that of observables, must be the union of subsets corresponding to regions $O \in B(M)$), and then also commutation relations, asymptotic behaviour, singularities at small distances, etc. Hence the objects endowed with both global and local properties of fields should be related in a definite way not only with the global algebra R but also with the local algebras $R(O)$. A natural class of such objects is provided by intertwining operators of representations of the algebras $R(O)$ in coherent sectors. Thus our first conclusion takes the following, more precise form: in the theory of observables fields should arise as operators intertwining local algebras of different superselection sectors. This hypothesis put forward by Borchers in 1965 [77] became the central idea for the whole new stage of the algebraic approach. It also provides a new point of view on the notion of field. It was traditionally believed (basing on obvious group-theoretical considerations and the unified spinor field theory by Heisenberg) that if there are some primary objects of which all fields and particles in the Universe are constructed, these objects can only be spinor fields. Contrary to this, the new idea suggested constructing any fields from observables, that is objects with boson statistics, and thus opened up so far unsuspected possibility of obtaining fermions from bosons. (It should be noted, however, that this idea is not fully realized even now. It was successfully verified in some models as well as for systems of fields with normal Bose–Fermi statistics (see Section 2.3). In the simplest cases the fields are indeed obtained as intertwining operators between definite Hilbert space representations of local algebras, or, equivalently, between finite-dimensional representations of the compact internal symmetry group G. In the general case, however, the usual apparatus of intertwining operators proves to be insufficient. In a system with an arbitrary nonabelian compact group G fields turn out to be intertwining objects of a more general kind: they intertwine between certain endomorphisms of a C^*-algebra. As a result, their construction is tantamount of the development of a certain nontrivial generalization of the classical Tannaka–Krein duality theory for compact groups. Work on this generalization is not yet completed; see [158-161].)

In constructing field-theoretical properties one uses the abstract description of the superselection structure as a certain family of representations of a quasilocal algebra $\mathfrak{A}$*. It is simply connected with the presented above concrete description of this structure as the central decomposition of a global algebra R. If we have a local quantum theory $\langle R, \mathcal{H}\rangle$ with the superselection structure $\mathcal{H} = \oplus_{\gamma \in \sigma} \mathcal{H}_\gamma$, then every coherent sector $\mathcal{H}_\gamma$ can be defined as a (factorial type I) representation–restriction of the concrete C^*-algebra $\mathfrak{A}$:

$$\pi_\gamma : \mathfrak{A} \to \mathfrak{A}_\gamma \equiv \mathfrak{A} \,|_{\mathcal{H}_\gamma} .$$

Similarly, one can define the local representations

$$\pi_\gamma^{(O)} = \pi_\gamma \,|_{R(O)} : R\,(O) \to R\,(O)\,|_{\mathcal{H}_\gamma} .$$

Since it is possible for regions $O'\,(O \in B\,(M))$ that $R\,(O') \not\subset \mathfrak{A}$, a special family of representations must be associated with unbounded regions:

$$\pi_\gamma^{(O')} = \pi_\gamma \,|_{\mathfrak{A}(O')} : \mathfrak{A}\,(O') \to \mathfrak{A}\,(O')\,|_{\mathcal{H}_\gamma}, \quad \mathfrak{A}\,(O') = \overline{\bigvee_{\tilde{O} \subset O'} R\,(\tilde{O})} .$$

If the two representations $\pi_\alpha^{(O)}$ and $\pi_\beta^{(O)}$ (or $\pi_\alpha^{(O')}$ and $\pi_\beta^{(O')}$), $\alpha, \beta \in \sigma$, are quasi- or unitarily equivalent, then there exist intertwining operators $T_{\alpha\beta}^{(O)}$ (respectively, $T_{\alpha\beta}^{(O')}$) which map $\mathcal{H}_\beta$ onto $\mathcal{H}_\alpha$ and also possess the following local properties:

$$T_{\alpha\beta}^{(O)} P_\beta \in R\,(O)', \qquad T_{\alpha\beta}^{(O')} P_\beta \in R\,(O')' .$$

Hence it is these families of operators that are natural candidates on the role of field objects. The nearest task is, therefore, to find the conditions of the existence of such operators, that is to study equivalence properties of coherent sectors.

Three sets of representations were introduced and on each set there are three equivalence relations: weak or physical equivalence, quasi-equivalence and unitary equivalence, denoted respectively by the symbols $\sim$, $\approx$ and $\simeq$ (Section 1.1). The superselection structure is thus characterized by nine kinds of equivalence properties of coherent sectors.

DEFINITION 2.2.1. Coherent sectors $\langle R_\alpha, \mathcal{H}_\alpha\rangle$ and $\langle R_\beta, \mathcal{H}_\beta\rangle$ are called:

(1) Physically equivalent, if $\pi_\alpha \sim \pi_\beta$;

(2, 3) Locally (strongly locally) physically equivalent, if

* An abstract generalization of the definition of the coherent sector can be formulated in several equivalent ways. If the theory (as is often the case, see Section 2.3) is specified not by an algebra of P-class and not an abstract algebra $\mathfrak{A}$, but by the quasilocal algebra $\mathfrak{A}_0$ of a certain singled out coherent sector $< R_0, \mathcal{H}_0 >$, then other sectors can be considered as (factorial and type-I) representations of $\mathfrak{A}_0$. On the other hand, if a factorial type-I representation of $\mathfrak{A}$ or $\mathfrak{A}_0$ is given, it is the same thing as if the unitary equivalence class of an irreducible representation is given. Thus such an equivalence class can also be taken as an abstract definition of the coherent sector.

$$\pi_\alpha^{(O)} \sim \pi_\beta^{(O)} \ (\pi_\alpha^{(O')} \sim \pi_\beta^{(O')}) ;$$

(4) Quasi-equivalent, if $\pi_\alpha \approx \pi_\beta$;

(5, 6) Locally (strongly locally) quasi-equivalent, if

$$\pi_\alpha^{(O)} \approx \pi_\beta^{(O)} \ (\pi_\alpha^{(O')} \approx \pi_\beta^{(O')}) ;$$

(7) Unitarily equivalent, if $\pi_\alpha \simeq \pi_\beta$;

(8, 9) Locally (strongly locally) unitarily equivalent, if

$$\pi_\alpha^{(O)} \simeq \pi_\beta^{(O)} \ (\pi_\alpha^{(O')} \simeq \pi_\beta^{(O')}) .$$

The problem whether these properties follow from the axioms of local quantum theory is solved by means of the trivial counterexample. Let $\langle R_i , \mathcal{H}_i \rangle_{i \in \mathcal{J}}$ be a certain family of Haag–Araki theories in which global algebras are factors of type I. Then one can construct a new Haag–Araki theory $\langle R, \mathcal{H} \rangle$, putting , by definition,

$$R(O) = \bigoplus_{i \in \mathcal{J}} R(O)_i , \qquad \mathcal{H} = \bigoplus_{i \in \mathcal{J}} \mathcal{H}_i$$

for all $O \in B(M)$. This is a theory of P-class with the coherent sectors $\langle R_i , \mathcal{H}_i \rangle$ and such that the projections P_i onto $\mathcal{H}_i$ lie in $R(O) \cap R'$. Due to this any P_i is the central support of the representation $\pi_i^{(O)}$ as well as the representations $\pi_i^{(O')}$ and π_i , whence it follows that in all these families of representations pairwise quasi-equivalence and unitary equivalence are excluded. Moreover. $P_i \notin \ker \pi_i^{(O)}$, $P_i \in \ker \pi_j^{(O)}, \forall\, i \neq j$, and similarly for the families $\pi_i^{(O')}$ and π_i , so that physical equivalence is excluded as well. As a result, we have an example of a Haag–Araki theory with the superselection structure of P-class (we shall call such theories *local theories of P-class*) satisfying none of the properties 1–9.

The properties 1–9 are connected by the following obvious relations: $1 \to 3 \to 2$, $4 \to 6 \to 5, 7 \to 9 \to 8, 7 \to 4 \to 1, 8 \to 5 \to 2, 9 \to 6 \to 3$. Besides, it is evident that the global properties 4 and 7 cannot hold since the central supports of the representations π_α and π_β (coinciding with P_α and P_β , respectively) are mutually orthogonal. Of the remaining properties, the only ones which do not follow from others, are 1 and 9, physical equivalence and strong local unitary equivalence. We turn to their analysis.

To study physical equivalence of the sectors we need information on the structure of ideals of the quasilocal algebra $\mathfrak{A}$ in local theory of P-class. In the general Haag-Araki theory the main properties of this structure are established by the following theorem due to Borchers.

THEOREM 2.2.1 [79]. *Let Axioms* I–III, V, VI *be valid and let* $\mathcal{J}$ *be a norm-closed two-sided ideal of* $\mathfrak{A}$. *Then*

(1) $\mathcal{J} \neq \{0\}$ *if and only if* $\mathcal{J} \cap Z_{\mathfrak{A}} \neq \{0\}$, *where* $Z_{\mathfrak{A}} = \mathfrak{A} \cap \mathfrak{A}'$ *is the centre of* $\mathfrak{A}$;

(2) *$\mathcal{J}$ is generated by the set $\mathcal{J} \cap Z_{\mathfrak{A}}$ i.e. $\mathcal{J} = \mathfrak{A}\,(\mathcal{J} \cap Z_{\mathfrak{A}})$ and is the smallest norm-closed two-sided ideal including $\mathcal{J} \cap Z_{\mathfrak{A}}$;*

(3) *the mapping $\mathcal{J} \to \mathcal{J} \cap Z_{\mathfrak{A}}$ is a one-to-one correspondence between norm-closed two-sides ideals of $\mathfrak{A}$ and ideals of $Z_{\mathfrak{A}}$.*

Proof relies on the following property of ideals of a C^*-algebra generated by a net of its W^*-subalgebras.

LEMMA 2.2.2. *Let $\mathfrak{A}$ be a* C**-algebra in a Hilbert space $\mathcal{H}$ which is the C^*-inductive limit of a sequence $\{R_i\}_{i=1}^{\infty}$ of von Neumann algebras:*

$$R_i \subset R_{i+1} \subset \dots \subset \mathfrak{A}\,, \quad \mathfrak{A} = \overline{\bigcup_{i=1}^{\infty} R_i}^{\,u} \quad .$$

Let also $\mathcal{J}$ be a norm-closed two-sided ideal of $\mathfrak{A}$. Then beginning from a certain n, $\mathcal{J} \cap R_n \neq \{0\}$. Moreover, $\overline{\mathcal{J} \cap (\cup_{i=1}^{\infty} R_i)}^{\,u} = \mathcal{J}$.

Proof of the Lemma. Let $A = A^* \in \mathcal{J}$, $\| A \| = 1$. Then there is a number n for which R_n includes an operator B such that $B = B^*$, $\| A - B \| \leq 1/8$. Approximating the spectral decomposition $B = \int_{-N}^{N} \lambda \, d E_\lambda$, $E_\lambda \in R_n$, we find that there exists a set of pairwise orthogonal spectral projections $E_k \in R_n$, $k = -4, -3, \dots, 3, 4$, such that

$$\left\| B - \sum_{k=-4}^{4} (k/4)\, E_k \right\| \leq 1/8 .$$

Hence taking account of $\| A - B \| \leq 1/8$ one has

$$\left\| A - \frac{1}{4} \sum_{k=-4}^{4} k\, E_k \right\| \leq 1/4 . \tag{2.1}$$

Since $\| A \| = 1$, not all E_k with $| k | \geq 3$ are equal to zero. Take now some faithful representation Π of $\mathfrak{A}/\mathcal{J}$ and denote by $\hat{T}$ the element in $\mathfrak{A}/\mathcal{J}$ corresponding to $T \in \mathfrak{A}$. Then $A \in \mathcal{J}$ entails $\hat{A} = \Pi\,(\hat{A}) = 0$, and so one has from (2.1):

$$\left\| \sum_{k=-4}^{4} k\, \Pi(\hat{E}_k) \right\| \leq 1 .$$

Since mutual orthogonality of E_k implies $\Pi\,(\hat{E}_k)\, \Pi\,(E_\ell) = \delta_{k\ell}\, \Pi\,(E_k)$ one has $\Pi\,(\hat{E}_k) = 0$ for $| k | > 2$. Π being faithful, this gives $\hat{E}_k = 0$. Hence $E_k \subset \mathcal{J} \cap R_n$ and $\mathcal{J} \cap R_n \neq \{0\}$.

To prove the last assertion note that the above argument implies the following property: for $A = A^* \in \mathcal{J}$ and any $\varepsilon > 0$ there is a number n such that one can find in R_n operators B, B_1 for which

$$\| A - B_1 \| \leq \varepsilon , \qquad \| B - B_1 \| \leq \varepsilon , \qquad B \in \mathcal{J} \cap R_n .$$

This implies that $A \in \overline{\mathcal{J} \cap (\cup_{i=1}^{\infty} R_i)}^{u}$. ❐

By virtue of the lemma one has for any norm-closed two-sided ideal $\mathcal{J}$ of $\mathfrak{A}$:

$$\mathcal{J} = \overline{\mathcal{J} \cap \mathcal{A}}^{u} . \tag{2.2}$$

Moreover, for some $O \in B(M)$ there is a $A = A^* = \int_{-M}^{M} \lambda \, dE_\lambda \in \mathcal{J} \cap R(O)$. Since $\mathcal{J} \cap R(O)$ is a two-sided ideal in $R(O)$, for any interval $\Delta \subset [-M, M]$ not including the origin one has

$$A E_\Delta = \int_\Delta \lambda \, d E_\lambda \in \mathcal{J} \cap R(O),$$

$$E_\Delta = (A E_\Delta) \int_\Delta \lambda^{-1} \, d E_\lambda \in \mathcal{J} \cap R(O). \tag{2.3}$$

Next, note the obvious property

$$E \in \mathcal{J} \cap R(O), \quad E \sim F \bmod R(O) \Rightarrow F \in \mathcal{J} \cap R(O). \tag{2.4}$$

By Proposition 1.3.11, for any region $O_1 >> O$ and any projection $E \in R(O)$ one can find a projection F such that

$$F \sim E \bmod R(O_1), \; F \in R(O_1) \cap R'.$$

Hence we obtain, taking account of (2.4), that the existence of a nontrivial projection $E_\Delta \in R(O) \cap \mathcal{J}$ implies the existence of a nontrivial projection $F \in \mathcal{J} \cap R(O_1) \cap R' \subset \mathcal{J} \cap Z_{\mathfrak{A}}$ which proves assertion 1.

To prove assertion 2 let us show first that the ideal $\mathcal{J} \cap R(O)$ in generated by the set $\mathcal{J}(O)^P$ of its projections. Indeed, due to (2.3) one has for any $\varepsilon > 0$, $\varepsilon < M$:

$$E_\varepsilon \equiv E_{(-M,-\varepsilon)} + (I - E_{(-M,-\varepsilon)}) \in \mathcal{J}(O)^P,$$

and AE_ε lies in the ideal generated by $\mathcal{J}(O)^P$. But $\| A - AE_\varepsilon \| \le \varepsilon$, and hence the closure of this ideal coincides with $\mathcal{J} \cap R(O)$. Owing to (2.2) this implies that $\mathcal{J}$ is generated by the set $\cup_{O \in B(M)} \mathcal{J}(O)^P$. Let now $\mathfrak{I}$ be the two-sided norm-closed ideal of $\mathfrak{A}$ generated by the ideal $\mathcal{J} \cap Z_{\mathfrak{A}}$ of $Z_{\mathfrak{A}}$. By definition, $\mathfrak{I} \subset \mathcal{J}$. As proved above, for every $E \in \mathcal{J}(O)^P$ there is a projection F equivalent to E modulo some $R(O_1)$, $O_1 >> O$, and lying in $\mathcal{J} \cap Z_{\mathfrak{A}}$. Clearly, this projection lies in $\mathfrak{I}$, whence $E \in \mathfrak{I}$ and $\cup_{O \in B(M)} \mathcal{J}(O)^P \subset \mathfrak{I}$. Since $\cup_{O \in B(M)} \mathcal{J}(O)^P$ generates $\mathcal{J}$, we obtain $\mathcal{J} \subset \mathfrak{I}$ and, therefore $\mathcal{J} = \mathfrak{I}$.

Finally, 3 can be obtained from 2 using that $\mathfrak{A}$ and $Z_{\mathfrak{A}}$ commute. If $\mathfrak{N}$ is an ideal in $Z_{\mathfrak{A}}$, then it generates in $\mathfrak{A}$ the two-sided norm-closed ideal $\mathfrak{A}\,\mathfrak{N}$. But due to 2 any such ideal $\mathcal{J}$ is of the form $\mathfrak{A}\,\mathfrak{N}$ with $\mathfrak{N} = \mathcal{J} \cap Z_{\mathfrak{A}}$, which implies that the mapping $\mathfrak{N} \to \mathfrak{A}\,\mathfrak{N}$ is surjective. Conversely, any ideal $\mathfrak{N}$ of $Z_{\mathfrak{A}}$ can be represented as $\mathfrak{N} = \mathcal{J} \cap Z_{\mathfrak{A}}$ where $\mathcal{J} = \mathfrak{A}\,\mathfrak{N}$ is an ideal in $\mathfrak{A}$. Hence, the mapping $\mathcal{J} \to \mathcal{J} \cap Z_{\mathfrak{A}}$ is also surjective. ❐

COROLLARY 1. *$\mathcal{J}$ is a maximal two-sided ideal in $\mathfrak{A}$ if and only if $\mathcal{J} \cap Z_{\mathfrak{A}}$ is a maximal ideal in $Z_{\mathfrak{A}}$.*

Proof follows from assertion 3.

COROLLARY 2. *In an arbitrary Haag–Araki theory the following conditions are equivalent:*

(1) *quasilocal algebra $\mathfrak{A}$ is simple;*

(2) $Z_{\mathfrak{A}} = C(\mathcal{H})$

and are implied by the condition

(3) *global algebra R is a factor.*

For the proof it is sufficient to recall that $C(\mathcal{H})$ is the only simple abelian C^*-algebra of $\mathcal{B}(\mathcal{H})$. As we shall see later, implication $2 \to 3$ is, indeed, not valid. It is violated (i.e. $Z_{\mathfrak{A}} = C(\mathcal{H}) \neq Z$), in particular, in any local theory of P-class having a nontrivial set of physically equivalent coherent sectors.

COROLLARY 3. *Let $\mathcal{J}$ be a nontrivial two-sided ideal of the algebra $\mathcal{A}$ of all local observables. Then $\mathcal{J}$ contains an nontrivial projection from the centre $\mathcal{A} \cap R'$ of this algebra.*

Proof is similar to that of assertion 1.

In local theory of P–class ideals of $\mathfrak{A}$ are closely connected with the physical equivalence classes of coherent sectors. In such a theory $\langle R, \mathcal{H} = \oplus_{\alpha \in \sigma} \mathcal{H}_\alpha \rangle$ the set σ of its coherent sectors splits uniquely into the subsets σ_i corresponding to classes of physical equivalence:

$$\sigma = \bigcup_{i \in \Delta} \sigma_i , \qquad \forall_{\alpha, \beta \in \sigma_i} \quad \ker \pi_\alpha = \ker \pi_\beta .$$

Denoting the projection onto the subspace $\mathcal{H}_i = \oplus_{\alpha \in \sigma_i} \mathcal{H}_i$ by Q_i, let us prove

PROPOSITION 2.2.3 [409]. *In any local theory of P-class the quasilocal algebra and its centre have the following form*

$$\mathfrak{A} = \bigoplus_{i \in \Delta} \mathfrak{A} Q_i , \qquad Z_{\mathfrak{A}} = \bigoplus_{i \in \Delta} C(\mathcal{H}_i) .$$

Moreover, the kernel of any representation π_α is a maximal two-sided ideal of $\mathfrak{A}$ having the form

$$\ker \pi_\alpha = (I - Q_i)\, \mathfrak{A} . \tag{2.5}$$

Proof. Assume that $Z_{\mathfrak{A}}$ contains at least one nontrivial projection F. Since $Z_{\mathfrak{A}} \subset Z$, one has $F = \Sigma_{\alpha \in \sigma(F)} P_\alpha$, where $\sigma(F)$ is such that

$$\forall_{\alpha \in \sigma(F)} \quad F \notin \ker \pi_\alpha , \qquad \forall_{\alpha \notin \sigma(F)} \quad F \in \ker \pi_\alpha .$$

Hence, $\sigma_i \cap \sigma(F) \neq \emptyset$ entails $\sigma_i \subset \sigma(F)$, so that

$$F = \sum_{i \in \Delta_F \subset \Delta} Q_i \ . \tag{2.6}$$

Now, let us associate with each $i \in \Delta$ the set

$$\mathcal{F}_i = \{ F \in Z_{\mathfrak{A}} \mid F \geq Q_i \},$$

which is always nonempty. It is a partially ordered set (with respect to the usual order relation for projections) in which any linearly ordered subset has the lower bound, according to (2.6). Hence by the Zorn Lemma, $\mathcal{F}_i$ has the minimal element $F_{m,i}$. One easily verifies that $F_{m,i}$ is a minimal projection in $Z_{\mathfrak{A}}$ and two minimal projections $F_{m,i}$ and $F_{m,k}$ are either equal or orthogonal to each other. Due to (2.6), the set of all such projections forms a partition of identity: $I = \Sigma_{i \in \Delta} Q_i = \Sigma_{i \in \Delta} F_{m,i}$. Obviously, the set $F = F_{m,i} \; Z_{\mathfrak{A}}$ is a minimal ideal in $Z_{\mathfrak{A}}$ (otherwise there exists an ideal $M \subset M_i$ and, by Theorem 2.2.1, it includes a projection $F < F_{m,i}$, which is impossible). Then $N_i = (I - F_{m,i}) \, Z_{\mathfrak{A}}$ is a maximal ideal in $Z_{\mathfrak{A}}$. (Otherwise there exists an ideal $N \supset N_i$ such that $F_{m,i} \neq 0$ (to exclude $N = N_i$). This implies $F_{m,i} N \subset N \cap M_i \neq \{0\}$ which is impossible, the ideal M_i being minimal.) Hence it follows due to Corollary 1 to Theorem 2.2.1 that $\mathcal{J} = (I - F_{m,i}) \, \mathfrak{A}$ is a maximal two-sided ideal in $\mathfrak{A}$, and, moreover, $\mathcal{J} P_\alpha = 0$ for all $\alpha \in \sigma(F_{m,i})$, i.e. $\mathcal{J} \subset \ker \pi_\alpha$. Since $\ker \pi_\alpha$ is a two-sided ideal in $\mathfrak{A}$, this implies $\mathcal{J} = \ker \pi_\alpha$ (and so $\ker \pi_\alpha$ is a maximal ideal). Therefore, we obtained that $\ker \pi_\alpha = \ker \pi_\beta = \mathcal{J}$ for all $\alpha, \beta \in \sigma(F_{m,i})$ and hence, for the corresponding $i \in \Delta$ one has $\sigma(F_{m,i}) \subset \sigma_i$, i.e. $F_{m,i} < Q_i$. Due to (2.6) this entails $F_{m,i} = Q_i$, and the second assertion of the proposition is proved.

At the same time, we have obtained $Q_i \in Z_{\mathfrak{A}}$ whence it follows that $\oplus_{i \in \Delta} C(\mathcal{H}_i) \subset Z_{\mathfrak{A}}$. To prove the inverse inclusion it suffices to show that

$$Z_{\mathfrak{A}} \big|_{\mathcal{H}_i} = C(\mathcal{H}_i)$$

for all $i \in \Delta$ (and then apply the trivial fact: for any *-subalgebra $\mathcal{B}$ of $\mathcal{B}(\mathcal{H})$ and any partition of identity $\{E_k\}_{k \in \mathcal{K}}$ in $\mathcal{B}'$ one has

$$\mathcal{B} \subset \bigoplus_{k \in \mathcal{K}} \mathcal{B}_k \, , \qquad \mathcal{B}_k \equiv \mathcal{B}\big|_{E_k \mathcal{H}} \, .$$

Let us consider the local P-class theory $\{R_{Q_i}, \mathcal{H} = \oplus_{\alpha \in \sigma_i} \mathcal{H}_\alpha\}$ in $\mathcal{H}_i$. All its sectors are physically equivalent. Moreover, the representations $\overset{(i)}{\pi}_\alpha$ of $\mathfrak{A}$ corresponding to these sectors are faithful since the formula (2.5) (already proved) implies that $\ker \overset{(i)}{\pi}_\alpha = \{0\}$. But $\ker \overset{(i)}{\pi}_\alpha$ is a maximal two-sided ideal in $\mathfrak{A}|_{\mathcal{H}_i}$ and hence, the algebra is simple. Corollary 2 in Theorem 2.2.1 implies then $Z_{\mathfrak{A}|_{\mathcal{H}_i}} = (Z_{\mathfrak{A}})|_{\mathcal{H}_i} = C(\mathcal{H}_i)$, and the proof is over since in the absence of nontrivial projections in $Z_{\mathfrak{A}}$ the proposition holds trivially. ❐

The important case when all coherent sectors are physically equivalent deserves to be considered in more detail.

THEOREM 2.2.4 [409]. *In local theory of P-class the following conditions are equivalent:*

(1) *all coherent sectors* $\langle R_\alpha, \mathcal{H}_\alpha \rangle$ *are physically equivalent;*

(2) $Z_{\mathfrak{A}} = C(\mathcal{H})$;

(3) *the algebra* $\mathfrak{A}$ *is simple;*

(4) *at least one of the representations* π_α *of* $\mathfrak{A}$ *is faithful;*

(5) *the algebra* $\mathcal{A}$ *of all local observables is simple;*

(6) *for all double cones* D *the local algebras are factors;*

(7) *extended locality holds: for all double cones* D ,

$$R(D) \cap R(D') = C(\mathcal{H}) .$$

Proof. Proposition 2.23 makes evident the relations $1 \leftrightarrow 2$ and $1 \leftrightarrow 4$; $2 \leftrightarrow 3$ is a part of Corollary 2 to Theorem 2.2.1; $2 \rightarrow 5$ follows from Corollary 3 to Theorem 2.2.1, and $5 \rightarrow 3$ from Lemma 2.2.2. $2 \rightarrow 6$ is obvious from $Z(D) \subset Z_{\mathfrak{A}}$ (see Proposition 1.3.6). $6 \rightarrow 7$ is trivial. $7 \rightarrow 5$: if $\mathcal{A}$ is not simple, then, by Corollary 3 to Theorem 2.2.1, $\mathcal{A} \cap R'$ has a nontrivial projection F lying in $R(D)$ for some double cone D. From $T(a) \in R$ and $F \in R(D) \cap R'$ one has for sufficiently large spacelike translations a :

$$F(a) = F \in R(D+a) \subset R(D') ,$$

whence $F \in R(D) \cap R(D') \neq C(\mathcal{H})$. ❐

Notice that we used for $2 \rightarrow 6$ the continuous isotony (Axiom Ia) not included usually in the list of axioms. Furthermore, condition 2 has a special physical meaning: since $Z_{\mathfrak{A}} = \mathfrak{A} \cap Z$, this condition is nothing else than the statement of 'global nature of superselection rules' : superselection operators and their spectral projections cannot lie in the quasilocal algebra, and so they are purely global observables. As pointed out in Section 2.1, this property is dictated by the physical interpretation of observables from Z as macroscopic quantities. Thus the relation $1 \leftrightarrow 2$ is an additional argument in favour of the requirement of physical equivalence. Finally, the fact that the conditions 1–7 include the factoriality of $R(D)$ but not $R(D')$ is due to the absence of the inclusion $Z(D') \subset Z_{\mathfrak{A}}$ and is just the first in the series of differences between the algebras $R(D)$ and $R(D')$.

Let us turn to the quasi-equivalence properties. To describe them means to find the algebraic structure produced by splitting of σ into the classes of local and strong local quasi-equivalence. We shall consider these properties for the representations $\pi_\alpha^{(\tilde{D})}$ with $\tilde{D} = D$ or D', because of factoriality of the latter which follows from Proposition 1.3.6 if the continuous isotony is assumed. We must also take into account the difference between the properties of quasi-equivalence on a separate region $\tilde{D}$ and on all such regions; the last property will be called the *full* (*local or strong local*) *quasi-equivalence*. Using the factoriality of all $\pi_\alpha^{(\tilde{D})}$ it is easy to show [409, Proposition 5] that the partition of identity $\{Q_i^{(\tilde{D})}\}_{i\in\Delta(\tilde{D})}$ formed by the projections $Q_i^{(\tilde{D})}$ on classes of local (or strong local) quasi-equivalence lies in the centre of $R(\tilde{D})$. Hence it follows that

$$R(\tilde{D}) = \bigoplus_{i\in\Delta(\tilde{D})} R_i(\tilde{D}), \qquad Z(\tilde{D}) = \bigoplus_{i\in\Delta(\tilde{D})} C(\mathcal{H}_i). \tag{2.7}$$

In its turn, this representation implies the following proposition:

PROPOSITION 2.2.5. *In local theory of P-class the following conditions are equivalent:*

(1) *all coherent sectors are quasi-equivalent on a region* $\tilde{D}$;

(2) *local algebra of the region* $\tilde{D}$ *is a factor;*

(3) *in some sector* $\mathcal{H}_\alpha$ *there is a vector* Φ_α *analytic in the energy and separating for* $R(\tilde{D})$;

and are implied by the condition:

(4) *in some sector* $\mathcal{H}_\alpha$ *there is a vector* Φ_α *the* $\mathcal{H}$*-image of which (as a state on* $R(\tilde{D})$*) intersects nontrivially with any other sector.*

Proof. $1 \leftrightarrow 2$ follows from (2.7) and $3 \to 2$ from the fact that Φ_α is cyclic for $Z(\tilde{D})'$ and belongs, due to (2.7), to an irreducible subspace of $Z(\tilde{D})'$. $2 \to 3$: for any $\Phi \in \mathfrak{N}$ one has

$$\mathcal{H}_\Phi^{R(\tilde{D})} = \mathcal{H}_\Phi^{R} \subset \mathcal{H}_\Phi^{R(\tilde{D})'},$$

whence

$$\mathcal{H}_\Phi^{Z(\tilde{D})'} = \overline{L\left\{R(\tilde{D})'\,\mathcal{H}_\Phi^{R(\tilde{D})}\right\}} = \mathcal{H}_\Phi^{R(\tilde{D})'}.$$

<u>$4 \to 1$</u>. For any $\Phi_\beta \in \mathfrak{M}_{\Phi_\alpha}(\tilde{D})$ one has

$$\pi_\beta^{(\tilde{D})} \approx \pi_{\Phi_\beta}^{(\tilde{D})} \simeq \pi_{\Phi_\alpha}^{(\tilde{D})} \approx \pi_\alpha^{(\tilde{D})},$$

where $\pi_{\Phi_{\alpha,\beta}}^{(\tilde{D})}$ are the representations of $R(\tilde{D})$ generated by the functionals $\omega_{\Phi_{\alpha,\beta}}^{(\tilde{D})}$ on $R(\tilde{D})$. ❐

The last two conditions of this proposition formulated in terms of the local $\mathcal{H}$-image $\mathfrak{M}_\Phi(\tilde{D})$ (recall that by (1.4) $\mathcal{H}_\Phi^{R(\tilde{D})'} = \overline{L\{\mathfrak{M}_\Phi(\tilde{D})\}}$) do not look too useful or physically meaningful. They become, however, of more interest in the important case when the set of coherent sectors includes a vacuum sector $\mathcal{H}_0$ with a vacuum state ω_Ω, $\Omega \in \mathcal{H}_0$ (by Proposition 1.3.47, such a state is unique in $\mathcal{H}_0$ and is pure). In this case the local $\mathcal{H}$-image $\mathfrak{M}_\Omega(\tilde{D})$ has a special physical interpretation.

DEFINITION 2.2.2 [285, 306]. A vector $\Phi \in \mathcal{H}$ and a state ω_Φ on R' are called strictly localized in a region $O \subset M$ if ω_Φ coincides with a vacuum state* on the local

* If a theory has different vacuum states ω_{Ω_i}, then each of these states generates its own set of

algebra of the region O':

$$\omega_\Phi |_{R(O')} = \omega_\Omega |_{R(O')} .$$

Let us denote by $SL(O)$ the set of all vectors strictly localized in O and put, by definition, $SL = \cup_{O \subset M} SL(O)$, $\mathcal{H}(O) = \overline{L\{SL(O)\}}$. For a local P-class theory $\langle R, \mathcal{H} = \oplus_{\alpha \in \sigma} \mathcal{H}_\alpha \rangle$ put also $SL(O)_\alpha = SL(O) \cap \mathcal{H}_\alpha$, $\mathcal{H}(O)_\alpha = \overline{L\{SL(O)_\alpha\}}$. It is immediately evident that for any $O \subset M$,

$$SL(O) = \mathfrak{M}_\Omega(O'), \quad \mathcal{H}(O) = \mathcal{H}_\Omega^{R(O')'} \tag{2.8}$$

and taking into account that $\Omega \in \mathfrak{N}$ we obtain:

PROPOSITION 2.2.6. *In local theory of P-class with a vacuum sector* $\langle R_0, \mathcal{H}_0 \rangle$ *the following conditions are equivalent:*

(1) *all coherent sectors are quasi-equivalent on a region* D';

(2) *the local algebra* $R(D')$ *is a factor;*

(3) $\mathcal{H}(D) = \mathcal{H}$,

and are implied by the condition

(4) *each sector contains at least one vector strictly localized in* D:

$$\forall_{\alpha \in \sigma} \; SL(D)_\alpha \neq \emptyset .$$

Thus the presence of a vacuum and strictly localized states induces certain differences between the properties of local and strong local quasi-equivalence. Similar differences take place between the properties of full local and full strong local quasi-equivalence.

The study of full quasi-equivalence properties represents a simple problem of the theory of nets of C^*-algebras developed by Haag *et al.* [234]. The problem is essentially to find out which subalgebra of $Z_{\mathfrak{A}}$ determines the decomposition into the classes of full local and strong local quasi-equivalence, in the same sense as $Z_{\mathfrak{A}}$ determines the decomposition into the classes of physical equivalence and the subalgebra $Z(\tilde{D}) \subset Z_{\mathfrak{A}}$ into classes of $\tilde{D}$-quasi-equivalence. Since it is possible that $Z(\tilde{D}_1) \neq Z(\tilde{D}_2)$ for $\tilde{D}_1 \neq \tilde{D}_2$, two different sets of classes arise, those of full local (strong local) quasi-equivalence and disjointness. They are defined as follows: in the former case sectors $\mathcal{H}_\alpha$ and $\mathcal{H}_\beta$ belong to the same class if $\pi_\alpha^{(\tilde{D})} \approx \pi_\beta^{(\tilde{D})}$ for all $\tilde{D}$, while in the latter case sectors $\mathcal{H}_\alpha$ and $\mathcal{H}_\beta$ belong to the same class if $\pi_\alpha^{(\tilde{D})} \approx \pi_\beta^{(\tilde{D})}$ for at least one $\tilde{D}$. The solution of the problem is simple to obtain (see [409, Section 2]). It can be directly checked that the partition into the classes of full local (strong local) quasi–equivalence is determined by the union of all the centres $Z(\tilde{D})$, i.e. by the W^*-algebras $Z^{\approx} = \{\cup_{\tilde{D} \subset M} Z(\tilde{D})\}$; whereas the partition into the classes of full local (strong local) disjointness is determined

---

strictly localized vectors. The sets corresponding to different ω_{Ω_i}, do not intersect and our further statements are valid for any of these sets.

by the intersection of all these centres, the W^*-algebra $\mathcal{Z}^{\flat} = \cap_{\tilde{D}\subset M} \mathcal{Z}(\tilde{D})$, which in the case $\tilde{D} = D'$ coincides with the algebra of asymptotic observables $\mathcal{Z}_{as} = \cap_{D\subset M} R\,(D')$ considered in Section 1.3.5. (In other words, the classes in question coincide with irreducible subspaces of $\mathcal{Z}^{\approx}$ and $\mathcal{Z}^{\flat}$, respectively.) Hence it follows, in particular, that all coherent sectors are in the same class of full strong local disjointness if and only if the algebra of asymptotic observables is trivial:

$$\forall_{\alpha,\beta\in\sigma} \quad \exists_{D\subset M} \quad \pi_{\alpha}^{(D')} \approx \pi_{\beta}^{(D')} \Leftrightarrow \mathcal{Z}^{\flat} = \bigcap_{D\subset M} R\,(D') = C(\mathcal{H}) .$$

On the other hand, due to $R\,(D) \subset \mathfrak{A}$, full local quasi-equivalence proves to be equivalent to physical equivalence: for any $\alpha, \beta \in \sigma$,

$$\forall_{D\subset M} \quad \pi_{\alpha}^{(D)} \approx \pi_{\beta}^{(D)} \Leftrightarrow \pi_{\alpha} \sim \pi_{\beta} , \tag{2.9}$$

which follows directly from Proposition 2.2.5 and condition 6 of Theorem 2.2.4. Clearly, this coincidence of equivalence classes means also the coincidence of the corresponding central subalgebras:

$$\mathcal{Z}^{\approx} = \left\{ \bigcup_{D\subset M} \mathcal{Z}\,(D) \right\}'' = \mathcal{Z}_{\mathfrak{A}} .$$

(2.9) implies also that full local quasi-equivalence of all coherent sectors is equivalent to all the conditions of Theorem 2.2.4, including the simplicity of $\mathfrak{A}$, extended locality and the global nature of superselection rules.

Finally, the properties of local (strong local) unitary equivalence are the direct enhancements of the corresponding quasi-equivalence properties. Therefore, criteria of their fulfillment consist of two parts:

(1) conditions of local (strong local) quasi-equivalence, and

(2) conditions under which quasi-equivalence entails unitary equivalence.

It is a natural division since conditions 1 determine when the algebras $R\,(\tilde{D})_{\alpha}$ and $R\,(\tilde{D})_{\beta}$ are isomorphic whereas conditions 2 determine when this isomorphism is spatial. Let us consider conditions 2. For arbitrary von Neumann algebras no necessary and sufficient conditions are known for their isomorphisms to be spatial, but only some necessary and some sufficient conditions. Nevertheless in the case of local algebras $R\,(\tilde{D})_{\alpha}$ in the theory of P-class such conditions can easily be found. The well known sufficient conditions under which an isomorphism of W^*-algebras is spatial concern the existence of bicyclic vectors for these algebras [142, p. 222], or the cardinality of partitions of identity in their commutants [142, p. 301]. In the case of local algebras $R\,(D)_{\alpha}$ such conditions, due to the Reeh–Schlieder theorem, are actually conditions on the global structure. Owing to this, they turn out to be not only sufficient, but necessary as well. As a result, we obtain the necessary and sufficient conditions of the transition from quasi-equivalence to unitary equivalence. They concern only global parameters of the theory (dimensions of its sectors $\mathcal{H}_{\alpha}$ and multiplicities κ'_{α} of the discrete factors R_{α})

and hence, they are the same for both the local and strong local properties. They are as follows:

(a) if $\kappa'_\alpha \le \aleph_0$, then $\kappa'_\beta \le \aleph_0$ and $\dim \mathcal{H}_\alpha = \dim \mathcal{H}_\beta$, (2.10)

(b) if $\kappa'_\alpha > \aleph_0$, then $\kappa'_\beta = \kappa'_\alpha$ and $\dim \mathcal{H}_\alpha = \dim \mathcal{H}_\beta$.

The derivation of (2.10) is straightforward [409, Theorem 3]: (b) is obtained from the condition [142, p. 301], whereas (a) from the condition [142, p. 222] and the following fact which is a simple corollary of Proposition 1.3.4: in local theory of P-class, sector algebras $R(D)_\alpha$ with $O' \ne \emptyset$ have bicyclic vectors if and only if $\kappa'_\alpha \le \aleph_0$.

As for conditions 1, they were considered above. As a result, the criteria of local unitary equivalence are given by Proposition 2.2.5 where we replace $\approx$ by $\simeq$ in condition 1 and add the conditions (2.10) to each of the conditions 2–4. Note that 4 becomes now equivalent to 1–3 because $\pi_\alpha^{(D)} \simeq \pi_\beta^{(D)}$ means the presence of an isometric intertwining operator which leaves all $\mathcal{H}$-images $\mathfrak{M}_\Phi^{(D)}$ invariant, according to (1.3.43). The criteria of strong local unitary equivalence are derived similarly, but in this case Proposition 2.2.6 provides an additional group of such criteria in terms of strictly localized vectors. Moreover, taking into account that

$$\mathcal{H}_0 = \overline{L\,\{R'\,\mathcal{H}_\Omega^R\}} = \overline{L\,\{R'\,\mathcal{H}_\Omega^{R(D)}\}} \subset \mathcal{H}_\Omega^{R(D')'}$$

implies $\mathcal{H}(D)_0 = \mathcal{H}_0$, and isometric operators leave total sets total, we find that condition 3 of Proposition 2.2.6 is enhanced and can be formulated now as the totality of each set $SL(D)_\alpha$ in the corresponding sector. As a result, we obtain the following proposition.

PROPOSITION 2.2.7. *In local theory of P-class with a vacuum vector* $\langle R_0, \mathcal{H}_0 \rangle$ *the following conditions are equivalent:*

(1) *all coherent sectors are unitarily equivalent on a region* D';

(2) *each sector contains a total set of vectors strictly localized in* D *and satisfies* (2.10) *with respect to the vacuum sector;*

(3) *each sector contains at least one vector strictly localized in* D *and satisfies* (2.10) *with respect to the vacuum sector.*

It can be easily verified that the presence of a total set of strictly localized vectors and hence, the property of strong local unitary equivalence imply definite restrictions on asymptotic behaviour of vector states.

PROPOSITION 2.2.8. *Let local theory of P-class have a vacuum sector and all the coherent sectors be strongly locally unitarily equivalent. Then any vector state* ω_Φ *enjoys the following asymptotic properties:*

(1) $$\lim_{\substack{|a| \to \infty \\ a^2 < 0}} \omega_\Phi(A(a)) = \omega_\Omega(A),$$

where ω_Ω *is a vacuum state;*

(2) $$\lim_{n \to \infty} \left\| \omega_\Phi^{(D'_n)} - \omega_\Omega^{(D'_n)} \right\| = 0 ,$$

where $\{D_n\}_{n=1}^{\infty}$ *is a sequence of infinitely expanding double cones:* $D_n \subset D_{n+1}$, $\cup_{n=1}^{\infty} D_n = M$;

(3) $$\lim_{\substack{|a| \to \infty \\ a^2 < 0}} \omega_\Phi (A(a) B) = \omega_\Omega (A)\, \omega_\Phi (B)$$

(generalized cluster property).

Proof. Properties 1–3 are the enhancements of the asymptotic and cluster properties of general Haag–Araki theory (see Theorem 1.3.42) which are derived easily using the presence in each sector of at least one strictly localized state. ❒

COROLLARY. *Local theory of P-class in which all the coherent sectors are strongly locally unitarily equivalent has not more than one vacuum state.*

Proof. Assuming that there exist two vacuum states ω_{Ω_1} and ω_{Ω_2}, we immediately derive from property 1 that $\omega_{\Omega_1} = \omega_{\Omega_2}$. ❒

Thus, the last and strongest type of equivalence of coherent sectors, that is strong local unitary equivalence, singles out the class of systems with a definite asymptotic behaviour and, more precisely, with the absence of long-range forces (because at large distances all states coincide with a vacuum). Accordingly, this equivalence property is typical of strong interaction theories and breaks down, for example, in quantum electrodynamics. Indeed, the electromagnetic field is a gauge field which obeys the Gauss law. For all such fields, the flux of the field through a sphere of a sufficiently large radius is not equal to zero in any state containing field sources, whereas in a vacuum state this flux is equal to zero. But the flux is an observable and the flux through an infinitely distant sphere is an observable of any region D'. Hence states with sources cannot satisfy condition 2 of Proposition 2.2.8 and are not strongly locally unitarily equivalent to the vacuum state. The differences in equivalence properties of various physical systems will be discussed in greater detail in the next section.

2.3. Program for Producing Field Theory by Means of Reconstructing its Charge Sectors

So far all we have found are the conditions for the existence of various intertwining operators between superselection sectors. These conditions do not show, however, any practical way to reconstruct field properties; they do not make clear how we could define spin, statistics or charge conjugation and introduce such objects as charges, currents, antiparticles – to say nothing of fields themselves. In other words, the practical reconstruction of a field theory on the basis of local quantum theory has not yet been started. But the ground for it is now well prepared, and we can easily proceed to outline its main steps (unfortunately, space does not permit a detailed exposition).

First of all, it became clear that the required field theory must be the one with superselection rules because the latter are not just specific features of some fields but belong to the core of the quantum field concept. Already this first conclusion is significant, being closely connected with the radical reorientation of quantum field theory which took place in the last decade. This period witnessed a radical change in deciding which types of relativistic quantum models are of major importance for describing interactions of elementary particles. Nowadays we believe that these types include, in the first place, models with gauge fields, phase transitions, topological charges and superselection rules of various kinds. For the most part, these models were known before but they were treated as very particular and rather exotic cases. Both in the Wightman axiomatics and in the early period of constructive theory the view was prevalent that all the basic elements relating to the essence of quantum relativistic interactions – at least strong ones – are already present in a simple theory of neutral scalar field.

This common view proved to be basically incorrect. It became clear that the above phenomena such as superselection rules, local gauge symmetries, etc., are very far from being just sources of technical complications and are extremely important for grasping the very nature of relativistic quantum dynamics. One must consider their presence a norm, almost a necessary property of any realistic model of elementary particles interaction. Many factors account for this conclusion, and the discovery of a close connection between superselection rules and the nature of quantum field is not the least of them.

This connection makes superselection rules different from other phenomena listed above. While any of them, be it solitons, phase transitions or gauge fields, characterize a certain class of quantum systems, superselection rules are common to all classes. Moreover, within each class they appear to be different, having some features specific to this class. This opens up the possibility for unified treatment of many systems completely unlike each other at first sight. Indeed, we can consider all these systems on the same footing as systems with superselection rules specified by a definite type of superselection structure. Thus the superselection structure stands out as a key characteristics of a quantum field system bringing forth a new classification principle for such systems. But all this is possible only if the superselection structure is taken fully into account, with both global (Section 2.1) and local (Section 2.2) properties. As we shall see, it is mainly the latter, i.e. local equivalence properties of coherent sectors, that carry the dynamical information. Quite similarly, in traditional quantum field theory the dynamics of a system is reflected in asymptotical and localization properties of states and fields.

The classification method is directly related to the problem of constructing realistic field theory: the ways and means of this construction turn out to be different for each class of systems. Let us therefore review it in brief.

I. 'Ordinary' systems with no gauge fields or zero-mass particles are characterized by superselection structures with strong local unitary equivalence of coherent sectors. This means that the family of representations π of quasilocal algebra corresponding to physical states is determined by the property

$$\pi^{(D')} \simeq \pi_0^{(D')} \tag{3.1}$$

(π_0 is a vacuum sector, D is a certain double cone depending on π). This class includes, in the first place, traditional models of strong interactions such as mesodynamics with the Yukawa Lagrangian.

In such systems there are fields and states of charged particles localized in bounded regions of Minkowski space because the equivalence of a sector or a state to a vacuum sector (state) on a region D' allows us to regard the interaction in this sector as localized in the bounded region $D'' = D$. Global structure in such a model is determined by one or several groups of the first kind gauge transformations which can be abelian (as the electric or baryon charge groups) as well as nonabelian (as the isospin group SU (2)). These groups generate the complete gauge group G for which sufficiently general assumptions (completeness of the set of asymptotic in-states and finite multiplicity of every irreducible component of the $\mathcal{P}_+^{\uparrow}$ representation in one-particle space) imply its compactness [149]. In the full space of states including all coherent sectors the unitary representation of G is acting and physical observables correspond to the operators invariant under this representation:

$$R = U(G)'.$$

The set of all coherent sectors is isomorphic to the spectrum $\hat{G}$ of the group G and is discrete since G is compact, so that R is a P-class algebra. Superselection operators having the coherent sectors as their eigenspaces coincide with the Casimir operators of G. Charge operators connected closely with superselection operators but different from them if G is nonabelian act as generators of one-parameter subgroups of G. And, of course, charged fields transform physical states from one coherent sector into another.

It is for such systems that an algebraic formalism has first been developed which produced the explicit description of field properties in terms of local observables. This was the formalism of localized morphisms by Doplicher–Haag–Roberts [149, 150], and it seemed at first that it is entirely dependent on the condition (3.1). However, later on it proved possible, by substantially complementing and generalizing this formalism, to rid it of this dependence (see below). As a result, a very versatile 'method of morphisms and cocycles' has been created on the basis of the Doplicher–Haag–Roberts technique. It is valid for reconstructing charge sectors of practically any quantum field system.

II. The study of systems having localized charged states* performed by Buchholz and Fredenhagen [111] provided a considerable extension of the Doplicher–Haag–Roberts formalism. It was found that the most general sufficient conditions for the existence of such states are weaker than the original Doplicher–Haag–Roberts condition which

* We mean here the generalized concept (discussed below) of the charged state assuming only its nonequivalence to vacuum state.

requires the coincidence of a localized state ω with the vacuum state on some region D' and thus ensures the localization of ω in the bounded region D. In the general case, the localization region of a charged state appears to be unbounded and have the form $D \cup S$ where S is a spacelike cone of any, however small, aperture angle, a 'thickening string', so to say. Moreover, direction of the cone S is not an observable quantity which means that the corresponding charge is topological. Thus the general form of a localized charged state is given not by states localized in double cones but by those localized in strings, and so the criterion (3.1) for the selection of physical states is now changed for a weaker condition:

$$\pi^{(S')} \simeq \pi_0^{(S')} \tag{3.2}$$

where S (for a given π) is an arbitrary spacelike cone. As a result, there arises a new class of superselection structures satisfying the condition (3.2) but not (3.1). Naturally, we must ask the question: what physical systems can be associated with such structures? Obvious arguments suggest that it should be systems with gauge fields: charged states in these systems cannot be isolated in bounded regions since they are sources of infinite lines ('strings') of the flux of a charged field. Furthermore, if the condition (3.1) is violated for a charged state, the corresponding charge produces a nonzero flux through a sphere of any sufficiently large radius and is therefore a gauge-type charge which must satisfy some analogue of the Gauss law. The same conclusion is provided also by the 'rejection argument': all the 'usual' strong interaction models without zero-mass particles satisfy the criterion (3.1). In addition, all theories with continuous abelian gauge groups must also be rejected: by a generalization of Swieca's theorem [110], charged states in such theories are possible only in the presence of massless particles while the whole analysis [111] as well as the Doplicher–Haag–Roberts formalism are only valid for purely massive theories. This leaves us the class of gauge theories which include massive particles only and possess the nonabelian or discrete gauge group. The first application of the [111] formalism to such theories was carried out in the recent work by Fredenhagen and Marcu [198]. The model analyzed here describes a gauge field on the lattice Z^d, $d \geq 2$, with the gauge group Z, interacting with a discrete analogue of the Higgs field. With the methods of Euclidean and constructive lattice field theory, charged states were explicitly constructed (for a certain region of the phase diagram) and their properties are in complete accordance with the general analysis in [111]. (As for the absence of charged states, 'strictly localized' in the sense of (3.1), it is demonstrated in [198] as a simple preliminary result.)

III. The next class, structurally rather close to the former one, includes abelian gauge theories without the charge confinement and, in the first place, quantum electrodynamics.

As is well known, in quantum electrodynamics there is the superselection rule in the electric charge Q which generates a one-parameter gauge group. Since the charge Q satisfies the Gauss law, the corresponding charge sectors cannot enjoy strong local unitary equivalence; but more general localization properties described in [111] still take place. Hence charged states in quantum electrodynamics (which are functionals on the quasilocal algebra generated by the observable operators $j_\nu(g^\nu)$ and $F_{\mu\nu}(f^{\mu\nu})$ and treated as an abstract C^*-algebra) are also localized in 'strings' S and their GNS representations satisfy the criterion (3.2). But then we discover important differences. The presence of massless bosons reveals itself in the fact that the asymptotic direction of the string S is now on observable quantity. Moreover, due to its asymptotic character, this observable is a 'classical' one, i.e. it commutes with all observables and therefore generates a certain superselection rule.

This is a new superselection rule in quantum electrodynamics, in addition to the well known superselection in the electric charge.

The opinion that numerous implications of the 'infrared problem' (i.e. the virtual presence of a cloud of low-energy photons in every physical state) should also include some superselection rule has been voiced repeatedly and on various grounds (see e.g. [199, 206]); the analysis by Buchholz is probably the most detailed [104]. It is shown there that the superselection observable is the quantity φ describing the asymptotic distribution of the electromagnetic field flux over all the directions n:

$$\varphi(n) = \lim_{r\to\infty} r^2 \ (n\, E\,(r\, n))\ ,$$

where E is the electric field strength. Any two different functions $\varphi_1(n)$, $\varphi_2(n)$ correspond to different superselection sectors ('φ-sectors'), with the only condition that the full flux $\int d\,\omega\,\varphi(n)$ must be a multiple of the electron charge. In accordance with the said above, this superselection rule can be interpreted as the superselection in the asymptotic direction of the string where the charged state creating the flux φ is localized (since the function φ can easily be related to this direction).

As a result, each charge sector (superselection sector in the charge Q) splits into an uncountable set of φ-sectors corresponding to all localization directions of states with a given value of Q. In other words, the superselection structure for systems of this type represents a family of sectors parametrized by two quantities: the value q of the charge Q and the direction of the localization cone S, in such a way that for any admissible q there is a sector $[\pi]$ with any given localization direction. Representations with a given localization cone S are singled out by the criterion:

$$\pi^{(S')} \simeq \pi_0^{(S')}\ , \tag{3.3}$$

where S is now not arbitrary for a given π, but fixed uniquely. A similar criterion can be formulated for picking up representations with a given charge value q.

In any theory where charged particles are not massless the charge value can be found by means of measurements in any forward light cone $\overset{+}{V}_x$, $x \in M$. Indeed, the world line of any massive particle being timelike, it will certainly get inside $\overset{+}{V}_x$ at sufficiently large times. Thus if two representations π_1, π_2 are unitarily equivalent on some cone $\overset{+}{V}_x$, they carry the same charge:

$$\underset{x \in M}{\exists} \quad \pi_1(V_x^+) \simeq \pi_2(V_x^+) \Rightarrow q^{(\pi_1)} = q^{(\pi_2)} .$$

Hence we can deduce under unrestrictive assumptions that the sector with a given charge q includes all those and only those representations for which

$$\pi^{(V^+)} \simeq \pi_q^{(V^+)} \tag{3.4}$$

where π_q is any representation with the charge q. We conclude that charges and fields in systems of this type should be constructed (by means of the universal technique mentioned in paragraph I) as objects associated with the family of representations determined by the selection criteria (3.3) and (3.4).

IV. The general superselection approach extends naturally to systems with topological charges. As was observed on many examples, nontrivial topological structure of a quantum system (induced either by objects like solitons, vortices and monopoles or by nontrivial topology of spacetime itself) leads to superselection rules and can be described by means of a superselection structure. Let us recall that already in class II the charge is topological because the direction of its localization cone is unobservable. However, in systems where topological phenomena are the only source of the superselection the distinctions from the most traditional class I are more essential. In such systems there are still (now topological) charge operators, but the theory now lacks gauge symmetry, in the general case, and these operators do not act as generators of gauge groups. This means, of course, that physical observables cannot be defined by the gauge invariance property any more. Instead, in systems of this class there is usually a certain family of local neutral boson fields which play the part of observables. In the class-I systems, charged fields φ were thought to be the only fields present and the observables were represented by their gauge-invariant constructions of the type $\varphi = \bar{\psi}\psi$. But now, however, quantities like φ appear as basic objects. They generate local algebras of a vacuum sector while higher sectors arise as representations of the corresponding quasilocal algebra and include above-mentioned topological objects: for instance, in soliton models, the set of sectors is labelled by the number of solitons, i.e. the Nth sector contains N-soliton states. Charged fields intertwining these sectors form another family of fields. In this situation, the reconstruction of the complete superselection structure is usually interpreted as the proof of the equivalence between two different quantum field models (with the fields φ and ψ, respectively) – rather than reconstruction of the fields ψ from the observables φ. In typical cases, when charged fields ψ are fermions, presenting them as intertwining operators between the representations of a boson field algebra can be considered as the reduction of the fermion system to the boson one (or vice versa): indeed, both systems have the same charged sectors, i.e. the same space of states. In this sense, the field φ is sometimes called the 'bosonization' of ψ while ψ is, in its turn, called the 'fermionization' of φ.

Let us give some basic examples. The most familiar one is the equivalence between the sine–Gordon and massive Thirring models discovered first by Coleman

[121] and studied rigorously (though not thoroughly so far) by Fröhlich [200–203] and Sushko and Pogrebkov [349, 406]. In this case, the system of neutral fields-observables φ is the model with the self-interaction $(\sin\varphi)_2$. Higher superselection sectors $\mathcal{H}_n$ $(n \in Z)$ of this model contain n-soliton states (which can be related to classical solitons, i.e. solutions of classical equations of motion not vanishing at infinity and having finite energy) and charged fields ψ intertwining soliton sectors coincide with fermion fields of the massive Thirring model. Another example is the Streater–Wilde model [394]. In this case φ is a free scalar neutral zero-mass field in two dimensional spacetime. Higher superselection sectors form an uncountable family parametrized by the values of two topological charges. Charged fields ψ satisfy some commutation relations of a more general form than the normal Bose–Fermi ones; however, for a certain subset of sectors these relations turn into anticommutators and the corresponding ψ are fermions. The construction of charge sectors in this model has been carried out by an original method according to which new inequivalent representations of quasilocal algebra $\mathfrak{A}$ of the vacuum sector were obtained by means of certain transformations of Weyl operators generating $\mathfrak{A}$ (cf. Section 3.3). This method is valid for systems described by algebras of canonical commutation of anticommutation relations and later on it was applied successfully to many more such systems (see, in particular [286, 351, 243]). For some three- and four-dimensional systems with topological charges the possibility of their description by means of superselection structure was demonstrated by Ezawa [186, 187] (though only 'on the physical level', i.e. ignoring mathematical correctness problems connected with the necessity of using distributions and unbounded operators). Similarly to the Streater–Wilde and $(\sin\varphi)_2$ models, the construction of superselection structure has been carried out in [186] for the abelian and nonabelian gauge models describing 'vortices', i.e. three-dimensional analogues of two-dimensional quantum solitons. Generalizing these results, it was (more briefly) shown in [187] that a similar structure, i.e. a family of superselection sectors and a set of operators intertwining the sectors and carrying a topological charge, is found in four-dimensional systems with Dirac monopoles and 'dyons', bound states of monopoles and isodoublet bosons. And finally, as mentioned above, topological charges followed by superselection rules may also arise owing to spacetime topology. This effect is very clearly demonstrated by the model studied in [55] and describing free electromagnetic field in the Universe with the Schwarzschield–Kruskal metrics. In this Universe the field has nontrivial electric and magnetic charges in spite of the absence of any sources. These two topological charges can take any real values and generate a two-parameter family of superselection sectors.

V. In studying charge sectors of any quantum field system we must always start with some vacuum sector assuming it to be given. But clearly, the analysis of the system cannot be considered complete if it does not include the study of the vacuum sector as well. Basically, this study must answer the following questions. Is the vacuum sector abelian? Are the vacuum sector and the vacuum state unique for a given system? And if the uniqueness is violated (which is quite likely, taking into account well known vacuum degeneration phenomena), then what is the structure of the set of all vacuum states? Some of these questions will be considered in the next section (Section 3.1), but now we mention only the answers. The most essential is that the structure of the set of all vacuum states is also described by a certain superselection rule. This 'vacuum superselection rule' is not related to any charges or charged fields. By its physical meaning, it shows the possibility for a quantum field system to be in different phases, like systems in statistical mechanics. Each phase is characterized by its ground state, and under certain conditions (see Section 3.1) different phases of a multiphase quantum field system are separated by the superselection rule. Both interpretation and studying of the vacuum superselection possess some specific features. Bearing no relation to any charges, vacuum superselection rules are less closely connected with the asymptotic behaviour and still less with the local structure of the system in question. The set of all vacuum sectors is determined not by criteria like (3.1)–(3.4) but with the help of Euclidean formulation of the model studied. This formulation associates with the general model a definite classical statistical system whose phase diagram provides the information about the set of ground states. Moreover, the interpretation of vacuum sectors as physical phases suggests that in this case neither observables nor fields can transform states from one sector into another. Hence in systems with vacuum superselection both algebra of observables and algebra of fields should be reducible. Taking all this into account, it is noteworthy that for a large class of systems the vacuum superselection is still described by the usual P-class global structure (although, alongside these systems, continuous vacuum superselection is also possible, when superselection sectors form a direct integral of Hilbert spaces).

Two conclusions can be drawn at once from this review of different classes of systems. Firstly, the superselection structure is really the core of the whole structure of a quantum field system; all principal dynamic and field properties of the latter are somehow related to or even determined by it. Hence, the problem of 'reconstructing a realistic field theory' in its first stage should be stated as the problem of reconstructing a superselection structure, i.e. a certain family of charge sectors. Secondly, starting the reconstruction of a field theory, we should not suppose a P-class global structure given beforehand. It should appear but at the end of the whole procedure as a result of constructing a physical state space. The global structure contains a whole range of field and dynamical characteristics which we can in no way take as known initially. On the contrary, we want to find all these characteristics starting with only the set of local observables. In other words, we cannot start with a given set of superselection sectors but must construct it with the help of some physical conditions like the selection criteria (3.1)–(3.4). Superselection structure is not to be given *a priori* but to be deduced and, in this sense, field theory reconstructed from its local observables is a 'quantum field theory predicting its own superselection rules', by the well known Streater formula [71].

The minimal input for reconstructing superselection structure is an abstract net of local algebras and a certain ground state or, equivalently, a net of local algebras of a vacuum coherent sector. Higher charge sectors are then to be obtained as certain representations of the vacuum sector quasilocal algebra. As said above, we should pick up these representations on the basis of some physical conditions. The very first of them must single out representations corresponding to arbitrary charged states irrespective of the type of charge. The following simple criterion, well prepared by all the 'superselection ideology' of this chapter provides a suitable condition which is more general than (3.1)–(3.4) and covers all kinds of charges.

DEFINITION 2.3.1 ('Borchers' criterion' [77]). The state ω of a system described by quasilocal algebra $\mathfrak{A}$ will be called a charged state if in the corresponding representation π_ω of $\mathfrak{A}$ all the Haag–Araki axioms hold but no vacuum state is present. The equivalence class $[\pi_\omega]$ of a charged state ω will be called the charge sector.

Specific to this definition is that it does not include any explicit conditions of charge existence and, as it seems, does not say anything at all about the charge. This is not accidental, of course. In fact, the ideas which we are now expounding constitute a new point of view on the structure of quantum field systems, according to which the concept of superselection is allotted the prime part. In the context of these ideas the charge appears as something secondary to superselection structure and its presence turns out to be an automatic consequence of the sectorial structure of the physical states set. Indeed, having a family of charge sectors, in the sense of Definition 2.3.1, we can step by step associate with this family various field objects and properties: fields, as intertwining operators of the sectors; spectrum of charge values, as the set of values of a parameter labeling the family, etc. – till it becomes possible to introduce an operator having the above set as its spectrum. With good reason this operator might be called a (generalized) charge.

One notices at once that all this logic is the direct opposite of that which is followed in building classical Lagrangian theory and which for a long time remained dominant in quantum field theory as well. As it is well known, in the Lagrangian formalism charges and currents are present from the very beginning, arising as soon as the Lagrangian itself is given. On the contrary, charge sectors are here quite a secondary object which in classical theory was hardly studied at all till recent works [343, 344] and in quantum theory is quite difficult to construct (because one needs here spectral decompositions of charge operators, very ill-defined as a rule). Invalidity of the Lagrangian formalism as a basis for rigorous unperturbative description of quantum field systems has been clearly seen long ago. But, quite understandingly, it took a considerable time to develop any sound alternative. Axiomatic and then constructive field theory can be considered as stages of the gradual liberation form structural patterns of the Lagrangian formalism. In the approach of 'local quantum physics' (as Haag and Wightman propose to call modern algebraic quantum field theory) this liberation is perhaps nearing its completion. Implementation of 'local quantum physics' ideas is provided by the technique of observable algebra morphisms and cocycles of symmetry groups [200–204, 362]. It is now developed rather thoroughly and its efficiency was demonstrated on models of all the types I–IV (the type V being not connected with charges, as we remember). We shall briefly describe its basic elements.

For all types of systems, the first stage of reconstructing field theory comprises the construction of a family of charge sectors, while at the next stages the explicit construction of field objects should be performed. Since the conditions on an arbitrary charge sector include only the fulfillment of the Haag–Araki axioms, the first stage has little to do with specific features of the system studied. Hence the technique of morphisms and cocycles is at this stage nearly universal, i.e. the same for all classes of systems. On the contrary, at the subsequent stages it differentiates to a great extent and takes a different form for each class. Here we shall restrict ourselves mainly to the universal elements.

The initial assumptions are as follows. The vacuum sector quasilocal algebra $\mathfrak{A}$ is irreducible; the net of its local subalgebras $R(O)$ satisfies duality (though this is used only at further stages); the sought sectors $[\pi_\alpha]$ should be locally unitarily equivalent (see Definition 2.2.1) to the vacuum sector. Due to the results of Sections 1.3.5 and 2.2, only the second of these assumptions means a loss of generality; however, it is physically justified in our problem: duality implies that the set of local observables is maximal and it is natural to expect that all the field properties of the system can only be reconstructed from the maximal set. As a first step, we start describing charge sectors by means of morphisms of the algebra $\mathfrak{A}$. Morphism ρ of an operator C^*-algebra $\mathfrak{A}$ in a Hilbert space $\mathcal{H}$ is a particular case of representation obtained when representation acts in the same space $\mathcal{H}$. Thus all the usual properties of representations are defined for morphisms, in particular, the notions of irreducibility (a morphism ρ is irreducible if $\rho(\mathfrak{A})'' = \mathcal{B}(\mathcal{H})$) and unitary equivalence. From every representation π or state ω, a morphism ρ makes it possible to obtain a new representation or state: $\pi' = \pi \circ \rho$; $\omega' = \omega \circ \rho$. In particular, when $\mathfrak{A}$ is the quasilocal algebra of a vacuum sector its identical representation π_{id} can be considered as corresponding to the vacuum state $\omega_0 = \omega_\Omega$ i.e. $\pi_{id} = \pi_{\omega_0}$. Then the morphisms ρ correspond to the states $\omega_0 \circ \rho$ and representations preserving the vacuum Ω as a cyclic vector and acting in subspaces

$$\mathcal{H}_\rho = \overline{\rho(\mathfrak{A})\,\Omega} \subset \mathcal{H}.$$

It is easy to check that any representation π_α locally unitarily equivalent to π_{id}, i.e. such that $\pi_\alpha^{(D)} = V_\alpha^{(D)} \cdot V_\alpha^{(D)*}$ (where D is a certain double cone and $V_\alpha^{(D)}$ is an isometry intertwining $\pi_{id}^{(D)}$ and $\pi_\alpha^{(D)}$) determines the morphism ρ_α equivalent to π_α by means of the formula

$$\rho_\alpha(A) = V_\alpha^{(D)*}\, \pi_\alpha(A) V_\alpha^{(D)}, \qquad \forall A \in \mathfrak{A}.$$

Since $\rho_\alpha \simeq \pi_\alpha$, then $[\rho_\alpha] = [\pi_\alpha]$ and so the construction of charge sectors is reduced to constructing morphisms of $\mathfrak{A}$. The morphisms must satisfy the Haag–Araki axioms; but morphisms being representations of $\mathfrak{A}$, only Axioms V and VI provide any restrictions on the choice of morphisms. Thus the only problem at this stage lies in the construction of morphisms satisfying the Poincaré covariance and spectrum condition.

The formulation of Poincaré covariance of a morphism ρ is provided by the definition of a covariant representation (Section 1.3.5): there is a continuous unitary representation U_ρ of the Poincaré group acting in $\mathcal{H}$ and such that

$$\rho\,(\alpha_{(a,\,\Lambda)}(A)) = U_\rho\,(a, \Lambda)\,\rho\,(A)\,U_\rho\,(a, \Lambda)^{-1}$$

for all $(a, \Lambda) \in \mathcal{P}_+^{\uparrow}$, $A \in \mathfrak{A}$. This condition can be formulated equivalently in terms of covariant states. Keeping in mind applications to other symmetry groups as well, we give this formulation in the general form: for an arbitrary locally compact group G, a morphism ρ of C^*-algebra $\mathfrak{A}$ is called *G-covariant* if the state $\omega \circ \rho$ is G-covariant for any G-covariant state ω. Clearly, the construction of G-covariant morphisms includes two problems: firstly, we should be able to construct new representations of G starting with the representation in the vacuum sector; secondly, we should be able to construct the morphisms of $\mathfrak{A}$, G-covariance of which is implemented by these representations. A way to the solution of these problems opens up if we consider not only representations but also cocycles of G.

DEFINITION 2.3.2. Let $(\mathfrak{A}, G, \alpha)$ be a C^*-dynamical system, ω a G-covariant state on $\mathfrak{A}$ and U_ω the corresponding representation of G in $\mathcal{H}_\omega$. The mapping

$$\Gamma : G \to U\,(\mathcal{H}_\omega)$$

is called the G-cocycle on $\mathcal{H}_\omega$ if $\Gamma\,(g)$ is strongly (or weakly) continuous and satisfies the cocycle identity:

$$\Gamma\,(g_1\, g_2) = \Gamma\,(g_1)\,U_\omega\,(g_1)\,\Gamma\,(g_2)\,U_\omega\,(g_1)^*\,.$$

This definition implies immediately the following:

PROPOSITION 2.3.1. *Let ω be a G-covariant state, U_ω the corresponding unitary representation of G in $\mathcal{H}_\omega$ and Γ G-cocycle in $\mathcal{H}_\omega$. Then the set of operators $V\,(G) = \Gamma\,(g)\,U_\omega\,(g)$ implements a strongly continuous unitary representation of G in $\mathcal{H}_\omega$.*

Thus the G-cocycle makes it possible to construct a new continuous unitary representation V from a given representation U_ω. The inverse correspondence is also obvious: any pair of continuous unitary representations V, U of the group G in a Hilbert space $\mathcal{H}$ determines the G-cocycle Γ defined by the formula

$$\Gamma\,(g) = V\,(g)\,U\,(g)^*\,.$$

It means that for every G-covariant morphism ρ there exists not only the representation V_ρ of G but also the G-cocycle $\Gamma_\rho\,(g) = V_\rho\,(g)\,U\,(g)^*$, where U is the representation of G corresponding to the vacuum state ω_0 (by definition, ω_0 is supposed to be G-covariant for all symmetry groups). In other words, the existence of the G-cocycle Γ_ρ is the necessary condition for G-covariance of ρ. However, sufficient conditions for G-covariance of morphisms are more needed to us. They can be formulated with the help of the notion of a transportable morphism.

DEFINITION 2.3.3. Morphism ρ of a C^*-algebra $\mathfrak{A}$ is called G-transportable, if there are a G-covariant state ω and a G-cocycle on $\mathcal{H}_\omega$ such that

$$\pi_\omega\left(\alpha_g^{-1} \circ \rho \circ \alpha_g (A)\right) = \Gamma_\rho (g)^* \, \pi_\omega (\rho (A)) \, \Gamma_\rho \ (g)$$

for all $g \in G, A \in \mathfrak{A}$.

We see that G-transportability of a morphism ρ implies the existence of a G-cocycle Γ_ρ and hence the existence of the representation $V_\rho (g) = \Gamma_\rho (g) \, U(g)$ of the group G connected with ρ in a certain way. In the general case, this connection may not be sufficient for the G-covariance of ρ since the representation V_ρ might lead out of the space $\mathcal{H}_\rho$. However, a number of simple conditions are known which secure the G-covariance of ρ when added to its G-transportability. The following assertion can be easily verified:

Let morphism ρ be G-transportable. Then each of the following conditions is sufficient for the G-covariance of ρ :

(1) $\mathcal{H}_\rho$ is an invariant subspace of the representation V_ρ ;

(2) ρ is an irreducible morphism;

(3) ρ is a localized morphism (i.e. $\rho\,|_{\mathfrak{A}(O')} = \mathrm{Id}$ for some region $O \in B(M)$. For a discussion of such morphisms, see below.)

Thus, the required G-covariant morphisms in a wide range of situations coincide with G-transportable ones, and to obtain the latter we only need cocycles Γ_ρ of the group G. Moreover, G is here some symmetry group of a quantum field system. Cocycles of these groups are relatively simple objects studied much better than C^*-algebra morphisms; and ways of constructing them are often available in concrete models. In constructing charge sectors the most needed ones are cocycles of the group $\hat{M}_{s,1}$ of translations in $(s+1)$-dimensional analogue of Minkowski space $M = M_{3,1}$. (Since the technique is applied to a wide range of models, including two-dimensional solitons, three-dimensional vortices, etc., it is natural to formulate it for an arbitrary dimension space.) Of course, these cocycles can only help to construct translationally covariant morphisms, but obtaining the latter is the main part of the problem: one can impose the spectrum condition on them and secure in such a way inequivalence of the states ω_0 and $\omega_0 \circ \rho$. And say, in quantum electrodynamics where the symmetry under the Lorentz group is spontaneously broken [205], translation covariance is the only symmetry required from charge sectors. For practical applications it is convenient to use a subclass of quasilocal translation cocycles which automatically give rise to certain morphisms of $\mathfrak{A}$.

DEFINITION 2.3.4. Translation cocycle $\Gamma(a)$, $a \in \hat{M}_{s,1}$, on a Hilbert space $\mathcal{H}$ is called quasilocal if it satisfies the following conditions:

(1) $\Gamma(a) A \Gamma(a)^* \in \mathfrak{A} \ \forall a \in \hat{M}_{s,1}$, $A \in \mathfrak{A}$;

(2) for all local observables $A \in \mathcal{A}$ and for all ways of approaching spacelike infinity in $\hat{M}_{s,1}$, $(|a| \to \infty$, $a^2 < 0)$ there exists the limit

$$\rho_{\Gamma}(a) = \underset{\substack{|a|\to\infty \\ a^2<0}}{w\text{-lim}} \; \Gamma(a) A \Gamma(a)^* \tag{3.5}$$

which is independent of the way chosen and defines a locally normal morphism of $\mathfrak{A}$ (i.e. $\omega_0 \circ \rho_{\Gamma}$ is a locally normal state);

(3) $$\underset{\substack{|a|,|a+b|\to\infty \\ a^2<0,(a+b)^2<0}}{w\text{-lim}} \alpha_{(-a)}(\Gamma(b)\,\alpha_{(a)}(A)\,\Gamma(b)^*) = A \quad \forall A \in \mathfrak{A} .$$

It is easy to check [203, 204] that the morphisms ρ_{Γ} introduced according to (3.5), are $\hat{M}_{s,1}$ -transportable and hence also $\hat{M}_{s,1}$ -covariant under any of the additional conditions given above.

Obtaining charged states with the help of (3.5) admits of a simple physical interpretation: clearly, relations of the (3.5) type express the idea (suggested first in [150]) that any charged state can be thought of as obtained from a vacuum state by means of charge transfer from infinity to some localization region. This property of a charged state does not depend on the type of charge being common to all of them. And in concrete applications, morphisms of the ρ_{Γ} form provide an effective tool for constructing charge sectors.

Now, to proceed further we need to be specific about the class of systems. As noted above, the method discussed is valid for all the classes I–IV: for any class appropriate kinds of morphisms and cocycles can be found which take account of its specific features. However, a detailed study and a wide enough range of results obtained are so far available only for the class 1.

The principal characteristic of the class 1 including systems without gauge fields and massless particles is the property (3.1), strong local unitary equivalence of superselection sectors. By using the duality condition (not used so far) it can be easily proved that each sector satisfying (3.1) is an equivalence class of so-called localized morphisms.

DEFINITION 2.3.5 [150]. A morphism ρ of C^*-algebra $\mathfrak{A}$ *is* called localized if a region $O \in B(M)$ can be found such that

$$\rho(R(O)) \subset R(O), \quad \rho|_{\mathfrak{A}(O')} = \mathrm{Id} .$$

The region O *is* called the support of ρ .

Any localized morphism can be obtained from a certain cocycle of the Poincaré group by means of an asymptotic relation of the (3.5) form. And this representation in terms of cocycles makes it possible to select the morphisms which are certainly inequivalent to the vacuum sector and hence carry a nontrivial charge [203, Theorem 2.16].

A large complex of results about the class I systems produced by the localized morphisms technique is sometimes called a 'field theory without fields': it comprises practically all properties of a realistic field theory and still they are all expressed without using field operators or field algebras.

Here we mention just the major results. Many of them are related to the cardinal feature of field theory according to which there must be an antiparticle for each particle. Quite a number of conditions on the formalism of the theory stem from this feature: each value of each charge must have its 'opposite' which means that a group structure on the set of all sectors must be possible; there must be some operation of charge conjugation transforming each sector to its 'opposite'; 'mutually opposite' sectors must have the same spectrum of masses and spins; etc. The definition of charge conjugation due to Doplicher–Haag–Roberts is based on the idea of charge transfer and asymptotic formulas of the (3.5) type. To construct a morphism corresponding to the 'opposite' charge the following hypothesis is used: if the transfer of a given charge is implemented by a sequence $\Gamma(a) \cdot \Gamma(a)^*$ in (3.5) then the similar transfer of the 'opposite' charge should be implemented by the inverse sequence $\Gamma(a)^* \cdot \Gamma(a)$.

Another group of results deals with introducing the notion of statistics for particles and sectors. The definition of the statistics of an arbitrary sector $[\rho]$ is given in terms of permutation properties of the composition of morphisms.

The type of statistics is proved to be determined uniquely by the value taken for a given sector by the 'statistics parameter' λ_ρ, spectrum of all λ_ρ values being $\{(0, \pm 1/d\}, d \in \mathbf{N}$. Sectors with $\lambda_\rho = \pm 1$ possess the normal Bose (Fermi) statistics and sectors with $\lambda_\rho = \pm 1/d$, $d > 1$, have para-Bose (para-Fermi) statistics of order d. The case $\lambda_\rho = 0$ is unphysical and its impossibility has recently been proved using locality and strong spectrum condition [192]. In concrete algebraic formalism, a sector with the statistics $\lambda_\rho = \pm 1/d$ has the global algebra R_ρ such that R_ρ' is a factor of type I_d.

Sectors with normal statistics have a number of special properties. Namely, according to [150] the following conditions are equivalent:

(1) sector $[\rho]$ has Bose-Fermi statistics;

(2) all the morphisms ρ_i of the sector $[\rho]$ are automorphisms, $\rho_i(\mathfrak{A}) = \mathfrak{A}$;

(3) ρ^2 is an irreducible morphism;

(4) the duality holds in the sector $[\rho]$, $\rho_i(\mathfrak{A}(D'))'' = \rho_i(R(D))'$.

The concluding results of the Doplicher–Haag–Roberts cycle establish for the class-I systems the spin-statistics theorem and scattering theory formalism. In the recent works by Doplicher *et al.* [148, 155] the existence of objects like charge density operators is proved using the split property. Let us mention, finally, that the problem of reconstructing field algebras has been solved (see [144, 350] for simplified reviews) for normal field systems only. The study of the general case started by Doplicher and Roberts was soon found to touch upon some large-scale mathematical problems such as reworking the abstract duality theory of compact groups and including it into the framework of C^*-algebras theory. Thus it resulted in a large work (see [158–161]) which has so far produced a number of quite sophisticated mathematical results but still no final physical synthesis.

In the class-II systems we find weaker localization properties. This implies, in particular, that morphisms satisfying the criterion (3.2) may not have the property $\rho(\mathfrak{A}) \subset \mathfrak{A}$ and so they must be defined as mappings from $\mathfrak{A}$ into $\mathcal{B}(\mathcal{H})$.

Nevertheless, it still proved possible [111] to extend to such systems the main bulk of the above results: the existence of an antiparticle for each particle (and hence the corpuscular interpretation of the formalism); Bose–Fermi statistics or finite order parastatistics for charge sectors and the existence of scattering theory. The results obtained for classes III and IV are more fragmentary* and were briefly reviewed above. It is also worth noting that the whole subject is developing rapidly and any attempt to fix up its present state can only be of short term value.

In conclusion a few words must be said about the class-V systems not covered by the method discussed. As noted above, the study of such systems is an indispensable complement to studying systems of classes I–IV: it provides information about the set of ground states which are used as the input in the method of morphisms. This method starts with some chosen ground state; but if the system in question has a whole family of such states, then, in principle, the method has to be applied to each of them. And one cannot exclude the possibility that the sets of charge sectors corresponding to different ground states may belong to different classes in our classification. In other words, the most general quantum field system might be conceived as a system having all the kinds of superselection rules. In this 'universal' system, the set of all coherent sectors forms a kind of two-dimensional structure: there is a series of vacuum sectors separated by a class V superselection rule and over each of them a hierarchy of charge sectors (belonging to one of classes I–IV) is constructed. Methods of analyzing the 'universal system' should comprise both the Euclidean theory (for constructing the 'horizon' of vacuum sectors) and the technique of morphisms and cocycles (for constructing 'towers' of charge sectors). A prototype of such a system and such an analysis can be seen in the study of the simplest φ_2^4 model carried out by Fröhlich [200–203]. This model has been found to possess two kinds of superselection rules, vacuum and soliton ones, and while soliton sectors are reconstructed by means of the morphisms technique, the vacuum ones are described by the Euclidean theory.

* A noticeable advance in the structural analysis of quantum electrodynamics over the recent period is due not so much to the morphisms method as to the so-called Strocchi–Wightman theory using indefinite metric spaces (see Appendix).

CHAPTER 3

Field Algebras and Their Applications

3.1. Op*-Algebras of Field Operators and Vacuum Superselection Rules

After the theory of superselection rules and the method of morphisms had clarified the differences between fields and observables, there appeared many investigations studying quantum field systems on the basis of field algebras. This approach presented some advantages over formalism of observable algebras: a concrete quantum field system is usually given in terms of some fields so that algebras generated by them provide the shortest way to their algebraic description (whereas to obtain observable algebras we need some additional operations, like forming gauge-invariant combinations of fields). Moreover, if our starting objects are quantized fields, it gives us richer information than that available in local quantum theory: above all, the relativistic covariance properties are now more detailed (see Section 1.3.3, 1.3.5 and 3.2.2). On the other hand, field algebras are not given beforehand but must be constructed from fields. The way of such a construction is not immediately obvious since all boson fields are represented by unbounded operators while all the apparatus above has been based so far on normed algebras. Hence the alternative arises: either to develop a new version of the algebraic approach able to include unbounded operators or to rid of the latter finding some way to describe systems of field operators by means of (local nets of) von Neumann algebras. We shall discuss the former in the present section and the latter in the next section.

It is fairly easy to represent a system of Wightman fields $A(f)$, $f \in S(\mathbf{R}^4)$ (we shall consider only neutral scalar theories) in the form of an algebraic structure. By the Wightman axioms, all fields are closable operators given on the common dense invariant domain $\mathcal{D}$ (the *Gårding domain* so-called by the analogy with the Lie groups theory; see [20, Vol. I, Chapter VIII, Notes]) and hence, generating the so-called polynomial algebra $\tilde{\mathcal{P}}$ – the set of all combinations of the form $\Sigma_{k=1}^{n} A(f_1) \ldots A(f_k)$ defined on $\mathcal{D}$. Since $A(f)$ are symmetric for $f = \bar{f}$, $\tilde{\mathcal{P}}$ is a *-algebra relative to the involution $A(f) \to A(f)^+ = A(f)^* |_{\mathcal{D}}$ and can be provided with a net of local subalgebras $\tilde{\mathcal{P}}(O)$, generated by all the operators $A(f)$ with $\operatorname{supp} f \subset O$.

However, a theory of the Wightman type possesses, in fact, deeper algebraic contents. It becomes evident as soon as we look closer at the correspondence $f \to A(f)$ between the test functions and field operators expressed by the reconstruction theorem. From the mathematical point of view, this well known theorem [4, Chapter 3, Section 3] states a simple and purely algebraic fact: the extension of the canonical GNS correspondence between positive functionals and cyclic representations of an algebra to a certain class of topological non-Banach *-algebras. The algebra in question here is the set of test functions endowed with algebraic structure, and the Wightman theory is considered as a representation of this algebra into the algebra of all unbounded operators in $\mathcal{H}$ with the domain $\mathcal{D}$. Systematic reworking of Wightman theory from this point of view produces a new algebraic formalism for quantum field theory called usually the *Borchers algebraic approach.* It was set up by Borchers in 1962 in the well known work [73] but the basic ideas were suggested at the same time by Uhlmann [415] and Maurin [318, 319].

Let us consider the graded set of test functions of Wightman theory:

$$\mathcal{A} = \bigoplus_{n=0}^{\infty} \mathcal{A}_n , \quad \mathcal{A}_0 = \mathbf{C}, \quad \mathcal{A}_k = S\,(\mathbf{R}^{4k}), \quad k = 1, 2, \ldots$$

and endow it with the structure of a topological *-algebra. Involution in $\mathcal{A}$ is defined by the following rule:

$$\begin{aligned} f &= \left(f_0 , f_1 (x_1) , \ldots , f_n (x_1 , \ldots , x_n), \ldots \right) \to f^* \\ &= \left(\bar{f}_0 , \bar{f}_1 (x_1) , \ldots , \bar{f}_n (x_n , \ldots , x_1), \ldots \right) ; \end{aligned}$$

addition and multiplication by elements of the field $\mathbf{C}$ are defined in a natural way. Identity is the element $f = (1, 0, 0, \ldots)$. Multiplication in $\mathcal{A}$ can be defined in various ways, both as a commutative operation (which is used in statistical mechanics (see [22]), and a noncommutative one. In the Borchers approach the following noncommutative multiplication is adopted:

$$\begin{aligned} f \times g &= \left(f_0\, g_0 , f_0\, g_1 (x_1) + f_1 (x_1)\, g_0 \ldots \right. \\ &\quad \left. \ldots \sum_{k=0}^{n} f_k (x_1 , \ldots , x_k)\, g_{n-k} (x_{k+1} , \ldots , x_n) , \ldots \right) . \end{aligned}$$

Topologies $\tau_{S,n}$ of spaces $S\,(\mathbf{R}^4)$ induce in the direct sum topology $\tau_{\mathcal{A}}$ defined as follows: a sequence $\{f^{(\gamma)}\} \subset \mathcal{A}$ is said to be converging to zero if for any $n > 0$ and $\gamma \to \infty$ $f_n^{(\gamma)} \to 0$ in the topology $\tau_{S,n}$ (and in the sense of convergence in $\mathbf{C}$ for $n = 0$) and, in addition, there exists $n_0 \in \mathbf{N}$, independent of γ and such that $f_n^{(\gamma)} \equiv 0$ for all γ and $n > n_0$. Topology $\tau_{\mathcal{A}}$ makes $\mathcal{A}$ a complete nuclear space and a topological *-algebra [18, Section 8] called the *Borchers algebra.* Each element of $\mathcal{A}$ has only a finite number of nonzero components. The algebra is semisimple; it has no minimal ideals and no idempotents except for 0 and I. Its centre is trivial $(Z_{\mathcal{A}} = \mathbf{C})$ and no element outside it has an inverse. The dual space $\mathcal{A}^*$ is also a complete nuclear space whose elements are sequences $(T_n)_{n=0}^{\infty}$, $T_n \in S\,(\mathbf{R}^4)'$, i.e. $\mathcal{A}^* = \Pi_{n=0}^{\infty} \mathcal{A}_n^*$. A number of important distinctions of $\mathcal{A}$ from a C^*-algebra concern properties of positive cones in $\mathcal{A}$ and $\mathcal{A}^*$. Like C^*-algebras, $\mathcal{A}$ has a generating cone $\mathcal{A}^+$ which defines an ordering on $\mathcal{A}$, i.e. any $f = f^* \in \mathcal{A}$ can be uniquely represented as $f = g - h$; $g, h \in \mathcal{A}^+$. However, the general form of the cone elements is now not $f = f^{(1)*} f^{(1)}$, but $f = \Sigma_{i=1}^{\infty} f^{(i)*} f^{(i)} \in \mathcal{A}$, and $\mathcal{A}^+$ has no inner points. Furthermore, as in the case of a C^*-algebra, $\mathcal{A}^+$ defines the dual cone $\mathcal{A}^{*+}$ in $\mathcal{A}^*$ formed by all $T \in \mathcal{A}^*$ such that $T(f) \geq 0$ for all $f \in \mathcal{A}^+$. By definition, elements of

$\mathcal{A}^{*+}$ are called positive functionals on $\mathcal{A}$. Again as for C^*-algebras, any positive functional on $\mathcal{A}$ is continuous though now not every Hermitian functional can be represented as a difference of positive functionals. (See [439] for simple proofs of these facts.)

According to the reconstruction theorem, a field theory with field operators $A(f)$ is obtained as a representation of the algebra $\mathcal{A}$ corresponding canonically to a certain state W on $\mathcal{A}$. This means that in the Borchers approach a physical system is given by a certain pair $(\mathcal{A}, W)$ and all its properties are expressed in terms of properties of this pair.

The algebra $\mathcal{A}$ determines only a class of test functions and so is affected only by a few physical properties such as renormalizability (in case of its violation, the Schwartz space has to be changed to some other space) or the absence of spinning particles (in their presence test functions must be the Lorentz group tensors). All other properties and, in the first place, the Wightman axioms, should be formulated as conditions on the state W. Formulations of the Wightman axioms in terms of W are easily obtained from those in terms of fields or Wightman functions.

(1) Relativistic covariance. On the algebra $\mathcal{A}$ an automorphic representation $\alpha : \mathcal{P}_+^\uparrow \to \mathrm{Aut}\,\mathcal{A}$ of the proper Poincaré group is defined which acts as follows:

$$\begin{aligned}\alpha_{(a,\Lambda)}(f) &= f_{(a,\Lambda)}\\ &= (f_0, f_1(\Lambda^{-1}(x-a)), \ldots, f_n(\Lambda^{-1}(x_1-a), \ldots, \Lambda^{-1}(x_n-a)), \ldots)\end{aligned}$$

The state W determining a physical theory is Poincaré-invariant: for all $f \in \mathcal{A}$, $(a, \Lambda) \in \mathcal{P}_+^\uparrow$

$$W(f_{(a,\Lambda)}) = W(f).$$

(2) Locality. There exists in $\mathcal{A}$ a two-side open Poincaré-invariant *-ideal

$$\begin{aligned}I_\ell = L\{f \in \mathcal{A} \mid f_n(x_1, \ldots, x_n) = g(x_1, \ldots, x_i, x_{i+1}, \ldots, x_n) -\\ - g(x_1, \ldots, x_{i+1}, x_i, \ldots, x_n);\ g(x_1, \ldots, x_n) = 0 \text{ for } (x_i - x_{i+1})^2 \geq 0\}\end{aligned}$$

called the *locality ideal*. The state W is such that its left kernel

$$N_W = \{f \in \mathcal{A} \mid W(f^* \times f) = 0\}$$

includes the locality ideal, $I \subset N_W$.

(3) Spectrum condition. There exists in $\mathcal{A}$ a left closed Poincaré-invariant *-ideal $I_{sp} = \{f \in \mathcal{A} \mid \tilde{f}_n(p_1, \ldots, p_n) = 0$ when

$$p_1 \in \overline{V^+},\ p_1 + p_2 \in \overline{V^+}, \ldots, p_1 + \ldots + p_n \in \overline{V^+}\}$$

($\tilde{f}_n$ is a Fourier-transform of f_n) called the *spectrum ideal*. The state W is such that its left kernel includes the spectrum ideal, $I_{sp} \subset N_W$. *

(4) Existence and cyclicity of vacuum (i.e., vacuum vector in the state space determined by the functional W). This property follows automatically from the Poincaré invariance of W.

(5) Uniqueness of vacuum. This condition (as in C^*-theory, see Section 1.3.5) is equivalent to the cluster property of W:

$$\lim_{\substack{\lambda \to \infty \\ a^2 < 0}} W(f \times g_{(\lambda a, 1)}) - W(f)\, W(g) = 0 .$$

It should be noted at once that the last condition is by no means as substantial as others in the list, and we shall study here more general theories which do not satisfy it.

The above formulations have been discussed many times (see e.g. [4, Chapter 3, Section 3.1]) and their derivation can be done as a simple exercise.

A state W on the Borchers algebra $\mathcal{A}$ satisfying the conditions (1)–(4) is called the *Wightman functional*. As said above, the description in terms of field operators is obtained by means of the GNS construction associating with W a *-representation Π of $\mathcal{A}$; W is then called the *generating functional* of Π. Due to the positivity of W, the space $\mathcal{H}_\Pi$ of Π is separable [27, Theorem 3 – 7]. Moreover, since $\mathcal{A}$ is not a B^*-algebra, operators $\Pi(f)$ of the representation Π are, in general case, unbounded operators in $\mathcal{H}_\Pi$ with the common dense domain $\mathcal{D} \subset \mathcal{H}_\Pi$, i.e. Π is the representation of $\mathcal{A}$ into the algebra $L_+(\mathcal{D})$ of all linear operators in $\mathcal{H}_\Pi$ with the common dense domain $\mathcal{D}$. Operators $\Pi(f)$ enjoy all the properties of Wightman fields and coincide exactly with them when $f = (0, f_1, 0, \dots)$. The values taken by W on products of such elements give the set of Wightman functions.

The image $\Pi(\mathcal{A}) \equiv \mathcal{P}$ of the algebra $\mathcal{A}$ is a *-algebra belonging to the class of so-called Op^*-algebras (see Definition 3.1.1 below). It can be provided with a net of local subalgebras related to regions $O \in B(M)$ and defined as images of the restrictions of Π to local subalgebras $\mathcal{A}(O) \subset \mathcal{A}$. The latter, in their turn, are defined as the sets of all $f \in \mathcal{A}$ such that each component $f_n(x_1, \dots, x_n)$ in each of its variables has the support in O:

$$\mathcal{A}(O) = \{ f \in \mathcal{A} \mid \operatorname{supp} f_n \subset O^{\times n},\ n = 1, 2, \dots \} ;$$

$$\mathcal{P}(O) = \{ \Pi(f),\ f \in \mathcal{A}(O) \}$$

($O^{\times n} = O \times O \times, \dots, \times O$ is the Cartesian product of n copies of O). Clearly, due to the presence of 'generalized field operators' $\Pi(f_n)$, the algebras $\mathcal{P}$ and $\mathcal{P}(O)$ differ from the algebras $\tilde{\mathcal{P}}$ and $\tilde{\mathcal{P}}(O)$ introduced above and generated directly by Wightman fields. This difference is insignificant since due to the nuclearity of Schwartz spaces

* Clearly, in conditions 2 and 3 only the latter part is a requirement, the former one just stating a fact.

generalized field operators are determined by products of ordinary fields: any continuous functional on product functions

$$f_n(x_1, \dots, x_n) = g^1(x_1) \dots g^n(x_n), \qquad g^i(x_i) \in S(\mathbf{R}^4)$$

determines a functional on the whole $S(\mathbf{R}^{4n})$. Owing to this, the Borchers algebra can be chosen equivalently in the following form:

$$\mathcal{A} = \bigoplus_{n=0}^{\infty} \bigotimes_{i=0}^{n} \mathcal{A}_i, \quad \mathcal{A}_0 = \mathbf{C}, \quad \mathcal{A}_i = S(\mathbf{R}^4), \qquad i = 1, 2, \dots .$$

A net of unbounded operator algebras $\{\mathcal{P}(O)\}_{O \in B(M)}$ presents the algebraic formulation of Wightman theory cognate to local quantum theory. It is easy to verify that, as a consequence of Wightman axioms 1–4, this net satisfies the following set of axioms similar to axioms I–VI of the algebraic approach:
isotony;
strong additivity:

$$\mathcal{P}(O_1 \cup O_2) = \mathcal{P}(O_1) \vee \mathcal{P}(O_2), \qquad \forall\, O_1, O_2 \in B(M);$$

locality: $\mathcal{P}(O_1)$ and $\mathcal{P}(O_2)$ with $O_1 \times O_2$ commute on $\mathcal{D}$;
spectrum condition;
existence and cyclicity (for the algebra $\mathcal{P}$) of a vacuum vector.

In this list of axioms only additivity is not obvious or resulting directly from conditions 1–4. Adopting the above form for the Borchers algebra and taking into account that every element $f \in \mathcal{A}$ has only a finite number of nonzero components f_n and the region $(O_1 \cup O_2)^{\times n}$ can be represented as a union of regions of the form $O_1^{\times k} \times O_2^{\times \ell}$, we obtain additivity for the net $\{\mathcal{A}(O)\}$:

$$\mathcal{A}(O_1 \cup O_2) = \mathcal{A}(O_1) \vee \mathcal{A}(O_2).$$

This property of the Borchers algebra is inherited by its representation Π and thus gives the required formula. The similar property (Axiom IIa) in local quantum theory implies weak additivity (Axiom II) which states that the set of all translations of any local algebra generates the global algebra. But in the Borchers approach weak additivity needs a special study because it is first necessary to understand how precisely local algebras $\mathcal{P}(O)$ generate the global algebra $\mathcal{P}$. For algebras $\mathcal{A}(O)$ the answer to the similar question is straightforward. For any n a subset of all functions with a compact support is dense in $S(\mathbf{R}^{4n})$ with respect to its topology. Hence, the union of all $\mathcal{A}(O)$, $O \in B(M)$, is dense in $\mathcal{A}$ with respect to the topology $\tau_{\mathcal{A}}$; and since there are partitions of unity in Schwartz spaces, the density holds also for the union of all translations of $\mathcal{A}(O)$ with a fixed $O \in B(M)$:

$$\mathcal{A} = \overline{\bigvee_{O \in B(M)} \mathcal{A}(O)}^{\tau_{\mathcal{A}}} = \overline{\bigvee_{a \in M} \mathcal{A}(O + a)}^{\tau_{\mathcal{A}}}.$$

However, to transport such a formula into the representation Π one needs a nontrivial analysis of continuity properties of Π and topologies on the algebra $\mathcal{P} = \Pi(\mathcal{A})$; and, as a result, the sought property has not been proved so far.

Fulfillment of the axioms for the net $\{\mathcal{P}(O)\}$ makes it possible to extend to this net many results of the algebraic approach not attached specifically to the category of C^* or W^* algebras. In particular, the Reeh–Schlieder theorem holds for $\mathcal{P}(O)$: although in the Haag–Araki theory its proof uses weak additivity, in the field theory the latter is not needed. Here lies one of the differences between field theory and theory of observables originating in the dependence of fields $A(f)$ upon the point functions $f(x)$. Due to this, the application of spectrum condition which in the Haag–Araki theory gives only

$$(\Phi, R(O)\,\Omega) = 0 \Rightarrow (\Phi,\ R(O+a)\,\Omega) = 0 \qquad \forall\, a \in M$$

(cf. p. 26) in the field theory gives at once 'the Reeh–Schlieder property':

$$(\Phi\,,\mathcal{P}(O)\,\Omega) = 0 \Rightarrow (\Phi\,,\ \mathcal{P}\Omega) = 0$$

(cf. [27, Theorem 4–2]). Hence the following equality takes place:

$$\overline{\mathcal{P}(O)\,\chi} = \overline{\mathcal{P}\,\chi}$$

for all regions $O \subset M$ and all vectors χ analytic in the energy and lying in the Gårding domain (i.e. $\chi \in \mathfrak{N} \cap \mathcal{D}$; as shown by Snellman [384] this set is dense in $\mathcal{H}$). This result needs a comment, however. In the Haag–Araki theory the Reeh–Schlieder theorem immediately implies a corollary which is often used: any local algebra $R(O)$ in an irreducible theory ($R = \mathcal{B}(\mathcal{H})$) has a dense set $\mathfrak{N}$ of cyclic vectors. The corollary is not directly transportable to $\mathcal{P}(O)$ since vectors from $\mathcal{D}$ in an irreducible Wightman theory are not necessarily cyclic for $\mathcal{P}$ (see Lemma 3.1.3). Nevertheless, for an irreducible Wightman theory with the strong spectrum condition and the unique vacuum vector Ω⋆ , the existence of a dense set of cyclic vectors for the algebras $\mathcal{P}$ and $\mathcal{P}(O)$ is established. It is a recent result by Driessler and Summers [170]: they have proved that the set

$$C_{\mathcal{P}} = \left\{ \Phi \in \mathcal{D} \mid \Phi = A\,\Omega\,,\ A \in \mathcal{P}\,,\ \underset{B=B(A)\in\mathcal{P}}{\exists} \quad B\,A\,\Omega = \Omega\,,\ B\,\Omega = 0 \right\}$$

is dense in $\mathcal{H}$ and each of its elements is cyclic for $\mathcal{P}$. Hence, each vector $\chi \in C_{\mathcal{P}} \cap \mathfrak{N}$ is cyclic for $\mathcal{P}(O)$ and the set $C_{\mathcal{P}} \cap \mathfrak{N}$ is also dense in $\mathcal{H}$, according to [170]. Thus the problem of cyclic vectors for local algebras of the Borchers approach has obtained a satisfactory solution.

Nevertheless, the nets $\mathcal{P}(O)$ rarely provide a convenient formalism for physical problems. Of course, this is chiefly because unbounded operator algebras are much more intricate and their theory is much less developed than that of C^* or W^* algebras. The formalism of von Neumann field algebras (Section 3.2) appears to be more efficient for concrete physical problems while the Borchers approach has so far concentrated more on its own mathematical problems: the structure of dual space of the Borchers algebra and Op^*-algebras, properties of tensor algebras of test functions, problem of topologization

⋆ Irreducibility of Wightman theory does not secure the uniqueness of vacuum. This is the 'Borchers anomaly' discussed in Section 3.1.2.

of Op^*-algebras (Leipzig school works; see e.g. [301, 302]). In the short list of physical problems to which this approach was applied, the following two are the more substantial ones: the existence of Wightman theory with a nontrivial interaction and the analysis of a vacuum structure of Wightman theory. The first one is the much discussed 'problem of a nontrivial example', one of the main problems for any axiomatic approach to quantum field theory. In studying this problem the Borchers approach has a certain advantage over local quantum theory because there exist in Wightman theory a great number of criteria distinguishing free and interacting systems (see [4] and Section 3.3). It proved possible to reformulate these criteria in terms of continuity properties of a Wightman functional [246]. But proving the existence of a positive functional, satisfying the axioms 1–3 and surely different from the Wightman functional of a free or generalized free field (the abstract solution of the nontrivial example problem) is an involved problem; it includes many questions about the spectrum and locality ideals for solving which there are no ready tools in the theory of algebras $\mathcal{A}$ and $\mathcal{P}$. We shall not discuss the partial results obtained so far; see [86, 442–4, 246, 60, 340].

On the contrary, the problem of vacuum structure will be considered in detail since the results are here more complete and the whole subject displays close ties with superselection rules (Section 2.3). Physically, this problem complements the description of the vacuum structure of local quantum theory in Section 1.3.5; it included some new aspects due to the correspondence between Wightman theory and quantum statistical mechanics. Indeed, as the Euclidean approach has established, Wightman functions are direct analogues of correlation functions of a statistical system. But each set of correlation functions determines a certain physical phase and thus the notion of a phase is transported to Wightman theory, together with all its concomitants like many-phase systems, phase diagrams, etc. Problems of studying phase diagrams and phase transitions in quantum field models are of great interest nowadays, and for all problems like these the formalism of Wightman functionals and algebras of Wightman fields should be used. Before starting the discussions of these problems we present the necessary mathematical information.

3.1.1. ELEMENTS OF OP*-ALGEBRAS

Basic information. Great many classes of unbounded operator algebras were introduced and studied in the mathematical literature: algebras of measurable and locally measurable operators [380, 24], EW^*-algebras by Inoue [259], V^*- and SV^*-algebras by Epifanio and Trapani [183, 33], partial *-algebras by Antoine *et al.* [34, 35, 36], etc. Of all these, the class of Op^*-algebras is singled out by a number of properties which make it a natural choice for a wide range of both mathematical and physical applications. Due to the presence of a common invariant dense domain for all their elements, it is Op^*-algebras that are images of Hilbert space representations of topological (unnormed) *-algebras. As for physics, all algebras generated by Wightman fields belong to this class, both in axiomatic theory and in concrete models. Let us give the exact definitions.

DEFINITION 3.1.1. A pair $(\mathcal{P}, \mathcal{D})$ where $\mathcal{P}$ is a set of linear operators in the Hilbert space $\mathcal{H}$ and $\mathcal{D}$ is a linear set (a lineal) dense in $\mathcal{H}$ is called *Op*-algebra if $\mathcal{P}$ is an unital algebra and for all its operators A the following conditions hold:

(1) $\operatorname{dom} A = \mathcal{D}$, $A\,\mathcal{D} \subset \mathcal{D}$.

If, in addition, the conditions

(2) $\operatorname{dom} A^* \supset \mathcal{D}$;

(3) $A^*|_{\mathcal{D}} \equiv A^+ \in \mathcal{P}$

hold, then $(\mathcal{P}, \mathcal{D})$ is called *Op**-algebra.

The symbol $\mathcal{D}$ will be omitted as a rule. Obviously, any *Op**-algebra $\mathcal{P}$ is a *-algebra with respect to the involution $A \to A^+$ and all its operators are closable. All *Op**-algebras with the same domain $\mathcal{D}$ can be considered as unital subalgebras of the *Op**-algebra $L_+(\mathcal{D})$ including all operators A such that $\operatorname{dom} A = \mathcal{D}$, $A\,\mathcal{D} \subset \mathcal{D}$, $A^*\,\mathcal{D} \subset \mathcal{D}$.

It is very often convenient to consider an *Op**-algebra as the image of a representation $\Pi : \mathcal{A} \to \mathcal{L}(\mathcal{H})$ of a certain *-algebra into the algebra of all linear operators in the Hilbert space $\mathcal{H}$. (Just so the *Op**-algebra of Wightman fields is the image of the Borchers algebra representation.) Since the definition of Π requires the existence of a dense lineal $\mathcal{D} = \mathcal{D}(\Pi)$ called the domain of Π and such that $\operatorname{dom} \Pi(f) = \mathcal{D}$ for all $f \in \mathcal{A}$, the image $\Pi(\mathcal{A})$ of any representation Π is an *Op**-algebra. If, in addition, Π satisfies the condition:

$$\Pi(f)^* \supset \Pi(f^*) \text{ for all } f \in \mathcal{A}$$

(such representations are called *Hermitian*), the image of Π is an *Op**-algebra.

A partial parallel exists between the theory of Hermitian representations and theory of Hermitian (symmetric) operators. Similar to extensions of a symmetric operator, a Hermitian representation possesses some canonical extensions. Let us consider the domains

$$\mathcal{D}(\overline{\Pi}) = \bigcap_{f \in \mathcal{A}} \operatorname{dom} \overline{\Pi(f)}, \quad \mathcal{D}(\Pi^*) = \bigcap_{f \in \mathcal{A}} \operatorname{dom} \Pi(f)^*,$$

where $\overline{\Pi(f)}$ is the closure of $\Pi(f)$. For any $f \in \mathcal{A}$ let us denote, by definition,

$$\overline{\Pi}(f) = \overline{\Pi(f)}\,|_{\mathcal{D}(\overline{\Pi})}, \quad \Pi^*(f) = \Pi(f^*)^*\,|_{\mathcal{D}(\Pi^*)}.$$

Let us consider also the domain

$$\mathcal{D}(\Pi^{**}) = \bigcap_{f \in \mathcal{A}} \operatorname{dom} \Pi^*(f)^*$$

and denote, by definition,

$$\Pi^{**}(f) = \Pi^*(f^*)^*\,|_{\mathcal{D}(\Pi^{**})}.$$

DEFINITION 3.1.2. Let Π be a Hermitian representation of a *-algebra $\mathcal{A}$. The mappings

$$\Pi : f \to \overline{\Pi}\,(f), \quad \Pi^* : f \to \Pi^*\,(f), \quad \Pi^{**} : f \to \Pi^{**}\,(f)$$

of $\mathcal{A}$ into $\mathcal{L}(\mathcal{H})$ determine the representations of $\mathcal{A}$ which we shall call, respectively, the closure, the adjoint and the second adjoint of Π.

The following relations are immediately implied by the definitions:

$$\Pi \subset \overline{\Pi} \subset \Pi^{**} \subset \Pi^* \tag{1.1}$$

where the symbol $\Pi_1 \subset \Pi_2$ for Hermitian representations Π_1, Π_2 means that Π_2 is an extension of Π_1, i.e. $\mathcal{D}(\Pi_1) \subset \mathcal{D}(\Pi_2)$ and $\Pi_1(f) \subset \Pi_2(f)$ for all $f \in \mathcal{A}$. The Op^*-algebra $\Pi_2(\mathcal{A})$ is called an extension of the Op^*-algebra $\Pi_1(\mathcal{A})$. Any of the inclusions (3.1.1) in general may not be an equality. It is easy to verify that the Hermiticity of Π implies the Hermiticity of both Π and Π^{**}. On the contrary, Π^*, is, in general, non-Hermitian and the following easily derived equalities [217, Theorem 2] hold for it:

$$\Pi^* = \overline{\Pi}^* = \Pi^{***}.$$

DEFINITION 3.1.3. A Hermitian representation Π (as well as the Op^*-algebra $\Pi(\mathcal{A})$) will be called*:

closed, if $\Pi = \overline{\Pi}$;

essentially self-adjoint, if $\Pi^{**} = \Pi^*$;

self-adjoint, if $\Pi = \Pi^*$.

As one easily sees, Π is closed if at least one of the operators $\Pi(f)$ is closed; Π is self-adjoint if at least one of $\Pi(f)$ is self-adjoint. Closed representations can also be characterized in terms of so-called graph topology on $\mathcal{D}(\Pi)$ much used in the theory of Op^*-algebras.

DEFINITION 3.1.4. Let Π be a representation of a *-algebra $\mathcal{A}$ in the Hilbert space $\mathcal{H}$ and S a finite subset of elements of $\mathcal{A}$. The topology on $\mathcal{D}(\Pi)$ generated by the basic neighbourhoods

$$\mathcal{N}(S, \varepsilon) = \left\{ \Psi \in \mathcal{D}(\Pi) \;\middle|\; \sum_{f \in S} \| \Pi(f)\Psi \| < \varepsilon \right\}$$

will be called graph topology. The corresponding closure of a set $\mathfrak{M} \subset \mathcal{D}(\Pi)$ will be denoted $\overline{\mathfrak{M}}^G$. A vector $\chi \in \mathcal{D}(\Pi)$ such that $\overline{\Pi(\mathcal{A})\chi}^G \subset \mathcal{D}(\Pi)$ as well as any representation having such vectors, will be called strongly cyclic.

By this definition, all operators $\Pi(f)$, $f \in \mathcal{A}$, are continuous with respect to graph topology. As shown by Powers [352], a representation Π is closed if and only if its domain is closed in the graph topology, which means

* There is a difference between our terminology and that of Powers [352] and Vassiliev [420]. In [352] a representation Π is called essentially self-adjoint if $\overline{\Pi} = \Pi^*$; in [420] Π is called closed if $\Pi^{**} = \Pi$.

$$\overline{\mathcal{D}(\Pi)}^{G} = \mathcal{D}(\Pi)\,. \tag{1.2}$$

Let us now define the strong Π'_s and weak Π'_w commutants of the Op^*-algebra $\Pi(\mathcal{A})$ with the domain $\mathcal{D}$:

$$\Pi'_s = \{T \in \mathcal{B}(\mathcal{H}) \mid AT \supset TA \;\; \forall A \in \Pi(\mathcal{A})\},$$

$$\Pi'_w = \{T \in \mathcal{B}(\mathcal{H}) \mid (A^*\Phi, T\Psi) = (T^*\Phi, A\Psi),\; \forall A \in \Pi(\mathcal{A}),\; \Phi, \Psi \in \mathcal{D}\}.$$

Over the last period various definitions of the commutant to $\Pi(\mathcal{A})$ as a set of unbounded operators are actively studied (though not much used yet); see, e.g. [183, 33, 49, 317, 260]. In particular, the following 'weak unbounded commutant' is the direct generalization of Π'_w:

$$\Pi'_\sigma = \{X \in C(\mathcal{D}, \mathcal{H}) \mid (A^*\Phi, X\Psi) = (X^*\Phi, A\Psi)\; \forall A \in \Pi(\mathcal{A}),\; \Phi, \Psi \in \mathcal{D}\}$$

where $C(\mathcal{D}, \mathcal{H})$ is the set of all closable operators B in $\mathcal{H}$ such that $\mathcal{D} \subset \operatorname{dom} B \cap \operatorname{dom} B^*$. Obviously, $\Pi'_w = \Pi'_\sigma \cap \mathcal{B}(\mathcal{H})$. The theory of unbounded commutants seems to us not yet ready for presentation in a book; however, in Section 3.1.2 we shall mention the first physical applications of the commutant Π'_σ. As for bounded commutants, it is easy to obtain for them the following basic properties:

(1) Π'_w is a weakly closed linear *-subset of $\mathcal{B}(\mathcal{H})$, including the identity;

(2) Π'_w is generated linearly by its positive elements;

(3) $\Pi'_w = \overline{\Pi}'_w$;

(1') Π'_s is a unital algebra;

(2') $\overline{\Pi}'_s$ is weakly closed;

(3') $\Pi'_s \subset \Pi'_w$ and if $\Pi = \Pi^*$ then $\Pi'_s = \Pi'_w$.

The properties $1, 3, 1', 3'$ are evident or verifiable directly. The property 2 is obtained taking into account that Π'_w is generated linearly by its Hermitian elements and if $T = T^* \in \Pi'_w$ then

$$\|T\| I \pm T \in \Pi'_w, \quad \|T\| I \pm T \geq 0, \quad T = \frac{1}{2}(\|T\| I + T) - \frac{1}{2}(\|T\| I - T).$$

To derive (2') let us take a directed set $\{T_\alpha\}$ from $\overline{\Pi}'_s$ converging weakly to a certain $T \in \mathcal{B}(\mathcal{H})$. For all $\mathcal{A} \in \Pi(\mathcal{A})$, $\Phi \in \mathcal{D}$ we have $T_\alpha A\Phi = A T_\alpha \Phi$ and hence the pair $(T_\alpha \Phi, T_\alpha A\Phi)$ belongs to the graph of A and converges weakly to $(T\Phi, TA\Phi)$ when $T_\alpha \to T$. But the weak and strong convergences coincide on linear subsets. Hence $(T\Phi, TA\Phi)$ belongs to the graph of $\bar{A}$ and consequently $T \in \overline{\Pi}'_s$.

A class of simple examples in which Π'_w is not an algebra was presented by Powers who has proved [352, Lemma 3.2] that for Hermitian representations Π of a free abelian *-algebra $\mathcal{A}$ generated by a single Hermitian generator, A_0 (i.e. $\mathcal{A} \ni A = \Sigma z_i A_0^i$), Π'_w is an algebra if and only if the operator $\Pi(A_0)$ is essentially self-adjoint. The condition that Π'_w is a von Neumann algebra distinguishes an important subclass of Op^*-algebras. As easily seen, it is larger than the subclass of self-adjoint ($\Pi = \Pi^*$) Op^*-algebras and includes, in particular, all Op^*-algebras for which $\Pi^{**} = \Pi^*$.

For a unital *-subalgebra of $\mathcal{B}(\mathcal{H})$, projections from its commutant are in one-to-one correspondence to its invariant subspaces and its subrepresentations acting in these subspaces. In case of an Op^*-algebra $\mathcal{P}$ this connection as well as the very notion of an invariant subspace become more complicated. The disposition of a subspace and the domain $\mathcal{D}$ of $\mathcal{P}$ can be different and due to this there exist a number of different kinds of invariant and reducing subspaces of $\mathcal{P}$. We shall consider the main ones (see [426, Section 2] for more detailed analysis). A lineal $\mathcal{M}$ in $\mathcal{H}$ is called invariant under an Op^*-algebra $\mathcal{P}$ if

$$\mathcal{M} \subset \mathcal{D}, \ \mathcal{P}\mathcal{M} \subset \mathcal{M}.$$

In this case all the operators $A \in \mathcal{P}$ have the restrictions $A|_{\mathcal{M}}$ the set of which, as it is easy to check, forms an Op^*-algebra $\mathcal{P}_{\mathcal{M}}$ acting in the Hilbert space $\overline{\mathcal{M}}$. In other words, if $\mathcal{P} = \Pi(\mathcal{A})$ then the Hermitian representation Π has the Hermitian subrepresentation $\Pi_{\mathcal{M}} : \Pi(f) \to \Pi(f)|_{\mathcal{M}} \ \forall f \in \mathcal{A}$. The first essential difference from the properties of *-algebras of bounded operators is that the projection $P_{\mathcal{M}}$ on $\overline{\mathcal{M}}$ may commute with the algebra $\mathcal{P}$ neither strongly nor weakly. The conditions $P_{\mathcal{M}} \in \Pi'_s$ or $P_{\mathcal{M}} \in \Pi'_w$ single out certain classes of invariant and reducing subspaces but may be violated for an arbitrary invariant subspace.

LEMMA 3.1.1. *Let* Π *be a Hermitian representation,* $\mathcal{D}_1$ *a lineal in* $\mathcal{H}$, P_1 *and* $P_1^{\perp}$ *projections on* $\mathcal{H}_1 = \overline{\mathcal{D}_1}$, *and* $\mathcal{H}_1^{\perp}$, *respectively. Then the following conditions are equivalent:*

(1) $P_1 \in \Pi'_s$;

(2) $P_1^{\perp} \in \Pi'_s$;

(3) $\mathcal{D}_1$ *'orthogonally reduces'* Π, *i.e.*

$$P_1\ \mathcal{D}(\Pi) = \mathcal{H}_1 \cap \mathcal{D}(\Pi);\ \Pi(\mathcal{A})\, P_1\ \mathcal{D}(\Pi) \subset P_1\ \mathcal{D}(\Pi);\ \Pi(\mathcal{A})\, P_1^{\perp}\, \mathcal{D}(\Pi) \subset P_1^{\perp}\, \mathcal{D}(\Pi).$$

Moreover, the following conditions are equivalent:

(1') $P_1 \in \Pi'_w$;

(2') $P_1^{\perp} \in \Pi'_w$;

(3') $\mathcal{D}_1$ *'weakly orthogonally reduces'* Π, *i.e.*

$$P_1\ \mathcal{D}(\Pi) \subset \mathcal{H}_1 \cap \mathcal{D}(\Pi);\ \Pi^*(\mathcal{A})\, P_1\ \mathcal{D}(\Pi) \subset P_1\ \mathcal{D}(\Pi),$$

$$\Pi^*(\mathcal{A})\, P_1^{\perp}\, \mathcal{D}(\Pi) \subset P_1^{\perp}\, \mathcal{D}(\Pi).$$

Proof. Relations $1 \leftrightarrow 2$ and $1' \leftrightarrow 2'$ follow from the presence of the identity in Π'_s and Π'_w. $1 \to 3$ and $1' \to 3'$ are obvious since $P_1 \in \Pi'_s$ entails $P_1\ \mathcal{D}(\Pi) \subset \mathcal{D}(\Pi)$ and $\Pi(\mathcal{A})\, P_1\ \mathcal{D}(\Pi) \subset P_1\ \mathcal{D}(\Pi)$ (i.e. the lineal $P_1\ \mathcal{D}(\Pi)$ is invariant under $\Pi(\mathcal{A})$), while $P_1 \in \Pi'_w$ entails $P_1\ \mathcal{D}(\Pi) \subset \mathcal{D}(\Pi^*)$ and $\Pi^*(\mathcal{A})\, P_1\ \mathcal{D}(\Pi) \subset P_1\ \mathcal{D}(\Pi)$ and similar relations hold for $P_1^{\perp}$. Let us prove $3 \to 1$. For all $\Phi, \Psi \in \mathcal{D}(\Pi)$ and $f \in \mathcal{A}$ one has

$$(\Pi(f)P_1\Phi,\Psi) = (\Pi(f)P_1\Phi,P_1\Psi)+(\Pi(f)P_1\Phi,P_1^{\perp}\Psi)$$
$$= (P_1\Pi(f)P_1\Phi,\Psi),$$

since due to Π-invariance of $P_1\mathcal{D}(\Pi)$, $P_1^{\perp}\Pi(f)P_1\Phi=0$. On the other hand, due to Π-invariance of $P_1^{\perp}\mathcal{D}(\Pi)$, one has

$$(P_1\Pi(f)\Phi,\Psi) = (P_1\Pi(f)P_1\Phi,\Psi)+(P_1\Pi(f)P_1^{\perp}\Phi,\Psi)$$
$$= (P_1\Pi(f)P_1\Phi,\Psi),$$

whence for all $\Phi\in\mathcal{D}(\Pi)$ and $f\in\mathcal{A}$ holds $\Pi(f)P_1\Phi=P_1\Pi(f)\Phi$, i.e. $P_1\in\Pi'_s$. $3'\to1'$ is derived identically. ❐

In the case of a self-adjoint representation $\Pi=\Pi^*$, when $\Pi'_s=\Pi'_w\equiv\Pi'$ is a von Neumann algebra, the condition $P_1\in\Pi'$ singles out the class of subspaces where self-adjoint subrepresentations of Π are acting. Such subspaces are also called self-adjoint.

PROPOSITION 3.1.2 [352]. *Let* Π *be a self-adjoint representation,* $\mathcal{H}_1$ *a subspace of* $\mathcal{H}$. *The following conditions are equivalent:*

(1) $P_1\in\Pi'$;

(2) $\mathcal{H}_1$ *is a self-adjoint subspace .*

When 1, 2 *hold, the self-adjoint representation acting in* $\mathcal{H}_1$ *is unique and its domain* $\mathcal{D}_1$ *is such that* $\mathcal{D}_1=P_1\mathcal{D}(\Pi)=\mathcal{H}_1\cap\mathcal{D}(\Pi)$.

Proof. 1 → 2. By Lemma 3.1.1, $P_1\in\Pi'$ implies that $P_1\mathcal{D}(\Pi)$ is a dense lineal in $\mathcal{H}_1$ invariant under $\Pi(\mathcal{A})$. Hence there is in $\mathcal{H}_1$ a Hermitian subrepresentation $\Pi_{\mathcal{D}_1}$ with the domain $\mathcal{D}_1=P_1\mathcal{D}(\Pi)$. To prove that $\Pi_{\mathcal{D}_1}$ is self-adjoint, let us choose $\Phi\in\mathcal{D}(\Pi^*_{\mathcal{D}_1})$ and show that $\Phi\in\mathcal{D}(\Pi_{\mathcal{D}_1})=P_1\mathcal{D}(\Pi)$. Since $\Phi\in\mathcal{D}(\Pi(f)|^*_{\mathcal{D}_1})$ for all $f\in\mathcal{A}$ there exists a constant K dependent on f and Φ and such that

$$|(\Phi,\Pi(f)\Psi)|\le K\|\Psi\| \qquad \forall\,\Psi\in\mathcal{D}_1 .$$

Taking into account condition 1 and $P_1\Phi=\Phi$ for all $\chi\in\mathcal{D}(\Pi)$ one has

$$|(\Phi,\Pi(f)\chi)|=|(\Phi,\Pi(f)P_1\chi)|\le K\|P_1\chi\|\le K\|\chi\| .$$

This means that $\Phi\in\mathcal{D}(\Pi(f)^*)$ for all $f\in\mathcal{A}$, whence $\Phi\in\mathcal{D}(\Pi^*)=\mathcal{D}(\Pi)$ and $P_1\Phi=\Phi$ give the required $\Phi\in P_1\mathcal{D}(\Pi)$.

2 → 1. First we show that the domain $\mathcal{D}_1$ of a self-adjoint representation $\Pi_{\mathcal{D}_1}$ in $\mathcal{H}_1$ coincides with $P_1\mathcal{D}(\Pi)$. Clearly, $\mathcal{D}_1\subset\mathcal{H}_1\cap\mathcal{D}(\Pi)\subset P_1\mathcal{D}(\Pi)$. To derive the inverse inclusion one takes $\Phi\in P_1\mathcal{D}(\Pi)$, i.e. $\Phi=P_1\Phi_1$, $\Phi_1\in\mathcal{D}(\Pi)$. For all $\Psi\in\mathcal{D}_1$ one has

$$|(\Phi,\Pi(f)\Psi)|=|(\Phi_1,\Pi(f)\Psi)|\le K\|\Psi\| . \tag{1.3}$$

Due to $\mathcal{D}(\Pi)\subset\mathcal{D}(\Pi^*)$, $\Phi_1\in\mathcal{D}(\Pi(f)^*)$ for all $f\in\mathcal{A}$. Hence there is a constant K dependent of f and Φ_1 and such that

$$|(\Phi_1\,,\,\Pi(f)\,\Psi)| \leq K\,\|\,\Psi\,\| \text{ for all } \Psi \in \mathcal{D}_1 \subset \mathcal{D}(\Pi)$$

(we can take, for example, $K = \|\,\Pi(f)^*\,\Phi_1\,\|$). Then taking (1.3) into account one has

$$|(\Phi, \Pi(f)\,\Psi)| = |(\Phi_1, \Pi(f)\,\Psi)| \leq K\,\|\,\Psi\,\|$$

for all $\Psi \in \mathcal{D}_1$ which implies that $\Phi \in \mathcal{D}\big(\Pi(f)\,|^{*}_{\mathcal{D}_1}\big)$ for all $f \in \mathcal{A}$, i.e. $\Phi \in \mathcal{D}\big(\overset{*}{\Pi}_{\mathcal{D}_1}\big) = \mathcal{D}(\Pi_{\mathcal{D}_1}) = \mathcal{D}_1$ since $\Pi_{\mathcal{D}_1}$ is self-adjoint. Thus we have obtained

$$\mathcal{D}_1 = P_1\,\mathcal{D}(\Pi)\,.$$

Using this, for all $\Phi, \Psi \in \mathcal{D}(\Pi)$ and $f \in \mathcal{A}$ one has

$$(\Pi(f)^*\,\Phi, P_1\,\Psi) = (\Phi, \Pi(f)\,P_1\,\Psi) = (\Phi, P_1\,\Pi(f)\,P_1\,\Psi)$$

and also

$$\begin{aligned}(\Phi, P_1\,\Pi(f)\,\Psi) &= (\Pi(f)^*\,P_1\,\Phi, \Psi) = \big(P_1\,\Pi(f^*)\,P_1\,\Phi, \Psi\big)\\ &= \big(\Pi(f^*)\,P_1\,\Phi, \Psi\big) = (\Phi, P_1\,\Pi(f)\,P_1\,\Psi)\,,\end{aligned}$$

whence it follows that $P_1 \in \Pi'_w = \Pi'$. ❐

Thus in case of a self-adjoint representation Π, there is a one-to-one correspondence between the set of all projections from Π' and the set of all self-adjoint subspaces (but not all invariant subspaces!).

Hermitian representations (and equally Op^*-algebras $\Pi(\mathcal{A})$) for which $\Pi'_w = C(\mathcal{H})$ are called *irreducible.* Their properties differ considerably from usual properties of irreducible representations of normed *-algebras. First of all, the standard equivalence between the irreducibility of a representation, the absence of closed invariant subspaces and the cyclicity of any vector in the representation space is now broken.

LEMMA 3.1.3 [217]. *Each of the following conditions implies the irreducibility of a self-adjoint representation* Π *:*

(2) Π *has no nontrivial self-adjoint subspaces;*

(3) Π *has no nontrivial invariant subspaces closed in the graph topology;*

(4) *any vector from* $\mathcal{D}(\Pi)$ *is strongly cyclic;*

(5) *any vector from* $\mathcal{D}(\Pi)$ *is cyclic.*

If the irreducibility of Π *is denoted by* (1), *then the following implications hold:*

$$\begin{array}{ccccc} 1 & \longleftrightarrow & 2 & \longleftarrow & 3 \\ \uparrow & & & & \updownarrow \\ 5 & \longleftarrow & & & 4 \end{array}$$

Proof. 3 → 2 and 4 → 5 are trivial. 2 ↔ 1 follows from Proposition 3.1.2.

3 → 4. For an arbitrary $\Phi \in \mathcal{D}(\Pi)$, $\Pi(\mathcal{A})\,\Phi$ is an nontrivial invariant lineal of Π. Let us prove that its graph closure (which lies in $\mathcal{D}(\Pi)$ since Π is closed) is an invariant lineal as well. For an arbitrary $\Psi \in \overline{\Pi(\mathcal{A})\,\Phi}^{\,G}$ there exists a directed set $\{\Psi_\alpha\}$ from $\Pi(\mathcal{A})\,\Phi$ which is graph-converging to Ψ. Hence for any $f_1\,, f_2 \in \mathcal{A}$ one

has

$$\Pi(f_1)\,\Pi(f_2)\,\Psi_\alpha = \Pi(f_1 f_2)\,\Psi_\alpha \to \Pi(f_1 f_2)\,\Psi = \Pi(f_1)\,\Pi(f_2)\,\Psi\,,$$

whence G-lim $\Pi(f_2)\,\Psi_\alpha = \Pi(f_2)\,\Psi$. Hence it follows that

$$\Pi(\mathcal{A})\,\overline{\Pi(\mathcal{A})\,\Phi}^{G} \subset \overline{\Pi(\mathcal{A})\,\Phi}^{G}\,,$$

i.e. $\overline{\Pi(\mathcal{A})\,\Phi}^{G}$ is a graph-closed invariant lineal. Due to 3, $\overline{\Pi(\mathcal{A})\,\Phi}^{G} = \mathcal{D}(\Pi)$, i.e. Φ is strongly cyclic.

4 → 3. If $\mathcal{D}_1$ is a graph-closed invariant lineal different from $\mathcal{D}(\Pi)$, then any $\Phi \in \mathcal{D}_1$ is not strongly cyclic.

5 → 1. Due to Proposition 3.1.2, a nontrivial self-adjoint subspace $\mathcal{H}_1$ corresponds to any nontrivial projection $P_1 \in \Pi'$. For any vector $\Phi \in P_1\,\mathcal{D}(\Pi)$ one has then $\overline{\Pi(\mathcal{A})\,\Phi} = \mathcal{H}_1$, i.e. Φ is not cyclic. ❒

Finally, let us describe the *Op**-generalization of the canonical GNS construction. In the case of normed *-algebras it establishes a reciprocal connection between states on an algebra and its cyclic representations. Now the place of cyclic representations is taken by strongly cyclic ones.

PROPOSITION 3.1.4. *Each state W on an unital *-algebra $\mathcal{A}$ corresponds canonically to a strongly cyclic Hermitian representation Π of $\mathcal{A}$ such that for all $f \in \mathcal{A}$*

$$W(f) = W_{\Phi_0}(f) = (\Phi_0\,,\Pi(f)\,\Phi_0)\,, \tag{1.4}$$

where Φ_0 is a strongly cyclic vector of Π. The representation Π generated by a state W is irreducible if and only if W is pure.

Proof. The construction of the Hilbert space $\mathcal{H}$ and operators $\Pi(f)$ of the representation Π is the same as in the theory of C^*-algebras and is given by the Wightman reconstruction theorem [4, Chapter 3, Section 3]. Usual formulations of this theorem do not include only the statement about strong cyclicity of Π. To prove it, let us note that Π is defined as the closure of the representation $\hat{\Pi}$ with the domain $\mathcal{D}(\hat{\Pi}) = \{\hat{\Pi}(f)\,\Phi_0\}_{f\in\mathcal{A}}$ where $\Phi_0 \in \mathcal{H}$ is the equivalence class of the identity in $\mathcal{A}$. Hence, according to (3.2), $\mathcal{D}(\Pi) = \overline{\mathcal{D}(\hat{\Pi})}^{G} = \overline{\Pi(\mathcal{A})\,\Phi_0}^{G}$.

The correspondence between strongly cyclic irreducible representations and pure states is also established in the same way as in the C^*-theory. If W is not pure then $W = \lambda W_1 + (1-\lambda) W_2$, where W_1 and W_2 are states on $\mathcal{A}$, $W_1 \neq W_2$, $\lambda \in (0, 1)$. By means of the state W_1 we introduce on $\{\Pi(\mathcal{A})\,\Phi_0\} \times \{\Pi(\mathcal{A})\,\Phi_0\}$ the bilinear form:

$$(\Pi(f)\,\Phi_0\,,\,\Pi(g)\,\Phi_0) = \lambda\,W_1(f^*g)\,.$$

The property $W(f^*f) > \lambda\,W_1(f^*f) > 0$ for all $f \in \mathcal{A}$ entails that the form $(\cdot\,,\cdot)$ defines uniquely a positive operator $0 < B < \lambda I$ such that for all $f,\ g \in \mathcal{A}$,

$$(\Pi(f)\,\Phi_0\,,\,\Pi(g)\,\Phi_0) = (\Pi(f)\,\Phi_0\,,\,B\,\Pi(g)\,\Phi_0)\,,$$

and for all $f, g_1, g_2 \in \mathcal{A}$,

$$\left(\Pi(g_1)\Phi_0, B\,\Pi(f)\,\Pi(g_2)\Phi_0\right) = \left(\Pi(f^*)\,\Pi(g_1)\Phi_0, B\,\Pi(g_2)\Phi_0\right). \tag{1.5}$$

Since operators $\Pi(f)$ and $\Pi(f^*)$ are continuous in the graph topology and $\Pi(\mathcal{A})\Phi_0$ is dense in $\mathcal{D}(\Pi)$, (1.5) still holds if we replace $\Pi(g_1)\Phi_0$ and $\Pi(g_2)\Phi_0$ by arbitrary $\Phi, \Psi \in \mathcal{D}(\Pi)$. Hence it follows

$$(\Phi, B\,\Pi(f)\Psi) = (\Pi(f^*)\Phi, B\Psi),$$

i.e. $B \in \Pi'_w$ and so Π is reducible.

Conversely, if $\Pi'_w \neq C(\mathcal{H})$, we can choose $B \in \Pi'_w$, $0 < B < 1$. Note that $(\Phi_0, B\Phi_0) > 0$, since otherwise

$$\left(\Pi(f)\Phi_0, B\,\Pi(f)\Phi_0\right) = \left(\Pi(f^*f)\Phi_0, B\Phi_0\right) = 0$$

for all $f \in \mathcal{A}$, which implies $B = 0$. For all $f \in \mathcal{A}$ one has

$$\left(\Phi_0, B\,\Pi(f^*f)\Phi_0\right) = \left(\Pi(f)\Phi_0, B\,\Pi(f)\Phi_0\right) > 0,$$

$$\left(\Phi_0, (I-B)\,\Pi(f^*f)\Phi_0\right) = \left(\Pi(f)\Phi_0, (I-B)\,\Pi(f)\Phi_0\right) > 0.$$

This means that $W_1(\cdot) \equiv (\Phi_0, B\Phi_0)^{-1}(\Phi_0, B\,\Pi(\cdot)\Phi_0)$ and $W_2(\cdot) \equiv (\Phi_0, (I-B)\Phi_0)^{-1}(\Phi_0, (I-B)\,\Pi(\cdot)\Phi_0)$ are states on $\mathcal{A}$ such that $W = \lambda W_1 + (1-\lambda) W_2$, where $\lambda = (\Phi_0, B\Phi_0)$ and $W_1 \neq W_2$ due to $B \neq cI$. □

Vector states on Op-algebras.* In studying vacuum structure of a Wightman theory with the field Op^*-algebra $(\mathcal{P}, \mathcal{D})$ and an arbitrary vacuum subspace $\mathcal{H}_0$ we centre on the Wightman functional which is a vector state $W_\Omega = (\Omega, \cdot\,\Omega)$ on $\mathcal{P}$ with $\Omega \in \mathcal{H}_0 \cap \mathcal{D} \cap S_1$. This brings forth the mathematical problem of investigating this class of states. It is most convenient to analyze vector states W_Φ on an Op^*-algebra $\mathcal{P}$ by comparing their properties with those of vector states on von Neumann algebras. To this end we shall associate with the algebra $\mathcal{P}$ the von Neumann algebra $R = (\mathcal{P}'_w)'$ and consider in parallel states W_Φ on $\mathcal{P}$ and ω_Φ on R with $\Phi \in \mathcal{D} \cap S_1$. The main characteristics of a state ω_Φ are the support (the projection $P^{R'}_\Phi$ or equivalently the cyclic subspace $\mathcal{H}^{R'}_\Phi$) and the $\mathcal{H}$-image $\mathfrak{M}^R_\Phi$; other characteristics include the projection P^R_Φ on the space $\mathcal{H}^R_\Phi$ of the GNS representation corresponding to ω_Φ and the central support of P^R_Φ and $P^{R'}_\Phi$, the projection $P^{Z'}_\Phi \in Z$. (The latter characteristic is of special interest in quantum field applications since the subspace $\mathcal{H}^{Z'}_\Phi$ for a pure state ω_Φ is interpreted as a coherent superselection sector.) Similar objects can be associated with the state W_Φ on $\mathcal{P} = \Pi(\mathcal{A})$, $\Phi \in \mathcal{D}(\Pi)$; the $\mathcal{H}$-image we can define as

$$\mathfrak{M}^{\mathcal{P}}_\Phi = \{\chi \in \mathcal{D}(\Pi) \mid W_\chi(A) = W_\Phi(A) \ \ \forall A \in \mathcal{P}\} \tag{1.6}$$

and the analogue of $\mathcal{H}^{Z'}_\Phi = \overline{L\{R R'\Phi\}}$ is the subspace $\overline{L\{\mathcal{P}'_w \mathcal{P}\Phi\}}$. But there are for W_Φ some other characteristics too: the following lineals are also related to it, for an arbitrary Hermitian representation Π :

$$\mathcal{D}_{\Phi}^{\mathcal{P}} = \mathcal{P}\Phi\,, \quad G_{\Phi}^{\mathcal{P}} = \overline{\mathcal{P}\,\Phi}^{G}, \quad \mathcal{H}_{\Phi}^{\mathcal{P}} = \overline{\mathcal{P}\Phi}\,, \quad \mathcal{H}_{\Phi}^{\mathcal{P}} \cap \mathcal{D}(\Pi)\,, \quad P_{\Phi}^{\mathcal{P}}\mathcal{D}(\Pi)\,,$$

and also $\mathcal{H}_{\Phi}^{\mathcal{P}} \cap \mathcal{D}(\Pi^*)$ and $P_{\Phi}^{\mathcal{P}}\mathcal{D}(\Pi^*)$. They are connected by easily checkable inclusions

$$\begin{array}{ccccccc} \mathcal{D}_{\Phi}^{\mathcal{P}} & \subset & \mathcal{H}_{\Phi}^{\mathcal{P}} \cap \mathcal{D}(\Pi) & \subset & P_{\Phi}^{\mathcal{P}}\mathcal{D}(\Pi) & & \\ \cap & & \cap & & \cap & & \\ G_{\Phi}^{\mathcal{P}} & \subset & \mathcal{H}_{\Phi}^{\mathcal{P}} \cap \mathcal{D}(\Pi^*) & \subset & P_{\Phi}^{\mathcal{P}}\mathcal{D}(\Pi^*) & \subset & \mathcal{H}_{\Phi}^{\mathcal{P}}\,. \end{array} \tag{1.7}$$

The projection $P_{\Phi}^{\mathcal{P}}$ may belong to the strong or weak commutant of $\mathcal{P}$ or in general, to none of them. In the last (and most general) case only a scarce list of properties can be obtained for W_{Φ}. As it was the case with invariant subspaces, we only get informative results under some additional conditions, of which the main ones concern the projection $P_{\Phi}^{\mathcal{P}}$.

DEFINITION 3.1.5. For a Hermitian representation Π (or equivalently its image $\mathcal{P}$) a unit vector $\Phi \in \mathcal{D}(\Pi)$ and state W_{Φ} will be called

(1) weakly regular, if $P_{\Phi}^{\mathcal{P}} \in \mathcal{P}_w'$;

(2) strongly regular, if $P_{\Phi}^{\mathcal{P}} \in \mathcal{P}_s'$;

(3) completely regular, if Φ is strongly regular and, moreover, $G_{\Phi}^{\mathcal{P}} \supset P_{\Phi}^{\mathcal{P}}\mathcal{D}(\Pi)$.

The sets of all such vectors will be denoted by $\mathfrak{N}_{wr}$, $\mathfrak{N}_{sr}$ and $\mathfrak{N}_{cr}$, respectively.

It is obvious that any cyclic vector of Π is strongly regular and any strongly cyclic vector is completely regular. Moreover, it is easy to see that if Π is closed then $P_{\Phi}^{\mathcal{P}}\mathcal{D}(\Pi) \subset G_{\Phi}^{\mathcal{P}}$ entails $P_{\Phi}^{\mathcal{P}}\mathcal{D}(\Pi) \subset \mathcal{D}(\Pi)$ whence it follows (see Proposition 3.1.5) that Φ is completely regular. Thus in this case the first condition in the definition of complete regularity can be omitted.

Now we formulate the criteria which determine the above classes of vectors.

PROPOSITION 3.1.5. *The following conditions are equivalent:*

(1) Φ *is weakly regular;*

(2) *the lineal* $P_{\Phi}^{\mathcal{P}}\mathcal{D}(\Pi)$ *is 'weakly reducing' for* Π, *i.e.*

$$P_{\Phi}^{\mathcal{P}}\mathcal{D}(\Pi) \subset \mathcal{D}(\Pi^*)\,, \quad \mathcal{P}^*\left(P_{\Phi}^{\mathcal{P}}\mathcal{D}(\Pi)\right) \subset P_{\Phi}^{\mathcal{P}}\mathcal{D}(\Pi)\,. \tag{1.8}$$

Moreover, the following conditions are equivalent:

(1') Φ *is strongly regular;*

(2') $P_{\Phi}^{\mathcal{P}}\mathcal{D}(\Pi) \subset \mathcal{D}(\Pi)$;

(3') *the restriction* $\mathcal{P}_{\Phi} = \mathcal{P}|_{P_{\Phi}^{\mathcal{P}}\mathcal{D}(\Pi)}$ *is an Op*-algebra in* $\mathcal{H}_{\Phi}$.

The following conditions are equivalent as well:

(1") Φ *is completely regular;*

(2") Φ *is strongly cyclic for the Op*-algebra* $\mathcal{P}_\Phi$ *(defined in* (3'))

(3") *the* GNS *representation of* Π *(or equivalently* $\mathcal{A}$*) corresponding to* W_Φ *coincides with the representation-restriction (reduction)* $\mathcal{P} \to \mathcal{P}_\Phi$ *or with its closure if it is not closed.*

Of these assertions, $1' \to 3' \to 2'$ and $1'' \to 2'' \to 3''$ are obvious; $1 \to 2$ follows from $1' \to 3'$ in Lemma 3.1.1. The derivation of $2 \to 1, 2' \to 1'$ and $3'' \to 1''$ is simple but cumbersome and so we refer for it to [426]. Let us add that the weak regularity is also equivalent to each of the conditions 1'–3' and the strong regularity to each of the conditions 1–3 in Lemma 3.1.1.

Let us consider the lineals (1.7) and $\mathcal{H}$-images $\mathfrak{M}_\Phi^{\mathcal{P}}$. As said above, the adequate statement of the problem is as follows: how are the properties of states W_Φ on $\mathcal{P}$ related to those of states ω_Φ on $R = (\mathcal{P}'_w)'$? To obtain these relationships we shall need some conditions not only on vectors Φ but on the algebra $\mathcal{P}$ as well. It is convenient to introduce the following 3 classes of *Op**-algebras:

(1) class I: $\mathcal{P}'_w$ is a von Neumann algebra;

(2) class II: $\mathcal{P}'_s = \mathcal{P}'_w$;

(3) class III: $\mathcal{P} = \mathcal{P}^*$.

Clearly, each condition here is stronger than preceding one(s) and each class is a subclass of preceding one(s). The difference between classes I and II is not very big: it can easily be proved [260] that any algebra of class I has an extension $\widetilde{\mathcal{P}} \supset \mathcal{P}$ which is of class II and has the same weak commutant, i.e. $\mathcal{P}'_w = \widetilde{\mathcal{P}}'_w = \widetilde{\mathcal{P}}'_s$. One more physically important class constitute *Op**-algebras generated by essentially self-adjoint operators. One can easily check [426] that such algebras belong to class I; for their study see [425]. This class includes *Op**-algebras of Wightman fields in all free, generalized free and constructive field models. For some simplest models even stronger properties of the field *Op**-algebra $\mathcal{P}$ were proved: according to [422], for free and generalized free fields as well as for the Wick square: φ^2: of a free field φ, $\mathcal{P}$ is a self-adjoint *Op**-algebra. (See also [97, 303].)

We shall now describe the properties of states W_Φ for algebras of class I and then of narrower classes II and III. Each algebra $\mathcal{P}$ of class I satisfies the relations:

$$\mathcal{H}_\Phi^{\mathcal{P}} \subset \mathcal{H}_\Phi^R,\ \overline{L\{\mathcal{P}'_w \mathcal{P}\Phi\}} \subset \mathcal{H}_\Phi^{Z'} \qquad \forall\, \Phi \in \mathcal{D}(\Pi) \tag{1.9}$$

where $Z = R \cap R'$. Here the first inclusion is implied by $P_\Phi^R \in R' = \mathcal{P}'_w$. Indeed, this entails

$$P_\Phi^R \mathcal{D}(\Pi) \supset P_\Phi^R \mathcal{P}\mathcal{D}(\Pi) \supset P_\Phi^R \mathcal{P}\Phi = \mathcal{P}^* P_\Phi^R \Phi = \mathcal{P}\Phi\,,$$

whence it follows that $\overline{\mathcal{P}\Phi} \subset P_\Phi^R \mathcal{D}(\Pi) \subset \overline{R\Phi}$.

The second inclusion is implied by the first, owing to

$$\overline{L\{\mathcal{P}'_w \mathcal{P}\Phi\}} = \overline{L\{\mathcal{P}'_w \overline{\mathcal{P}\Phi}\}} = L\{\mathcal{P}'_w \overline{\mathcal{P}\Phi}\} \tag{1.10}$$

Weakly regular vectors (and hence also strong and completely regular) have the useful property that the inclusions in (1.9) turn for them into equalities, the equality in the first of the formulas (1.9) being the necessary and sufficient condition of weak regularity.

PROPOSITION 3.1.6. *For each Op*-algebra $\mathcal{P}$ of class I the following conditions are equivalent:*

(1) $\Phi \in \mathfrak{N}_{wr}$;

(2) $\mathcal{P}'_w \Phi \cap \mathcal{D}(\Pi) \subset \mathfrak{N}_{wr}$;

(3) $\mathcal{H}^{\mathcal{P}}_{\Phi} = \mathcal{H}^{R}_{\Phi}$,

and imply

(4) $\overline{L\{\mathcal{P}'_w \mathcal{P}\Phi\}} = \mathcal{H}^{Z'}_{\Phi}$.

Proof. $\underline{2 \to 1}$ is trivial and $3 \to 4$ obvious from (1.10). $3 \to 2$. For any $B \in \mathcal{P}'_w$ such that $B\Phi \subset \mathcal{D}(\Pi)$ one has $B\mathcal{P}\Phi = \overline{B\,\mathcal{P}\Phi} = \overline{\mathcal{P}B\,\Phi}$. Due to this it follows from 3 for such B that

$$\mathcal{H}^{\mathcal{P}}_{B\Phi} = B\,\mathcal{H}^{\mathcal{P}}_{\Phi} = B\,\mathcal{H}^{R}_{\Phi} = \mathcal{H}^{R}_{B\Phi},$$

i.e. $P^{\mathcal{P}}_{B\Phi} \in R' = \mathcal{P}'_w$ and $B\Phi \in \mathfrak{N}_{wr}$.

$\underline{1 \to 3}$. With $P^{\mathcal{P}}_{\Phi} \in R'$ one has

$$\mathcal{H}^{\mathcal{P}}_{\Phi} = P^{\mathcal{P}}_{\Phi}\, R\,\mathcal{H} \supset P^{\mathcal{P}}_{\Phi}\, R\,\Phi = R\,P^{\mathcal{P}}_{\Phi}\,\Phi = R\,\Phi,$$

whence $\mathcal{H}^{R}_{\Phi} \subset \mathcal{H}^{\mathcal{P}}_{\Phi}$. ❐

Both in the theory and applications one often needs to consider restrictions of Op^*-algebras to their cyclic subspaces, i.e. algebras $\mathcal{P}_{\Phi} = \mathcal{P}|_{P^{\mathcal{P}}_{\Phi}\mathcal{D}(\Pi)}$ (defined correctly when $P^{\mathcal{P}}_{\Phi} \in \mathcal{P}'_s$, by Proposition 3.1.5). For such objects one has the following proposition.

PROPOSITION 3.1.7. *For each Op*-algebra $\mathcal{P}$ of class I and $\Phi \in \mathfrak{N}_{sr}$, $\mathcal{P}_{\Phi}$ is also the class* I *algebra. Moreover,*

$$(\mathcal{P}_{\Phi})'_w = R'_w, \qquad R_{\Phi} \equiv R_{P^{R}_{\Phi}}.$$

Proof. By definition, for any $B \in (\mathcal{P}_{\Phi})'_w$ and $f \in \mathcal{A}$ one has

$$(\mu, B\,\Pi(f)\,\nu) = (\Pi(f^*)\,\mu, B\,\nu) \qquad \forall\,\mu, \nu \in P^{\mathcal{P}}_{\Phi}\,\mathcal{D}(\Pi). \tag{1.11}$$

On the other hand, with $P^{\mathcal{P}}_{\Phi} \in \mathcal{P}'_s$ one has

$$(P^{\mathcal{P}}_{\Phi}\chi, B\,\Pi(f)\,P^{\mathcal{P}}_{\Phi}\Psi) = (\chi, P^{\mathcal{P}}_{\Phi}\,B\,P^{\mathcal{P}}_{\Phi}\,\Pi(f)\,\Psi) \qquad \forall\,\chi, \Psi \in \mathcal{D}(\Pi)$$

$$\left(\Pi(f^*)\,\overset{\mathcal{P}}{P_\Phi}\chi\,,\; B\,\overset{\mathcal{P}}{P_\Phi}\Psi\right)=\left(\Pi(f^*)\,\chi\,,\;\overset{\mathcal{P}}{P_\Phi}\,B\,\overset{\mathcal{P}}{P_\Phi}\Psi\right).$$

Together with (1.11) this gives us $\overset{\mathcal{P}}{P_\Phi}\,B\,\overset{\mathcal{P}}{P_\Phi}\in\mathcal{P}'_w=R'$, i.e.

$$\overset{\mathcal{P}}{P_\Phi}\,(\mathcal{P}_\Phi)'_w\,\overset{\mathcal{P}}{P_\Phi}\subset R',$$

whence due to $\overset{\mathcal{P}}{P_\Phi}=\overset{R}{P_\Phi}$ one has $(\mathcal{P}_\Phi)'_w\in R'_\Phi$.

To derive the inverse inclusion we take $B\in\mathcal{P}'_w$. Then due to $\overset{\mathcal{P}}{P_\Phi}\in\mathcal{P}'_s$ one has for all $\chi\,,\;\Psi\in\overset{\mathcal{P}}{P_\Phi}\,\mathcal{D}(\Pi)\subset\mathcal{D}(\Pi)$:

$$(\chi\,,\;B\,\Pi(f)\,\Psi)=\left(\chi\,,\;\overset{\mathcal{P}}{P_\Phi}\,B|_{\mathcal{H}^{\mathcal{P}}_\Phi}\,\Pi(f)\,\Psi\right),$$

$$(\Pi(f^*)\,\chi\,,\;B\,\Psi)=\left(\Pi_\Phi(f^*)\,\chi\,,\;\overset{\mathcal{P}}{P_\Phi}\,B|_{\mathcal{H}^{\mathcal{P}}_\Phi}\,\Psi\right).$$

In these equalities the left-hand sides are equal to each other because $B\in\mathcal{P}'_w$. The equality of the right-hand sides implies then $\overset{\mathcal{P}}{P_\Phi}\,B|_{\mathcal{H}^{\mathcal{P}}_\Phi}\in(\mathcal{P}_\Phi)'_w$ whence due to $\overset{\mathcal{P}}{P_\Phi}=\overset{R}{P_\Phi}$ one has $R'_\Phi\subset(\mathcal{P}_\Phi)'_w$. ❒

One often needs also to use the functionals dominated by a given state $(W\prec W_\Phi)$. In the case of a W^*-algebra R we have seen in Section 2.1 that

$$\omega\prec\omega_\Phi\Leftrightarrow\omega=\omega_{T\Phi}\,,\;T\in R',$$

for any $\Phi\in\mathcal{H}$. In the case of an Op^*-algebra we are able to obtain a similar result only for a certain subclass of vectors.

PROPOSITION 3.1.8. *For each Op^*-algebra of the class* I *and* $\Phi\in\mathfrak{N}_{cr}$ *one has*

$$W\prec W_\Phi\Leftrightarrow W=W^{(\mathcal{P}^*)}_{T\Phi},\qquad T\in\mathcal{P}'_w\,,\tag{1.12}$$

(where $W^{(\mathcal{P}^*)}$ denotes a functional on $\mathcal{P}^*$). If $\mathcal{P}$ is of class II *then*

$$W\prec W_\Phi\Leftrightarrow W=W_{T\Phi}\,,\qquad T\in\mathcal{P}'\,.$$

Proof. Let us take first an arbitrary Op^*-algebra $\mathcal{P}$ with a strongly cyclic vector χ and show that in this case $W\prec W_\chi$ is equivalent to the existence of a positive operator B such that $W(A)=(\chi\,,B\,A\,\chi)$ for all $A\in\mathcal{P}$. Sufficiency of this condition is obvious. To show the necessity we follow the proof of a similar fact for B^*-algebra [18, Section 19.1] defining on the dense domain $\mathcal{P}\chi$ a bilinear form

$$q(\Phi,\Psi)=W(A_1^*A_2)\,,\qquad\Phi=A_1\chi\,,\qquad\Psi=A_2\chi\,,\;A_1\,,A_2\in\mathcal{P}\,.$$

Due to $W\prec W_\chi$ the form q is bounded and by Riesz's Lemma, $q(\Phi,\Psi)=(\Phi,B\,\Psi)$ where B is a bounded operator. Taking a vector $\xi=A_3\chi\,,A_3\in\mathcal{P}$, one has from the equality $(A_1^*A_2)^*A_3=A_2^*(A_1A_3)$ valid on $\mathcal{D}(\Pi)$:

$$(A_1^*\Psi,\;B\,\xi)=(\Psi,\;B\,A_1\,\xi)\qquad\forall\,\Psi,\;\xi\in\mathcal{P}\chi\,.\tag{1.13}$$

A_1, A_1^* being continuous with respect to the graph topology, one can extend this equality from the domain $\mathcal{P}\chi$ to its graph closure. Hence, owing to the strong cyclicity of χ, (1.13) is valid for all $\Psi, \xi \in \mathcal{D}(\Pi)$ which means $B \in \mathcal{P}'_w$. Positivity of B follows from the fact that W is positive and symmetric.

Coming back to the assertion (1.12) we see that only the necessity is not evident in it. If $\Phi \in \mathfrak{N}_{cr}$ and $W \prec W_\Phi$ then Φ is strongly cyclic for $\mathcal{P}_\Phi$ and hence, as we have just proved,

$$W(f) = (B_\Phi \Phi, \Pi_\Phi(f)\Phi), \ B_\Phi \in (\mathcal{P}_\Phi)'_w, \ B_\Phi \geq 0.$$

Defining the operator $B \equiv P_\Phi^{\mathcal{P}} B_\Phi P_\Phi^{\mathcal{P}}$ one has $B \in \mathcal{P}'_w$ due to $P_\Phi^{\mathcal{P}} = P_\Phi^R$ and Proposition 3.1.7; in addition, $W(f) = (B\Phi, \Pi(f)\Phi)$ and, $\mathcal{P}'_w$ being a von Neumann algebra, we obtain $B = T^*T$, $T \in \mathcal{P}'_w$. Taking finally into account that $\mathcal{P}'_w = \mathcal{P}'_s$ leaves $\mathcal{D}(\Pi)$ invariant we have the last assertion of our Proposition. ❐

Denote by $PV(\mathcal{P})$ the set of all pure vector states on $\mathcal{P}$. One has the following proposition.

PROPOSITION 3.1.9. *For each Op*-algebra $\mathcal{P}$ of the class I and $\Phi \in \mathfrak{N}_{wr}$,*

$$W_\Phi \in PV(\mathcal{P}) \Rightarrow \omega_\Phi \in PV(R).$$

For all $\Phi \in \mathfrak{N}_{cr}$ the converse is also true.

Proof. For $\Phi \in \mathfrak{N}_{wr}$ it is sufficient to prove that $R'_{P_\Phi^R} = C(\mathcal{H}_\Phi^R)$ if $W_\Phi \in PV(\mathcal{P})$, i.e. the GNS representation Π_Φ^{GNS} corresponding to W_Φ is irreducible,

$$(\Pi_\Phi^{GNS})'_w = C(\mathcal{H}_\Phi^R) \qquad (1.14)$$

Π_Φ^{GNS} is defined as $\mathcal{P} \to \mathcal{P}|_{G_\Phi^{\mathcal{P}}}$. Denoting by $\hat{\Pi}_\Phi$ the representation $\mathcal{P} \to \mathcal{P}|_{\mathcal{P}_\Phi}$ one has:

$$\overline{\hat{\Pi}_\Phi} = \Pi_\Phi^{GNS}, \ (\hat{\Pi}_\Phi)'_w = (\Pi_\Phi^{GNS})'_w. \qquad (1.15)$$

Taking $B \in \mathcal{P}'_w$ we then proceed like in proving the second inclusion in Proposition 3.1.7 (but now with $\Phi \in \mathfrak{N}_{wr}$ instead of $\Phi \in \mathfrak{N}_{sr}$). This gives us

$$R'_{P_\Phi^R} \subset (\hat{\Pi}_\Phi)'_w$$

whence the sought result follows due to (1.14) and (1.15).

Choosing now $\Phi \in \mathfrak{N}_{cr}$, let us prove the converse, i.e. the purity of ω_Φ implies the purity of W_Φ. We have $R'_{P_\Phi^R} = C(\mathcal{H}_\Phi^R)$ and, moreover, by Proposition 3.1.7,

$$(\mathcal{P}_\Phi)'_w = R'_{P_\Phi^R}.$$

But for $\Phi \in \mathfrak{N}_{cr}$ the representation $\mathcal{P} \to \mathcal{P}_\Phi$ (or its closure having the same weak commutant) coincides with Π_Φ^{GNS}. Due to $\mathcal{H}_\Phi^{\mathcal{P}} = \mathcal{H}_\Phi^R$ this implies (1.14), i.e. $W_\Phi \in PV(\mathcal{P})$. ❐

Let us now find the relation between unitary equivalence of states W_Φ and states ω_Φ. One should take into account that if a *-algebra $\mathcal{A}$ is unnormed, intertwining operators T between its representations Π_1 and Π_2 must satisfy the additional condition $T\mathcal{D}(\Pi_1) \subset \mathcal{D}(\Pi_2)$.

PROPOSITION 3.1.10. *For each Op*-algebra* $\mathcal{P}$ *of class* II *and* $\Phi, \Psi \in \mathfrak{N}_{cr}$,

$$\omega_\Phi \simeq \omega_\Psi \Leftrightarrow W_\Phi \simeq W_\Psi .$$

Proof. Taking states $\omega_\Phi \simeq \omega_\Psi$ and an isometric intertwining operator

$$U : \overset{R}{\mathcal{H}}_\Phi \to \overset{R}{\mathcal{H}}_\Psi , \; U\overset{R}{P}_\Phi \in R' ,$$

one has for all $\chi \in \mathcal{D}(\Pi)$, $f \in \mathcal{A}$ $\left(\text{due to } \overset{\mathcal{P}}{G}_\Phi = \overset{R}{P}_\Phi , \; \mathcal{D}(\overline{\Pi})\, \overset{\mathcal{P}}{G}_\Psi = \overset{R}{P}_\Phi \, \mathcal{D}(\overline{\Pi})\right)$

$$\begin{aligned} U \overset{\text{GNS}}{\Pi}_\Phi (f) \overset{R}{P}_\Phi \chi &= U \overline{\Pi} \; (f) \overset{R}{P}_\Phi \chi \\ &= U \overset{R}{P}_\Phi \; \overline{\Pi} (f) \chi = \overline{\Pi} (f)\, U \overset{R}{P}_\Phi \chi = \overset{\text{GNS}}{\Pi}_\Psi (f)\, U \overset{R}{P}_\Phi \chi , \end{aligned}$$

which means that U is intertwining $\overset{\text{GNS}}{\Pi}_\Phi$ and $\overset{\text{GNS}}{\Pi}_\Psi$. Checking in the same way that U^* is intertwining $\overset{\text{GNS}}{\Pi}_\Psi$ and $\overset{\text{GNS}}{\Pi}_\Phi$, we obtain $W_\Phi \simeq W_\Psi$.

Conversely, if $W_\Phi \simeq W_\Psi$ and U an isometric intertwining operator then it is sufficient to check that $U\overset{R}{P}_\Phi \in \mathcal{P}'$. This property follows from the chain of equalities:

$$\begin{aligned} U\overset{\mathcal{P}}{P}_\Phi \overline{\Pi} (f) \chi &= U\overset{\mathcal{P}}{P}_\Phi \overline{\Pi} (f) \left(\overset{\mathcal{P}}{P}_\Phi \chi + (\overset{\mathcal{P}}{P}_\Phi)^\perp \chi\right) = U \overline{\Pi}_\Phi (f) \overset{\mathcal{P}}{P}_\Phi \chi \\ &= \overline{\Pi}_\Psi (f)\, U\overset{\mathcal{P}}{P}_\Phi \chi = \overline{\Pi} (f)\, U \overset{\mathcal{P}}{P}_\Phi \chi , \end{aligned}$$

for all $\chi \in \mathcal{D}(\Pi)$ and $f \in \mathcal{A}$. Thus U is an intertwining operator for ω_Φ and ω_Ψ and $\omega_\Phi \simeq \omega_\Psi$.

In conclusion we consider $\mathcal{H}-$ images of states W_Φ. ❐

PROPOSITION 3.1.11.

(A) *For* $\Phi \in \mathcal{D}(\Pi) \cap S_1$ *and an algebra* $\mathcal{P}$ *of class* I,

$$\overset{\mathcal{P}}{\mathfrak{M}}_\Phi \supset \overset{R}{\mathfrak{M}}_\Phi \cap \mathcal{D}(\Pi) . \tag{1.16}$$

If $\mathcal{P}$ *is of class* II *then, in addition,*

$$\overset{\mathcal{P}}{\mathfrak{M}}_\Phi \supset \overset{R}{\mathfrak{M}}_\Phi . \tag{1.17}$$

(B) *For* $\Phi \in \mathfrak{N}_{cr}$ *and an algebra* $\mathcal{P}$ *of class* I,

$$\overset{\mathcal{P}}{\mathfrak{M}}_\Phi \subset \mathcal{P}'_w \cap \mathcal{D}\,(\Pi) \cap S_1 . \tag{1.18}$$

If $\mathcal{P}$ *is of class* II *then, in addition,*

$$\overset{\mathcal{P}}{\mathfrak{M}}_\Phi \subset \mathcal{P}'\Phi \cap S_1 . \tag{1.19}$$

If $\mathcal{P}$ is of class III *then*

$$\mathfrak{M}_{\Phi}^{\mathcal{P}} = \left\{ \Psi = T\,\Phi \mid T \in \mathcal{P}',\ T^*T = P_{\Phi}^{\mathcal{P}},\ T\,T^* = P_{\Psi}^{\mathcal{P}} \right\} \tag{1.20}$$

and, in addition,

$$\mathfrak{M}_{\Phi}^{\mathcal{P}} \subset \mathfrak{N}_{\mathrm{cr}},\ \mathfrak{M}_{\Phi}^{\mathcal{P}} = \mathfrak{M}_{\Phi}^{R}. \tag{1.21a, b}$$

Proof. The formula (1.3.43) for $\mathcal{H}$-image $\mathfrak{M}_{\Phi}^{R}$ implies that if $\Psi \in \mathcal{D}(\Pi) \cap \mathfrak{M}_{\Phi}^{R}$ then $\Psi = T\,\Phi$, $T \in \mathcal{P}_W'$. Whence it follows for all $f \in \mathcal{A}$,

$$T\,\Pi\,(f)\,\Phi = \Pi^*\,(f)\,T\,\Phi = \Pi\,(f)\,T\,\Phi$$

and we obtain (1.16):

$$W_{\Psi}\,(f) = (T\,\Phi, \Pi\,(f)\,T\,\Phi) = (T^*T\,\Phi, \Pi\,(f).\Phi) = W_{\Phi}\,(f)\,.$$

(1.17) follows simply from (1.16). To derive (1.18) it is convenient to use $\mathcal{H}$-images $\mathfrak{M}_{\Phi}^{\mathcal{P}^*}$ of vector states on $\mathcal{P}^*$ defined as

$$\mathfrak{M}_{\Phi}^{\mathcal{P}^*} = \left\{ \Psi \in \mathcal{D}\,(\Pi^*) \mid W_{\Phi}^{(\mathcal{P}^*)} = W_{\Psi}^{(\mathcal{P}^*)} \right\}$$

for all $\Phi \in \mathcal{D}\,(\Pi^*)$. It is easy to see that

$$\mathfrak{M}_{\Phi}^{\mathcal{P}} = \mathfrak{M}_{\Phi}^{\mathcal{P}^*} \cap \mathcal{D}\,(\Pi) \tag{1.22}$$

for all $\Phi \in \mathcal{D}\,(\Pi)$. For $\Phi \in \mathfrak{N}_{\mathrm{cr}}$ (1.12) entails

$$\mathfrak{M}_{\Phi}^{\mathcal{P}^*} \subset \mathcal{P}_W' \cap S_1\,.$$

Hence, taking account of (1.22), we have (1.18) from which (1.19) easily follows.

Finally, for a class III algebra it is convenient to derive first (1.21a). Take $\Psi \in \mathfrak{M}_{\Phi}^{\mathcal{P}}$, i.e. $W_{\Phi} \simeq W_{\Psi}$. GNS representations of these functionals are unitarily equivalent and, Φ being completely regular, one has $P_{\Phi}^{R} \in \mathcal{P}'$. By Proposition 3.1.2 this implies self-adjointness of $\Pi_{\Phi}^{\mathrm{GNS}}$. Hence $\Pi_{\Psi}^{\mathrm{GNS}}$ is also self-adjoint, due to $\Pi_{\Phi}^{\mathrm{GNS}} \simeq \Pi_{\Psi}^{\mathrm{GNS}}$. Next, $\mathcal{P}$ being closed, one has $G_{\Phi}^{\mathcal{P}} \subset \mathcal{D}\,(\Pi)$, i.e. $\Pi_{\Psi}^{\mathrm{GNS}}$ is a subrepresentation of Π. Again, by Proposition 3.1.2, self-adjointness of $\Pi_{\Psi}^{\mathrm{GNS}}$ entails $P_{\Psi}^{R} \in \mathcal{P}'$ which, in its turn, entails self-adjointness of the representation-restriction $\pi_{\Psi} : \mathcal{P} \to \mathcal{P}_{\Psi}$. Due to the closedness of $\mathcal{P}$, $\Pi_{\Psi}^{\mathrm{GNS}} \subset \pi_{\Psi}$ whence $\Pi_{\Psi}^{\mathrm{GNS}} = \pi_{\Psi}$, i.e. $\mathcal{D}\left(\Pi_{\Psi}^{\mathrm{GNS}}\right) = G_{\Psi}^{\mathcal{P}} \subset \mathcal{D}\,(\pi_{\Psi}) = P_{\Psi}^{\mathcal{P}}\,\mathcal{D}\,(\Pi)$ and $\Psi \in \mathfrak{N}_{\mathrm{cr}}$. Let us prove (1.20). Since we have proved that $\pi_{\Phi} \simeq \pi_{\Psi}$, there exists an isometry $U : \mathcal{H}_{\Phi}^{\mathcal{P}} \to \mathcal{H}_{\Psi}^{\mathcal{P}}$ such that

$$U\,\Phi = \Psi,\ U\,\mathcal{D}\,(\pi_{\Phi}) = \mathcal{D}\,(\pi_{\Psi}),\ U\,\pi_{\Phi}\,(A)\,\chi = \pi_{\Psi}\,(A)\,U\,\chi \tag{1.23}$$

for all $A \in \mathcal{P}$ and $\chi \in \mathcal{D}\,(\pi_{\Phi})$. The operator $T = P_{\Psi}^{\mathcal{P}}\,U\,P_{\Phi}^{\mathcal{P}}$ has all the necessary properties. Indeed, since both $P_{\Psi}^{\mathcal{P}}$ and $P_{\Phi}^{\mathcal{P}}$ are in $\mathcal{P}'$, one has from (1.23): $T \in \mathcal{P}'$, $T.T^* = P_{\Psi}^{\mathcal{P}}\,U\,P_{\Phi}^{\mathcal{P}}\,U^*P_{\Psi}^{\mathcal{P}} = P_{\Psi}^{\mathcal{P}}$ and, in the same way, $T^*\,T = P_{\Phi}^{\mathcal{P}}$.

Conversely, if $\Psi = T\Phi$, T having all the properties from (1.20), then due to $\overset{\mathcal{P}}{P}_{\Phi}$, $\overset{\mathcal{P}}{P}_{\Psi} \in \mathcal{P}'$ one has for all $A \in \mathcal{P}$:

$$W_{\Psi}(A) = (T\Phi,\ AT\Phi) = (T\Phi,\ TA\Phi) = (T^{*}T\Phi,\ A\Phi) = W_{\Phi}(A)\,.$$

Finally, (1.21b) follows from (1.20) taking account of the formulas (1.3.43), (1.17) and the equality $\overset{\mathcal{P}}{P}_{\Phi} = \overset{R}{P}_{\Phi}$ valid for all $\Psi \in \overset{\mathcal{P}}{\mathfrak{M}}_{\Phi} \subset \mathfrak{N}_{\mathrm{cr}}$. ❐

COROLLARY . *If $\mathcal{P}$ is of class* I *and Φ its pure completely regular vector, then*

$$\overset{\mathcal{P}}{\mathfrak{M}}_{\Phi} = \mathcal{P}'_{w}\,\Phi \cap \mathcal{D}(\Pi) \cap S_1\,. \tag{1.24}$$

If $\mathcal{P}$ is of class II *then*

$$\overset{\mathcal{P}}{\mathfrak{M}}_{\Phi} = \mathcal{P}'_{w}\,\Phi \cap\ S_1 = \overset{R}{\mathfrak{M}}_{\Phi}\,. \tag{1.25}$$

Proof. (1.24) follows from (1.16) and (1.18) taking account of Proposition 3.1.9 and Lemma 2.1.7. (1.25) follows from (1.24). ❐

3.1.2. VACUUM STRUCTURE OF QUANTUM FIELD THEORY

Now we come back to our physical problem. It is parallel to that discussed in Section 1.3.5: there the vacuum structure was studied for a system of local observables while here we should study it for a system of quantized fields. Mathematically, this corresponds to the transition from C^*-dynamical systems to those with 'Op^*-dynamics' given by an Op^*-algebra and a group of its translation automorphisms. The new problem is more involved and the results are not complete as yet. However, we have already a clear view of both the general situation (according to which all vacuum states are grouped into sectors separated by a superselection rule, and the decomposition into such sectors is directly related to the decomposition over pure thermodynamical phases) and the main effect distinguishing the Op^* and C^* types of dynamics (it is the so-called 'Borchers anomaly': possibility of vacuum degeneration in irreducible field theory).

Let us consider a general Wightman theory given as a Poincaré covariant representation of the Borchers algebra. It is described by a Hilbert space $\mathcal{H}$, field Op^*-algebra $(\mathcal{P}, \mathcal{D})$ and unitary representation $U(a, \Lambda)$ of the Poincaré group. Vacuum structure of the theory is determined by the 'vacuum subspace' $\mathcal{H}_0$ which includes all translation invariant vectors in $\mathcal{H}$ and may be, in general, of an arbitrary dimension d_0. Any Wightman theory can be supposed to be canonically generated by some Wightman functional which is, by the reconstruction theorem (Proposition 3.1.4), a vector state W_{Ω} on $\mathcal{P}$ corresponding to a strongly cyclic translation invariant vector $\Omega \in \mathcal{H}_0 \cap S_1$. Thus, without loss of generality, $\mathcal{H}_0$ can be supposed to include at least one strongly cyclic vector. Furthermore, vacuum structure is also characterized by physical vacuum states. By definition, any physical state is a functional on the observable algebra of the system studied, i.e. quasilocal algebra $\mathfrak{A}$ or global algebra $\pi(\mathfrak{A})''$ of some chosen representation of $\mathfrak{A}$. Hence we must define these objects, and in our framework it should be done in terms of the algebra $\mathcal{P}$ and Wightman functional W_{Ω}.

In scalar neutral theories which we consider, C^* and W^* algebras of observables coincide with C^* and W^* field algebras (while in the case of fields with nontrivial quantum numbers we have seen in Chapter 2 that observable algebras are typically defined as gauge-invariant subalgebras of field algebras; e.g., the global algebra being typically of the form $R = F \cap U(G)'$, with F a von Neumann algebra of fields and U a unitary representation of the group G of all internal symmetries of the system studied.) It is not difficult to associate such algebras with our Op^*-algebra. As seen above, the von Neumann algebra $R = (\mathcal{P}_w')'$ is naturally associated with every Op^*-algebra $\mathcal{P}$. We shall accept this formula for R as the definition of the global observable algebra in a scalar neutral Wightman theory. In physical models another definition is in standard use: when field operators $A(f)$, $f \in S(\mathbf{R}^4)$, $\mathrm{Im} f = 0$, are essentially self-adjoint, the bounded functions of them are correctly defined, and the field W^*-algebra is introduced as the von Neumann algebra generated by all these functions or, equivalently, by the exponents $\exp\left[i\overline{A(f)}\right]$. This definition, in contrast to the former one, needs certain conditions on field operators. It is easy to show (see Section 3.2) that if these conditions are met and so both ways of defining W^* field algebras are possible for a Wightman theory, they produce the same algebras. Thus, we define vacuum states as vector and normal translation invariant states ω on R.

Like in the C^*-theory (Section 1.3.5), the problem of vacuum structure has the closest connection with the global properties of the system, needs sometimes an account of asymptotical properties and practically does not depend upon local behaviour. To exploit asymptotical properties we need local algebras, however. In particular, the important property of existence and unicity of a vacuum state in coherent superselection sector (Proposition 1.3.47) has been proved with the help of local net structure. This property will be used below in a number of cases; and hence in these cases we actually suppose Wightman theory to be endowed not only with a global observable algebra R but also with a local net $\{R(O)\}$. Such situation is described by the following definition.

DEFINITION 3.1.6. Wightman theory given by the field Op^*-algebra $(\mathcal{P}, \mathcal{D})$ and a unitary representation $U(a, \Lambda)$ of the Poincaré group will be said to have an associated Haag–Araki theory if the algebra $R = (\mathcal{P}_w')'$ serves as the global observable algebra for the Haag–Araki theory with the same representation $U(a, \Lambda)$ of the Poincaré group.

In contrast to the existence of a global algebra R, the existence of a local net $R(O)$ puts some restrictions on the Wightman theory. They will be analyzed in Section 3.2.

Let us turn to describing existing types of vacuum structure, proceeding from simpler to more complicated ones. The axioms of translation invariance and spectrum condition will be always supposed valid; the other ones will be specially mentioned when used.

Irreducible Field Systems. Borchers' Anomaly.

We start with the simplest case corresponding to the standard Wightman theory with an irreducible system of fields and a unique strongly cyclic vacuum:

$$d_0 = 1, \ \mathcal{H}_0 = \{c\,\Omega\}_{c \in \mathbf{C}}, \ \overline{\mathcal{P}\Omega}^G = \mathcal{D}.$$

This case is fully characterized by the following elementary proposition.

PROPOSITION 3.1.12. *Let Wightman theory be given with the class* II *Op^*-algebra* $\mathcal{P}$ *and a strongly cyclic vacuum vector* Ω. *Then the condition*

(1) $d_0 = 1$

implies the conditions

(2) W_Ω *is pure;*

(3) ω_Ω *is pure;*

(4) $\mathcal{P}'_w = \mathbf{C}(\mathcal{H})$,

which are equivalent to one another and, in their turn, imply the condition

(1') $d_0 = 1$ or ∞.

If an associated Haag–Araki theory exists, the conditions 2–4 *imply* 1.

Proof. Note a simple fact: in Wightman theory with an arbitrary algebra $\mathcal{P}$, translation operators $T(a)$ implement a group of unitary automorphisms of the von Neumann algebra R and these automorphisms can be supposed inner without the loss of generality; i.e. $T(a) \in R$ (whence $P_0 \in R$ too). Indeed, one has for any Op^*-algebra $(\mathcal{P}, \mathcal{D})$ and any unitary operator $\mathcal{U}$ such that $\mathcal{U}\mathcal{D} \subset \mathcal{D}$, $\mathcal{U}^* \mathcal{D} \subset \mathcal{D}$:

$$(\mathcal{U}\mathcal{P}\mathcal{U}^*)'_w = \mathcal{U}\mathcal{P}'_w\,\mathcal{U}^*.$$

This formula can be checked directly. Being applied to our algebra $\mathcal{P}$ with the group of automorphisms $T(a)\cdot T(-a)$ it implies the first statement, $T(a)\cdot T(-a)$ are automorphisms of R. The second statement follows from the spectrum condition and Theorem 1.3.43.

Due to the above fact, the $\mathcal{H}$-image $\mathfrak{M}^R_\Omega$ of the state ω_Ω belongs to $\mathcal{H}_0$. Hence, if $d_0 = 1$, this $\mathcal{H}$-image is one-dimensional and ω_Ω is pure. Moreover, Ω being cyclic for R (due to $\mathcal{H}^{\mathcal{P}}_\Omega \subset \mathcal{H}^R_\Omega$ for the class I algebra $\mathcal{P}$), $R = B(\mathcal{H})$ and we have obtained $1 \to 3 \to 4$. The equivalence $4 \leftrightarrow 2$ is well known and $4 \to 3$ is trivial since $R' = \mathcal{P}'_w$. $2 \to 1'$ follows from Theorem 3.1.19 below (which uses the class II assumption for $\mathcal{P}$). The last assertion is obvious since in the Haag–Araki theory the vacuum vector is unique if it is cyclic and pure (the equivalence $4 \leftrightarrow 6$ on p. 107). ❒

We see that already in this simplest situation the vacuum structure of field theory displays an important distinction from that of local quantum theory: an irreducible field theory may have, in the general case, not only one-dimensional but also infinite-dimensional vacuum subspace. This phenomenon was discovered by Borchers in 1965 [75] and is called sometimes *Borchers' anomaly.*

Let us consider it in more detail. Proposition 3.1.12 tells us next to nothing about it: here the Borchers anomaly appears just as a formal possibility which we cannot exclude being not able to prove the implication (2–4) $\to$ 1. But, in fact, the roots of this anomaly are not difficult to trace down (at least on the level of intuitive arguments). In the Haag–Araki theory we have seen (Proposition 1.3.46 using the locality axiom) that R_{P_0} is an abelian von Neumann algebra; and if, moreover, $\mathcal{H}_0$ is cyclic for Z' then R'

is isomorphic to R'_{P_0} so that the irreducibility of R is equivalent to that of R_{P_0}. But the only irreducible abelian von Neumann algebra is $\mathcal{B}(\tilde{\mathcal{H}}) = C(\tilde{\mathcal{H}})$ in one-dimensional space $\tilde{\mathcal{H}}$ – and so we get the classical result that in a local quantum theory the global algebra is irreducible iff the vacuum is unique and cyclic. As for the Wightman theory, we have seen that $P_0 \in R = (\mathcal{P}'_w)'$ and hence if $\mathcal{H}_0$ is cyclic for Z' we conclude again that R' and R'_{P_0} are isomorphic and the irreducibility of R is equivalent to that of R_{P_0}. Hence, if an associated Haag–Araki theory exists, we come to the former conclusion about the unicity of vacuum in an irreducible theory: $R = \mathcal{B}(\mathcal{H}) \Leftrightarrow d_0 = 1$. Suppose, however, that for our field theory the existence of an associated Haag–Araki theory is not established. Then, having no local W^*-algebras, we cannot prove that R_{P_0} is abelian. But we have instead local Op^*-algebras $\mathcal{P}(O)$, and by the same arguments as in the Haag–Araki theory we can expect that now some other algebra will be abelian, namely, the reduction of Op^*-algebra $\mathcal{P}$, i.e. the algebra $\mathcal{P}_{P_0} = P_0 \mathcal{P} P_0|_{\mathcal{H}_0}$ (of course, if we can prove that it is well defined). And if, in addition, the irreducibility of R_{P_0} is equivalent to that of $\mathcal{P}_{P_0}$ (in the same way as the irreducibility of R is equivalent to that of $\mathcal{P}$), for which it is sufficient that

$$(\mathcal{P}_{P_0})'_w = R'_{P_0} \tag{1.26}$$

then we come to the following 'Wightman analogue' of the connection between the irreducibility and vacuum structure: the irreducibility of $\mathcal{P}$ is equivalent to that of the abelian Op^*-algebra $\mathcal{P}_{P_0}$. But it is well known (see, e.g., [352]) that an abelian Op^*-algebra can be irreducible not only in one-dimensional but also in infinite-dimensional space! The final conclusion will be then as follows: the Borchers' anomaly is present iff the representation-restriction of the abelian Op^*-algebra: $P_0 \mathcal{P} P_0 \to \mathcal{P}_{P_0}$ is irreducible and infinite-dimensional.

Rigorous realization of this scheme brings forth a number of refinements and assumptions due to which the final results turn out to be noticeably weaker. The starting problem is, as we have seen, the definition of a certain 'vacuum Op^*-algebra' acting in $\mathcal{H}_0$. This turns out to be the most complicated problem. We shall not be able to define correctly the 'reduction' $\mathcal{P}_{P_0}$ and shall construct instead of this a somewhat different Op^*-algebra in $\mathcal{H}_0$. Let $(\mathcal{P}_{\text{loc}}, \mathcal{D})$ be the Op^*-subalgebra of $(\mathcal{P}, \mathcal{D})$ consisting of all the operators $\Pi(f)$ with $f \in \mathcal{A}_{\text{loc}} \equiv \cup_{O \in B(M)} \mathcal{A}(O) = \mathbf{C} \oplus_{n=1}^{\infty} \mathcal{D}(\mathbf{R}^4)$. $\mathcal{A}_{\text{loc}}$ may not be closed even if $\mathcal{P}$ is closed, i.e. in general,

$$\mathcal{D}_{\text{loc}} \equiv \mathcal{D}(\overline{\mathcal{P}}_{\text{loc}}) = \bigcap_{f \in \mathcal{A}_{\text{loc}}} \overline{\Pi(f)} \supset \mathcal{D}.$$

To start with, we prove the following proposition.

PROPOSITION 3.1.13 [75]. *Let a Wightman theory be given with the field Op^*-algebra $(\mathcal{P}, \mathcal{D})$ and vacuum subspace $\mathcal{H}_0$ including a strongly cyclic vector Ω. Let also the axioms of translation covariance, locality and spectrum condition hold. Then the region $\mathcal{D}_{0,\text{loc}} \equiv \mathcal{D}_{\text{loc}} \cap \mathcal{H}_0$ is dense in $\mathcal{H}_0$.*

Proof is based on the results by Borchers [74] concerning the analytic properties of

Wightman functions. By Theorem 2 of [74], for arbitrary $\Psi \in \mathcal{D}_0 \equiv \mathcal{D} \cap \mathcal{H}_0$ and f, g from the Borchers algebra $\mathcal{A}$, the expectations

$$F(x, y) = \left(\Psi, \Pi(f^*) T(x) \Pi(g^* g) T(y) \Pi(f) \Psi\right)$$

can be continued analytically into the future tube, $\mathrm{Im}\, x \in V^+$, $\mathrm{Im}\, y \in V^+$. Let us take $\Pi(f) \in \mathcal{P}(O_1)$, $\Pi(g) \in \mathcal{P}(O_2)$ for some $O_1, O_2 \in B(M)$, and note that, due to spectrum condition, for complex translations y with the imaginary part in the light cone, not only weak but also strong convergence holds in the asymptotic formula for translation operators (Proposition 1.3.36):

$$\underset{\lambda \to \infty}{s\text{-lim}}\; T(\pm \lambda y) = P_0 , \quad \mathrm{Im}\, y \in V^{\pm} . \tag{1.27}$$

Choosing $x \in (O_1 \cup O_2)'$ and $\mathrm{Im}\, y \in V^+$ or $\mathrm{Im}\, y = 0$ one has

$$| F(x, y) | \leq \| \Pi(f) T(-x) \Pi(g) \Psi \| \cdot \| \Pi(f) \Psi \| .$$

Hence, due to the translation invariance of Ψ and locality,

$$| F(x, y) | \leq \| \Pi(f) \Psi \| \cdot \| \Pi(f^* f) \Psi \|^{1/2} \cdot \| \Pi(g^* g) \Psi \|^{1/2} .$$

Thus the modulus of $F(x, y)$ is uniformly bounded in the following region in $M + iM$:

$$\{x \in (O_1 \cup O_2)', \ \mathrm{Im}\, y \in V^+ \text{ or } \mathrm{Im}\, y = 0\}$$
$$\cup \{x \in (O_1 \cup O_2)', \ \mathrm{Im}\, x \in V^+ \text{ or } \mathrm{Im}\, x = 0\}.$$

Fixing a vector $x \in (O_1 \cup O_2)'$, let us consider $F(x, y)$ for the arguments $x = (\lambda_1 u_1 \tilde{x}^0, \lambda_1 \tilde{\mathbf{x}})$, $y = (\lambda_2 u_2 \tilde{x}^0, \lambda_2 \tilde{\mathbf{x}})$ where λ_1, λ_2 are real and u_1, u_2 complex parameters. Then we obtain the function $f_{\lambda_1 \lambda_2}(u_1, u_2)$ whose modulus at $|\lambda_1|$, $|\lambda_2| \geq 1$ is uniformly bounded in the region

$$\{\mathrm{Im}\, \lambda_1 u_1 > 0, |u_2| \leq 1, \mathrm{Im}\, u_2 = 0\} \cup \{\mathrm{Im}\, \lambda_2 u_2 > 0, |u_1| \leq 1, \mathrm{Im}\, u_1 = 0\}.$$

By Lemma 8 of [74], the modulus of $f_{\lambda_1 \lambda_2}(u_1, u_2)$ will be then uniformly bounded in the region

$$\{\mathrm{Im}\, \lambda_1 u_1 > 0, \mathrm{Im}\, \lambda_2 u_2 > 0, \ |u_1| < 1, \ |u_2| < 1\}.$$

Hence, it follows that at fixed values of $\mathrm{Im}\, u > 0$, $|u| < 1$ the set of norms

$$n_\lambda = \| \Pi(g) T(\lambda u \tilde{x}^0, \lambda \tilde{\mathbf{x}}) \Pi(f) \Psi \|$$

is bounded on the interval $\lambda > 1$. Taking also into account that due to (1.27) the sequence of scalar products $\left(\Phi, \Pi(g) T(\lambda u \tilde{x}^0, \lambda \tilde{\mathbf{x}}) \Pi(f) \Psi\right)$ is convergent at $\lambda \to \infty$ for all Φ from the dense set $\mathrm{dom}\, \Pi(g)^*$, we obtain the weak convergence of the sequence of vectors $\Pi(g) T(\lambda u \tilde{x}^0, \lambda \tilde{\mathbf{x}}) \Pi(f) \Psi$ at $\lambda \to \infty$. Moreover, due to (1.27),

$$T(\lambda u \tilde{x}^0, \lambda \tilde{\mathbf{x}}) \Pi(f) \Psi \xrightarrow[\lambda \to \infty]{} P_0 \Pi(f) \Psi .$$

Thus, $\overline{\Pi(g)}$ being a closed operator, we have obtained that $P_0 \Pi(f) \Psi \in \operatorname{dom} \overline{\Pi(g)}$ for all $\Pi(f) \in \mathcal{P}(O_1)$, $\Pi(g) \in \mathcal{P}(O_2)$ and arbitrary compact regions O_1, O_2 in M. Hence, it follows

$$\mathcal{P}_0 \mathcal{P}_{\text{loc}} \mathcal{D}_0 \subset \mathcal{D}_{0,\text{loc}} . \tag{1.28}$$

But, by the Reeh–Schlieder theorem for the net $\{\mathcal{P}(O)\}$, there exists a cyclic vector for $\mathcal{P}_{\text{loc}}$ in $\mathcal{D}_0$. Hence, $\mathcal{P}_{\text{loc}} \mathcal{D}_0$ is dense in $\mathcal{H}$ and the left-hand side of (1.28) is dense in $\mathcal{H}_0$. □

Now we can introduce an abelian Op^*-algebra in $\mathcal{H}_0$.

PROPOSITION 3.1.14 [75]. *Under the assumptions of Proposition* 3.1.13, *the set of operators*

$$\mathcal{P}_{0,\text{loc}} \doteq \vee \left\{ P_0 \overline{\Pi} P_0 \big|_{\mathcal{D}_{0,\text{loc}}}, \ \Pi \in \mathcal{P}_{\text{loc}} \right\}$$

is an abelian Op^*-algebra in $\mathcal{H}_0$ with the domain in $\mathcal{D}_{0,\text{loc}}$.

Proof. Obviously, the operators $P_0 \overline{\Pi} P_0 |_{\mathcal{D}_{0,\text{loc}}}$ as well as their products are well defined on $\mathcal{D}_{0,\text{loc}}$ and $\mathcal{P}_{0,\text{loc}}$ is an Op-algebra. It is also an Op^*-algebra since we check directly, first for the elements of the form $P_0 \overline{\Pi} P_0 |_{\mathcal{D}_{0,\text{loc}}}$ and then for an arbitrary $\Pi_0 \in \mathcal{P}_{0,\text{loc}}$, that

$$\operatorname{dom} \Pi_0^* \supset \mathcal{D}_{0,\text{loc}} , \quad \Pi_0^* \big|_{\mathcal{D}_{0,\text{loc}}} \in \mathcal{P}_{0,\text{loc}} .$$

Finally, to prove the abelianness of $\mathcal{P}_{0,\text{loc}}$ it is sufficient to show that for all $\Phi, \Psi \in \mathcal{D}_{0,\text{loc}}$ and all $\Pi(f) \in \mathcal{P}(O_1)$, $\Pi(g) \in \mathcal{P}(O_2)$ with arbitrary $O_1, O_2 \in B(M)$ one has

$$\left(\Phi , \overline{\Pi(f)}\, P_0\, \overline{\Pi(g)}\, \Psi\right) = \left(\Phi , \overline{\Pi(g)}\, P_0\, \overline{\Pi(f)}\, \Psi\right) . \tag{1.29}$$

Fixing again $\tilde{x} \in (O_1 \cup O_2)'$ let us consider vectors of the form $x_{\lambda,u} = (\lambda u \tilde{x}^0, \lambda \tilde{\mathbf{x}})$, where λ is a real and u complex parameter, and define the function

$$f_\lambda(u) = \left(\Phi , \overline{\Pi(f)}\ T(x_{\lambda,u})\, \overline{\Pi(g)}\, \Psi\right), \quad |u| < 1, \lambda > 1.$$

By asymptotic formulae (1.27) one has for $\operatorname{Im} u > 0$;

$$\lim_{\lambda \to \infty} f_\lambda(u) = \left(\Phi , \overline{\Pi(f)}\, P_0\, \overline{\Pi(g)}\, \Psi\right) . \tag{1.30}$$

For $\operatorname{Im} u < 0$, checking by means of graph topology that if two elements of some Op^*-algebra are commuting, their extensions to the domain of its closure are also commuting, we have, due to the locality axiom:

$$\lim_{\lambda \to \infty} f_\lambda(u) = \left(\Phi , \overline{\Pi(g)}\, P_0\, \overline{\Pi(f)}\, \Psi\right) . \tag{1.31}$$

Since

$$f_\lambda(u) = \int e^{-\lambda \operatorname{Im} u\tilde{x}^0 p_0} \; e^{i\lambda(\operatorname{Re} u\tilde{x}^0 p_0 - \tilde{x}p)} \left(\Phi, \overline{\Pi\ (f)}\, dE(p)\, \overline{\Pi\ (f)}\, \Psi\right),$$

$f_\lambda(u)$ is uniformly bounded for $\lambda > 0$ and $|u| < 1$. Hence, by the Vitali theorem [9, IV, 10.9], (1.30) and (1.31) entail (1.29).

Finally, the desired connection between the irreducibility of $\mathcal{P}$ and $\mathcal{P}_{0,\mathrm{loc}}$ is provided by the elementary proposition 3.1.15.

PROPOSITION 3.1.15. *Under the assumptions of Proposition* 3.1.13,

$$R'_{P_0} \subset [(\mathcal{P}_{0,\mathrm{loc}})'_w]'' . \tag{1.32}$$

If, moreover, $\mathcal{P}$ *is the class* I *algebra, then*

$$R'_{P_0} \subset (\mathcal{P}_{0,\mathrm{loc}})'_w . \tag{1.33}$$

Proof. For $T \in \mathcal{P}'_w$ and any $\Phi, \Psi \in \mathcal{D}_{0,\mathrm{loc}}$, $\Pi \in \mathcal{P}_{\mathrm{loc}}$ one has due to $P_0 \in R$:

$$(\Phi, T\overline{\Pi}\Psi) = \left(\Phi, T|_{\mathcal{H}_0} P_0 \overline{\Pi} P_0 \Psi\right).$$

On the other hand, since T commutes weakly with $\overline{\Pi}$,

$$(\Phi, T\overline{\Pi}\Psi) = (\Pi^* \Phi, T\Psi) = \left(P_0 \Pi^+ P_0 \Phi, T|_{\mathcal{H}_0} \Psi\right),$$

i.e. $T|_{\mathcal{H}_0}$ commutes weakly with the generating set of elements of $\mathcal{P}_{0,\mathrm{loc}}$ and hence, with all the elements of this algebra. This means that

$$\mathcal{P}'_w|_{\mathcal{H}_0} \subset (\mathcal{P}_{0,\mathrm{loc}})'_w . \tag{1.34}$$

If $\mathcal{P}$ is the class I algebra, the left-hand side here is equal to R'_{P_0} and we obtain (1.33). In the general case, because of $P_0 \in R$,

$$R' = (\mathcal{P}'_w)'' , \quad R'_{P_0} = (\mathcal{P}'_w|_{\mathcal{H}_0})'' ,$$

which implies (1.32). □

These results bring us to our main goal, to show the possibility of the Borchers anomaly by finding for it some sufficient conditions. Indeed, by (1.32) the irreducibility of $\mathcal{P}_{0,\mathrm{loc}}$ implies that of $\mathcal{P}$ and we conclude: in general Wightman theory with a vacuum subspace cyclic for Z', the Borchers anomaly takes place in all cases when $\mathcal{P}_{0,\mathrm{loc}}$ is an irreducible abelian *Op**-algebra in the infinite-dimensional space $\mathcal{H}_0$. But it is clear, too, that this conclusion is weaker than our preliminary formulations and, in particular, (1.32) is weaker than the expected (1.26). However, a comprehensive analysis of the Borchers anomaly is absent thus far, and we cannot say whether an abelian *Op**-algebra in $\mathcal{H}_0$ can be found, the weak commutant of which coincides exactly with R'_{P_0}. Let us only mention that, besides (still incomplete) general results, there is an explicit model example of the Borchers and Yngvason [91]. In this model the state space of the Wightman theory is taken in the form $\mathcal{H}_1 \otimes \mathcal{H}_0$, where $\mathcal{H}_1$ is the tensor product of $2n$ Fock spaces corresponding to free fields of different nonzero masses while in $\mathcal{H}_0$ there acts an infinite-dimensional irreducible representation of a free abelian algebra with

n generators, $n > 2$. Such representations are known to exist and their construction originates in the Hilbert example of a positive polynomial which cannot be represented as a sum of squares (see, e.g. [11, 352]).

Let us stress that the Borchers anomaly is 'anomalous' not only from the mathematical point of view (as a property of Op^*-dynamics having no analogues in the behaviour of C^*-dynamical systems) but physically as well: in its presence the theory has pure Wightman functionals (ergodic states on $\mathcal{P}$) which are not clustering and so do not determine a pure phase. Hence, it is desirable to know some criteria of its absence which should be as simple and clear as possible, to be suitable for practical checking in models. For this reason no small attention was paid to looking for such criteria and a number of them have been found. Of course, there is one obvious condition, namely, the existence of an associated Haag–Araki theory. But it is not easily checkable and also, may be, too restrictive. Furthermore, the necessary and sufficient condition of the uniqueness of vacuum vector, and hence, the sufficient condition of absence of the Borchers anomaly, is the cluster property of the Wightman functional W_Ω. Based on this, the following condition was proved [92, Section 6] to be necessary and sufficient for absence of the Borchers anomaly: the functional $W_\Omega^{(0)}$ on the abelian Op^*-algebra $\mathcal{P}_0$ in $\mathcal{H}_0$, corresponding to the vacuum Ω, should be a product state, i.e. Hermitian character on $\mathcal{P}_0$. By [441, Theorem 4.3] this condition is tantamount to existence of self-adjoint extensions of symmetric operators generating $\mathcal{P}_0$; these extensions are acting, in general, in an extended Hilbert space $\hat{\mathcal{H}} \supset \mathcal{H}_0$ and preserving commutativity and positivity on the Borchers algebra elements of the form $f^* \times f$. Another criterion found recently [254] does not refer to operators or algebras in $\mathcal{H}_0$ and uses the weak unbounded commutant $\mathcal{P}'_{\mathrm{loc},\sigma}$ of the algebra $\mathcal{P}_{\mathrm{loc}}$. The following representation of vacuum subspace of a general Wightman theory has been obtained:

$$\mathcal{H}_0 = \overline{\mathcal{P}'_{\mathrm{loc},\sigma}\, \Omega}$$

(where $\mathcal{P}_{\mathrm{loc}}$ is supposed to have $\mathcal{P}_{\mathrm{loc}}\,\Omega$ as its domain). It easily implies that in any irreducible Wightman theory the Borchers anomaly is absent iff $\mathcal{P}'_{\mathrm{loc},\sigma} = \mathcal{P}'_w$. Some conditions which are only sufficient have been found as well. The first of them [202, 441] requires that Schwinger functions (analytical continuations of Wightman functions to imaginary time values) should be the moments of a Euclidean-invariant measure satisfying a certain positivity condition. Two more conditions have been obtained by Driessler and Summers [171]. One of them weakens the condition of existence of an associated Haag–Araki theory: at least one field operator $A(f)$, $f \in S(\mathbf{R}^4)$ should exist such that for some $O \in B(M)$ the Poincaré transformations of the von Neumann algebra generated by the polar decomposition of $A(f)$ commute with this algebra if the corresponding Poincaré transformations in M bring O into its spacelike complement (such field operators are called *intrinsically local*; Wightman theories possessing them were studied in [428, 173]). Another condition is the fulfillment of so-called 'generalized H-bound': for every test function $f \in Z(\mathbf{R}^4)$ (the space of the Fourier transforms of functions with compact support) there exists a monotone decreasing and strictly positive function F_f on $(0, \infty)$ such that

$F_f\,(H)\,\mathcal{H} \subset \text{dom}\,A\,(f)$ and $\|A\,(f)\,F_f(H)\| < \infty$ where $H = \hat{p}_0$. Estimates of this kind were studied in the Wightman theory [197] and checked in constructive field models. According to the arguments presented in [171], one can expect the validity of the generalized H-bound even for a general field theory with strong spectrum condition; moreover, this bound is implied by the Driessler–Fröhlich condition (see Section 3.2.1).

Systems With a Unique Vacuum Sector.

Having described the simplest vacuum structure, which other kinds of structure should we turn to now? To describe vacuum structure of a general Wightman theory means, above all, to obtain the decomposition of its generating functional W_Ω over extremal vacuum states on the Borchers algebra (*Op**-states, for the sake of brevity). In addition, a comprehensive analysis should obtain the decomposition of the corresponding physical vacuum state ω_Ω over ergodic states on the global algebra R and find the structure of this global algebra. Hence, it is natural to classify vacuum structures according to the form of the above-mentioned decompositions. Then the next kind of vacuum structure which we have to consider is that possessing a unique (up to the equivalence) extremal vacuum *Op**-state. This structure is fully enough described by the following proposition.

PROPOSITION 3.1.16 [426]. *In Wightman theory with a field Op*-algebra* $\mathcal{P}$ *of class* II *and strongly cyclic vacuum* Ω *the following conditions are equivalent:*

(1) $W_\Omega = \Sigma_{i=1}^{\kappa'} c_i\, W_{\Omega_i}$ *where* W_{Ω_i} *are pure Wightman functionals unitarily equivalent to one another;*

(2) $\omega_\Omega = \Sigma_{i=1}^{\kappa'} c_i\, \omega_{\Omega_i}$ *where* ω_{Ω_i} *are pure vacuum states unitarily equivalent to one another;*

(3) R *is a factor of type* I *and multiplicity* κ' ;

(4) $\mathcal{H} = L\,\overline{\{\mathcal{P}\,\mathcal{P}'\,\Omega_0\}}\,,\ \omega_{\Omega_0} \in PV\,(R)$.

Proof. The simplest way is to prove the equivalence of (3) to each of the conditions 1, 2, and 4. If (3) holds, then in every irreducible subspace $\mathcal{H}_1$ of R there is a pure vacuum vector $\Omega_0 = P_1\,\Omega$. Taking an orthonormal basis $\{\Omega'_i\}_{i=1}^{\kappa'} \subset \mathfrak{M}^{R}_{\Omega_0}$ in κ'-dimensional subspace $\mathcal{H}^{R'}_{\Omega_0}$ one has

$$\mathcal{H} = \bigoplus_{i=1}^{\kappa'} \mathcal{H}^{R}_{\Omega'_i}\ . \tag{1.35}$$

In each $\mathcal{H}^{R}_{\Omega'_i}$ we choose a vacuum vector $\Omega_i = P^{R}_{\Omega'_i}\Omega/\|P^{R}_{\Omega'_i}\Omega\|$; then $\Omega = \Sigma_{i=1}^{\kappa'} a_i\, \Omega_i\ ,\ a_i\ \|P^{R}_{\Omega'_i}\Omega\|$. Since $P^{R}_{\Omega'_i} \equiv \hat{P} \in \mathcal{P}'$ then $G^{\mathcal{P}}_{\hat{P}\Omega} = \hat{P}\, G^{\mathcal{P}}_{\Omega}$, and the cyclicity of Ω implies

$$\mathcal{H}^{\mathcal{P}}_{\hat{P}\Omega} = \hat{P}\ \mathcal{H}^{\mathcal{P}}_{\Omega} = \hat{P}\ \mathcal{H}\,,$$

i.e. $\hat{P} = P^{\mathcal{P}}_{\hat{P}\Omega}$, and due to $G^{\mathcal{P}}_{\Omega} \supset \mathcal{D}\,(\Pi)$ one has

$$G^{\mathcal{P}}_{\hat{P}\Omega} = \hat{P}\, G^{\mathcal{P}}_{\Omega} \supset \hat{P}\, \mathcal{D}(\Pi) = P^{\mathcal{P}}_{\hat{P}\Omega}\, \mathcal{D}(\Pi)$$

which means that Ω_i are completely regular vectors. Hence by Proposition 3.1.6 and due to $\mathfrak{N}_{cr} \subset \mathfrak{N}_r$,

$$\mathcal{H}^{R}_{\Omega_i} = \mathcal{H}^{\mathcal{P}}_{\Omega_i},\ \mathcal{H}^{Z'}_{\Omega_i} = \overline{L\ \{\mathcal{P}\,\mathcal{P}'\,\Omega_i\}} = \mathcal{H},\ i = 1\,, \dots\,, \kappa'\,. \tag{1.36}$$

The states ω_{Ω_i} are pure and unitarily equivalent and so are W_{Ω_i} (due to Propositions 3.1.9 and 3.1.10). Hence (1.36) entails $3 \to 4$ and (1.36) together with (1.35) entails $3 \to 1$ and $3 \to 2$ (with $c_i = |a_i|^2$).

<u>$4 \to 3$</u>. Due to $\overline{\mathcal{P}\,\Omega_0} \subset \overline{R\,\Omega_0}$ it follows from 4 that

$$\mathcal{H} = \overline{L\ \{\mathcal{P}\,\mathcal{P}'\,\Omega_0\}} = \mathcal{H}^{Z'}_{\Omega_0}\,.$$

But, if $\omega_{\Omega_0} \in PV(R)$ then $R_{P^{Z'}_{\Omega_0}} = \mathcal{B}\left(\mathcal{H}^{R}_{\Omega_0}\right)$ and due to $P^{Z'}_{\Omega_0} = I$, R is a discrete factor.

<u>$1 \to 3$</u>. Let us prove for the vectors Ω_i in 1 that

$$\omega_{\Omega_i} \in PV(R)\,,\ \omega_{\Omega_i} = \omega_{\Omega_j}\,,\ i\,,\ j = 1, \dots\, \kappa'\,. \tag{1.37}$$

Since $W_{\Omega_i} \prec W_{\Omega}$, $\Omega \in \mathfrak{N}_{cr}$, one can take $\Omega_i = T_i$, $T_i \in R'\,\Omega$, whence $\Omega_i \subset \mathfrak{N}_{cr}$ and $\omega_{\Omega_i} \in PV(R)$ (Propositions 3.1.8, 3.1.6, 3.1.9). It suffices to prove that a nontrivial intertwining operator exists for pure states ω_{Ω_i} and ω_{Ω_j} [143, Sections 5.3.3; 5.3.6]. Denoting by U_{ij} an isometry intertwining $\Pi^{GNS}_{\Omega_i}$ and $\Pi^{GNS}_{\Omega_j}$ one has for all $f, g \in \mathcal{A}$:

$$U_{ij}\, T_i\ \Pi(f)\, \Pi(g)\, \Omega = U_{ij}\, \Pi^{GNS}_{\Omega_j}(f)\, \Pi(g)\, \Omega_i = \Pi^{GNS}_{\Omega_j}(f)\, U_{ij}\, \Pi(f)\, \Omega_i$$

$$= \overline{\Pi}(f)\, U_{ij}\, T_i\, \Pi(g)\, \Omega\,,$$

whence due to $\mathcal{P}$-cyclicity of Ω, $U_{ij}\, T_i \in R'$ and $T_{ij} \equiv U_{ij}\, T_i\, P^{R}_{\Omega_i} \in R'$. Hence $\tilde{T}_{ij} \equiv T_{ij}|_{\mathcal{H}^{R}_{\Omega_i}}$ is an intertwining operator between ω_{Ω_i} and ω_{Ω_j}, and $\tilde{T}_{ij} \neq 0$ since $\tilde{T}_{ij}\, \Omega_i = U_{ij}\, T_i^2\, \Omega \neq 0$ (Ω being separating for R'). Furthermore, due to $\Omega_i, \Omega_j \in \mathfrak{N}_r$ one has from (1.37):

$$\mathcal{H}^{Z'}_{\Omega_i} = \mathcal{H}^{Z'}_{\Omega_j} = \overline{L\ \{\mathcal{P}\,\mathcal{P}'\,\Omega_i\}} = \overline{L\ \{\mathcal{P}\,\mathcal{P}'\,\Omega_j\}}\,,\ i, j = 1, \dots\, \kappa'\,.$$

It is easy to get from here that $\mathcal{H}^{Z'}_{\Omega_i} = \mathcal{H}$ for all $i = 1, \dots\, \kappa'$. If $\mathcal{H} = \mathcal{H}^{Z'}_{\Omega_i} \oplus \mathcal{H}^{\perp}$ then $\Omega = P^{Z'}_{\Omega_i}\, \Omega + \Omega_\perp$ (both terms in the sum lying in $\mathcal{D}$, clearly) and, denoting $P^{Z'}_{\Omega_i}\, \Omega = \Omega_Z$ one has due to 1:

$$W_{\Omega} = W_{\Omega_Z} + W_{\Omega_\perp} = \sum_{i=1}^{\kappa'} c_i\, W_{\Omega_i}\,. \tag{1.38}$$

$W_{\Omega_i} \prec W_{\Omega}$ implies $W_{\Omega_i} = W_{T_i \Omega}$, $T \in \mathcal{P}'$, and due to $P^{Z'}_{\Omega_i} \in Z$,

$T_i\,\Omega = \Omega_i \in \mathcal{H}^{Z'}_{\Omega_i}$ one has $T_i\,\Omega = T_i\,\Omega_Z$. Since $\Omega_Z \in \mathfrak{N}_{cr}$, it means that $W_{\Omega_i} \prec W_{\Omega_Z}$ whence $\Sigma^{\kappa'}_{i=1} c_i\, W_{\Omega_i} \prec W_{\Omega_Z}$, whereas (1.38) entails $W_{\Omega_Z} \prec \Sigma^{\kappa'}_{i=1} c_i\, W_{\Omega_i}$. Hence it follows that $W_{\Omega_\perp} = 0$ and $\Omega_\perp = 0$ too, $\mathcal{P}$ being unital. Thus $\mathcal{H}^{Z'}_{\Omega_i} = \mathcal{H}$, i.e. R is a discrete factor and it is easy to check that its multiplicity equals κ'.

<u>2 → 3</u> being elementary, the equivalence of 1–4 is proved.

Finally, the assertion that W_{Ω_i} in 1 are Wightman functionals follows by noting that relativistic covariance of W_{Ω_i} is evident while locality and spectrum condition are implied by the fact that, any W_{Ω_i} being majorized by the Wightman functional W_Ω, its left kernel includes the left kernel of the latter. □

Let us discuss physical interpretation of this vacuum structure. First of all, its connection with superselection structure is quite clear: our Wightman theory represents precisely a coherent superselection sector, i.e. a relativistic quantum system whose global observable algebra is a type I factor (though, having just one sector, we cannot speak about the presence of a superselection rule). At the same time, we can see the difference between this structure and a vacuum coherent sector in the Haag–Araki theory which always contains a unique and pure vacuum state and is abelian whenever there is a cyclic vacuum vector (Proposition 1.3.47). The difference lies in the following two properties:

(1) irreducible subspaces $\mathcal{H}^{R}_{\Omega_i}$ of $R' = (\mathcal{P}'_w)'$ may have the Borchers anomaly, i.e. $\dim\left(\mathcal{H}_0 \cap \mathcal{H}^{R}_{\Omega_i}\right) = \infty$;

(2) irrespective of the Borchers anomaly it is possible a priori that R' is not abelian despite the presence of a cyclic vacuum vector, and vacuum state is not unique, but instead there is a whole set of vacuum states including both pure and mixed ones.

These properties are of different nature, however. The Borchers anomaly was proved by the example to be possible in field systems satisfying all Wightman axioms, and thus it represents a real distinction between Op^* field dynamics and C^* observable dynamics. On the contrary, the second property, the possibility of non-abelian R' and non-unique (even in the absence of the Borchers anomaly) vacuum state in theories with a cyclic vacuum, turns out to be excluded, if we take into account locality and spectrum condition (not used in Proposition 3.1.16).

THEOREM 3.1.17. *Let a Wightman theory be given having a cyclic vacuum* $\Omega \in \mathcal{D} \cap \mathcal{H}_0$ *and satisfying translation covariance, locality and spectrum condition. Then any von Neumann algebra lying in the weak commutant of the field* Op^**-algebra* $(\mathcal{P}, \mathcal{D})$ *is abelian.*

Proof. Let us consider the vacuum Op^*-algebra $(\mathcal{P}_{0,\mathrm{loc}}, \mathcal{D}_{0,\mathrm{loc}})$ and define in $\mathcal{H}_0$ on a dense lineal $\mathcal{P}_{0,\mathrm{loc}}\,\Omega$ ($\mathcal{P}_{0,\mathrm{loc}}\,\Omega \subset \mathcal{D}_{0,\mathrm{loc}}$ due to (1.28)) an operator

$$j : \Pi\,\Omega \to \Pi^*\,\Omega\,,\quad \Pi \in \mathcal{P}_{0,\mathrm{loc}}\,.$$

Using the abelianness of $\mathcal{P}_{0,\mathrm{loc}}$, we check trivially that j is antiisometric on its initial domain:

$$(j\Phi, j\chi) = (\chi, \Phi) \ \forall \chi, \Phi \in \mathcal{P}_{0,\mathrm{loc}}\Omega,$$

and so it can be extended to an antiunitary operator in $\mathcal{H}_0$, for which the same symbol will be used. Similarly, the following properties are checked:

$$j = j^*, j T j = T^* \ \forall T \in (\mathcal{P}_{0,\mathrm{loc}})'_w .$$

(Clearly, j is an analogue of the modular involution $J = S = F$ in the trivial modular theory for an abelian von Neumann algebra with a cyclic vector; this analogy is confirmed by other properties of j.) *-Invariant lineal $(\mathcal{P}_{0,\mathrm{loc}})'_w$ may not be an algebra, in general; but any *-algebra $N_0 \subset (\mathcal{P}_{0,\mathrm{loc}})'_w$ must be abelian. Indeed, due to $j^2 = I$ on has for all $T_1, T_2 \in N_0$:

$$T_1 T_2 = j T_1^* jj T_2^* j = j T_1^* T_2^* j = T_2 T_1 .$$

Let M be a von Neumann algebra from $\mathcal{P}'_w$. Ω being cyclic for $\mathcal{P}$, it is separating for $\mathcal{P}'_w$ and hence separating for M and cyclic for M'. Since, moreover, $P_0 \in R \subset M'$, the mapping $M \to M_{P_0}$ is an induction and an isomorphism. Due to (1.34) one has

$$M_{P_0} \subset \mathcal{P}'_w|_{\mathcal{H}_0} \subset (\mathcal{P}_{0,\mathrm{loc}})'_w ,$$

whence it follows that M_{P_0} and therefore M as well, are abelian von Neumann algebras. □

COROLLARY. *Under the assumptions of Theorem* 3.1.17, *if in a Wightman theory* $\mathcal{P}'_w$ *is an algebra , it is an abelian von Neumann algebra.*

These results establish the Wightman analogue of the classical result by Araki [41]: in local quantum theory the cyclicity of vacuum implies abelianness of R'. Owing to them, of all the structures described by Proposition 3.1.16, only irreducible ones are allowed by the Wightman axioms (i.e. $\kappa' = 1$ and $R = \mathcal{B}(\mathcal{H})$). In other words, all Wightman theories with a cyclic vacuum and the unique equivalence class of vacuum states belong, in fact, to the class of irreducible systems described before. And the Borchers anomaly stands out as the only distinction of principle between the *Op** and *C** types of quantum dynamics.

Important aspects are added to the physical picture of Wightman theory by its description in terms of pure phases. In Section 2.1 we have seen the analogy between the coherent sector in local quantum theory and the pure phase in statistical mechanics. But within that framework this analogy could be considered as just a formal parallel which did not yet justify transporting concepts of phase and pure phase to local quantum theory. In field theory the situation is different. Wightman functions being analytical continuations of correlation functions, each system of Wightman fields is directly connected with a certain system of statistical mechanics. Hence, phase and pure phase should be among its characteristics, and structural analysis of a Wightman theory should include obtaining its decomposition over pure phases. Physical states corresponding to a pure phase are extremal equilibrium states having definite values of macroscopic observables (which implies that these states are factorial as noted in Section 2.1); but, in addition, they should have the (weak) cluster property. (See, e.g. [22, 200].) In the case of C^*-dynamics we have seen in Section 1.3.5 that extremal invariant states of asymptotically abelian systems are always weakly clustering, and if the spectrum condition holds, they are factorial. Thus, they are pure phases. In the case of Op^*-dynamics this may not be so. For both physical vacuum state ω_Ω and Op^*-state W_Ω weak cluster property is equivalent to the uniqueness of a vacuum ray in the space of the corresponding representation and hence, it is broken for pure Wightman functionals whose representation space enjoys the Borchers anomaly. We conclude: a Wightman theory described by Proposition 3.1.16 has a unique vacuum coherent sector but, in general, not a unique pure phase. In such theory pure phase is unique if and only if the Borchers anomaly is absent (which can be guaranteed by existence of an associated Haag–Araki theory or by any condition on p. 196).

Discrete Vacuum Superselection.
The most general vacuum structure is that of a system with an arbitrary set of (equivalence classes of) extremal vacuum Op^*-states. Basing on Proposition 3.1.16, we can expect that the structure of such a system should look like the decomposition into sectors, each of which collects states from the same equivalence class and has a discrete factor as its global algebra. By Theorem 3.1.17, the sectors should be abelian if locality is taken into account. The decomposition can be discrete (into a direct sum of algebras and Hilbert spaces) or continuous (into a direct integral). Let us consider the discrete case first.

THEOREM 3.1.18 [426]. *For any Op^*-algebra $(\mathcal{P}, \mathcal{D})$ of class* II *with a strongly cyclic vector Ω, the following conditions are equivalent:*

(1) $W_\Omega = \sum_{i\in I} c_i W_i$ *where* W_i *are pure Op^*-states on* $\mathcal{P}$, $c_i > 0$, $\sum_{i\in I} c_i = 1$;

(2) $\omega_\Omega = \sum_{i\in I} c_i\, \omega_i$ *where* ω_i *are pure states on* R;

(3) R *is a P-class algebra:*

$$\mathcal{H} = \bigoplus_{j\in J} \mathcal{H}_j, \qquad R = \bigoplus_{j\in J} R_j, \qquad R_j \text{ are discrete factors;}$$

(4) $\mathcal{H} = \bigoplus_{j\in J} \overline{L\{\mathcal{P}\mathcal{P}'\Omega_j\}}, \qquad \omega_{\Omega_j} \in PV(R).$

If $(\mathcal{P}, \mathcal{D})$ is a field Op^*-algebra of Wightman theory and Ω a vacuum vector, then in conditions 1–4 *W_i are Wightman functionals, ω_i are vacuum states, $I = J$, $R_j = \mathcal{B}(\mathcal{H}_j)$ and*

$$\mathcal{H}_0 = \bigoplus_{j\in J} (\mathcal{H}_0 \cap \mathcal{H}_j), \qquad \dim(\mathcal{H}_0 \cap \mathcal{H}_j) = 1 \text{ or } \infty. \tag{1.39}$$

Proof follows the same scheme as that of Proposition 3.1.16. If 3 holds, an obvious analogue of (1.35) takes place:

$$\mathcal{H} = \bigoplus_{j\in J} \bigoplus_{i=1}^{\kappa'} \mathcal{H}^{R}_{\Omega_{ij}} , \tag{1.40}$$

where κ'_j is the multiplicity of the discrete factor R_j, $\mathcal{H}^{R}_{\Omega_{ij}}$ are irreducible subspaces of R and Ω_{ij} normed projections of the strongly cyclic vacuum Ω which are completely regular vacuum vectors due to $P^{R}_{\Omega_{ij}} \in \mathcal{P}'$. Hence

$$\mathcal{H}^{R}_{\Omega_{ij}} = \mathcal{H}^{\mathcal{P}}_{\Omega_{ij}} ; \quad \mathcal{H}^{Z'}_{\Omega_{ij}} = \overline{L\ \{\mathcal{P}\,\mathcal{P}'\,\Omega_{ij}\}} = \mathcal{H}_j , \qquad i = 1, \ldots , \kappa'_j \tag{1.41}$$

which means 3 → 4. The states $\omega_{\Omega_{ij}}$ are pure and $\omega_{\Omega_{ij}}$ with the same value of j are unitarily equivalent; the same is valid for W_{ij}. Hence, taking into account (1.40) and the first formula in (1.41), one has 3 → 1 and 3 → 2.

4 → 3. Due to $\mathcal{P}\Omega_j \subset R\,\Omega_j$ one has $\overline{L\ \{\mathcal{P}\,\mathcal{P}'\,\Omega_j\}} \subset \mathcal{H}^{Z'}_{\Omega_j}$ for all $j \in J$. So, all subspaces $\mathcal{H}^{Z'}_{\Omega_j}$ being mutually orthogonal, one has

$$\mathcal{H} = \bigoplus_{j\in J} \overline{L\ \{\mathcal{P}\,\mathcal{P}'\,\Omega_j\}} = \bigoplus_{j\in J} \mathcal{H}^{Z'}_{\Omega_j} .$$

Since $\omega_{\Omega_j} \in PV(R)$ this means that R is a P-class algebra, $R = \oplus_{j\in J} R_j$, where R_j are discrete factors in subspaces $\mathcal{H}^{Z'}_{\Omega_j}$. (Clearly, we have also obtained $\overline{L\ \{\mathcal{P}\,\mathcal{P}'\,\Omega_j\}} = \mathcal{H}^{Z'}_{\Omega_j}$ for all j, and Ω_j can be assumed to be completely regular since otherwise completely regular vectors $\Omega'_j = P^{R'}_{\Omega_j}\ \Omega$ can be taken instead of Ω_j.)

1 → 3. $W_i \prec W_\Omega$ entails $W_i = W_{\Omega_i}$ where Ω_i is a strongly (which is the same as weakly in our case) regular vector. Let us rewrite the sum in 1, grouping together classes of unitary equivalence:

$$W_\Omega = \sum_{j\in J} \sum_{i=1}^{\kappa'_j} c_{ij}\ W_{\Omega_{ij}} .$$

Similarly to deriving 1 → 3 in Proposition 3.1.16, we obtain that $\omega_{\Omega_{ij}}$ are also pure and unitarily equivalent at fixed values of j, and $\overline{L\ \{\mathcal{P}\,\mathcal{P}'\,\Omega_{ij}\}} = \mathcal{H}^{Z'}_{\Omega_{ij}}$. And using again the *ad absurdum* proof, it is easy to see that if

$$\bigoplus_{j\in J} \bigoplus_{i=1}^{\kappa'_j} \mathcal{H}^{Z'}_{\Omega_{ij}} \subsetneqq \mathcal{H} ,$$

then the decomposition of W_Ω into pure states includes additional terms, besides W_i.

2 → 3 is derived identically. The last assertion of the theorem follow from Proposition 3.1.16 and Theorem 3.1.17 while (1.39) follows from Proposition 3.1.12 and $P_0 \in R$. ❒

This theorem makes completely clear the close connection between vacuum and superselection structures. We see that vacuum structure is determined (up to Borchers anomalies) by the central decomposition of global algebra R: the generating functional W_Ω of a Wightman theory has a discrete decomposition into pure states iff it is a direct sum of discrete factors. In the light of Section 2.1 we can say that in this case a discrete superselection rule is present with superselection sectors $\langle R_j, \mathcal{H}_j \rangle$. This rule will be called the *vacuum superselection* and the components $\langle R_j, \mathcal{H}_j \rangle$ *vacuum superselection sectors*. By the condition 4 of Theorem 3.1.18, the sectors $\mathcal{H}_j$ are invariant subspaces of the field Op^*-algebra $\mathcal{P}$. If each sector includes a single pure phase (i.e. in the absence of Borchers anomalies) any two pure phases are separated by the vacuum superselection and the decomposition into vacuum superselection sectors is at the same time the decomposition into pure phases.

The following result is a useful complement to Theorem 3.1.18 showing that the discrete vacuum superselection is always present in a large class of Wightman theories including all those with a finite-dimensional vacuum subspace ($1 < d_0 < \infty$).

THEOREM 3.1.19 [426]. *Let Op^*-algebra $(\mathcal{P}, \mathcal{D})$ of class* II *be given having a strong cyclic vector Ω and such that the representation-induction $\pi'_0 : R' \to R'_{P_0}$ (where P_0 is a projection from R such that $P_0\mathcal{H} = \mathcal{H}_0 \ni \Omega$) has a discrete decomposition into irreducibles, $\pi'_0 = \oplus_{j\in J}\, \kappa'_j\, \pi'_j$, κ'_j being the multiplicity of the component π'_j. The properties* 1–4 *from Theorem* 3.1.18 *hold for $(\mathcal{P}, \mathcal{D})$. If, in addition, $(\mathcal{P}, \mathcal{D})$ is the field Op^*-algebra of a Wightman theory, Ω a vacuum vector and $\mathcal{H}_0$ the vacuum subspace, then all the properties from Theorem* 3.1.18 *hold and the theory has the discrete vacuum superselection with the set J of abelian superselection sectors. If $d_0 < \infty$, then each sector has one and only one vacuum ray.*

Proof. We shall show that the set of irreducible representations π'_j of the algebra R' is in one-to-one correspondence with a certain set of irreducible representations π_j of R, and this correspondence preserves disjointness and unitary equivalence. Due to this, the decomposition of π'_0 implies discrete decomposition into irreducibles for the representation $\pi_0 : R \to R_{P^R_{\mathcal{H}_0}}$ of the algebra R, i.e. of this algebra itself, because $P^R_{\mathcal{H}_0} = I$. Then it will be easy to check any of the conditions 1–4. Thus, let us choose in the decomposition

$$\pi'_0 = \bigoplus_{j\in J} \kappa'_j\, \pi'_j$$

with the corresponding decomposition of the representation space

$$\mathcal{H}_0 = \bigoplus_{j\in J} \bigoplus_{i=1}^{\kappa'_j} \mathcal{H}_{0i} \tag{1.42}$$

an arbitrary component $\pi'_i : R' \to R'_{P^{R'}_{\Omega'_i}}$ (Ω'_i being any unit vector in $\mathcal{H}_{0i} = P^{R'}_{\Omega'_i}\mathcal{H}_0$). π'_i being irreducible, $P^{R'}_{\Omega'_i}$ is a minimal projection in R which implies that $P^{R}_{\Omega'_i}$ is a minimal projection in R'. Hence $\pi_i : R \to R_{P^R_{\Omega'_i}}$ is an irreducible representation, $\mathcal{H}^R_{\Omega'_i}$ is an irreducible subspace of R and the state $\omega_{\Omega'_i}$ is pure. Since central supports of $P^{R'}_{\Omega'_i}$ and $P^{R}_{\Omega'_i}$ coincide, one has

$$\pi'_i \simeq \pi'_j \Leftrightarrow \pi_i \simeq \pi_j \ ,$$

$$\pi'_i \between \pi'_j \Leftrightarrow \pi_i \between \pi_j$$

where $\between$ denotes the disjointness relation). Thus, we have obtained the set of pure states $\omega_{\Omega'_i}$ which can be transformed in a standard way (cf. the proof of Proposition 3.1.16), without changing corresponding representations π_i, into a set of pure and completely regular states. Grouping these states into unitary equivalence classes we obtain that the set of these classes corresponds to the set of mutually orthogonal 'sectors' $\mathcal{H}^{Z'}_{\Omega_j} = \overline{L\{\mathcal{P}\mathcal{P}'\Omega_j\}}$ But every subspace $\mathcal{H}_{0i}$ in (1.42) belongs to some sector in this set, whence one has

$$\mathcal{H}_0 \subset \bigoplus_{j \in J} \mathcal{H}^{Z'}_{\Omega_j} \ , \quad \mathcal{H}^R_{\mathcal{H}_0} = \mathcal{H} \subset \bigoplus_{j \in J} \mathcal{H}^{Z'}_{\Omega_j} \ .$$

This proves condition 4. The other assertions are evident. (To prove the last one, one notes that if $R = \mathcal{B}(\mathcal{H})$ and $d_0 < \infty$, $\pi'_0 = d_0 \pi'$, π' being one-dimensional representation of $C(\mathcal{H})$. Then the application of Proposition 3.1.16 gives contradiction if $d_0 \neq 1$.) □

As proved in [425], both theorems 3.1.18 and 3.1.19 remain valid for Wightman theories with essentially self-adjoint field operators.

Continuous Vacuum Superselection.
Having the same algebraic structure as usual superselection rules, the vacuum superselection has, of course, many physical properties common with them. For instance, like them, it is a principally global phenomenon: of all its main properties, only the abelianness of vacuum sectors depends on the locality axiom. On the other hand, both its origins and physical interpretation are markedly different; in the terminology of Section 2.1, it belongs to neither kinematical nor dynamical superselection rules. This difference is reflected, among other things, in the problem of continuous superselection. We have argued in Section 2.1 that basic physical properties of superselection rules, both kinematical and dynamical, exclude so-called continuous superselection rules characterized by decompositions of the state space $\mathcal{H}$ into a direct integral of superselection sectors. But a large part of these properties does not hold for the vacuum superselection (it is important, in particular, that there are no superselection operators in this case) – and, as a consequence, our conclusion that continuous superselection is physically unacceptable is not valid anymore. Decomposition of $\mathcal{H}$ into vacuum superselection sectors corresponds to the decomposition of generating functional W_Ω into pure states. The latter decomposition is determined by so-called 'quantum measure' obtained in the Euclidean formulation, and it is quite possible for it to be continuous. Let us give a simple sufficient condition for this.

Consider a Wightman theory with the class I algebra $\mathcal{P}$ and a unique physical vacuum state, but a non-unique vacuum ray:

$$d_0 > 1 \ , \qquad \mathcal{V}_1 = \mathcal{H}_0 \cap S_1 = \mathfrak{M}^R_\Omega \ . \tag{1.43}$$

Combination of these conditions with the strong cyclicity of Ω creates a number of specific features of this vacuum structure: in particular, it turns out to be possible only when $d_0 = \infty$; vacuum vectors cannot correspond to pure states on R as well as on $\mathcal{P}$; decompositions into pure phases and pure Wightman states are of the form of direct integrals with no discrete part; and hence, if there is any vacuum superselection in such a theory, it is continuous.

Let us derive these properties. By Proposition 3.1.11, $\Omega \in \mathfrak{N}_{cr}$ implies $\overset{\mathcal{P}}{\mathfrak{M}}_\Omega \subset \mathcal{P}'_w\ \Omega \cap S_1 \cap \mathcal{D} \subset \mathcal{V}_1 \cap \mathcal{D}$, the second inclusion following from $T(a) \in R$. Since $\overset{\mathcal{P}}{\mathfrak{M}}_\Omega \supset \overset{R}{\mathfrak{M}}_\Omega \cap \mathcal{D}$, one has

$$\overset{\mathcal{P}}{\mathfrak{M}}_\Omega = \overset{R}{\mathfrak{M}}_\Omega \cap \mathcal{D} = \mathcal{V}_1 \cap \mathcal{D},$$

i.e. the vacuum Op^*-state is unique as well. Next, both states ω_Ω and W_Ω cannot be pure. Indeed, in the case of ω_Ω its uniqueness and purity combined with the cyclicity of Ω imply $d_0 = 1$. Purity of W_Ω is equivalent to that of ω_Ω due to $\Omega \in \mathfrak{N}_{cr}$. Thus, $\mathcal{H}$ has no pure vectors either for R or $\mathcal{P}$.

Hence, it follows that the decomposition of ω_Ω into pure states includes no discrete part: if $\omega = \omega_1 + \omega_2$ with $\omega_1 \in PV(R)$, then $\omega_1 \prec \omega_2$ and $\omega_1 = \omega_{T\Omega}$, $T\Omega \in R'\Omega \subset \mathcal{H}_0$, i.e. $T\Omega$ is a pure vacuum vector. By Theorem 3.1.19, when $1 < d_o < \infty$, ω_Ω has a discrete decomposition into pure states. Hence, the structure (1.43) implies $d_0 = \infty$.

As for the decomposition of W_Ω into pure states, it may, in general, include a discrete part but the latter cannot include vector functionals. Indeed, the equality

$$W_\Omega = W_1 + W_2,\ W_1 \in PV(\mathcal{P}), \tag{1.44}$$

according to Proposition 3.1.8 implies only

$$W_1(f) = (T\Omega, T\Pi(f)\Omega),\ T \in \mathcal{P}'_w,\ T\Omega \in \mathcal{V}_1 \cap \mathcal{D}(\mathcal{P}^*).$$

If $\mathcal{D}(\mathcal{P}^*) \supsetneqq \mathcal{D}$ and $T\Omega \notin \mathcal{V}_1 \cap \mathcal{D}$, then W_1 is not a vector functional on $\mathcal{P}$ and (1.44) does not contradict (1.43). But if W_1 is a vector functional, (1.44) is impossible since $\mathcal{V}_1 \cap \mathcal{D} = \overset{\mathcal{P}}{\mathfrak{M}}_\Omega$ and $T\Omega \in \mathcal{V}_1 \cap \mathcal{D}$ entail $W_1 = W_\Omega$. For algebras $\mathcal{P}$ of the class II, W_1 is always a vector functional (Proposition 3.1.8) and so the decomposition of W_Ω cannot include any discrete part.

Thus in our example the decompositions of W_Ω (as well as ω_Ω) into pure invariant states must have the form of integrals over measures with the support having no discrete part. General theory of decompositions of functionals on unbounded operator algebras, with special applications to Wightman functionals on the Borchers algebra, was developed by Borchers and Yngvason [91, 92, 442]. According to this theory, decomposition of an Op^*-state W on an algebra $(\mathcal{P}, \mathcal{D})$ in a Hilbert space $\mathcal{H}$ over extremal Op^*-states can be constructed by means of a certain extension $\hat{\mathcal{P}}$ of the algebra $\mathcal{P}$ acting, in the general case, in an extended Hilbert space $\hat{\mathcal{H}} \supset \mathcal{H}$. It is of the form

$$W(\Pi) = \int^{\oplus} W_\xi(\Pi)\, d\nu_\xi,\ \Pi \in \mathcal{P} \tag{1.45}$$

where $d\nu_\xi$ is a finite positive Borel measure on $\mathbf{R}$ produced by the Gelfand isomorphism [8, Section 1.4] for a certain abelian von Neumann algebra acting in $\hat{\mathcal{H}}$ and lying in the strong commutant of $\hat{\mathcal{P}}$. Op^*-states W_ξ on $\mathcal{P}$ are pure almost everywhere in the measure ν. If W is a Wightman functional, so are W_ξ. In general, they admit of the Borchers anomaly, i.e. may have an infinite-dimensional vacuum subspace, be not clustering and not corresponding to a pure phase. However, in very many theories, in particular, all theories with the algebra $\mathcal{P}$ of the class I, there is no need in using the complicated construction of the extension $\hat{\mathcal{P}}$. In these theories there exists [171, 254] the decomposition of W_Ω over pure Wightman functionals, the measure in which is determined by the abelian (by Theorem 3.1.17) algebra R', if $\mathcal{P}'_w$ is an algebra, or by a certain maximal abelian von Neumann algebra from R', completely in parallel with the theory of decomposition over ergodic states for C^*-dynamical systems [5]. But so far we lack the exact description of the class of theories possessing such decompositions and thus, it is not quite clear to which extent the general Borchers–Yngvason theory is of a real necessity in physical situations.

Now we have seen the examples of Wightman theories possessing the discrete or continuous vacuum superselection. Obviously, the most general vacuum structure might combine both cases. Indeed, by Lemma 2.1.3 one has for the global algebra $R = (\mathcal{P}'_w)'$ of a general Wightman theory:

$$R = R_p \oplus R_m\,, \qquad \mathcal{H} = \mathcal{H}_p \oplus \mathcal{H}_m \tag{1.46}$$

where R_p is a P-class algebra and R_m does not have pure vector states. Due to the presence of a strongly cyclic vacuum, (1.46) implies decomposition of the Wightman functional:

$$W_\Omega = W_d + W_c\,, \qquad W_d = W_{\Omega_d} = \sum_{i\in I} c_i\, W_{\Omega_i}\,, \qquad W_c = W_{\Omega_c}\,,$$

and similarly

$$\omega_\Omega = \omega_d + \omega_c\,, \qquad \omega_d = \omega_{\Omega_d} = \sum_{i\in I} c_i\, \omega_{\Omega_i}\,, \qquad \omega_c = \omega_{\Omega_c}\,,$$

where $\Omega_d = P_p\,\Omega$, $\Omega_c = P_m\,\Omega$, and $P_p\,, P_m$ are the projections on $\mathcal{H}_p$ and $\mathcal{H}_m$, respectively. The state W_d determines a Wightman theory with discrete vacuum superselection while W_c determines the Wightman theory in $\mathcal{H}_m$ having no pure vector states either for $\mathcal{P}$ or R and the decomposition over pure invariant states in the form of a direct integral (1.45). If (all or some) components of this direct integral are pure phases, the Wightman theory $\langle W_c\,, \mathcal{H}_m\rangle$ describes a continuous vacuum superselection.

Finally, let us stress again that the complete parallel between our description of vacuum superselection and the theory of usual superselection rules (Section 2.1) is only mathematical. By its physical meaning, vacuum superselection, in contrast to usual superselection rules, means reducibility and sectorial structure of a field algebra F rather than observable algebra R, although in the theories considered both algebras coincide.

3.2. Construction and Properties of Von Neumann Field Algebras

3.2.1. METHODS OF CONSTRUCTION AND THE LOCALITY PROBLEM

A certain amount of results obtained in the framework of field Op^*-algebras does not change the general state of things according to which unbounded operator algebras still are too cumbersome and little-studied apparatus and normed algebras are much preferable as a basis for the algebraic technique of quantum theory. It is also desirable to have algebraic formalism for field theory as close as possible to the Haag–Araki theory, so that the technique and results of the latter could be used. For these reasons the initial problem in constructing an algebraic description of a quantum field system can be formulated as follows:

> To associate with the set of field operators $A(f)$ or, equivalently, the net of field Op^*-algebras $\mathcal{P}(O)$, a net of von Neumann field algebras $F(O)$ satisfying the basic postulates of relativistic quantum theory.

This problem consists of two parts: we have to define or construct a certain net of von Neumann algebras based on $A(f)$ or $\mathcal{P}(O)$ and check that this net satisfies the physical postulates (or choose out of all possible nets that – or those – satisfying the postulates). Clearly, there is a wide range of different ways by which the first stage can be performed, and hence, in this stage the solution of our problem is highly non-unique. Many ways are known how to associate a von Neumann algebra with a system of unbounded closable operators; another group of ways makes it possible to proceed to von Neumann algebras from an Op^*-algebra $\mathcal{P}(O)$, and, finally, some more variants of defining $F(O)$ arise if we consider additional properties due to physical interpretation of $A(f)$ and $\mathcal{P}(O)$: for instance, it is possible to define $F(O)$ basing on $\mathcal{P}(O')$ instead of $\mathcal{P}(O)$. It is hardly possible nor is there much need to point out all the the existing methods. But we still shall give a wide range of examples, wider than practical needs dictate, in order to demonstrate the diversity of available variants.

Thus we start with possible definitions of $F(O)$ directly in terms of Wightman fields. Scalar Wightman theory will be usually taken from the most general class determined by the reconstruction theorem. Such a theory has an arbitrary field Op^*-algebra $\mathcal{P}$ and arbitrary vacuum subspace including at least one cyclic vector of $\mathcal{P}$. Test function space is assumed to be $S(\mathbf{R}^4)$ although (as it was the case in Section 3.1) the extension to strictly localized theories is not difficult.

(1) Since all $A(f)$ are closable, there is the polar decomposition:

$$A(f) = V_f H_f, \qquad V_f \in \mathcal{B}(\mathcal{H}), \qquad H_f \geq 0, \qquad H_f = \int_0^\infty \lambda \, dE(f, \lambda).$$

Hence the set of all fields $A(f)$, $\operatorname{supp} f \subset O$, for an arbitrary $O \in B(M)$ generates the von Neumann algebra

$$F_1(O) = \{V_f, E(f, \lambda) \mid f \in S(\mathbf{R}^4), \ \operatorname{supp} f \subset O\}''.$$

Clearly, this definition does not require imposing any conditions on the field system.

(2) Assuming for any symmetric operator $A(f)$, $f \in S_R(\mathbf{R}^4)$ ($S_R(\mathbf{R}^4)$ is the subspace of all real functions in $S(\mathbf{R}^4)$) the existence of at least one self-adjoint extension $\tilde{A}(f)$, we can define the algebra of fields in $O \in B(M)$ as the von Neumann algebra generated by bounded functions of $A(f)$, $\operatorname{supp} f \subset O$. As is well known, for a given system of self-adjoint operators $\tilde{A}(f)$, the same von Neumann algebra is generated by the following sets of bounded operators:

all bounded functions of $\tilde{A}(f)$;

exponents $\exp[i\tilde{A}(f)]$;

resolvents $\tilde{R}(f, \mu) = (\tilde{A}(f) - \mu)^{-1}$, $\operatorname{Im} \mu \neq 0$.

Thus, we have a field algebra

$$F_2(O) = \{e^{i\tilde{A}(f)}\}'' = \{\tilde{E}(f, \lambda)\}'' = \{\tilde{R}(f, \mu), \operatorname{Im} \mu \neq 0\}''.$$

(3) Symmetrical operator $A(f)$, $f = \bar{f}$, has generally a set (possibly, an empty one) of self-adjoint extensions and so a set of different algebras $F_2(O)$ exists for a given Wightman field. One sufficient condition of the existence and uniqueness of $F_2(O)$ is the essential self-adjointness of $A(f)$. In this case $\tilde{A}(f) = \overline{A(f)} = A(f)^* = A(f)^{**}$ and $F_2(O)$ takes the form

$$F_3(O) = \{\exp(i\overline{A(f)}) \mid f \in S_R(\mathbf{R}^4),\ \operatorname{supp} f \subset O\}''.$$

Next, let us point out a few ways to define $F(O)$ in terms of $\mathcal{P}(O)$.

(4) The most straightforward way is that used in Section 3.1: without any restrictions on $\mathcal{P}(O)$, we can always associate with it the von Neumann algebra

$$F_4(O) = (\mathcal{P}(O)'_w)'.$$

(5, 6) Again without any restrictions on $\mathcal{P}(O)$, the following sets are always von Neumann algebras:

$$F_5(O) = \{\mathcal{P}(O)'_s \cup (\mathcal{P}(O)'_s)^*\}', \quad F_6(O) = \{\mathcal{P}(O)'_s \cap (\mathcal{P}(O)'_s)^*\}'.$$

(7) For an arbitrary set $\mathfrak{R}$ of closed operators von Neumann [335] has introduced one more commutant which is, in general, different from both weak and strong commutants. Namely, the following set of bounded operators will be called the *von Neumann commutant* of the set $\mathfrak{R}$:

$$\mathfrak{R}'_N = \{T \in \mathcal{B}(\mathcal{H}) \mid TA \subset AT,\ TA^* \subset A^*T\ \forall A \in \mathfrak{R}\}.$$

This is the smallest of all known types of the commutant to a set of unbounded operators; obviously, for an arbitrary *Op**-algebra $\mathcal{A}$, $\mathcal{A}'_N \subset \mathcal{A}'_s \subset \mathcal{A}'_w$, where $\mathcal{A}'_N$ is defined, allowing for the possible unclosedness of operators in $\mathcal{A}$:

$$\mathcal{A}'_N = \{T \in \mathcal{B}(\mathcal{H}) \mid T\bar{A} \subset \bar{A}T,\ TA^* \subset A^*T\ \forall A \in \mathcal{A}\}. \tag{2.1}$$

Still von Neumann himself has proved that $\mathfrak{R}'_N$ is a 'ring', i.e. what we call today von Neumann algebra. Hence the von Neumann field algebra of a region O can be defined as

$$F_7(O) = (\mathcal{P}(O)'_N)' .$$

Finally, let us recall that $A(f)$ and $\mathcal{P}(O)$ possess a local net structure with respect to Minkowski space. Fields in a region O commute with the fields in O' and, obviously, this property must be inherited by the field algebra $F(O)$, i.e. we should have

$$F(O) \subset \mathcal{P}(O')'_{(?)}$$

(of course, the type of the commutant is not determined by this intuitive argument).

(8) Thus, if we want to choose the largest field algebra (which is preferable for it provides the maximal information), we must take the maximal von Neumann algebra lying in $\mathcal{P}(O')'_{(?)}$. Out of various types of the commutant, the weak one produces the largest $F(O)$. Hence we can define the von Neumann field algebra as

$$F_8(O) = {}_{\mathrm{alg}}\mathcal{P}(O')'_w ,$$

where for an arbitrary set $\mathcal{K} \subset \mathcal{B}(\mathcal{H})$,

$$ {}_{\mathrm{alg}}\mathcal{K} = \{T \in \mathcal{K} \,|\, T\,\mathcal{K} \subset \mathcal{K} \supset \mathcal{K}T\} . \tag{2.2}$$

(9) The last definition can be modified in several ways. Note, for instance, that locality, by its direct meaning, should require the commutativity of physical operators relating to a region O not with any elements of $\mathcal{P}(O')$ but only with the fields $A(f)$, $\operatorname{supp} f \subset O'$. For this reason the condition (2.2) can be replaced by

$$F(O) \subset L(O')'_N ,\ L(O') = \{A(f),\ \operatorname{supp} f \subset O'\} ,$$

and, again, by the maximal information argument, the von Neumann field algebra can be defined as

$$F_9(O) = L(O')'_N .$$

(10) Finally, by the same argument of possibility to replace $\mathcal{P}(O)$ by its subset $L(O)$, we can modify the definition (7) as well and choose the von Neumann field algebra as

$$F_{10}(O) = L(O'_N)' .$$

Each of the introduced sets of algebras generated in the standard way quasilocal and global field algebras:

$$\mathcal{F}_i = \overline{\bigcup_{O \in B(M)} F_i(O)}^{\,u} ,\quad F_i = \Big\{ \bigcup_{O \in B(M)} F_i(O) \Big\}'' , \qquad i = 1, \dots , 10 .$$

The above set of definitions shows clearly the principles of choosing von Neumann field algebras as well as the degree of ambiguity in this choice; the reader could easily add to our list a few more variants. Since all the algebras introduced have the same destination and contain largely the same information, there must be various connections between them. Again, they can be obtained easily by the reader himself; we shall give here just the basic ones.

Thus the following relationships hold:

$$F_1(O) = F_{10}(O) \subset F_7(O);$$

$$F_1(O) = F_2(O) = F_3(O) = F_4(O) = F_{10}(O)$$

(if $F_3(O)$ is well defined);

$$F_4(O) \subset F_5(O) \subset F_6(O) \supset F_7(O)$$

(for self-adjoint algebras, $\mathcal{P}(O) = \mathcal{P}(O)^*$, this changes to

$$F_4(O) = F_5(O) = F_6(O) = F_7(O));$$

$$F_4(O') \subset F_8(O)';$$

$$F_4(O) \subset F_{10}(O); \tag{2.3}$$

$$F_9(O) = F_{10}(O')' \subset F_8(O).$$

They all follow from the definitions of $F_i(O)$, without using the Wightman axioms. The derivation of most of them is evident, for others brief comments will suffice. The equality in the first formula follows if we recall that for any closed operator $A = V\int_0^\infty \lambda dE_A(\lambda)$ and a von Neumann algebra N, the properties $V \in N, E_A(\lambda) \in N$ are equivalent to $A \,\eta\, N$. Due to this, the equality expresses the well known property of the von Neumann commutant: for any set of closed operators, the commutant to its von Neumann commutant is the smallest von Neumann algebra, to which all the operators are affiliated. In the second formula it is sufficient to derive the third and the fourth equalities. The third one is implied by the simple property: $\mathcal{P}(O)$ being the polynomial algebra for the set of fields $A(f)$, $\operatorname{supp} f \subset O$, its weak commutant coincides with that of this set, i.e.

$$\mathcal{P}(O)'_w = L(O)'_w. \tag{2.4}$$

(Indeed, one has for any $B \in L(O)'_w$, all $f_i \in S_R(\mathbf{R}^4)$, $\operatorname{supp} f_i \subset O$, $i = 1, \dots, n$, and all $\Phi, \Psi \in \mathcal{D}$:

$$(\Phi, BA(f_1) \dots A(f_n)\Psi) = (A(f_1)\Phi, BA(f_2) \dots A(f_n)\Psi)$$

$$\dots = (A(f_n) \dots A(f_1)\Phi, B\Psi) = ((A(f_1) \dots A(f_n))^* \Phi, B\Psi),$$

which means that $B \in \mathcal{P}(O)'_w$.) (2.4) implies $F_3(O) = F_4(O)$, since $L(O)'_w = F_3(O)'$. (Hence it follows that in all theories with self-adjoint operators $\mathcal{P}(O)'_w$ and $\mathcal{P}'_w$ are von Neumann algebras, as mentioned before.) The fourth equality follows by taking into account the easily checkable fact: if $\mathcal{A}$ is a set of essentially self-adjoint operators given on a common core, then $\mathcal{A}'_w = \mathcal{A}'_N = \mathcal{A}'_s$. Indeed, due to this equality, essential self-adjointness of $A(f)$, $f \in S_R(\mathbf{R}^4)$ on $\mathcal{D}$ entails

$$L(O)'_w = L(O)'_s = L(O)'_N \quad ,$$

which implies the sought for (but, note, does not, in general, imply $\mathcal{P}(O)'_w = \mathcal{P}(O)'_s$).

In the third of the formulas (2.3) only the last inclusion is not evident. It follows from the simple properties of the von Neumann commutant:

$$\mathcal{A}'_N = \bar{\mathcal{A}}'_N = \bar{\mathcal{A}}'_s \cap (\bar{\mathcal{A}}'_s)^* \supset \mathcal{A}'_s \cap (\mathcal{A}'_s)^* \tag{2.5}$$

when $\mathcal{A}$ is an arbitrary Op^*-algebra. These properties can be derived by the reader as a useful exercise; the derivation uses the following facts: if T is a bounded and A closable operator, then $TA \subset AT$ entails $T\bar{A} \subset \bar{A}T$; if, in addition, A belongs to a closed Op^*-algebra and T to the strong commutant of the latter, the converse is also true; finally, $(TA)^* = A^*T^*$, but the inclusion $(AT)^* \supset T^*A^*$ may not be the equality. By virtue of (2.5), self-adjoint Op^*-algebras provide another example when all three commutants coincide, $\mathcal{B}'_w = \mathcal{B}'_s = \mathcal{B}'_N$ for any such algebra $\mathcal{B}$. This gives us the fourth of the formulas (2.3). The next to the last formula follows by taking into account (2.4), the last one uses the obvious corollary of (2.4):

$$\mathcal{P}(O')'_w = L(O')'_w \supset L(O')'_N = F_9(O) .$$

Let us mention here one more general property [253]: in typical cases, when $F_i(O)$ are type III factors and $F_i(O) \supset F_4(O)$, $F_i(O)$ do not leave the Gårding domain $\mathcal{D}$ invariant and so do the global algebras F_i whenever they are P-class algebras and $F_i \supset F_4$.

It will be seen below that, although not all the algebras introduced are equally natural and useful, none of them is preferable to all others in all respects. In different problems different kinds of field algebras are the most suitable. For instance, in simple models, like free and generalized free fields (see Section 3.3) where one succeeded in proving 'good' properties of the field operators, such as the essential self-adjointness on the Gårding domain, we come naturally to the algebra $F_2(O)$ and $F_3(O)$. Algebras $F_4(O)$, as shown above, are the direct extensions of $F_3(O)$ to the case of an arbitrary Wightman theory. Algebras $F_8(O)$ and $F_9(O)$, as will be shown in Section 3.2.2, are of a distinct status in the following important respect: for them it is possible to prove the local commutativity. For algebras $F_{10}(O)$ the cyclicity of vacuum holds, for $F_9(O)$ the duality holds (under a certain assumption). Finally, there are certain arguments in favour of $F_1(O)$ too: as we have seen, it is the smallest von Neumann algebra to which all the fields $\overline{A(f)}$ and $A(f)^*$, $\operatorname{supp} f \subset O$, are affiliated; and the condition that the fields are affiliated to the field von Neumann algebra seems to be justified. A number of other algebras $F_i(O)$ and their commutants also turn out to be 'affiliating' von Neumann algebras for various sets of Wightman fields. For example, using again the fact that the smallest 'affiliating' von Neumann algebra of a set of closed operators is equal to the commutant of the von Neumann commutant of this set, we see: $F_7(O)$ is the minimal von Neumann algebra to which all the closures and adjoints of operators from $\mathcal{P}(O)$ are affiliated; $F_9(O)'$ is the similar algebra for operators from

$L(O')$, $F_8(O)$ is also connected with certain affiliation properties of field operators, see [66, Lemma 10].

Of course, the main requirement to any of the nets $\{F_i(O)\}$ is fulfillment of the axioms of relativistic quantum theory. The fulfillment of isotony for all nets $\{F_i(O)\}$, $i = 1, \dots, 10$, is obvious from their definitions. To check relativistic covariance one needs the following fact. Let $(\mathcal{A}, \mathcal{D})$ be the set of closed operators with a common dense domain $\mathcal{D}$ such that $A\,\mathcal{D} \subset \mathcal{D}$, $A^*\mathcal{D} \subset \mathcal{D}$ for all $A \subset \mathcal{A}$. Then for any unitary or antiunitary operator U such that $U\,\mathcal{D} \subset \mathcal{D}$, $U^*\mathcal{D} \subset \mathcal{D}$ one has

$$(U\mathcal{A}U^*)'_c = U\mathcal{A}'_c U^*, \quad {}_{\mathrm{alg}}(U\mathcal{A}U^*)'_w = U_{\mathrm{alg}}\mathcal{A}'_w U^*, \tag{2.6}$$

the index c denoting any of the three commutants, weak, strong and von Neumann's. The derivation of (2.6) is straightforward; in the case of von Neumann's commutant one uses that

$$\overline{U\,A\,U^*} = U\,\bar{A}\,U^*, \qquad (U\,A\,U^*)^* = U\,A^*\,U^*.$$

Formulae (2.6), as can be easily seen, secure relativistic covariance of all the nets $\{F_i(O)\}$ except $\{F_2(O)\}$ (clearly, if there exist self-adjoint extensions of a Wightman field, their relativistic covariance is not guaranteed and may require additional conditions). Thus, if $U(a, \Lambda)$ is a unitary representation of the Poincaré group implementing relativistic covariance of the Wightman theory, then

$$U(a,\Lambda)\,F_i(O)\,U(a,\Lambda)^{-1} = F_i(\Lambda^{-1}(O-a)), \qquad i = 1, \dots, 10, i \neq 2.$$

Hence, it follows that the spectrum condition is also satisfied for each of these nine nets.

The situation is different with weak additivity. As shown in Section 3.1, the nets $\{\mathcal{P}(O)\}$ satisfy strong additivity:

$$\mathcal{P}(O_1 \cup O_2) = \mathcal{P}(O_1) \vee \mathcal{P}(O_2).$$

However, the definitions of $\{F_i(O)\}$ are such that neither this property nor weak additivity has been proved for them up to now. The negative result, the breaking of weak additivity, has not been proved either, and thus, the problem stays open. This entails an important consequence: we cannot guarantee that the nets $\{F_i(O)\}$ satisfy the conditions of the Reeh–Schlieder theorem. As a result, although Wightman theory has a vacuum Ω cyclic for both $\mathcal{P}$ and $\mathcal{P}(O)$, we cannot prove the cyclicity of Ω for $F_i(O)$.

In the case of $F_4(O)$ there is another way of proving the cyclicity of Ω, besides the Reeh–Schlieder theorem: we can use the formulae (1.9) which give us

$$\overline{F_4(O)\,\Omega} \supset \overline{\mathcal{P}(O)\,\Omega} = \mathcal{H} \tag{2.7}$$

whenever $\mathcal{P}(O)'_w$ is a von Neumann algebra. If (2.7) holds, Ω will be cyclic for all $F_i(O)$, $i = 1, 3, 5-7, 10$ as well. This remark is, however, of little help since we know of no direct ways how to prove that $\mathcal{P}(O)'_w$ is a von Neumann algebra. Thus, whenever we need the cyclicity of vacuum for $F_i(O)$, we have to adopt it as an additional condition.

The algebras $F_2(O)$ and $F_{10}(O)$ are an exception, however. It can be easily proved that all vectors cyclic for $\mathcal{P}(O)$ remain cyclic for these algebras (and hence, due to (2.3), for $F_i(O)$, $i = 1, 3, 7$, as well).

LEMMA 3.2.1. *All vectors analytic in the energy and cyclic for $\mathcal{P}$ are cyclic for $F_i(O)$, $i = 1, 2, 3, 7, 10$ and separating for $F_k(O)$, $k = 8, 9$, for any open nonempty region $O \subset M$.*

Proof. Consider $F_{10}(O)$ first. Let us recall the simple fact: any cyclic vector χ of an Op^*-algebra $\mathcal{A}$ is separating for $\mathcal{A}'_w$. (Indeed, since

$$(A_1 \chi, T A_2 \chi) = (A_2^* A_1 \chi, T\chi)$$

for all $A_1, A_2 \in \mathcal{A}$ and $T \in \mathcal{A}'_w$, then, due to the density of sets $\{A_i \chi, A_i \in \mathcal{A}\}$, $i = 1, 2$, $T\chi = 0$ implies $T = 0$.) Hence, by the Reeh–Schlieder theorem, any vector χ analytic in the energy and cyclic for $\mathcal{P}$ is separating for $\mathcal{P}(O')'_w$. Due to the inclusions

$$F_9(O) \subset F_8(O) \subset \mathcal{P}(O')'_w,$$

χ is separating for $F_{8,9}(O)$ as well. Hence, taking account of (2.3), χ is cyclic for $F_{10}(O)$, where O is any region of M such that $O' \neq \emptyset$.

The cyclicity of χ for $F_3(O)$ is implied then by (2.3); but the algebras $F_2(O)$ must be given a special consideration. Let $\tilde{A}(f)$ be a self-adjoint extension of the Wightman field $A(f)$ having the spectral decomposition

$$\tilde{A}(f) = \int_{-\infty}^{\infty} \lambda \, d\tilde{E}(f, \lambda).$$

Let O be an arbitrary region in M, $\{f_i\}_{i=1}^{\infty}$ a family of test functions with $\operatorname{supp} f_i \subset O$ and $\Delta_\lambda = (-\lambda, \lambda)$ an interval on $\mathbf{R}$. Consider bounded operators

$$A_i(\lambda) = \tilde{A}(f_i)\, \tilde{E}(f_i, \Delta_\lambda) = \int_{-\lambda}^{\lambda} \mu \, d\tilde{E}(f_i, \mu) \in F_2(O).$$

The spectral decomposition being strongly convergent for all $\Phi \in \mathcal{D}$, one has

$$\lim_{\lambda \to \infty} A_i(\lambda)\Phi = \tilde{A}(f_i)\Phi = A(f_i)\Phi \in \overline{F_2(O)\Phi} \tag{2.8}$$

and, in particular,

$$\Phi_1 \equiv A(f_1)\chi \in \overline{F_2(O)\chi} \cap \mathcal{D},$$

χ being any vector analytic in the energy and cyclic for $\mathcal{P}$. Hence, the operators from $F_2(O)$ being bounded, it follows from (2.8) in the same way:

$$\Phi_2 \equiv A(f_2)A(f_1)\chi = \lim_{\lambda \to \infty} A_2(\lambda)\Phi_1 \in \overline{F_2(O)\Phi_1} \subset \overline{F_2(O)\chi} \cap \mathcal{D}.$$

Repeating the procedure, one has for any natural n

$$A(f_n) \dots A(f_1)\chi \in \overline{F_2(O)\chi} \cap \mathcal{D},$$

whence the result follows:

$$L\,\overline{\{A\,(f_n)\,\dots\,A\,(f_1)\,\chi\}} = \overline{\mathcal{P}(O)\,\chi} = \mathcal{H} \subset \overline{F_2\,(O)\,\chi}\;. \qquad \square$$

Another remark is that, by the converse Reeh–Schlieder theorem (Proposition 1.3.2), the Reeh–Schlieder property $\left(\overline{F_i\,(O)\,\chi} = \overline{F_i\,\chi}\right.$ for all vectors χ analytic in the energy$\left.\right)$ is a sufficient condition of weak additivity. Hence, whenever this property is proved or adopted as an additional condition, this axiom is fulfilled. This does not mean, however, that Lemma 3.2.1 implies the weak additivity of the nets $\{F_i\,(O)\}$, $i = 1-3, 7, 10$, since by this Lemma we have the Reeh–Schlieder property only for those analytic in the energy vectors which lie in $\mathcal{D}$ and are cyclic for $\mathcal{P}$. (Still, taking into account that the set of such vectors is dense in $\mathcal{H}$ [170], one could try to obtain the corresponding generalization of Proposition 1.3.2.)

The main problem in constructing the formalism of von Neumann field algebras is to ensure the locality axiom. As was the case with additivity, our definitions of the nets $\{F_i\,(O)\}$ do not permit us to check their local commutativity directly, basing on the locality for the net $\{\mathcal{P}(O)\}$. What is more, simple arguments show that, in general, the locality of Wightman fields may not imply locality of $F_i\,(O)$ and we may need some additional conditions to ensure it. Indeed, let us assume that the fields $A\,(f)$ are essentially self-adjoint. Then the operators $A\,(f)$ and $A\,(g)$ with $\operatorname{supp} f = O_1 \times O_2 = \operatorname{supp} g$ are essentially self-adjoint on the common core $\mathcal{D}$, leave this domain invariant and are permutable on it. But the well known Nelson example [20, Vol. I, Section VIII.5] proves that two operators satisfying all this set of conditions, still may have non-commuting bounded functions. Hence, if this possibility is realized for some pair $A\,(f), A\,(g)$, the corresponding algebras $F_3\,(O_1)$ and $F_3\,(O_2)$ do not commute, and due to (2.3), locality will be violated for the nets $\{F_i\,(O)\}$, $i = 1, 4-7, 10$ as well. And in a general theory, with still less conditions on $A\,(f)$, such a loss of locality in the transition from $A\,(f)$ to $F\,(O)$ all the more cannot be excluded.

Thus the problem of locality of the algebras $F_i\,(O)$ is formulated as the problem of finding additional conditions (on $A\,(f)$ or $\mathcal{P}(O)$) which could ensure locality. Of course, both necessary and sufficient conditions are of interest. But one more example from the operator theory tells us that there are no necessary conditions for our problem: two essentially self-adjoint operators can have commuting bounded functions but at the same time be defined on cores with trivial intersection, so that no kind of commutativity properties holds for the operators themselves. (In other words, pairs of operators A, B exist such that $\overline{A} = A^*, \overline{B} = B^*$, $[\exp\,(it\,\overline{A}\,),\ \exp\,(it\,\overline{B}\,)] = 0$, but $\operatorname{dom} AB = \operatorname{dom} BA = \{0\}$). As a result, the problem reduces to finding sufficient conditions. Several of them are known now, the first and the simplest being the 'postulate of the analyticity of vacuum' put forward by Borchers and Zimmermann in 1964 [94] and later on studied and generalized by a number of authors (see, e.g. [207, 208, 63, 175]).

The original condition by Borchers and Zimmermann states that the cyclic vacuum Ω should be an analytic vector for all the field operators $A\,(f)$, $f \in S_R\,(\mathbf{R}^4)$.

Adding this condition to the Wightman axioms, one easily obtains the following result [94, Theorem 2]: for any $A(f)$, $f=\bar{f}$, $\operatorname{supp} f \subset O$, all vectors of the form $\Phi = \Pi_{i=1}^{n} A(g_i)\,\Omega$, $\operatorname{supp} g_i \subset O'$, $i = 1, 2, \dots$, are also analytic; and the set of all such vectors being total in $\mathcal{H}$ due to the Reeh–Schlieder theorem, $A(f)$ is essentially self-adjoint on the Gårding domain, by the well known Nelson theorem. Thus in any theory satisfying the Borchers–Zimmermann postulate the net $\{F_3(O)\}$ is well defined. But the locality of this net is still to be proved (and the warning should be made that the original proof [94, Theorem 3] has a gap: it is not taken into account that the analytic vectors of $A(f)$ may not be analytic for the restriction of $A(f)$ to the domain $\mathcal{D}' = (A(g) - \lambda)\,\mathcal{P}(O_2)\,\Omega$, $\operatorname{supp} f \subset O_2 \subset O'$, since this domain is not invariant under $A(f)$; cf. [20, Corollary 2 and Example 1 to Theorem X.39]). The need for this proof can be avoided, however, if we use, instead of the Nelson theorem, the theorem by Nussbaum [337] which gives a certain criterion of essential self-adjointness as well as a certain condition of commutativity of bounded functions of two operators satisfying the criterion and commuting on a common core. This way, facilitating the proof of locality, produces also its less restrictive condition because the Nussbaum criterion is formulated in terms of quasianalytic (instead of analytic) vectors. As a result, the improved sufficient condition takes the form of the 'postulate of the quasianalyticity of vacuum':

> The cyclic vacuum Ω of Wightman theory is an quasianalytic vector for all the field operators $A(f)$, $f \in S_R(\mathbf{R}^4)$, i.,e. the following condition holds: $\sum_{n=1}^{\infty} \| A(f)^n \Omega \|^{-1/n} = \infty$.

Application of this condition, as already said, is based on Nussbaum's theorem [337]. Let A and B be symmetric operators and $\mathcal{D}_q$ the set of their common quasianalytic vectors lying in the domain of their polynomial algebra. Let A and B as well as all their powers commute on $\mathcal{D}_q$. Then, if $\mathcal{D}_q$ is dense in $\mathcal{H}$, A and B are essentially self-adjoint and their spectral projections are commuting. Using this theorem we can easily prove the following proposition.

PROPOSITION 3.2.2 [175]. *Let Wightman theory satisfy the postulate of the quasi-analyticity of vacuum. Then field operators* $A(f)$, $f \in S_R(\mathbf{R}^4)$, *are essentially self-adjoint on the Gårding domain and their bounded functions generate the local net* $\{F_3(O)\}$.

Proof. Choosing arbitrary regions O_1, $O_2 \subset O'_1$, $O \subset (O_1 \cup O_2)'$, one has for test functions $f_i \in S_R(\mathbf{R}^4)$, $\operatorname{supp} f_i \subset O_i$, $i = 1, 2$, and $g_k \in S(\mathbf{R}^4)$, $\operatorname{supp} g_k \subset O$, $k = 1, \dots, m$, and any natural n:

$$\| A(f_i)^n A(g_1) \dots A(g_m)\,\Omega \|^2$$

$$= (A(g_1) \dots A(g_m)\,\Omega,\ A(f_i)^{2n}\, A(g_1) \dots A(g_m)\,\Omega)$$

$$= (A(g_m)^* \dots A(g_1)^* A(g_1) \dots A(g_m)\,\Omega,\ A(f_i)^{2n}\,\Omega)$$

$$\le K\, \| A(f_i)^{2n}\,\Omega \|, \qquad i = 1, 2,$$

where

$$K = \| A(g_m)^* \dots A(g_1)^* A(g_1) \dots A(g_m) \Omega \| .$$

Hence for any $\Phi \in \mathcal{D}(O) \equiv \mathcal{P}(O)\Omega$

$$\| A(f_i)^n \Phi \|^2 \leq K(\Phi) \| A(f_i)^{2n} \Omega \| , \; i = 1, 2,$$

and using the quasi-analyticity of Ω one has

$$\sum_{n=1}^{\infty} \| A(f_i)^n \Phi \|^{-1/n} \geq C(\Phi) \sum_{n=1}^{\infty} \| A(f_i)^{2n} \Omega \|^{-1/2n} = \infty ,$$

since $u_n = \| A(f_i)^n \Omega \|^{1/n}$ is an increasing positive sequence and the divergence of the series $\Sigma_{n=1}^{\infty} u_n^{-1}$ is equivalent to that of $\Sigma_{n=1}^{\infty} u_{2n}^{-1}$. Thus all the vectors in the domain $\mathcal{D}(O)$, which is dense by the Reeh–Schlieder theorem, are quasi-analytic for $A(f_1)$ and $A(f_2)$. Due to the locality of Wightman theory, these operators and all their natural powers commute on $\mathcal{D}(O)$, and so all the conditions of the Nussbaum theorem are fulfilled. ❐

Since every analytic vector is quasi-analytic but the converse is, in general, not true, the quasi-analyticity of Ω is a less restrictive condition than the analyticity. But still it is restrictive enough to exclude some well known models: as pointed out repeatedly (see, e.g. [4, Appendix A]) the quasi-analyticity of vacuum is valid for free and generalized free fields but is broken for the Wick monomial $:A^n:$ in the free field A for odd powers n.

A more realistic condition has been found by Driessler and Fröhlich [168]. It is adjusted to the framework of Wightman theory obtained by the Osterwalder–Schrader reconstruction theorem form a certain set $\{S_n\}_{n=1}^{\infty}$ of Schwinger functions. As a result, it has two formulations: the initial one is a condition on the functions S_n imposed in the original Euclidean theory; after the transition to relativistic theory this formulation induces a certain condition on the Wightman field $A(f)$.

The main part of the Euclidean Driessler–Fröhlich condition is of the same kind as the (quasi) analyticity of vacuum, i.e. it puts some restriction on the growth of the sequence $\{S_n\}$ with the order n of the correlation function. Namely, the sequence $\{S_n\}$ is required to satisfy the estimate

$$|S_n(f_1, \dots, f_n)| \leq (n!)^{1/2} \prod_{i=1}^{n} \| f_i \|_s \; \forall f_1, \dots, f_n \in S(\mathbf{R}^4) , \tag{2.9}$$

where $\| \cdot \|_s$ is some continuous norm on $S(\mathbf{R}^4)$. (Let us point out for the comparison that the analyticity of vacuum, as easily shown in [94], is equivalent to the estimate

$$|W_n(f, \dots, f)| = |(\Omega, A(f)^n \Omega)| \leq n! \, c(f)^n$$

for all $f \in S(\mathbf{R}^4)$.) However, after the transition to relativistic theory (2.9) entails the condition of a different kind. Since it exploits the spectrum condition, its formulation includes the positive generator $H = \hat{p}_0$ of time translations. The condition requires the following inequality to hold for quadratic forms of the operators $A(f)$, H:

$$|(\Psi, A(f)\Psi)| \leq |f| (\Psi, (H + I)\Psi) \tag{2.10}$$

where Ψ belongs to the domain of the quadratic form of H, $f \in S_R(\mathbf{R}^4)$ and $|f|$ is some translation-invariant norm on $S(\mathbf{R}^4)$. Similarly to the vacuum quasi-analyticity postulate, the use of (2.10) as a sufficient condition for the existence and locality of a net $\{F(O)\}$ is based on a certain 'commutator theorem', stating sufficient conditions of the essential self-adjointness of a pair of symmetric operators and the commutativity of bounded functions of these operators. In connection with (2.10) Driessler and Fröhlich prove the following commutator theorem:

COMMUTATOR THEOREM. *Let H be a positive self-adjoint operator, A, B closed symmetric operators weakly commuting on some core $\mathcal{D}_0 \subset \mathcal{D}(A) \cap \mathcal{D}(\mathcal{B})$ for the quadratic form H^2:*

$$(A\,\Phi, B\,\Psi) = (B\,\Phi, A\,\Psi) \;\; \forall\, \Phi, \Psi \in \mathcal{D}_0 .$$

Let there be a constant K such that on some core $\mathcal{D}_1$ for the quadratic form H, A and B as quadratic forms are bounded by H in the following sense:

$$|(\chi, A\,\chi)| \le K\,(\chi, (H+I)\,\chi),\; |(\chi, B\,\chi)| \le K\,(\chi, (H+I)\,\chi) .$$

$$|(\chi, i\,[H, A\,]\,\chi)| \le K\,(\chi, (H+I)\,\chi) .$$

$$|(\chi, i\,[H, B\,]\,\chi)| \le K\,(\chi, (H+I)\,\chi) .$$

for all $\chi \in \mathcal{D}_1$. Then A and B are self-adjoint operators and their bounded functions commute.

With the aid of this theorem it is easy to prove the following Proposition:

PROPOSITION 3.2.3 [130]. *If a Wightman theory satisfies the condition* (2.10), *the field operators $A(f)$, $f \in S_R(\mathbf{R}^4)$, are essentially self-adjoint on the Gårding domain as well as on any core of the Hamiltonian H. The bounded functions of fields generate a local net of von Neumann algebras $F_3(O)$.*

Proof. Clearly, (2.10) is adjusted precisely to the conditions of the commutator theorem. Applying the theorem to $A = \overline{A\,(f)}$, $B = \overline{A\,(g)}$, $f, g \in S_R(\mathbf{R}^4)$, $\operatorname{supp} f \times \operatorname{supp} g$, we see that the locality of Wightman fields ensures the fulfillment of the first assumption of the theorem while (2.10) the fulfillment of the first two bounds (and, the Gårding domain being a core for both the operator H and its quadratic form, one can choose $\mathcal{D}_0 = \mathcal{D}_1 = \mathcal{D}$). Then, taking account of $i\,[H, A\,(f)] = -A\,(\partial f/\partial t)$, we note that the fulfillment of (2.10) implies that of a similar condition obtained from it by replacing $A\,(f)$ by $i\,[H, A\,(f)]$ and $|f|$ by $|\partial f/\partial t|$. This means that the last two bounds hold too, and, all the assumptions of the commutator theorem being fulfilled, the fields are essentially self-adjoint on the Gårding domain and their bounded functions are locally commuting. Clearly, all the assumptions will also be fulfilled if the fields are taken initially not on $\mathcal{D}$ but on any core $\mathcal{D}_1 \subset \mathcal{D}$ of the form H. □

Study of the Euclidean version (2.9) of the Driessler–Fröhlich condition shows that in a large class of Euclidean and constructive models the following situation is typical: whenever the model admits of the construction of correct dynamics, it is usually found to satisfy (2.9) and (2.10). The class includes models with a mass gap and finite field strength renormalization, in particular, φ_d^4-models in dimensions $d = 2, 3, 4$. But, although the Driessler–Fröhlich condition is less restrictive and better fit for checking in concrete models than, say, the vacuum quasi-analyticity postulate, we cannot claim that it is already optimal; also, it is not very helpful in proving other properties of field algebras, besides locality. Thus the study of the ways of constructing local algebras $F(O)$ went on and still continues (see [173] for the latest analysis of the problem). For the most part it is based now on the so-called Bisognano–Wichmann formalism to which we proceed.

3.2.2. ALGEBRAS OF WEDGE REGIONS: THE BISOGNANO–WICHMANN FORMALISM

Significant advance in the problem of constructing and analyzing von Neumann field algebras was achieved when some special properties of (repeatedly used above) wedge regions were discovered and exploited [65, 66]. These properties make it possible to obtain very detailed information about the algebras $F(W_i)$, $i = R, L$: not just find for them new locality conditions but prove the locality for $F_8(W_i)$ and $F_9(W_i)$ without any additional assumptions as well as establish a number of deeper structural facts. The latter include, above all, the duality and representations for modular involution and modular automorphisms in terms of operators of spacetime transformations. Part of these facts is then extended to algebras of other regions, in particular, double cones.

All these results are based on the careful analysis of group-theoretical and analytic properties of physical objects related to the regions W_i. Some facts about these properties have already been presented in Section 1.3.3; we shall now recall and augment them.

Everything is rooted in the specific geometry of the pair (W_R, W_L). As said before, these regions are spacelike complements to each other:

$$W_R' = W_L, \qquad W_L' = W_R,$$

and there are three subgroups of the Poincaré transformations which leave both W_R and W_L invariant:

(1) translations along the x^2 and x^3 axes;

(2) rotations $R(e^1, \varphi)$, $\varphi \in [0, 2\pi]$ in the plane (x^2, x^3);

(3) Lorentz's boosts (velocity transformations) $\Lambda(e^1, t)$, $t \in \mathbf{R}$, in the plane (x^0, x^1).

There are also a number of transformations mapping W_R onto W_L, in particular,

$$R(e^2, \pi)\ (x^0, x^1, x^2, x^3) = (x^0, -x^1, x^2, -x^3),$$

$$R(e^3, \pi)\ (x^0, x^1, x^2, x^3) = (x^0, -x^1, -x^2, x^3),$$

$$G(x^0, x^1, x^2, x^3) = (-x^0, -x^1, x^2, x^3).$$

Here the first two transformations belong to $\mathcal{P}_+^\uparrow$, but, in contrast to them, G is a discrete transformation (inversion in the plane (x^0, x^1)) which belongs to the complexification of the subgroup $\Lambda(e^1, t)$: since

$$\Lambda(e^1, t) = \begin{pmatrix} \text{ch } t & \text{sh } t & & 0 \\ \text{sh } t & \text{ch } t & & \\ & & 1 & \\ & 0 & & 1 \end{pmatrix},$$

one has $G = \Lambda(e^1, i\pi)$.

Due to relativistic covariance of Wightman theory, the subgroups (1)–(3) are implemented in $\mathcal{H}$ by subgroups of the unitary group $U(\mathcal{P}_+^\uparrow)$ (the subgroup $U(0, \Lambda(e, t))$ will be denoted by $V(t)$ for brevity) and they are also implemented by automorphisms of Op^*-algebras $\mathcal{P}(W_i)$ as well as von Neumann algebras $F_k(W_i), k = 1, \ldots, 10$. Similarly, $R(e^2, \pi)$ and $R(e^3, \pi)$ are also implemented by unitary operators in $\mathcal{H}$ which, in their turn, implement the isomorphisms of corresponding algebras, both Op^* and W^*:

$$U(0, R(e^i, \pi))\, \mathcal{P}(W_R)\, U(0, R(e^i, \pi))^{-1} = \mathcal{P}(W_L),$$

$$U(0, R(e^i, \pi)) F_k(W_R)\, U(0, R(e^i, \pi))^{-1} = F_k(W_L), \quad i = 2, 3, k = 1, \ldots, 10.$$

But the covariance of these algebras under the inversion G, i.e. the existence of an antiunitary operator $U(G)$ in $\mathcal{H}$ such that

$$U(G)\, F(W_R)\, U(G) = F(W_L),$$

is not included in the relativistic covariance axiom. Also not included in this axiom but ensured by CPT-theorem [27, Section 4.3] is the existence of CPT-operator which is an antiunitary operator with the properties

$$\theta A(f)\theta = A^*(f_{PT}), \; f_{PT}(x) = f(-x), \; \theta\Omega = \Omega,$$

$$\theta \mathcal{P}(O)\theta = \mathcal{P}(-O), \; \theta U(a, \Lambda)\theta = U(-a, \Lambda), \tag{2.11}$$

implying, due to (2.6), CPT-covariance of all nets $F_k(O)$:

$$\theta F_k(O)\theta = F_k(-O), \qquad k = 1, \ldots, 10. \tag{2.12}$$

The existence of θ is of prime importance for studying the algebras $F(W_i)$. As pointed out in Section 1.3, in general local quantum theory CPT-theorem may not hold. Exactly because of this in field theory one can obtain significant enhancement of Proposition 1.3.24 concerning properties of the observable algebras $R(W_i)$ in the Haag–Araki theory. The results of this proposition are incomplete in the following respects:

(1) the condition (1.3.27) was not proved, and hence the conclusion that $V(t)$ is the modular group of $R(W_R)$ was based not only on the axioms but on this additional condition as well;

(2) the modular involution J, in contrast to modular automorphisms, did not get any physico-geometrical interpretation.

As we shall see, in the case of $F(W_l)$ both shortcomings can be eliminated.

The formula $G = \Lambda(e^1, i\pi)$ and Proposition 1.3.24 suggest that transformations mapping algebras of the region W_R into those of W_L may be connected with the complexification of the unitary group $V(t)$. As is well known [13], this complexification $V(t) \to V(t + i\lambda)$ is well defined for all $\tau = t + i\lambda \in \mathbf{C}$, and $V(\tau)$ with $\operatorname{Im}\tau \neq 0$ are closed unbounded operators whose domains can be readily characterized with the help of operator functional calculus. Let K be the self-adjoint generator of $V(t)$ and $E(K, s)$ the spectral measure of K. Then for any function $c(s)$ with compact support on the real axis, there exists a bounded operator

$$c(K) = \int_{-\infty}^{\infty} c(s)\, dE(K, s) = \int_{-\infty}^{\infty} V(t)\, \tilde{c}(t)\, dt\,,$$

$\tilde{c}(t)$ being the Fourier transform of $c(s)$. Since the function $\exp(-i\tau s)\, c(s)$ is in $\mathcal{D}(\mathbf{R})$ for any $\tau \in \mathbf{C}$, there exists also a bounded operator

$$V(\tau)\, c(K) = \int_{-\infty}^{\infty} e^{-i\tau s}\, c(s)\, dE(K, s)\,,$$

which is an entire function of τ. Hence the domain of any $V(\tau)$ includes the lineal $L\{c(K)\mathcal{H} \mid c(s) \in \mathcal{D}(\mathbf{R})\}$. Since the space $\mathcal{Z}(\mathbf{R})$ of the Fourier transforms of compact support functions includes sequences converging to delta function, this lineal is dense in $\mathcal{H}$. Moreover, for any dense lineal Q, the lineal

$$\mathcal{D}_c = L\{c(K)\, Q \mid c(s) \in \mathcal{D}(\mathbf{R})\}\,,$$

which also belongs to the domain of any $V(\tau)$, is dense, too. One can check that $\mathcal{D}_c$ is a core for $V(\tau)$ and the latter acts on the former as

$$V(\tau)\,\chi = \int_{-\infty}^{\infty} e^{-i\tau s}\, c(s))\, dE(K, s)\,\xi \quad \forall\, \chi = c(K)\,\xi\,, \qquad \xi \in Q\,. \tag{2.13}$$

Obviously, closing the operator $V(\tau)$ from the domain $\mathcal{D}_c$ means complementing the set of functions $c(s)$ with such functions which are not in $\mathcal{D}(\mathbf{R})$ anymore but still produce the convergent integral in (2.13). Hence it follows that the domain of $V(\tau)$ depends on $\operatorname{Im}\tau = \lambda$ only and, moreover, denoting $\mathcal{D}(V(\tau)) \equiv \mathcal{D}(\lambda)$, one has

$$\mathcal{D}(\lambda_1) \subset \mathcal{D}(\lambda_2) \text{ for } 0 \le \lambda_2/\lambda_1 \le 1\,, \tag{2.14}$$

and $\mathcal{D}(\lambda_1)$ is a core for all $V(\tau)$ with $0 \le (\operatorname{Im}\tau)/\lambda_1 \le 1$. For all $\Phi \in \mathcal{D}(\lambda_1)$ the vector-valued function $V(\tau)\,\Phi$ is analytic in the open strip $0 < (\operatorname{Im}\tau)/\lambda_1 < 1$ and continuous on its boundary. Finally, $\mathcal{D}_c$ being a core for $V(\tau)$, one has from (2.13):

$$V(\tau)^* = V(-\bar{\tau})\,.$$

Hence all $V(i\lambda)$, $\lambda \in \mathbf{R}$, are self-adjoint operators. The operator $V(i\pi)$ needs a special consideration because it is expected (see (1.3.27)) to transform the objects related to W_R to those related to W_L. Let us define a family of open bounded subregions $O_n \subset W_R$, $n = 1, 2, \ldots$ in the following way. Let O_1 be an arbitrary nonempty open subset of W_R and a vector $x_0 \in W_R$ be such that $(x - x_0) \in W_L$ for $x \in \overline{O}_1$. Then for any integer $n > 1$ we put

$$O_n = \{x \in M \mid x = y + (n-1)\, x_0 \,,\ y \in O_1\} \,.$$

It is easy to see that all O_n belong to W_R and are mutually spacelike. Using this family, consider the action of $V(i\pi)$ on Wightman fields in W_R.

LEMMA 3.2.4. *Let a Wightman theory be given with the Gårding domain* $\mathcal{D}$ *and cyclic vacuum* Ω. *Then for each family of test functions* $f_k(x)$, $k = 1, \ldots, n$, *such that* $\operatorname{supp} f_k \subset O_n$ *and each function* $c(s) \in \mathcal{D}(\mathbf{R})$ *one has*

$$V(i\pi)\, c(K)\, A(f_1) \ldots A(f_n)\, \Omega = c(K)\, A\left(f^j_1\right) \ldots \left(f^j_n\right) \Omega \,, \tag{2.15}$$

where $f^j_k(x) = f_k(Gx)$.

Proof. (2.15) will be derived by means of a chain of equalities. At first, we shall prove that

$$V(i\pi/2)\, c(K)\, A(f_1) \ldots A(f_n)\, \Omega$$

$$= \operatorname*{s-lim}_{t \to +0}\ V(i\pi/2)\, c(K)\, T(ite^0)\, A(f_1) \ldots T(ite^0)\, A(f_n)\, \Omega \,, \tag{2.16}$$

where $e^0 = (1, 0, 0, 0)$, $t \in \mathbf{R}$. It suffices to show that the vector-valued function in the right-hand side is strongly continuous in t for $t \geq 0$. To this end we shall apply to it the following representation:

$$T(z_1)\, A(f_1) \ldots A(f_n)\, \Omega$$

$$= \int \prod_{k=1}^{n} d^4 x_k\, A(x_1) \ldots A(x_n)\, f(x_1, \ldots, x_n; z_1, \ldots, z_n)\, \Omega \,, \tag{2.17}$$

where

$$(z_1, \ldots, z_n) \in \mathcal{T}^+_n = \{(z_1, \ldots, z_n) \in \mathbf{C}^{4n} \mid \operatorname{Im} z_i \in V^+,\ i = 1, \ldots, n\} \,,$$

test function f is the Fourier transform in the variables $p_1, \ldots, p_n$ of the function

$$\tilde{f}(p_1, \ldots, p_n; z_1, \ldots, z_n) = \prod_{k=1}^{n} \tilde{f}_k(p_k)\, E\left(\sum_{r=k}^{n} p_r \,;\, z_k\right) \tag{2.18}$$

and the function E is defined as

$$E(p; z) = u_0(p^2)\, u_0\left((p^2)^2\right) e^{ipz} \,, \quad u_0 \in C^\infty(\mathbf{R}), \quad u_0(s) = \begin{cases} 1, s \geq 0 \\ 0, s \leq 1 \end{cases} . \tag{2.19}$$

One can derive (2.17) by induction in n from the standard formulae (see, e.g. [15, Chapter 4]) expressing translation covariance of Wightman theory in the complex region: for $z \in \overline{\mathcal{T}^+} = \mathbf{R}^4 + i\overline{V^+}$ one has

$$T(z)\, A(f_1) \ldots A(f_n)\, \Omega = \int \prod_{k=1}^{n} d^4 x_k\, A(x_1) \ldots A(x_n)\, f(x_1, \ldots, x_n; z)\, \Omega \,,$$

where f is the Fourier transform in the variables $p_1, \ldots, p_n$ of a function which is equal to

$$\tilde{f}\,(p_1\,,\ldots,p_n\,;z)=\tilde{f}_1\,(p_1)\ldots\tilde{f}_n(p_n)\,e^{\,iz\sum_{r=1}^{n}p_r}$$

when

$$(p_1\,,\ldots,p_n)\in K_n\equiv\left\{(p_1,\,\ldots,\,p_n)\in \mathbf{R}^{4n}\,\middle|\,\sum_{r=k}^{n}p_r\in\overline{V^+},\;k=1,\,\ldots,\,n\right\}$$

(2.18) and (2.19) show that $\tilde{f}\,(p_1\,,\ldots,p_n\,;z_1\,,\ldots,z_n)$ is continuous in $z_1\,,\ldots,z_n$ on $\overline{\mathcal{T}_n}$ (as a $S\,(\mathbf{R}^{4n})$-valued function), and hence $f\,(x_1\,,\ldots,x_n\,;z_1\,,\ldots,z_n)$ is continuous too. Finally, the vector-valued tempered distribution in the right-hand side of (2.17) is continuous in the test function f and so this right-hand side is strongly continuous in $z_1\,,\ldots,z_n$ on $\overline{\mathcal{T}_n}$. Hence, the operator $V\,(i\,\pi/2)\,c\,(K)$ being bounded, the function in the right-hand side of (2.16) is strongly continuous in t for $t\geq 0$ ($i\,t\,e^0\in\overline{\mathcal{T}^+}$ for $t\geq 0$).

Next equality

$$\begin{aligned}&\underset{t\to+0}{s\text{-lim}}\;V\,(i\,\pi/2)\,c\,(K)\,T\,(i\,t\,e^0)\,A\,(f_1)\ldots T\,(i\,t\,e^0)\,A\,(f_n)\,\Omega\\&=\underset{t\to+0}{s\text{-lim}}\;V\,(i\,\pi/2)\,c\,(K)\int\prod_{k=1}^{n}d^4x_k f_1\,(x_1)\ldots f_n\,(x_n)\\&\quad\times A\,(i\,t\,e^0+x_1)\,A\,(i\,t\,e^0+x_2-x_1)\ldots A\,(i\,t\,e^0+x_n-x_{n-1})\,\Omega\end{aligned}\tag{2.20}$$

is a simple application of the convolution operation for tempered distributions, and we can proceed to proving the following equality:

$$\begin{aligned}&\underset{t\to+0}{s\text{-lim}}\;V\,(i\,\pi/2)\,c\,(K)\int\prod_{k=1}^{n}d^4x_k f_1\,(x_1)\ldots f_n\,(x_n)\,A\,(i\,t\,e^0+x_1)\\&\quad\times A\,(i\,t\,e^0+x_2-x_1)\ldots A\,(i\,t\,e^0+x_n-x_{n-1})\,\Omega\\&=\int\prod_{k=1}^{n}d^4x_k f_1\,(x_1)\ldots f_n\,(x_n)\;\underset{t\to+0}{s\text{-lim}}\;V\,(i\,\pi/2)\,c\,(K)\,A\,(i\,t\,e^0+x_1)\\&\quad\times A\,(i\,t\,e^0+x_2-x_1)\ldots A\,(i\,t\,e^0+x_n-x_{n-1})\,\Omega\,.\end{aligned}\tag{2.21}$$

The operator $V\,(i\,\pi/2)\,c\,(K)$ being bounded, its transport is allowed. Possibility of interchanging the strong limit in t and integration over x_k can be established by means of the following formula:

$$\begin{aligned}&\underset{t\to+0}{s\text{-lim}}\;V\,(i\,\pi/2)\,A\,(x_1+i\,t\,e^0)\ldots A\,(x_n+i\,t\,e^0)\,\Omega\\&=\underset{t\to+0}{s\text{-lim}}\;V\,(-i\,\pi/2)\,A\,(G\,x_1+i\,t\,e^0)\ldots A\,(G\,x_n+i\,t\,e^0)\,\Omega\\&=A\,(z_1)\ldots A\,(z_n)\,\Omega,\end{aligned}\tag{2.22}$$

where $z_k = (i \overset{1}{x}_k , i \overset{0}{x}_k ; i \overset{2}{x}_k , i \overset{3}{x}_k)$, $k = 1, \dots , n$.

To derive (2.22) let us check first that for $z_1 , \dots, z_n$ $(z_j = x_j + i\, y_j , j = 1, \dots , n)$ such that $\overset{2}{y}_k = \overset{3}{y}_k = 0$, $\overset{0}{y}_k > |\overset{1}{y}_k|$, $k = 1, \dots , n$ the vector $A(z_1) \dots A(z_n)\,\Omega$ is in the domain of $V(i\pi/2)$ when $x_1 , \dots , x_n \in W_R$ and in the domain of $V(-i\pi/2)$ when $x_1 , \dots , x_n \in W_L$; moreover, for all $\lambda \in [0, \pi/2]$ and $x_1 , \dots , x_n \in W_R$ as well as for all $\lambda \in [-\pi/2, 0]$ and $x_1 , \dots , x_n \in W_L$ one has

$$V(i\lambda) A(z_1) \dots A(z_n)\,\Omega = A(z_1(i\lambda)) \dots A(z_n(i\lambda))\,\Omega , \tag{2.23}$$

where $z_k(\tau) = \Lambda(e^1, \tau)\, z_k$. (2.23) is proved in a standard way: it suffices to show that for all vectors χ from some core for $V(i\lambda)^*$ one has

$$(V(i\lambda)^*\chi , A(z_1) \dots A(z_n)\,\Omega) = (\chi , A(z_1(i\lambda)) \dots A(z_n(i\lambda))\,\Omega) . \tag{2.24}$$

Choosing the core to be the (discussed above) lineal $L\,\{c(K)\,\mathcal{H} \mid c(s) \in \mathcal{D}(\mathbf{R})\}$, we have that the left-hand side of (2.23) is an entire analytic function of λ . One can check by the direct calculation that for the points $(z_1 , \dots , z_n)$ considered in (2.23) one has $(z_1(i\lambda), \dots , z_n(i\lambda)) \in \overset{+}{\mathcal{T}}_n$ for all $x_1 , \dots , x_n \in W_R$ and $\lambda \in [0, \pi/2]$. Hence, by the well known analytic properties of Wightman fields, the right-hand side of (2.24) is also an analytic function of λ in a domain including (since $\overset{+}{\mathcal{T}}_n$ is an open set) a certain open neighbourhood $\mathcal{N}$ of the segment $[0, \pi/2]$. Since the two functions coincide in a certain real neighbourhood of $\lambda = 0$, they coincide in the whole $\mathcal{N}$ as well, which implies (2.24), implying, in its turn, (2.23). The assertions about W_L are proved identically, taking into account that one has $(z_1(i\lambda) , \dots , z_n(i\lambda)) \in \overset{+}{\mathcal{T}}_n$ for all $x_1 , \dots , x_n \in W_L$ and $\lambda \in [-\pi/2, 0]$.

Returning to the proof of (2.22), one has from (2.23) for $x_1 , \dots , x_n \in W_L$ and $t > 0$:

$$V(i\pi/2) A(x_1 + i\,t\,e^0) \dots A(x_n + i\,t\,e^0)\,\Omega = A(z_1') \dots A(z_n')\,\Omega , \tag{2.25}$$

where $z_k' = z_k'(t) = z_k - (0, t, 0, 0)$, $k = 1 , \dots , n$ and z_k are the same as in (2.22). Since $x_1 , \dots , x_n \in W_R$ implies $G\,x_1 , \dots , G\,x_n \in W_L$, we get from (2.23) when $x_1 , \dots , x_n \in W_R$, and $t > 0$:

$$V(-i\pi/2) A(G\,x_1 + i\,t\,e^0) \dots A(G\,x_n + i\,t\,e^0)\,\Omega = A(z_1'') \dots A(z_n'')\,\Omega \tag{2.26}$$

where $z_k'' = z_k''(t) = z_k + (0, t, 0, 0)$, $k = 1, \dots , n$. Again, by the analyticity properties of Wightman fields [15], the vector-valued functions in the right-hand sides of (2.25) and (2.26) are analytic in the open tube $\overset{+}{\mathcal{T}}_n$ and strongly continuous in the closed tube $\overline{\overset{+}{\mathcal{T}}}_n$ (this can be proved simply by the same reasoning as that used in deriving (2.16) and (2.17)). Thus, the functions in the right-hand sides of (2.25) and (2.26) have strong limits at $t \to +0$, and by taking them we obtain both equalities in (2.22).

In its turn, (2.22) readily implies (2.21): since $\operatorname{supp} f_k \subset O_k \subset W_R$ in (2.21), the equalities (2.22) can be applied to the integrand in (2.21) and imply that it has the strong limit at $t \to +0$, approaching it uniformly with respect to $x_1 , \dots , x_n$. On the other hand, convergence of the integral in the left-hand side of (2.21) is uniform with respect to t. Thus interchanging the limit $t \to +0$ and integration over $x_1, \ldots, x_n$ is allowed.

Now it is not difficult to obtain all the remaining equalities which we need:

$$\int \prod_{k=1}^{n} d^4 x_k f_1(x_1) \ldots f_n(x_n) \underset{t \to +0}{\text{s-lim}} V(i\pi/2)\, c(K)\, A(i t e^0 + x_1)$$
$$\times A(i t e^0 + x_2 - x_1) \ldots A(i t e^0 + x_n - x_{n-1})\, \Omega$$
$$= \int \prod_{k=1}^{n} d^4 x_k f^j_1(x_1) \ldots f^j_n(x_n) \underset{t \to +0}{\text{s-lim}} V(-i\pi/2)\, c(K)$$
$$\times A(i t e^0 + x_1)\, A(i t e^0 + x_2 - x_1) \ldots A(i t e^0 + x_n - x_{n-1})\, \Omega$$
$$= \underset{t \to +0}{\text{s-lim}} V(-i\pi/2)\, c(K)\, T(i t e^0) A(f^j_1)\, T(i t e^0) \ldots T(i t e^0) A(f^j_n)\, \Omega$$
$$= V(-i\pi/2)\, c(K)\, A(f^j_1) \ldots A(f^j_n)\, \Omega . \tag{2.27}$$

Here the first equality is obtained from (2.22) by the change of the integration variables $x_k \to G x_k$. The second equality: by the same arguments as in deriving (2.21), we can again interchange in the second term in (2.27) the integration over $x_1, \ldots, x_n$ and limit at $t \to +0$. Then, taking the bounded operator $V(-i\pi/2)\, c(K)$ out of the integrand and using the equality similar to (2.20) we obtain the third term in (2.27). Finally, the third equality is obtained exactly like (2.26).

Summing up, the whole chain of equalities (2.16), (2.20), (2.21) and (2.27) gives us

$$V(i\pi/2)\, c(K)\, A(f_1) \ldots A(f_n)\, \Omega = V(-i\pi/2)\, c(K)\, A(f^j_1) \ldots A(f^j_n)\, \Omega .$$

Since the operators $V(\tau)$ form a representation of $\mathbf{C}$ as a group by addition, and $c(K)\, \mathcal{H} \subset \mathcal{D}(V(\tau))\ \forall \tau \in \mathbf{C}$, we can act by the operator $V(i\pi/2)$ on this equality. Thus, (2.15) follows. ☐

The results of Section 1.3.3 suggest that the group $V(\tau)$ and its complexification are still insufficient for the complete characterization of modular structure of the algebras $F(W_i)$. Even if we prove the fulfillment of (1.3.27), we can only conclude that $V(t)$ is the modular group of $F(W_R)$ and $V(-t)$ the modular group of $F(W_L)$. But the second main object of modular theory, the Tomita involution, still remains unrelated to physical properties and geometrical structure of Wightman theory. Being antiunitary, it cannot be one of the operators $V(\tau)$. Thus, to describe it we need some new objects independent of the groups $V(t)$ and $V(\tau)$.

We have already enough facts at our disposal to discern the basic properties of such an object and understand the reasons for its choice in the Bisognano–Wichmann formalism. Indeed, according to Proposition 1.3.24, the condition (1.3.27) ensures not only the coincidence of the group $V(t)$ and the modular group of $F(W_R)$ but also the duality for this algebra, $F(W_R)' = F(W_L)$. This means that the modular involution J is an antiunitary operator transforming $F(W_R)$ into $F(W_L)$. So the unitary operators $U(0, R(e^2, \pi))$ and $U(0, R(e^3, \pi))$ cannot be candidates for it, and geometrically the involution J should be connected not with the rotations $R(e^3, \pi), R(e^3, \pi) \in \mathcal{P}_+^\uparrow$, but with the inversion G. The Haag–Araki theory may not include operators implementing such a transformation (which explains why in Section 1.3.3 we have not advanced as far as constructing the modular involution); however, in Wightman theory they are always present due to *CPT*-theorem. Indeed, the antiunitary *CPT*-operator includes the complete inversion *PT* while G can be represented as the composition of *PT* and the inversion in the plane (x^2, x^3) which is the Euclidean rotation by π about the x^1 axis. Thus we can expect the operator

$$J = U(0, R(e^1, \pi))\,\theta \tag{2.28}$$

to be a good candidate for the modular involution of $F(W_R)$ This hypothesis turns out to be true.

Let us consider basic properties of J. Taking account of (2.11), we get easily:

$$J^2 = I, \qquad J\Omega = \Omega, \qquad J\,U(a, \Lambda) = U(Ga, G\Lambda G). \tag{2.29}$$

In particular, at $a = 0, \Lambda = \Lambda(e^1, t)$ we obtain

$$J\,V(t)\,J = V(t) \ \ \forall t \in \mathbf{R}, \tag{2.30}$$

whence it follows due to the antiunitarity of J:

$$J\,\mathcal{D}(K) = \mathcal{D}(K), \qquad J K J = -K, \qquad J\,\mathcal{D}(\lambda) = \mathcal{D}(-\lambda), \qquad J\,V(\tau)\,J = V(\bar{\tau}). \tag{2.31}$$

Using these properties, we can obtain the enhancement of Lemma 3.2.4 where (2.15) takes the form closer to (1.3.27).

LEMMA 3.2.5. *Let* $\{O_n\}_{n=1}^{\infty}$ *be the same family of* W_R *subregions as in Lemma 3.2.4 and* $\tilde{\mathcal{P}}(W_R) \subset \mathcal{P}(W_R)$ *the linear hull of the identity operator and all operators of the form* $A = A(f_1) \dots A(f_n)$, $\operatorname{supp} f_k \subset O_k$, $k = 1, \dots n$. *Then* $\tilde{\mathcal{P}}(W_R)$ *is a* **-set with respect to the involution in* $\mathcal{P}$ *and* $\tilde{\mathcal{P}}(W_R)\,\Omega$ *is dense in* $\mathcal{H}$. *For all* $A \in \tilde{\mathcal{P}}(W_R)$ *and* $c(s) \in \mathcal{D}(\mathbf{R})$,

$$V(i\pi)\,c(K)\,A\,\Omega = c(K)\,J\,A^*\,\Omega\,. \tag{2.32}$$

Proof. Density of $\tilde{\mathcal{P}}(W_R)\,\Omega$ follows from the 'Wightman analogue' of the corollary to Proposition 1.3.1 (the Reeh–Schlieder theorem). The fact that $A \in \tilde{\mathcal{P}}(W_R)$ entails $A^+ = A^+|_{\mathcal{D}} \in \tilde{\mathcal{P}}(W_R)$ is evident since $A(f)^*|_{\mathcal{D}} = A(\bar{f})$ and due to $O_i \times O_k$, $i, k = 1, \dots, n, i \neq k$, we have

$$(A(f_1) \dots A(f_n))^*|_{\mathcal{D}} = A(\bar{f}_n) \dots A(\bar{f}_1) = A(\bar{f}_1) \dots A(\bar{f}_n) \in \tilde{\mathcal{P}}(W_R). \tag{2.33}$$

To derive (2.32) note that (2.11) entails $J\mathcal{D} = \mathcal{D}$ and (2.11) together with (2.29) entail

$$J\,A(f)\,J = A(f^j)\,. \tag{2.34}$$

Hence, due to (2.33) one has for all $A \in \tilde{\mathcal{P}}(W_R)$ of the form $A = A(f_1) \dots A(f_n)$:

$$J A^+ J = A\left(f_1^j\right) \dots A\left(f_n^j\right).$$

Putting this into (2.15) and taking into account $A^* \Omega = A^+ \Omega$, (2.32) follows. ❐

In the last of the preparatory lemmas we collect all properties of the 'modular objects', $J,\ V(t),\ V(\pm i\pi)$ with respect to *Op**-algebras $\mathcal{P}(W_i)$.

LEMMA 3.2.6. *In any Wightman theory with cyclic vacuum* Ω *the following properties of Op*-algebras* $\mathcal{P}(W_i)$ *hold:*

(1) $V(t) \cdot V(t)^{-1}$ *is the group of automorphisms of* $\mathcal{P}(W_i)$;

(2) $J\ \mathcal{P}(W_R)\, J = \mathcal{P}(W_L)$;

(3) $\mathcal{P}(W_R)\,\Omega \subset \mathcal{D}(\pi), \qquad \mathcal{P}(W_L)\,\Omega \subset \mathcal{D}(-\pi)$;

(4) *for all* $A \in \mathcal{P}(W_R),\ B \in \mathcal{P}(W_L)$,

$$J\, V(i\pi)\, A\, \Omega = A^*\, \Omega, \qquad J\, V(-i\pi)\, B\, \Omega = B^*\, \Omega. \tag{2.35}$$

Proof. The property 1 is evident and so is 2, taking account of (2.34). The first formulas in 3 and 4 can be proved simultaneously in the standard way, choosing some core $\mathcal{D}_+$ for the self-adjoint operator $V(i\pi)$ and showing that

$$(A\,\Omega, V(i\pi)\,\chi) = (J\,A^*\,\Omega, \chi) \tag{2.36}$$

for all $A \in \mathcal{P}(W_R),\ \chi \in \mathcal{D}_+$. Take $\mathcal{D}_+ = L\,\{c(K)\,\tilde{\mathcal{P}}(W_R)\,\Omega \mid c(s) \in \mathcal{D}(\mathbf{R})\}$. The density of $\tilde{\mathcal{P}}(W_R)\,\Omega$ in $\mathcal{H}$ ensures that it is a core for $V(i\pi)$. Then in (2.36) $\chi = c(K)\,T\,\Omega,\ c(s) \in \mathcal{D}(\mathbf{R}),\ T \in \tilde{\mathcal{P}}(W_R)$ and we obtain

$$\begin{aligned}(A\,\Omega, V(i\pi)\,\chi) &= (A\,\Omega, V(i\pi)\,c(K)\,T\,\Omega) = (A\,\Omega, c(K)\,J\,T^*\,\Omega)\\ &= (A\,\Omega, c(-K)^*\,T^*\,\Omega) = (c(-K)^*\,T^*\,\Omega, J\,A\,\Omega),\end{aligned} \tag{2.37}$$

where the second equality follows from (2.32), the third from (2.31) and the last from the antiunitarity of J. Continuing this chain of equalities with the aid of the representation

$$c(K) = \int_{-\infty}^{\infty} V(t)\,\tilde{c}(t)\,dt, \quad \tilde{c}(t) = \frac{1}{2\pi}\int_{-\infty}^{\infty} ds\, e^{its}\, c(s), \tag{2.38}$$

we get

$$\begin{aligned}(c(-K)^*\,T^*\,\Omega, J\,A\,\Omega) &= \int_{-\infty}^{\infty} dt\,\tilde{c}(t)\,(V(t)\,T^*\,V(t)^{-1}\,\Omega,\ J\,A\,J\,\Omega)\\ &= \int_{-\infty}^{\infty} dt\,\tilde{c}(t)\,((J\,A\,J)^*\,\Omega, V(t)\,T\,V(t)^{-1}\,\Omega)\\ &= (J\,A^*\,\Omega, c(K)\,T\,\Omega) = (J\,A^*\,\Omega, \chi).\end{aligned} \tag{2.39}$$

Here the first equality uses that the scalar product in $\mathcal{H}$ is antilinear in the first argument; the second follows from the commutativity on $\mathcal{D}$ of the operators

$V(t)\, T^*\, V(t)^{-1}$ and $(J\, A\, J)^*$; the third and the fourth are evident. (2.37) and (2.39) taken together entail (2.36) and the statements relating to $\mathcal{P}(W_R)$ and $V(i\,\pi)$ are proved. The statements relating to $\mathcal{P}(W_L)$ and $V(-i\,\pi)$ follow from them, taking account of 2 and $J\, V(i\,\pi) = V(-i\,\pi)$. ❐

Now we can proceed to the main results.

THEOREM 3.2.7.

(A) *In any Wightman theory with cyclic vacuum* Ω *the following properties of the von Neumann algebras* $F_n(W_i)$, $n = 8, 9$; $i = R,\ L$, *hold:*

(1) *locality is fulfilled,* $F_n(W_R) \subset F_n(W_L)'$;

(2) $F_n(W_R) = J\, F_n(W_L)\, J$; (2.40)

(3) $F_n(W_R)\,\Omega \subset \mathcal{D}(\pi)$, $F_n(W_L) \subset \mathcal{D}(-\pi)$ *and for all* $A \in F_n(W_R)$, $B \in F_n(W_L)$,

$$J\, V(i\,\pi)\, A\,\Omega = A^*\,\Omega,\ J\, V(-i\,\pi)\, B\,\Omega = B^*\,\Omega\ . \tag{2.41}$$

(B) *If, moreover,* Ω *is cyclic for* $F_n(W_R)$, *the following properties hold:*

(4) Ω *is bicyclic for* $F_n(W_i)$;

(5) *duality is fulfilled,* $F_n(W_R)' = F_n(W_L)$;

(6) *the modular structure of* $F_n(W_i)$ *is described by the following objects:*

(a) $V(\pm t),\ t \in \mathbf{R}$, *is the modular group for the pair* $(F_n(W_R), \Omega)$ *and* $(F_n(W_L), \Omega)$, *respectively;*

(b) $\Delta^{\pm 1} = \exp(\pm 2\,\pi\, K)$ *are the modular operators for the said pairs;*

(c) $J\, V(i\,\pi) = S_R$, $J\, V(-i\,\pi) = S_L$ *are the involutions in the achieved generalized Hilbert algebras* $F_n(W_R)\,\Omega$ *and* $F_n(W_L)\,\Omega$, *respectively; the domains* $\mathcal{D}(\pm\, i\,\pi)$ *are the spaces* $\mathcal{D}^\#$ *and* $\mathcal{D}^b$ *connected with these algebras;*

(d) J *is the modular involution,* $J\, F_n(W_i)\, J = F_n(W_i)'$.

(C) *If, moreover,* Ω *is the unique vacuum vector in* $\mathcal{H}$, *then* $F_n(W_i)$ *are factors of type* III_1 .

Proof shall be given first for $F_9(W_i)$ and then we shall check that it is valid for $F_8(W_i)$ as well, with only the minor changes. The statements in part (A), except locality are derived simply, using Lemma 3.2.6. Indeed, (A2) is analogous to (2) in the lemma and readily follows from it taking account of (2.6). (A3) are analogous to (3) and (4) in the lemma and are proved in the same way, i.e. we should (in the case of $F_9(W_i)$) derive (2.36) for all χ from some core for $V(i\,\pi)$ and all $A \in F_9(W_R)$. At first we obtain an auxiliary equality:

$$(J\, V(t)\,\eta\ ,\ J\, A^*\,\Omega) = (V(i\,\pi)\, J\, V(t)\,\eta\ ,\ A\,\Omega) \tag{2.42}$$

for all $A \in F_9(W_R)$ and $\eta \in \mathcal{P}(W_L)\,\Omega$. Note that there is $B \in \mathcal{P}(W_L)$ such that

$V(t)\eta = B_t \Omega$ where $B_t = V(t) B V(t)^{-1} \in \mathcal{P}(W_L)$. Taking this into account we transform the left-hand side of (2.42) to the right-hand side with the aid of (2.31) and (2.35):

$$(JB_t \Omega,\ JA^* \Omega) = (A^* \Omega, B_t \Omega) = (B_t^* \Omega, A \Omega)$$
$$= (JV(-i\pi) B_t \Omega, A \Omega) = (V(i\pi) JV(t)\eta,\ A\Omega)$$

Hence, it follows by (2.38) for an arbitrary $c(s) \in \mathcal{D}(\mathbf{R})$:

$$(Jc(K)\eta,\ JA^* \Omega) = \int_{-\infty}^{\infty} dt\, \tilde{c}(t)\, (JV(t)\eta,\ JA^* \Omega)$$
$$= \int_{-\infty}^{\infty} dt\, \tilde{c}(t)\, (V(i\pi) JV(t)\eta,\ A\Omega)$$
$$= (V(i\pi) J c(K)\eta,\ A\Omega).$$

Since the domain $\mathcal{D}_- = L\{c(K)\mathcal{P}(W_L)\Omega \mid c(s) \in \mathcal{D}(\mathbf{R})\}$ is a core for $V(-i\pi)$, and hence, due to (2.31), $J\mathcal{D}_-$ is a core for $V(i\pi)$, we get what we need: for all $A \in F_9(W_R)$

$$A\Omega \in \mathcal{D}(\pi), \quad JV(i\pi) A\Omega = A^* \Omega .$$

It can be proved identically that for all $B \in F_9(W_L)$

$$B\Omega \in \mathcal{D}(-\pi), \quad JV(-i\pi) B\Omega = B^* \Omega .$$

Let us turn to the proof of locality. We show first that $F_9(W_R)$ and $F_9(W_L)$ commute on the vector Ω:

$$[A, B]\Omega = 0, \ \forall A \in F_9(W_R), \quad B \in F_9(W_L) . \tag{2.43}$$

Due to (2.40) it is equivalent to

$$[A, JCJ]\Omega = 0, \ \forall A, C \in F_9(W_R) . \tag{2.44}$$

Since $V(t) \cdot V(t)^{-1}$ are the automorphisms of $F_9(W_R)$, for all $t \in \mathbf{R}$ one has $AV(t) C^* V(t)^{-1} \in F_9(W_R)$. Hence, for any bounded operator Y such that $Y\Omega \in \mathcal{D}(-\pi)$, $JV(-i\pi) Y\Omega = Y^* \Omega$, one has, using (2.41):

$$(Y\Omega, AV(t) C^* V(t)^{-1} \Omega) = (Y\Omega, JV(i\pi) V(t) CV(t)^{-1} A^* \Omega)$$
$$= (Y\Omega, V(-i\pi) JV(t) CV(t)^{-1} A^* \Omega) = (V(-i\pi) Y\Omega, JV(t) CV(t)^{-1} A^* \Omega)$$
$$= (V(-i\pi - t) Y\Omega, JCV(-t) A^* \Omega) = (V(-i\pi - t) Y\Omega, JCJV(i\pi - t) A\Omega),$$

whence it follows that for all real t,

$$f(t) \equiv (Y\Omega, AV(t) C^* \Omega) - (V(-t - i\pi) Y\Omega, JCJV(i\pi - t) A\Omega) = 0. \tag{2.45}$$

Consider the vector-valued analytic functions in this formula. Since

$F_9(W_R)\,\Omega \subset \mathcal{D}(\pi)$ and $Y\Omega \in \mathcal{D}(-\pi)$, the functions $AV(\tau)\,C^*\,\Omega, JCJV(i\pi-\tau)A\,\Omega$ and $V(-\bar{\tau}-i\pi)\,Y\Omega$, due to (2.14), are well defined and strongly continuous in the closed strip $0 \le \operatorname{Im}\tau \le \pi$. Moreover, the first two functions are strongly analytic in the corresponding open strip while the third one is strongly analytic in $\bar{\tau}$ in the open strip $0 > \operatorname{Im}\tau > -\pi$. Hence $f(t)$ can be analytically continued to a function, analytic in the open strip $0 < \operatorname{Im}\tau < \pi$ and continuous in its closure (we continue the function $V(-t-i\pi)\,Y\Omega$ in (2.45) as $V(-\bar{\tau}-i\pi)\,Y\Omega$). Thus $f(\tau)=0$ on the set $\operatorname{Im}\tau = 0$ implies $f(\tau)=0$ on the whole closed strip $0 \le \operatorname{Im}\tau \le \pi$. Now, having $f(i\pi)=0$ and taking account of (2.41) and $V(-\bar{\tau}-i\pi)=I$ at $\tau = i\pi$ we get

$$(Y\Omega,\ [A, JCJ]\,\Omega) = 0\,. \tag{2.46}$$

Let us prove that the set of all Y in (2.46),

$$\mathcal{A}(W_L) \doteq \{Y \in \mathcal{B}(\mathcal{H}) \mid Y\Omega \in \mathcal{D}(-\pi)\,,\ JV(-i\pi)\,Y\Omega = Y^*\,\Omega\}$$

produces the dense set of vectors $\mathcal{A}(W_L)\,\Omega$. To this end we associate with each vector $\eta \in \mathcal{D}(-\pi)$ a bounded operator $Y(\eta)$ defined (in the Dirac bracket notations) as*

$$Y(\eta) = |\,\eta\,\rangle\langle\,\Omega\,| + |\,\Omega\,\rangle\langle\,JV(-i\pi)\,\eta\,| - \langle\,\Omega\,|\,\eta\,\rangle\langle\,\Omega\,|\,\Omega\,\rangle\,.$$

The advantage of this notation is that it gives immediately the adjoint operator:

$$Y(\eta)^* = |\,\Omega\,\rangle\langle\,\eta\,| + |\,JV(-i\pi)\,\eta\,\rangle\langle\,\Omega\,| - \langle\,\eta\,|\,\Omega\,\rangle\,|\,\Omega\,\rangle\langle\,\Omega\,|\,.$$

In the usual notation, taking account of $(\Omega\,,\ \eta) = (JV(-i\pi)\,\eta\,,\ \Omega)$, one has for all $\Phi \in \mathcal{H}$:

$$Y(\eta)\,\Phi = (\Omega, \Phi)\,\eta + (JV(-i\pi)\,\eta\,,\ (I - P_\Omega)\,\Phi)\,\Omega\,,$$

$$Y(\eta)^*\,\Phi = (\eta, \Phi)\,\Omega + (\Omega, \Phi)\,[JV(-i\pi) - P_\Omega]\,\eta\,,$$

P_Ω being the projection on $\{c\,\Omega\}$. In particular,

$$Y(\eta)\,\Omega = \eta,\ Y(\eta)^*\,\Omega = JV(-i\pi)\,\eta \qquad \forall\,\eta \in \mathcal{D}(-\pi)\,.$$

These formulas imply immediately $Y(\eta) \in \mathcal{A}(W_L)$ and $\mathcal{A}(W_L)\,\Omega = \mathcal{D}(-\pi)$. Hence the set of all vectors $Y\Omega$ in (2.46) is dense. Then (2.46) entails (2.44) together with its equivalent (2.43).

Taking into account that $(a, I)\,W_R \subset W_R$ for all $a \in W_R$, it follows from the relativistic covariance of $F_9(O)$ that $A(a) \in F_9((a, I)\,W_R) \subset F_9(W_R)$ for all $A \in F_9(W_R)$. Hence (2.43) entails

* The mapping $Y : \mathcal{D}(-\pi) \to \mathcal{B}(\mathcal{H})$ is akin to the representation $\pi' : \mathcal{D}^b \to \mathcal{L}(\mathcal{H})$ in the Tomita-Takesaki theory [28, Section 3]: according to the statement (B6) in our theorem, $\mathcal{D}(-\pi)$ is the space $\mathcal{D}^b$ for the pair $(F(W_R), \Omega)$ and $JV(-i\pi)$ is the involution in the Hilbert algebra $\mathcal{U}' = F_9(W_L)\,\Omega$. It does not coincide with it, however, since $\pi'(\eta)$ are bounded only for $\eta \in \mathcal{U}'$ but not for all $\eta \in \mathcal{D}^b$; moreover, the mapping Y is not a representation at all and its image $\mathcal{A}(W_L)$ is larger than $F_9(W_L)$ and not an algebra. Identically, the mapping $X : \mathcal{D}(\pi) \to \mathcal{B}(\mathcal{H})$ can be introduced whose image $\mathcal{A}(W_R)$ includes $F_9(W_R)$ and is not an algebra [65, Lemma 12].

$$[A(a), B]\,\Omega = 0 \;\; \forall A \in F_9(W_R), B \in F_9(W_L)\,. \tag{2.47}$$

Consider $O = W_L \cup (a, I)\, W_R$; then $A(a) \in F_9(O), B \in F_9(O)$ for the operators in (2.47). $F_9(O)$ being an algebra and $F_9(O) \subset F_8(O)$, one has

$$[A(a), B] \in F_9(O) \subset F_8(O) \subset \mathcal{P}(O')'_w\,.$$

By the definition of weak commutant and due to (2.47) it implies

$$(S_1\,\Omega, [A(a), B]\, S_2\,\Omega) = (S_2^*\, S_1\,\Omega, [A(a), B]\,\Omega) = 0 \tag{2.48}$$

for all $S_1, S_2 \in \mathcal{P}(O')$. Since $O' = W_R \cap (a, I)\, W_L$ is an open nonempty set for any $a \in W_R$, by the Reeh–Schlieder theorem, the vectors $S_1\,\Omega$ and $S_2\,\Omega$ in the left-hand side of (2.48) run over the dense set $\mathcal{P}(O')\,\Omega$, which implies $[A(a), B] = 0$ for all $a \in W_R$. The representation $U(a, \Lambda)$ being strongly continuous, the operator-valued functions $A(a)$ and $[A(a), B]$ are strongly continuous in a, whence $[A(a), B] = 0$ for all $a \in \overline{W_R}$ as well. Since $a = 0 \in \overline{W_R}$, we obtain the locality for $F_9(W_R)$.

The proof of remaining statements about $F_9(W_i)$ is simple. Consider part (B). Ω is cyclic for both $F_9(W_R)$ and $F_9(W_L)$ due to (2.40) and bicyclic for them since they commute. Thus these algebras and the group $V(t)$ satisfy all the conditions of the Rigotti theorem (see Section 1.3.3; in particular, the basic condition (3) follows from (2.41) due to (2.40)). This proves the following facts: $V(\pm t)$ is the modular group for the pair $(F_9(W_{R,L})\,\Omega)$ and $\Delta^{\pm 1} = \exp(\pm 2\pi K)$ the modular operator; the duality holds. The duality together with (2.40) and $J\,\Omega = \Omega$ implies that J is the modular involution for the said pair. (6c) is evident owing to (2.41).

Finally, the statements about the factoriality and type of $F_9(W_i)$ in part (C) we obtain, taking into account that W_i are lightlike monotone regions: it is easily seen that the proofs of Proposition 1.3.6 and 1.3.12 use only the facts proved or assumed here for $\{F_9(O)\}$.

Now we have to extend the proof to the algebras $F_8(W_i) = {}_{\rm alg}\mathcal{P}(W_i')'_w$.

(a) It is easy to see that the proofs of (2.40) and (2.41) are, in fact, valid for any set of bounded operators commuting weakly with $\mathcal{P}(W_L)$ and such that $V(t) \cdot V(t)^{-1}$, $t \in \mathbf{R}$, are its automorphisms. Hence the statements (A2) and (A3) are valid for the sets $\mathcal{P}(W_L)'_w$ and therefore the algebras $F_8(W_i)$.

(b) Hence all the proof of (2.44) is still valid if we take $A \in F_8(W_R)$, $C \in \mathcal{P}(W_L)'_w$. Indeed, the only property required of these operators, besides (2.40) and (2.41), is

$$AV(t)\, C^*\, V(t)^{-1} \in F_8(W_R),$$

which follows from the definition of $F_8(O)$.

(c) Note that what we used in proceeding from (2.44) and (2.43) to (2.48) was actually not $[A, B] \in F_9(O)$, but only $[A, B] \in \mathcal{P}(O')'_w$. This holds if $B \in \mathcal{P}(W_R)'_w \subset \mathcal{P}(O')'_w$ and, in particular, for all $B \in F_8(W_L)$. The operator A can also be taken from $F_8(W_R)$: the net being relativistically covariant, we shall have $A(a) \in F_8(W_R)$ for all $a \in W_R$, and the property $[A, B] \in \mathcal{P}(O')'_w$ sufficient for (2.48) will be preserved. Thus the locality of $F_8(W_i)$ follows.

All the remaining part of the proof does not need any change. The additional condition

$$\overline{F_8(W_R)\,\Omega} = \mathcal{H} \tag{2.49}$$

together with the locality for $F_8(W_i)$ and the condition (2.41) with $A \in F_8(W_R)$, $B \in F_8(W_L)$ imply that $F_8(W_i)$ satisfies all the assumptions of the Rigotti theorem whence the statements of Part (B) follow. Assuming the uniqueness of vacuum, we get the statements of part (C). ❐

The results obtained deserve some discussion. At first, the theorem has the simple corollary.

COROLLARY 1. *In any Wightman theory with cyclic vacuum the following relations hold:*

$$F_9(W_i) \subset F_8(W_i) \subset \mathcal{P}(W_j)'_w \subset F_8(W_j)' \subset F_9(W_j)' = F_{10}(W_i)\,,$$

$$F_8(W_i) \subset F_4(W_i) \subset F_8(W_j)'\,, \qquad i\,,j = R\,,L\,,i \neq j\,. \tag{2.50}$$

Proof. The second formula (2.50) is the commutant to the middle part of the first formula. In the first formula, due to (2.3), all the inclusions are evident, except the third one, which is a by-product of the locality proof for $F_8(W_i)$. Indeed, in the last stage of this proof we have noticed that the property $[A, B] = 0$ proved first for $A \in F_9(W_R)$, $B \in F_9(W_L)$ by exactly the same reasoning extends to $A \in F_8(W_R)$, $B \in \mathcal{P}(W_R)'_w$. ❐

Furthermore, the condition (2.49) is, in general, weaker than the cyclicity of Ω for $F_9(W_R)$. Accordingly, it entails a smaller set of properties. If the two conditions are different effectively, i.e. the class of Wightman theories such that

$$F_9(W_i) \subsetneqq F_8(W_i),\ \overline{F_9(W_R)\,\Omega}\ \subsetneqq\ \overline{F_8(W_R)\,\Omega} = \mathcal{H}$$

is nonempty, then in theories of this class the algebras $F_8(W_i)$ satisfy locality and duality while $F_9(W_i)$ satisfy locality and violate duality (the latter follows from Corollary 2 below); moreover, the operators $V(\pm t), V(\pm i\pi), J$ describe the modular structure of $F_8(W_i)$ but not $F_9(W_i)$, in general. If Ω is cyclic for $F_9(W_R)$ as well, then, obviously, $F_8(W_i)$ and $F_9(W_i)$ coincide and also a number of other properties hold.

COROLLARY 2. *In a Wightman theory with cyclic vacuum* Ω *the following conditions are equivalent:*

(1) *Ω is cyclic for $F_9(W_R)$;*

(2) *Ω is cyclic for all $F_k(W_i)$, $\quad k = 1, \ldots, 10$;*

(3) *Ω is separating for $F_{10}(W_i)$;*

(4) *the locality for $F_{10}(W_i)$;*

(5) *the duality for $F_9(W_i)$;*

(6) *$F_4(W_i) = F_8(W_i) = F_9(W_i) = F_{10}(W_i) = \mathcal{P}(W_j)'_w$; $\quad i, j = R, L;\ i \neq j$, and imply*

(7) *the duality for* $F_4(W_i)$.

If, moreover, the fields are essentially self-adjoint on the Gårding domain, then

(8) $F_k(W_i) = F_\ell(W_i)$, $\quad k, \ell = 1\text{–}4, 8\text{–}10$.

All the algebras in the equalities (6) *and* (8) *satisfy locality and duality.*

Proof. Check the chain $1 \to 5 \to 4 \to 3 \to 1$.

$\underline{1 \to 5}$ is the part of the theorem.

$\underline{5 \to 4}$. By the last formula in (2.3), the duality for $F_9(W_i)$ entails $F_9(W_i) = F_{10}(W_i)$ so that $F_{10}(W_i)$ satisfy both locality and duality.

$\underline{4 \to 3}$. Ω being cyclic for $F_{10}(W_L)$ (Lemma 3.2.1), $F_{10}(W_L) \subset F_{10}(W_R)'$ entails 3.

$\underline{3 \to 1}$. If Ω is cyclic for $F_{10}(W_R)'$, (2.40) and the last formula in (2.3) imply 1. Thus, the conditions 1, 3 – 5 are equivalent.

Now, we have seen (5) to imply $F_9(W_i) = F_{10}(W_i)$, which entails (6), taking account of (2.50). (7) follows from (5) and (6). (2) is evident for $k = 1 - 3, 8 - 10$; for $k = 4 - 7$ it follows from (6), (1.9) and (2.3): due to (6), $\mathcal{P}(W_R)'_w$ is a von Neumann algebra, and hence by (1.9) and the Reeh–Schlieder theorem one has

$$\mathcal{H} = \overline{\mathcal{P}(W_R)\,\Omega} \subset \overline{F_4(W_R)\,\Omega} \subset \overline{F_k(W_R)\,\Omega}\,, \qquad k = 5, 6, 7,$$

the last inclusion following from (2.3). The last assertion follows from (6) and (2.3). ❐

Theorem 3.2.7 together with its implications is central for the whole formalism of field algebras. Its main results are twofold: they clarify the conditions of the existence of wedge algebras satisfying locality and duality and they provide a geometrical interpretation of the modular structure of these algebras. Let us discuss the second fact first. We have found that the operators $V(t), V(\pm i\pi), J$ related to relativistic symmetries of a quantum system are at the same time structural characteristics of the wedge algebras describing fully their modular structure. This result by Bisognano and Wichmann is a new nontrivial consequence of the relativistic covariance of the Wightman theory. As already stressed above, it is based on specific features of field theory and could not be obtained in general theory of local observables. But it should be noted that even in field theory the results about the modular structure are not strictly axiomatic: their derivation needs the Reeh–Schlieder property $\overline{F_9(W_R)\,\Omega} = \mathcal{H}$ (or, equivalently, some of the conditions in Corollary 2). Moreover, these results, being so closely connected with the specific properties of wedge regions, cannot be extended to other regions, except those obtained from W_i by means of Poincaré transformations (in the last case, i.e. for the regions $(a, \Lambda)\, W_i$, $(a, \Lambda)\, W'_i$, $(a, \Lambda) \in \mathcal{P}_+^\uparrow$, Theorem 3.2.7 and its corollaries hold automatically). For this reason, it is unclear whether similar – or any, for that matter – geometrical interpretation takes place for the modular structure of field algebras of non-wedge regions. Up to now the results of such kind were obtained only for free massless fields, and here they are not completely independent of the wedge results: due to the conformal symmetry of such fields, the transformations in Minkowski space mapping double cone D and forward light cone V^+ onto W_i turn out to be unitarily implemented in $\mathcal{H}$, and so the algebras $F(D)$ and $F(V^+)$ are now spatially isomorphic to $F(W_i)$. Hence the geometrical interpretation for the modular structure of $F(D)$ and $F(V^+)$ follows; in particular, for $F(V^+)$ the modular group is now connected with the dilatations in M while for $F(D)$ with the generalized linear fractional transformations in the (x^0, x^1)-plane. See [103, 245, 311].

In the problem of locality and duality the extension to other regions is the matter of necessity since having these properties just for one pair of regions, of course, does not mean the locality and duality of the theory. The satisfactory solution of the locality problem should give the proof of locality at least for some base set of regions $O_\alpha \in B(M)$. If we want to obtain such a proof basing on the locality for $F(W_i)$, we should be able to reduce the study of $F(O_\alpha)$ to that of $F(W_i)$. It can really be done.

Consider the set of all wedge-like regions

$$\mathcal{W} \doteq \{W \subset M \mid W = (a\,\Lambda)\,W_i\,,\ i = R,\ L\,,\ (a\,\Lambda) \in \mathcal{P}_+^\uparrow\}$$

and associate with the net $F(O)$ (which can be any of $F_k(O), k = 1, \ldots, 10$) an auxiliary net $\{F^w(D)\}$ defined for double cones and their spacelike complements by the formulae

$$F^w(D) = \bigcap_{\mathcal{W} \ni W \supset D} F(W)\,,\quad F^w(D') = \Big\{ \bigcup_{\mathcal{W} \ni W \subset D'} F(W) \Big\}''\,. \tag{2.51}$$

LEMMA 3.2.8. *Let $\{F(O)\}$ be a net of von Neumann field algebras of some Wightman theory. Then its auxiliary net $\{F^w(\tilde{D})\}_{\tilde{D}=D,D'}$ defined according to* (2.51) *enjoys the following properties:*

(1) *if the algebras $F(W)$, $W \in \mathcal{W}$, satisfy the conditions of isotony, relativistic covariance and CPT covariance, then $\{F^w(\tilde{D}\}$ is a net satisfying these conditions, too;*

(2) *if the algebras $F(W_i)$, $i = R, L$, satisfy locality or duality, $\{F^w(\tilde{D}\}$ also satisfies, respectively, locality or duality and locality holds both for pairs of double cones $D_1 \times D_2$ and for pairs of the form D, D';*

(3) $$F(D) \subset F^w(D)\,,\ F(D') \supset F^w(D')\,, \tag{2.52}$$

and for any triplet (D_1, D_2, W), where $W \in \mathcal{W}$ and D_1, D_2 double cones such that $D_1 \subset W \subset D'_2$ one has

$$F^w(D_1) \subset F(W) \subset F^w(D'_2)\,. \tag{2.53}$$

Proof. (1) and (3) are evident from (2.51). If $\{F(O)\}$ is a local net, for any double cone D and wedge $W \subset D'$ one has due to (2.52) and (2.53):

$$F(W) \subset F(W')' \subset F^w(D)'\,,$$

whence

$$F^w(D') = \left\{ \bigcup_{W \subset D'} F(W) \right\}'' \subset F^w(D)'.$$

Locality for the pairs $D_1 \times D_2$ follows from here due to the isotony of the auxiliary net. If $\{F(O)\}$ satisfies duality,

$$F^w(D')' = \left\{ \bigcup_{W \subset D'} F(W) \right\}' = \bigcap_{W \subset D'} F(W)'$$

$$= \bigcap_{W \subset D'} F(W') = \bigcap_{\tilde{W} \supset D} F(\tilde{W}) = F^w(D). \qquad \square$$

COROLLARY. *In a Wightman theory with the vacuum* Ω *cyclic for each net* $\{F_k(O)\}$, $k = 1, \ldots, 10$, *the nets* $F_k^w(\tilde{D})$ *enjoy isotony, relativistic covariance and CPT-covariance. The nets* $F_n^w(\tilde{D})$, $n = 8, 9$, *enjoy locality. If, in addition, the theory satisfies the condition* $\overline{F_9(W_R)\,\Omega} = \mathcal{H}$ *(or any of the equivalent conditions in Corollary* 2), *then*

$$F_4^w(\tilde{D}) = F_8^w(\tilde{D}) = F_9^w(\tilde{D}) = F_{10}^w(\tilde{D}),$$

and all these nets enjoy duality and the Reeh–Schlieder property (Ω *is bicyclic for* $F_k^w(\tilde{D})$, $\tilde{D} = D, D'$).

Proof. Only the last assertion is not immediately obvious; but it follows from (2.52), cyclicity of Ω for $F_{10}(D)$ (Lemma 3.2.1) and locality of $F_k^w(\tilde{D})$. $\square$

Thus, the auxiliary nets provide a rather satisfactory description of a field system, guaranteeing the fulfillment of basic axioms as well as *CPT*-covariance, duality and Reeh–Schlieder property (the latter is again not a theorem but an independent condition, weak additivity being not proved). However, these nets have no visible connection with Wightman fields and so it is desirable to use them as intermediate objects only, returning ultimately to $F_k(O)$. Such a return proves possible since one can find a set of operators which is left invariant under the mapping $F_k(\tilde{D}) \to F_k^w(\tilde{D})$.

LEMMA 3.2.9. *In an arbitrary Wightman theory one has*

$$\mathcal{P}(D')'_w = \mathcal{P}^w(D')'_w \equiv \bigcap_{\mathcal{W} \ni W \subset D'} \mathcal{P}(W)'_w . \tag{2.54}$$

Proof. The inclusion from left to right being evident, it is sufficient to prove for an arbitrary operator T in the right-hand side of (2.54) that $T \in \mathcal{P}(D')'_w$. Due to (2.4), to this end it is sufficient to check that T commutes weakly with all $A(f)$, $\operatorname{supp} f \subset D'$. Choose a compact $O \subset D'$. For any $x \in O$ one can choose $\rho_x > 0$ such that

$$B(x, 2\rho_x) \subset W \subset D',$$

where $W \in \mathcal{W}$ and $B(x, \rho)$ is the ball of the radius ρ and with the centre in x. The set $\{B(x, \rho_x) \mid x \in O\}$ is a covering of O and, O being compact, we can choose a finite subcovering. This means that there is a finite set of points $x_1, \ldots, x_n \in O$ and wedge regions $W_1, \dots, W_n \in \mathcal{W}$ such that

$$O \subset \bigcup_{k=1}^{n} B(x_k, \rho_{x_k}); \quad B(x_k, 2\rho_{x_k}) \subset W_k \subset D'. \tag{2.55}$$

For the subcoverings (2.55) there exists a family of functions $f_1, \dots, f_n \in \mathcal{D}(\mathbf{R}^4)$ such that

$$\operatorname{supp} f_k \subset B(x_k, 2\rho_{x_k}), \qquad \sum_{k=1}^{n} f_k(x) = 1 \qquad \forall x \in O.$$

Since f_k are multiplicators in $S(\mathbf{R}^4)$, for any $A(f) \in \mathcal{P}(O)$ one has

$$A(f) = \sum_{k=1}^{n} A(ff_k), \qquad \operatorname{supp} ff_k \subset W_k,$$

i.e. $A(ff_k) \in \mathcal{P}(W_k)$, $W_k \subset D'$. But any T in the right-hand side of (2.54) is such that $T \in \mathcal{P}(W_k)'_w$ which implies that T commutes weakly with $A(ff_k)$ and $A(f)$:

$$(A(f)\Phi, T\Psi) = (T^*\Phi, A(f)^+\Psi) \;\; \forall \Phi, \Psi \in \mathcal{D}. \tag{2.56}$$

Since O is an arbitrary compact subregion of D', compact support functions are dense in $S(\mathbf{R}^4)$ with respect to its topology and the matrix elements in (2.56) are tempered distributions, (2.56) holds for all $A(f)$ with $\operatorname{supp} f \subset D'$. Due to (2.4), this proves the lemma, giving $T \in L(D')'_w = \mathcal{P}(D')'_w$. ❒

Note that the proof cannot be repeated for weak commutants $\mathcal{P}(D')'_w$ for bounded regions $O \in B(M)$.

Using (2.54), it is not difficult to find the connection between $F_k(O)$ and $F_k^w(D)$.

COROLLARY. *In a Wightman theory with cyclic vacuum* Ω *satisfying the condition* $F_9(W_R)\Omega = \mathcal{H}$ *(or any equivalent condition in Corollary* 2 *to Theorem* 3.2.7) *one has for* $k = 4, 8$–10:

$$F_k^w(D) = F_8(D) = \mathcal{P}(D')'_w \supset F_{10}(D) \supset F_9(D), \tag{2.57}$$

$$F_k^w(D') \subset F_9(D') \subset F_8(D') \subset \mathcal{P}(D)'_w. \tag{2.58}$$

Proof. Due to (2.54), $F_8(D) = F_8^w(D)$ and, by the Corollary to Lemma 3.2.8, we get the first equality in (2.57). The second equality follows from (2.54) and Corollary 2 to Theorem 3.2.7:

$$\mathcal{P}(D')'_w = \mathcal{P}^w(D')'_w = F_8^w(D).$$

Since $F_{10}(D) \subset F_{10}^w(D)$, due to (2.52), the first inclusion in (2.57) holds. To get the second inclusion, note that (2.52) together with locality for $F_9^w(\tilde{D})$ imply locality for $F_9(D)$; then use (2.3). In (2.58) the first inclusion follows from (2.52) and the second from (2.3). ❒

(We see that the coincidence of some auxiliary nets, $F_k^w(\tilde{D}) = F_\ell^w(\tilde{D})$, may not imply the coincidence of the original nets $F_k(\tilde{D})$ and $F_\ell^w(\tilde{D})$.)

Now we can formulate the final results about the properties of the algebras $F_k(O)$ without using $F_k^w(\tilde{D})$.

PROPOSITION 3.2.10. *In any Wightman theory with the cyclic vacuum* Ω, *the von Neumann field algebras* $F_k(O), k = 1, \dots, 10$, *satisfy the conditions of isotony, relativistic covariance, CPT-covariance and spectrum conditions. The algebras* $F_8(D) = {}_{\mathrm{alg}}\mathcal{P}(D')'_w$ *and* $F_9(D) = L(D')'_N$ *satisfy locality and* $F_{10}(D) = F_1(D) = (L(D)'_N)'$ *possess the Reeh–Schlieder property* (Ω *is bicyclic for all* $F_{10}(\tilde{D}), \tilde{D} = D, D'$). *If, moreover, the theory satisfies the condition* $\overline{F_9(W_R)\,\Omega} = \mathcal{H}$ *(or another equivalent condition in Corollary* 2 *to Theorem* 3.2.7), *the following properties hold:*

- *the nets* $F_k(\tilde{D}), k = 1, 4, 8\text{–}10$, *satisfy locality;*
- *the net* $F_8(\tilde{D})$ *satisfies duality, and* $F_8(D) = \mathcal{P}(D')'_w$ *;*
- *the net* $F_8(\tilde{D})$ *possesses the Reeh–Schlieder property.*

Proof. All statements concerning the general Wightman theory were already proved. Duality for $F_8(\tilde{D})$ follows from (2.57) and duality for $F_8^w(\tilde{D})$ (Corollary to Lemma 3.2.8). Locality for $F_{10}(\tilde{D})$ follows from $F_{10}(D) \subset F_8(D)$, locality for $F_4(\tilde{D})$ from $F_4(D) \subset F_{10}(D)$. The Reeh–Schlieder property for $F_8(\tilde{D})$ follows from (2.57). ❒

Let us make a few conclusive remarks. Since $F_9(O)$ is defined as the algebra commuting with fields in O', locality for $F_9(O)$, like essential duality, can be interpreted as expressing transitivity of locality: a net which is local relative to the local system of fields $\{L(O)\}$ turns out to be local, too. Let us also recall that both the duality and Reeh–Schlieder property for the net $\{F_8(D)\}$ are valid under weaker additional condition $\overline{F_8(W_R)\,\Omega} = \mathcal{H}$. Furthermore, it should be noted that in a Wightman theory with essentially self-adjoint field operators our results ensure only locality but not duality of the 'natural' net $\{F_3(D)\}$ and leave open the possibility that $F_3(D) \subsetneq F_8(D)$, the net $\{F_8(D)\}$ satisfying both locality and duality; the relation between the two nets was considered 'mysterious' in [130]. Finally, the development of the Bisognano–Wichmann formalism has lead to its detailization in which nets of von Neumann algebras associated with Wightman fields belong to various classes of so-called AB-systems, i.e. nets including only the algebras of the regions W, D and D' [428, 173]. Based on the notions of AB-systems and intrinsically local operator (see Section 3.1.2), in [173] a new analysis of the relationship between Wightman fields and von Neumann field algebras has been performed producing more detailed results, in particular, about different forms of the duality condition.

Besides its main goal, to clarify the relation between quantum fields and nets of von Neumann algebras, over the most recent periods the Bisognano–Wichmann formalism has found a number of other promising applications of which we briefly mention the following three.

(1) It was used for constructing local nets of bounded operator algebras (not von Neumann algebras anymore) in quantum gauge field theory in the indefinite metric (Krein) state space ([127], see Appendix).

(2) The properties of wedge algebras, and in particular, duality of the latter, play an important part in studies of Bell's inequalities and their violation in quantum field theory [404, 405, cf. 297].

(3) The formalism provides the basis for rigorous treatment of some topical problems in cosmology and gravitation. Above all, it proves suitable for constructing the theory of the Hawking–Unruh thermalization effect, according to which in certain accelerated frames – in particular, those connected with black holes – states of quantum fields systems are 'thermalized', i.e. became states of nonzero temperature. The ability of the formalism to describe such effects is almost evident. Indeed, in the corresponding accelerated frame the group $V(t)$ of Lorentz boosts becomes the group of proper time translations; and due to the KMS condition, the coincidence of this group with the modular group of the pair $(F(W_R), \Omega)$ means that in the said frame the state $\omega_\Omega^{(W_R)}$ is an equilibrium state with nonzero temperature. The number of works devoted to this and related subjects is rapidly increasing [382, 383, 61, 278-80, 193, 236, 196].

3.3. Free and Generalized Free Fields

As we have already said, space does not permit the inclusion in this book of the modern treatment of concrete models by means of the method of local morphisms. However, the algebraic formalism for the simplest systems, free and generalized free neutral fields, having no charged sectors, is not connected with this method. It is based on the usual formalism of field algebras (the latter now coinciding with the observable algebras) and we shall now consider it in order to give an explicit example of local quantum theory, thereby completing our introduction to this theory.

A wide range of different methods is known for the algebraic treatment of free and generalized free fields: relativistically covariant description [40], fixed-time description [40, 42], description in the formalism of canonical systems [273, 395] and on the basis of infinite tensor products [135]. Such a diversity is well justified: all these ways proved to be useful in studying free systems and often also in extending the technique and results of these studies to less trivial models, like those of constructive field theory. As for our aims, we want to define, for the free field, a system of local algebras $\{R(O)\}_{O \in B(M)}$ to verify that this system is a concrete example of the Haag–Araki theory and check it for the main structural properties studied in Section 1.3. The traditional Araki formalism [39, 40, 42] remains the most convenient for these aims.

So let us consider the well known model of a free scalar neutral field of a non-zero mass $m > 0$. The model is assumed to be given in its Wightman formulation [4, Chapter 3, Section 4] so that there is an operator-valued distribution $A(f)$, $f \in S(\mathbf{R}^4)$, satisfying all Wightman axioms and, in addition, the Klein–Gordon equation:

$$A\,(Kf) = 0,\;\; K = \Box + m^2 .$$

The state space $\mathcal{H}$ of the field $A\,(f)$ is endowed with the structure of Fock space, i.e. the direct sum of symmetrized tensor powers (see [64] or [4, Chapter 2, Section 5]) of the one particle space $\mathcal{H}_1$:

$$\mathcal{H} = \mathcal{H}_F\,(\mathcal{H}_1) = \bigoplus_{n=0}^{\infty} \text{sym}\ \mathcal{H}_1^{\otimes n} .$$

$\mathcal{H}_1$ is the Hilbert space obtained from the test function space by its completion with respect to some L^2-norm. There are a great many different Hilbert norms on Schwartz spaces, and this freedom of choice is widely exploited in the algebraic description of the free field. The standard one-particle space for the Wightman theory of the field $A\,(f)$ is the complex Hilbert space $L\,(\mathbf{R}^4) = \mathcal{H}_{m,0}$ of the irreducible representation $[m, 0]$ of the Poincaré group $\mathcal{P}_+^{\uparrow}$. It is obtained as follows: introduce on the real space $S_R\,(\mathbf{R}^4)$ the complex inner product

$$(f, g)_L = 2\,i \int d^4 x\ d^4 y\, f\,(x)\, \Delta^+\,(x-y)\, g\,(y) = \int \frac{d^3 p}{\tilde{\omega}}\ \overline{\tilde{f}\,(\omega, \mathbf{p})}\ \tilde{g}\,(\tilde{\omega}, \mathbf{p}),$$

where $\tilde{\omega} = (p^2 + m^2)^{1/2}$ and $\Delta^+\,(x)$ is the Pauli–Jordan positive frequency function:

$$\Delta^+\,(x) = -\,i\,(2\,\pi)^{-3} \int d^4 p\ e^{-ipx}\, \theta\,(p_0)\, \delta\,(p^2 - m^2)\,,$$

take the quotient space by the subspace

$$S_0\,(\mathbf{R}^4) = \{f \in S_R\,(\mathbf{R}^4) \mid \; \|f\|_L = 0\}$$

and complete with respect to $(\cdot, \cdot)_L$. Thus

$$L\,(\mathbf{R}^4) = \overline{S_R\,(\mathbf{R}^4) \,/\, S_0\,(\mathbf{R}^4)}^{\,\|\cdot\|_L} .$$

Multiplication by i for elements of $S_R\,(\mathbf{R}^4) \,/\, S_0\,(\mathbf{R}^4)$ (equivalence classes of real functions) is defined by means of the Fourier transformation to that $f' = i f$ means $\tilde{f}'\,(p) = i\,\tilde{f}\,(p)$ for $p \in \overset{+}{V}_{(m)}$.

The relativistically covariant Araki formalism uses as the basic one-particle space for the field $A\,(f)$ the real Hilbert space $\hat{L}\,(\mathbf{R}^4)$ obtained by means of endowing $L\,(\mathbf{R}^4)$ with the real inner product

$$(f, g)_{\hat{L}} \doteq \text{Re}\,(f, g)_L = -\int d^4 x\ d^4 y\, f\,(x)\, \Delta_1\,(x-y)\, g\,(y)\,,$$

where

$$\Delta_1\,(x) = (2\,\pi)^{-3} \int d^4 p\ e^{-ipx}\ \delta\,(p^2 - m^2)$$

is the even Pauli–Jordan function. Since i is not an element of the field of scalars for $\hat{L}$, the operation $f \to i f$ is now interpreted as the action of a 'complex structure operator' β such that $\beta^* = -\beta,\ \beta^2 = -I$. One has then

$$(f, \beta\, g)_{\hat{L}} = -\operatorname{Im}\ (f, g)_L = -\int d^4x\ d^4y\, f(x)\, \Delta(x-y)\, g(y)\,,$$

where

$$\Delta(x) = -i\,(2\pi)^{-3} \int d^4p\ e^{-ipx}\ \varepsilon(p_0)\, \delta(p^2 - m^2)$$

is the Pauli–Jordan function. The state space is assumed to be the Fock space $\mathcal{H}_F(\hat{L})$ over the one-particle space $\mathcal{H}_1 = \hat{L}$. With the operators in $\mathcal{H}_1$ one associates canonically (see [3] or [10, Section 1.3]) their 'second quantizations', i.e. operators in $\mathcal{H}_F$. In such a way one defines in $\mathcal{H}_F$ unitary operators $U(a, \Lambda)$, implementing a strongly continuous representation of the Poincaré group. The fields $A(f)$, $f \in \hat{L}$, are acting in $\mathcal{H}_F(\hat{L})$ as essentially self-adjoint operators with the common invariant dense domain

$$\mathcal{D} = L\,\{A(f_1) \ldots A(f_n)\,\Omega,\ n = 0, 1, 2, \ldots\}$$

where Ω is the unique vacuum vector in $\mathcal{H}_F$ (the Fock vacuum). In this realization $A(f)$ has the commutation relations

$$[A(f), A(g)] = -i\,(f, \beta\, g)_{\hat{L}}$$

on $\mathcal{D}$. Due to essential self-adjointness of $A(f)$, for any lineal $H_1 \subset \hat{L}$ we can define the von Neumann algebra

$$R(H_1) = \left\{ e^{i\overline{A(f)}} \,|\, f \in H_1 \right\}'' = \{E_\Delta(f) \mid \Delta \subset \mathbf{R},\ f \in H_1\}'' \tag{3.1}$$

$E_\Delta(f)$ being spectral projections of $A(f)$. By the Araki–Woods lemma [52, Lemma 2.3], the operators $W(f) = \exp(i\,\overline{A(f)})$ are strongly continuous (with respect to the strong topology on $\hat{L}$) and therefore the von Neumann algebras corresponding to a lineal and its closure coincide. Hence it suffices to consider the algebras $R(H_1)$ for subspaces of $\hat{L}$ only. For any $O \in B(M)$ we associate with the set $S(O)$ of all functions from $S_R(\mathbf{R}^4)$ with the support in O a certain lineal in $\hat{L}$ and its closure, the subspace $\hat{S}(O) \subset \hat{L}$.

DEFINITION 3.3.1. Von Neumann algebras $R(O) = R(\hat{S}(O))$ will be called local algebras (of fields or, which is the same, observables) of the free massive field model.

PROPOSITION 3.3.1. *The net of local algebras of the free field* $A(f)$ *satisfies all the Haag–Araki axioms, additivity and spectrum conditions being valid in their strong forms* IIa *and* VIa, *respectively. In the state space* $\mathcal{H} = \mathcal{H}_F(\hat{L})$ *there is a unique Poincaré-invariant vector* Ω *and the global algebra is irreducible.*

Proof. To prove the strong additivity it is enough to check that $W(f) \in \overline{\vee_i R(O_i)}^{\,w}$ for any function $f \in L\,\{\overline{\cup_i \hat{S}(O_i)}\}$. It is clearly so for $f \in \cup_i \hat{S}(O_i)$. For $f = f_1 + f_2$ with $f_1, f_2 \in \cup_i \hat{S}(O_i)$ the Weyl–Segal commutation relations:

$$W(h_1)\, W(h_2) = W(h_1 + h_2) \exp\left\{ -\frac{i}{2}(h_1, \beta\, h_2)_{\hat{L}} \right\} \tag{3.2}$$

for all $h_1, h_2 \in \hat{L}$, imply

$$W(f_1), W(f_2) \in \overline{\bigvee_i R(O_i)}^w \Rightarrow W(f_1 + f_2) \in \overline{\bigvee_i R(O_i)}^w .$$

Hence, it follows

$$W(f) \in \overline{\bigvee_i R(O_i)}^w \quad \forall f \in L\left\{\bigcup_i \hat{S}(O_i)\right\} .$$

Finally, if $f = \lim_{n\to\infty} f_n$, $f_n \in L\{\cup_i \hat{S}(O_i)\}$, strong continuity of $W(h)$ implies again $W(f) \in \overline{\vee_i R(O_i)}^w$ and IIa is proved.

Locality, the proof of which was the main problem in the general theory of field algebras (Section 3.2), is now evident from (3.2) since the index of the exponent in (3.2) vanishes for $\operatorname{supp} h_1 \times \operatorname{supp} h_2$. It follows also from the fact that the Fock vacuum Ω is an analytic vector for $A(f)$: it is easy to check [94, Section 2] that the condition of the analyticity of Ω for $A(f)$ in terms of truncated Wightman functions W_n^{tr} has the form

$$\left| \int \prod_{i=1}^{n} (d^4 x_i f(x_i))\, W_n^{\mathrm{tr}}(x_1, \dots, x_n) \right| \le n!\, c(f)^n ,$$

while for the free field $W_n^{\mathrm{tr}}(x_1, \dots, x_n) \equiv 0$ for all $n > 2$.

The fulfillment of IV follows from the fact that $A(f)$ satisfies the Klein–Gordon equation and hence, is determined by its initial data. Relativistic covariance is fulfilled due to one of the basic properties of the Fock formalism [3, 10, Chapter 1, Theorem 1]: if u is a unitary operator in $\hat{L}$ and $U(u)$ its second quantization,

$$U(u) = \bigoplus_{n=0}^{\infty} u^{\otimes n}, \quad u^{\otimes 0} = I ,$$

then for all $f \in \hat{L}$,

$$W(uf) = U(u)\, W(f)\, U(u)^{-1} .$$

Operators implementing the irreducible representation of $\mathcal{P}_+^{\uparrow}$ in $\hat{L}(\mathbf{R}^4)$ are induced by the representation of $\mathcal{P}_+^{\uparrow}$ in $S_R(\mathbf{R}^4)$:

$$f(x) \to f(\Lambda^{-1}(x - a)) \quad \forall (a, \Lambda) \in \mathcal{P}_+^{\uparrow} .$$

In particular, taking translation operators

$$T(a) = \bigoplus_{n=0}^{\infty} u(a, I)^{\otimes n} = \int e^{ipa}\, dE(p)$$

and noting that for all equivalence classes $[f] \in S_R(\mathbf{R}^4)/S_0(\mathbf{R}^4)$ one has $u(a, I)[\tilde{f}(p)] = [e^{ipa}\tilde{f}(p)]$, we obtain VIa. Finally, the uniqueness of vacuum and irreducibility of the set of field operators are the standard properties of free massive fields verified by means of expansions over creation-annihilation operators. ❐

COROLLARY 1. *The net* $R(O)$ *enjoys the Reeh–Schlieder property: each vector analytic in the energy is cyclic for all* $R(O)$, $O \subset M$ *and separating for all* $R(O)$ *with* $O' \neq \emptyset$.

COROLLARY 2. *The quasilocal algebra* $\mathfrak{A}$ *of the net* $R(O)$ *is simple.*

Proof. See Corollary 2 to Theorem 2.2.1. ❐

Further study of the model is not based on the net $\{R(O)\}$ anymore since no efficient technique was found for its straightforward analysis. All the existing methods for investigating algebraic structure of the free field use a different net whose algebras are assigned to regions of the Euclidean space $\mathbf{R}^3$. This space is treated as the hyperplane $S_0 \subset M$ and, accordingly, the description in terms of $\{R(O)\}_{O \in B(M)}$ is called the covariant formulation and that in terms of $\{R(B)\}_{B \subset \mathbf{R}^3}$ is called the equal-time or fixed-time formulation of the free field model. Of course, an arbitrary Wightman field (and even generalized free field, see below) lacks, in general. the equal-time formulation; its existence is entirely due to the fact that $A(f)$ satisfies the Klein–Gordon equation and so has well defined 'canonical fields' φ, π i.e., the field itself and its time derivative taken at the initial time moment. Canonical fields generate the algebras $R(B)$ exactly in the same way as $A(f)$ generates $R(O)$, and the known connection between $A(f)$ and φ, π makes it possible to establish the connection between the nets $\{R(B)\}$ and $\{R(O)\}$. Then, the information about the former can be converted into that about the latter. And to study $\{R(B)\}$ the powerful technique of canonical systems theory is at our disposal.

(It is worthwhile recalling here that the existence of canonical variables is characteristic of not just free systems but a considerably larger class of canonical or Hamiltonian systems. Thus, in all cases when a relativistic quantum system can be brought correctly to the Hamiltonian form – i.e. in models of constructive field theory – the algebraic formalism arises which is very close to that presented below and has nearly the same complex of results, including the connection between the 'covariant' and 'canonical' nets and duality for local algebras but not including thus far the theorems about the type of latter. See [215, Sections 8–10, 167, 403].)

By definition, the canonical field and canonical momentum of the covariant field $A(f)$ are given by the formulae

$$\varphi(h_f) = A(f), \quad \pi(h_f) = A(\dot{f}), \tag{3.3}$$

where $\dot{f} = \partial f / \partial x^0$, $\tilde{h}_f(p) = \tilde{f}(\tilde{\omega}, p)$. If $f \in S_R(\mathbf{R}^4)$, then $h_f \in S_R(\mathbf{R}^3)$; in other words, the formulae (3.3) define two different mappings of $S_R(\mathbf{R}^4)$ into $S_R(\mathbf{R}^3)$, having the following form in the momentum representation

$$(\widetilde{\delta_0 f})(p) = \tilde{f}(\tilde{\omega}, p), \qquad (\widetilde{\delta_0 f})(p) = (i\tilde{\omega})^{-1} \tilde{f}(\tilde{\omega}, p). \tag{3.4}$$

To make the definition of operator-valued distributions $\varphi(h), \pi(g)$ complete, these mappings should be extended from $S_R(\mathbf{R}^4)$ to all of $\hat{L}(\mathbf{R}^4)$. To this end we split $S_R(\mathbf{R}^4)$ into subspaces of even and odd functions in the variable x^0:

$$S_R(\mathbf{R}^4) = S^+(\mathbf{R}^4) \dotplus S^-(\mathbf{R}^4), \tag{3.5}$$

where

$$S^{\pm}(\mathbf{R}^4) = \{f \in S_R(\mathbf{R}^4) \mid f(x) = g_{\pm}(x) \equiv \frac{1}{2}\left(g(x^0, x) \pm g(-x^0, x)\right),\ g \in S_R(\mathbf{R}^4)\}$$

and the notation $\dot{+}$ means that every $f \in S_R(\mathbf{R}^4)$ can be uniquely represented in the form $f = f_+ + f_-$, $f_{\pm} \in S^{\pm}(\mathbf{R}^4)$. Since the distribution Δ_1 is even and one has

$$\tilde{f}_{\pm}(p_0, p) = \frac{1}{2}\left(\tilde{f}(p_0, p) \pm \tilde{f}(p_0, -p)\right)$$

in the momentum representation, (3.5) induces the orthogonal decomposition

$$\hat{L}(\mathbf{R}^4) = \hat{L}^+(\mathbf{R}^4) \oplus \hat{L}^-(\mathbf{R}^4),$$

where $\hat{L}^- = \beta \hat{L}^+$. Functions f being real, one has

$$\tilde{f}_{\pm}(p_0, p) = \pm \tilde{f}_{\pm}(-p_0, p), \tag{3.6}$$

which implies that the mappings (3.4) will be one-to-one modulo $S_0(\mathbf{R}^4)$, if we define them on any one of the components S^+ or S^-. It is convenient to define δ_0 on S^+ and δ_1 on S^-, and then to extend them to $\hat{L}^+$ and $\hat{L}^-$, respectively. Let us endow $S_R(\mathbf{R}^3)$ with the inner products

$$(h, g)_{\varphi} = (\omega^{-1/2} h, \omega^{-1/2} g) = \int \frac{d^3 p}{\tilde{\omega}}\ \overline{\tilde{h}(p)}\, \tilde{g}(p)$$

$$(h, g)_{\pi} = (\omega^{1/2} h, \omega^{1/2} g) = \int d^3 p\, \tilde{\omega}\ \overline{\tilde{h}(p)}\, \tilde{g}(p).$$

Here $\omega^t = (-\Delta + m^2)^{t/2}$ is the positive operator in $\hat{L}$ defined as the Fourier transform of the multiplication operator $\tilde{\omega}^t = (p^2 + m^2)^{t/2}$; clearly, its domain includes $S_R(\mathbf{R}^3)$. By (3.6) one has for $f_1, f_2 \in \hat{L}^+$ or $f_1, f_2 \in \hat{L}^-$:

$$(f_1, f_2)_{\hat{L}} = (h_{f_1}, h_{f_2})_{\varphi}.$$

Hence, it follows that δ_0 and δ_1 can be extended to (denoted by the same symbols) isometrical mappings

$$\delta_0 : \hat{L}^+(\mathbf{R}^4) \to L_{\varphi}(\mathbf{R}^3) \equiv \overline{S_R(\mathbf{R}_3)}^{\|\cdot\|_{\varphi}},$$

$$\delta_1 : \hat{L}^-(\mathbf{R}^4) \to L_{\pi}(\mathbf{R}^3) \equiv \overline{S_R(\mathbf{R}^3)}^{\|\cdot\|_{\pi}}.$$

Here the spaces L_{φ} and L_{π} belong to the well studied class of the Sobolev spaces $\mathcal{L}^p_{\alpha}$ of fractional index α and coincide with $\mathcal{L}^2_{-1/2}(\mathbf{R}^3)$ and $\mathcal{L}^2_{1/2}(\mathbf{R}^3)$, respectively, (see [26]). The operator ω considered in the space $L_{\varphi} \oplus L_{\pi}$ maps isometrically L_{φ} onto L_{π}. The extended mappings δ_0, δ_1 let us define the canonical fields φ, π as self-adjoint operators in the same state space $\mathcal{H}_F(\hat{L})$:

$$\forall_{h \in L_{\varphi}(\mathbf{R}^3)}\quad \varphi(h) = \overline{A(\delta_0^{-1} h)}, \qquad \forall_{g \in L_{\pi}(\mathbf{R}^3)}\quad \pi(g) = \overline{A(\delta_1^{-1} h)}.$$

The inverse relationship, as implied by these definitions, has the form

$$\forall_{f \in \hat{L}(\mathbf{R}^4)} \quad \overline{A(f)} = \varphi(\delta_0 f_+) + \pi(\delta_1 f_-), \tag{3.7}$$

implying, in its turn, that the set of φ, π is irreducible. Taking into account that h, g are real functions, δ_0, δ_1 are isometric and one has $\delta_0 \beta \delta_1^{-1} = -\omega$, $\delta_1 \beta \delta_0^{-1} = \omega^{-1}$, we obtain the commutation relations

$$[\varphi(h_1), \varphi(h_2)] = [\pi(g_1), \pi(g_2)] = 0, \tag{3.8}$$

$$[\varphi(h), \pi(g)] = (h, g).$$

Self-adjoint operators $\varphi(h), \pi(g)$ produce the Weyl forms $U(h) = \exp[i\varphi(h)]$, $V(g) = \exp[i\pi(g)]$ strongly continuous with respect to topologies of L_φ and L_π, respectively. Thus, for each pair of subspaces $K_\varphi \subset L_\varphi$ and $K_\pi \subset L_\pi$ we can define, similarly to Definition 3.3.1, the Weyl algebra over the one-particle space L_φ:

$$R(K_\varphi, \omega K_\pi) = \{U(h), V(g) \mid h \in K_\varphi, \omega g \in \omega K_\pi\}''.$$

Now, to define an algebra for a region $B \subset \mathbf{R}^3$, it is enough to associate with this region a pair of 'local subspaces' of the K_φ, K_π form. In the case of nonrelativistic second quantization formalism (obtained when $\omega \equiv 1$) over one-particle space $L^2(\mathbf{R}^3, d^3x)$, such pair reduces to one subspace $K_\varphi = K_\pi = L^2(B)$ generating a very simple structure:

$$L^2(\mathbf{R}^3) = L^2(B) \oplus L^2(B^\perp), L^2(B^\perp) = L^2(B)^\perp, \; B^\perp \equiv \mathbf{R}^3 \setminus B. \tag{3.9}$$

If $\omega \neq 1$, K_φ, K_π are subspaces of the fractional index Sobolev spaces L_φ, L_π. Local subspaces in such spaces are constructed as follows. Let B be a region in $\mathbf{R}^3$ with piecewise smooth boundary, and let canonical embeddings be given

$$j_1 : L_\pi(\mathbf{R}^3) \to L^2(\mathbf{R}^3), \qquad j_2 : L^2(\mathbf{R}^3) \to L_\varphi(\mathbf{R}^3).$$

Then we define

$$L_\varphi(B) \doteq \overline{j_2 L^2(B)}^{\|\cdot\|_\varphi}, \qquad L_\pi(B) \doteq j_1^{-1}(j_1 L_\pi(\mathbf{R}^3) \cap L^2(B)).$$

It is easy to see these definitions to be equivalent to

$$L_\pi(B) = \{g \in L_\pi(\mathbf{R}^3) \mid \operatorname{supp} g \subset \operatorname{int} B\},$$

$$L_\varphi(B) = \{h \in L_\varphi(\mathbf{R}^3) \mid (\omega^{-1/2} h, \omega^{1/2} g) = 0 \quad \forall g \in L_\pi(B^\perp)\}.$$

The following important relations are also easy to verify:

$$\omega L_\pi(B)^\perp = L_\varphi(B^\perp), \qquad \omega L_\pi(B^\perp) = L_\varphi(B)^\perp, \tag{3.10}$$

though, ω being a nonlocal operator, the structure (3.9) is not valid for L_φ and L_π, and $\omega L_\pi(\mathbf{R}^3) = L_\varphi(\mathbf{R}^3)$ does not imply $\omega L_\pi(B) = L_\varphi(B)$. In fact, as Segal and

Goodman [381] have found, the operator ω is even 'antilocal', in the sense that for any $g \in L_\pi(B)$ the function ωg vanishes in no full neighbourhood of a point $x \in B^\perp$. Hence, it follows that $\omega L_\pi(B) \cap L_\varphi(B) = \{0\}$. This relation, together with (3.10), turns out to be one of the key properties determining the structure of local algebras.

DEFINITION 3.3.2. For any Lebesque measurable region $B \subset \mathbf{R}^3$ let us define the local algebras of $A(f)$ as the Weyl algebra

$$R_{\varphi,\pi}(B) \doteq R(L_\varphi(B), \omega L_\pi(B)) .$$

The system of algebras $R_{\varphi,\pi}(B)$ enjoys the properties of isotony, strong additivity and locality in the noncovariant form:

$$R_{\varphi,\pi}(B_1) \subset R_{\varphi,\pi}(B_2)' \quad \text{for} \quad B_1 \cap B_2 = \emptyset \tag{3.11}$$

and the corresponding global algebra is irreducible:

$$R_{\varphi,\pi} \equiv \left\{ \bigcup_{B \subset \mathbf{R}^3} R_{\varphi,\pi}(B) \right\}'' = R_{\varphi,\pi}(\mathbf{R}^3) = B(\mathcal{H}) .$$

These properties follow easily from (3.7), (3.8) and (3.10); the strong additivity is verified in the same way as for the net $R(O)$.

PROPOSITION 3.3.2. *Let $B \subset \mathbf{R}^3$ be a measurable region with piecewise smooth boundary such that* $\operatorname{int}\bar{B} = B$ *and $C(B)$ its hull convex with respect to the light cone* in M:*

$$C(B) = \{x \in M \mid (x-y)^2 < 0 \ \ \forall\, y \in (M \setminus B) \cap S_0\} .$$

Then

$$R_{\varphi,\pi}(B) = R(C(B)) .$$

Proof.

(1) For an arbitrary $O \subset M$ we define an extension $\hat{R}(O)$ of $R(O)$ as the projective limit of $R(\tilde{O})$ with $\tilde{O} \supset O$:

$$\hat{R}(O) = \bigcap_{\tilde{O} \supset O} R(\tilde{O})$$

and prove that

$$R_{\varphi,\pi}(B) \subset \hat{R}(B) \subset \hat{R}(C(B)) . \tag{3.12}$$

The second inclusion follows from isotony. To derive the first we check that $R_{\varphi,\pi}(B)$ is generated by the subset of operators $U(h), V(g)$ with $h, \omega^{-1} g \in \mathcal{D}_R(B)$. By the strong continuity of $U(h)$, $V(g)$, the assertion holds if $j_2\, \mathcal{D}_R(B)$ is dense in $L_\varphi(B)$ and $j_1^{-1}\, \mathcal{D}_R(B)$ is dense in $L_\pi(B)$. The first

* Cf. [30, Section 28], where different expressions for $C(B)$ are also given. If B is bounded, $C(B)$ is a symmetrical double cone, the base of which is the circumscribed ball of B. If B is unbounded, $C(B)$ is an unbounded region in M the boundary of which is formed by hyperplanes tangent to light cones. In particular, it is easy to check that for $B = \{x \in \mathbf{R}^3 \mid x^1 > 0\}$ one has $C(B) = W_R$.

property is evident. To derive the second we approximate arbitrary $g \in L_\pi(B)$ by compact support functions $g_n \in L_\pi(B)$. Take

$$p(t) = \begin{cases} 1, t \geq 1 \\ 0, t \leq 0 \end{cases} \quad p' \in \mathcal{D}_R(\mathbf{R}),$$

and put for any $\lambda \in \mathbf{R}$, $\boldsymbol{a} \in \mathbf{R}^3$

$$f_{\lambda,\boldsymbol{a}}(\boldsymbol{x}) \equiv p((\boldsymbol{x}\,\boldsymbol{a}) - \lambda).$$

Verifying that the norm $\| f_{\lambda,\boldsymbol{a}}\, g \|_\pi$ is a bounded function of λ for any $g \in L_\pi(B)$ and for any $u(x) \in S_R(\mathbf{R}^3)$ one has $\lim_{\lambda\to\infty}(u, f_{\lambda,\boldsymbol{a}}\, g) = 0$, we obtain, taking into account that $S_R(\mathbf{R}^3)$ is dense in $L_\pi(\mathbf{R}^3)$:

$$L_\pi\text{-}\lim_{\lambda\to\infty} (1 - f_{\lambda,\boldsymbol{a}}(\boldsymbol{x}))\, g(\boldsymbol{x}) = g(\boldsymbol{x}).$$

The support properties of $f_{\lambda,\boldsymbol{a}}(\boldsymbol{x})$ are such that, although $\operatorname{supp}(1 - f_{\lambda,\boldsymbol{a}})$ is not compact, there exists a finite set $(\boldsymbol{a}_1, \ldots, \boldsymbol{a}_n)$ such that

$$\operatorname{supp} \prod_{j=1}^{n} (1 - f_{\lambda_j}, \boldsymbol{a}_j) \subset \bigcap_{j=1}^{n} \operatorname{supp}(1 - f_{\lambda_j}, \boldsymbol{a}_j)$$

is compact. Putting

$$g_{\lambda_1 \ldots \lambda_n}(\boldsymbol{x}) = \prod_{j=1}^{n} (1 - f_{\lambda_j}, \boldsymbol{a}_j(\boldsymbol{x}))\, g(\boldsymbol{x}),$$

we have the desired approximation,

$$g = \underset{\lambda_1, \ldots, \lambda_n \to \infty}{L_\pi\text{-}\lim} g_{\lambda_1 \ldots \lambda_n}$$

where $g_{\lambda_1 \ldots \lambda_n}$ are compact support functions from $L_\pi(B)$.

Now, let us show that for any pair of functions $h, g \in \mathcal{D}_R(B)$ there exist sequences $\{h_n\}, \{g_n\} \subset \hat{L}(\mathbf{R}^4)$ such that

$$U(h) = \underset{n\to\infty}{s\text{-}\lim}\, W(h_n), \qquad V(g) = \underset{n\to\infty}{s\text{-}\lim}\, W(g_n). \tag{3.13}$$

If $h_n \in \hat{L}^+$, $g_n \in \hat{L}^-$, one has by (3.7)

$$W(h_n) = U(\delta_0 h_n), \qquad W(g_n) = V(\delta_1 g_n).$$

Hence (3.12) holds if

$$\delta_0 h_n \to h, \ \delta_1 g_n \to g \tag{3.14}$$

in some topology, with respect to which $U(h)$ and $V(g)$ are continuous. Check that the sequences of the form $h_n = h_n^0(x^0)\, h(\boldsymbol{x})$, $g_n = g_n^0(x^0)\, g(\boldsymbol{x})$ will do, where the functions h_n^0, g_n^0 are such that $|\tilde{h}_n^0(p_0)| \leq 1$, $|\tilde{g}_n^0(p_0)| \leq |p_0|$ and at $n \to \infty$ one has $\tilde{h}_n^0(p_0) \to 1$, $\tilde{g}_n^0(p_0) \to 1$ in the topology of uniform convergence on compacts. Indeed, it is possible to choose $h_n \in \hat{L}^+, g_n \in \hat{L}^-$, since in the

x-representation we must have $h_n^0(x^0) \to \delta(x^0)$, $g_n^0(x^0) \to \delta'(x^0)$ and the distributions δ and δ' enjoy necessary symmetry properties; due to (3.4), such h_n, g_n will satisfy (3.14).

It follows from the properties of h_n, g_n that for any $\varepsilon > 0$ there is a number $N = N(\varepsilon)$ such $W(h_n), W(g_n) \in R(B_\varepsilon)$ for all $n \geq N$, where $B_\varepsilon = \{x \in M \mid |x^0| < \varepsilon, \boldsymbol{x} \in B\}$. Therefore, it follows from (3.13) that $U(h)$, $V(g) \in \hat{R}(B)$ for all h, $g \in \mathcal{D}_R(B)$. But the set of such operators generates $R_{\varphi,\pi}(B)$ and (3.12) is proved.

(2) We prove the enhancement of (3.12):

$$R_{\varphi,\pi}(B) \subset R(C(B)). \tag{3.15}$$

To this end one verifies that both the covariant and equal-time nets are continuous from the inside, i.e $R(O)$ and $R_{\varphi,\pi}(B)$ are the W^*-inductive limits of algebras of their proper subregions

$$R(O) = \Big\{ \bigcup_{\bar{O}_\alpha \subset O} R(O_\alpha) \Big\}'' , \quad R_{\varphi,\pi}(B) = \Big\{ \bigcup_{\bar{B}_\alpha \subset B} R_{\varphi,\pi}(B_\alpha) \Big\}'' , \tag{3.16 a,b}$$

Here only the inclusions from left to right are not trivial. For $R(O)$ this inclusion follows from the fact that $\overline{\operatorname{supp} f} \subset O$ for any $f \in S_R(O)$ and hence, there is an open set O_α such that $\overline{O_\alpha} \subset O$, $\operatorname{supp} f \subset O_\alpha$, i.e. $W(f) \in R(O_\alpha)$. For $R_{\varphi,\pi}(B)$ note that it is generated by $U(h)$, $V(g)$ with h, $g \in \mathcal{D}_R(B)$ and, the boundary of B being piecewise smooth, one has

$$\mathcal{D}_R(B) = \overline{\bigcup_{\bar{B}_\alpha \subset B} \mathcal{D}_R(B_\alpha)} .$$

Hence it follows that $R_{\varphi,\pi}(B)$ is generated by $U(h), V(g)$ with h, $g \in \mathcal{D}_R(B_\alpha)$, and (3.16) is proved.

By the definition of $\hat{R}(O)$ and (3.16a), one has $\hat{R}(O_1) \subset R(O_2)$ for any pair of regions O_1, $O_2 \subset M$ such that $\bar{O}_1 \subset O_2$. This fact, together with (3.12), entails for any region B_α with $\bar{B}_\alpha \subset B$:

$$R_{\varphi,\pi}(B_\alpha) \subset \hat{R}\left(C(B_\alpha)\right) \subset R(C(B)) .$$

This formula, together with (3.16b), gives (3.15).

(3) Now we have only to prove the opposite inclusion to (3.15). To do this, consider how the localization regions of a function $f \in \hat{L}$ and its images $\delta_0 f_+$, $\delta_1 f_-$ are related. Associating with $f \in S_R(\mathbf{R}^4)$ a smooth solution of the Klein–Gordon equation:

$$F_f(x) = \int d^4 y \, f(y) \, \Delta(x - y) ,$$

we see that

$$F_f(0, \boldsymbol{x}) = (\delta_1 f_-)\, \boldsymbol{x}), \quad \left(\frac{\partial F_f}{\partial x^0} \right)(0, \boldsymbol{x}) = -(\delta_0 f_+)(\boldsymbol{x}) .$$

Since the support of $\Delta(x-y)$ is in the light cone, which implies $F_f(x)=0$ for all $x \in (\mathrm{supp} f)'$, we obtain the required connection between the localization regions:

$$\left(f \in S_R(\mathbf{R}^4),\ \mathrm{supp} f \subset D\right) \Rightarrow \begin{cases} \mathrm{supp}\ \delta_0 f_+ \subset (M \setminus D') \cap S_0 \\ \mathrm{supp}\ \delta_1 f_- \subset (M \setminus D') \cap S_0 \end{cases}$$

(D is a double cone). Hence, taking into account that the property

$$\left(M \setminus C(B_\alpha)'\right) \cap S_0 = C(B_\alpha) \cap S_0 = B_\alpha$$

holds only if $\mathrm{int}\ \overline{B_\alpha} = B_\alpha$, we obtain: if $\mathrm{supp}\ f \subset C(B_\alpha)$ and $\mathrm{int}\ \overline{B_\alpha} = B_\alpha$, then $\mathrm{supp}\ \delta_0 f_+$, $\mathrm{supp}\ \delta_1 f_- \subset B_\alpha$. Since it follows from (3.7) that for all $f \in S_R(\mathbf{R}^4)$,

$$W(f) = U(\delta_0 f_+)\, V(\delta_1 f_-) \exp\left\{\frac{1}{2}\left(j_2^{-1}\, \delta_0 f_+,\ j_1\, \delta_1 f_-\right)\right\},$$

we obtain

$$R\left(C(B_\alpha)\right) \subset R_{\varphi,\pi}(B)$$

for any open B such that $\overline{B_\alpha} \subset B$. But (3.16a) implies that if $\mathrm{int}\ \overline{B} = B$, the algebras $R\left(C(B_\alpha)\right)$ with $\mathrm{int}\ \overline{B_\alpha} = B_\alpha$ generate $R(C(B))$. ❐

COROLLARY. *The net $R_{\varphi,\pi}(B)$ enjoys the Reeh–Schlieder property: each vector analytic in the energy is bicyclic for all $R_{\varphi,\pi}(B)$ with B satisfying the assumptions of Proposition* 3.3.2.

Thus our problem is reduced essentially to studying the net $\{R_{\varphi,\pi}(B)\}$. We must investigate the type and the centre of $R_{\varphi,\pi}(B)$ and check the fulfillment of causal relations, of which the main ones are duality and causal independence. It can be expected from Chapter 1 that the problem of the centre presents the least difficulties. And, indeed, in our model this problem is solved trivially.

PROPOSITION 3.3.3. *For any measurable regions $B \subset \mathbf{R}^3$, the algebras $R_{\varphi,\pi}(B)$ and $R(C(B))$ are factors.*

Proof. By virtue of locality, the algebras $R_{\varphi,\pi}(B)$ and $R_{\varphi,\pi}(B^\perp)$ are commuting; by virtue of strong additivity they generate $R_{\varphi,\pi}(\mathbf{R}^3) = \mathcal{B}(\mathcal{H})$. Hence, they form a factorization and, therefore, are factors [18, Section 34.4]. ❐

Let us make two remarks. Firstly, in case of the covariant net we have obtained the factoriality of $R(O)$ only for the regions of the form $O = C(B)$ or $O = (a, \Lambda)\, C(B)$. For other regions the possibility $\mathcal{Z}(O) \neq C(\mathcal{H})$ stays quite open. Secondly, factoriality may also not hold for the Weyl algebra $R(K_1, K_2) = \{U(h), V(g) \mid h \in K_1, \omega g \in K_2\}''$ if the pair of subspaces K_1, K_2 is not of the form $L_\varphi(B), \omega L_\pi(B)$. It is not difficult to find the centre of $R(K_1, K_2)$ [39, Section 5]: taking pairwise intersections in the quartet of subspaces K_1, K_2, $K_1^\perp$, $K_2^\perp$ and representing the one-particle space as the direct sum of these intersections, we obtain in a standard way the representation of the Fock space as the tensor product in which each multiplier corresponds to one of the terms in the direct sum. The Weyl algebra then also takes the form of a tensor product in which the centre of each multiplier is easily calculable. In this way we find that the centre of $R(K_1, K_2)$ is equal to $R\left(K_1 \cap K_2^{\perp}, K_2 \cap K_1^{\perp}\right)$ whence it follows that $R(K_1, K_2)$ is a factor iff

$$K_1 \cap K_2^{\perp} = K_2 \cap K_1^{\perp} = \{0\}.$$

On the contrary, the problem of type is the most tricky. Of course, the final answer is simple and well known: $R_{\varphi,\pi}(B)$ is of the type III. However, both existing proofs of this result [42, 135] are quite involved and use the technique of infinite tensor products (ITP) not discussed in our book so far. On the other hand, as a sort of compensation, in the formalism of ITP duality appears as a simple by-product of results concerning the type. And the technique of ITP (used for a long time in studies of canonical systems) has repeatedly been described in physical literature (see, e.g. [10, Chapter 3; 5, Section 2.7.2]).

To define ITP of Hilbert spaces $\mathcal{H}_\gamma$ ($\gamma \in \Gamma$, card $\Gamma \geq \aleph_0$), one starts with a set $\Omega = \{\Omega_\gamma\}_{\gamma \in \Gamma}$ of unit vectors $\Omega_\gamma \in \mathcal{H}_\gamma$. This set determines the set of all sets $\xi = \{\xi_\gamma\}_{\gamma \in \Gamma}$, $\xi_\gamma \in \mathcal{H}_\gamma$ such that $\xi_\gamma = \Omega_\gamma$ for all $\gamma \in \Gamma$ except finite number of values. The latter set is a vector space with respect to operations

$$c_1 \xi + c_2 \eta = \{c_1 \xi_\gamma + c_2 \eta_\gamma\}_{\gamma \in \Gamma}, \qquad c_1, c_2 \in \mathbf{C}.$$

Introducing the notation $\xi = \otimes_{\gamma \in \Gamma} \xi_\gamma$ for its elements and endowing it with the inner product

$$\left(\bigotimes_{\gamma \in \Gamma} \xi_\gamma, \bigotimes_{\gamma \in \Gamma} \eta_\gamma\right) \doteq \prod_{\gamma \in \Gamma} (\xi_\gamma, \eta_\gamma),$$

we complete it with respect to the corresponding norm thus obtaining a Hilbert space which is denoted by $\otimes^{\Omega}_{\gamma \in \Gamma} \mathcal{H}_\gamma$ and called the ITP (or incomplete direct product, in von Neumann's terminology) of Hilbert spaces $\mathcal{H}_\gamma$.

Let then $\{R_\gamma\}_{\gamma \in \Gamma}$ be a set of von Neumann algebras acting in $\mathcal{H}_\gamma$. Consider the set of all sets $A = \{A_\gamma\}_{\gamma \in \Gamma}$ where $A_\gamma \in R_\gamma$ and $A_\gamma = I_\gamma$ for all $\gamma \in \Gamma$ except a finite number of values (I_γ is the identity operator in $\mathcal{H}_\gamma$). Each element of this set determines uniquely an operator in $\mathcal{H} = \otimes^{\Omega}_{\gamma \in \Gamma} \mathcal{H}_\gamma$, acting on elements $\otimes_{\gamma \in \Gamma} \xi_\gamma$ by the law

$$A \bigotimes_{\gamma \in \Gamma} \xi_\gamma \doteq \bigotimes_{\gamma \in \Gamma} A_\gamma \xi_\gamma$$

and extending to the whole of $\mathcal{H}$ by continuity. Denoting such operators by $\otimes_{\gamma \in \Gamma} A_\gamma$ we endow their set by the structure of *-algebra, putting by definition

$$\left(\bigotimes_{\gamma \in \Gamma} A_\gamma\right)\left(\bigotimes_{\gamma \in \Gamma} B_\gamma\right) \doteq \bigotimes_{\gamma \in \Gamma} A_\gamma B_\gamma, \quad \left(\bigotimes_{\gamma \in \Gamma} A_\gamma\right)^* = \bigotimes_{\gamma \in \Gamma} A_\gamma^*.$$

Taking the closure of this algebra with respect to weak topology of $\mathcal{B}(\mathcal{H})$, we obtain a von Neumann algebra in $\mathcal{H}$ which is denoted by $\otimes^{\Omega}_{\gamma \in \Gamma} R_\gamma$ and called the ITP of von

Neumann algebras R_γ. Properties of this algebra depend crucially on the choice of the vector $\Omega = \otimes_{\gamma \in \Gamma} \Omega_\gamma$ (and we shall see it in the examples below); but if all R_γ are factors, $R = \otimes^{\Omega}{}_{\gamma \in \Gamma} R_\gamma$ is a factor as well.

Returning to our model we now formulate the central result.

THEOREM 3.3.4. *For any measurable region $B \subset \mathbf{R}^3$ the algebra $R_{\varphi,\pi}(B)$ is a hyperfinite factor of the type* III_1 *and satisfies the duality condition:*

$$R_{\varphi,\pi}(B) = R_{\varphi,\pi}(B^{\perp})'.$$

These properties are also valid for the algebras $R(D)$ of the covariant formulation.

Proof. Adopting again the convenient notations by Araki,

$$R_{\varphi,\pi}(B) = R(K_1, K_2), \qquad K_1 = L_\varphi(B), \qquad K_2 = \omega L_\pi(B),$$

let us derive, as promised, a representation of $R(K_1, K_2)$ in the form of an ITP. The standard procedure is as follows. Define on the subspace $K_1 \subset K \equiv L_\varphi$ an operator

$$Y : K_1 \to K_1^{\perp} \quad \underset{\eta \in K_2}{\forall} \quad Y P_1 \eta = (I - P_1)\eta,$$

where P_1 is the projection on K_1. Y is defined on the set of all K_1-projections of vectors from K_2 and transforms the K_1-projection of a vector $\eta \in K_2$ into its $K_1^{\perp}$–projection. If the graph of an operator A is identified with the set of vectors $\{\xi + A\xi, \xi \in \mathcal{D}(A)\}$, the graph of Y is K_2. Properties of Y are determined by geometry of the pair K_1, K_2. In our case K_1 and K_2 are in a generic position, i.e. all the pairwise intersections in the quartet K_1, K_2, $K_1^{\perp}$, $K_2^{\perp}$ are trivial. Indeed, $K_1 \cap K_2^{\perp} = \{0\}$ due to (3.10), while $K_1 \cap K_2 = K_1^{\perp} \cap K_2^{\perp} = \{0\}$ due to antilocality of the operator ω. This geometrical structure is of a prime importance for the algebraic structure; it lies in the roots of the factoriality of $R_{\varphi,\pi}(B)$ (as we have seen), as well as other structural properties. As for the most immediate implications, it determines the properties of Y: it is easy to check that

due to $K_2 \cap K_1^{\perp} = \{0\}$, Y is well defined and closed;

due to $K_1 \cap K_2^{\perp} = \{0\}$, the domain of Y is dense in K_1;

due to $K_1^{\perp} \cap K_2^{\perp} = \{0\}$, the range of Y is dense in $K_1^{\perp}$;

due to $K_1 \cap K_2 = \{0\}$, the kernel of Y is trivial.

Hence it follows that there exists the adjoint operator Y^* with the dense domain in $K_1^{\perp}$, the operator $\alpha = Y^*Y$ is positive and Y has the polar decomposition $Y = U\alpha^{1/2}$, where U is an isometry from K_1 into $K_1^{\perp}$ and $\alpha^{1/2}$ a positive operator in K_1 ($\alpha^{1/2} > 0$).

In the remaining part we follow the proof by Dell'Antonio [135, Section 9]. We apply to $\alpha^{1/2}$ the following lemma by von Neumann:

LEMMA. *Each self-adjoint operator A in a separable Hilbert space can be represented* *in the form* $A = A_d + A_c$ *where* A_d *is a self-adjoint operator with purely discrete spectrum coinciding with* A *on the discrete spectrum subspace of* A, *and* A_c *is a Hilbert–Schmidt operator.*

By this lemma, $\alpha^{1/2} = \alpha_d^{1/2} + \alpha_c^{1/2}$ and there exists in K_1 an orthonormal basis $\{h_i\}_{i\in \mathbf{N}}$, the elements of which are eigenvectors of $\alpha_d^{1/2}$. Denote by $\{\lambda_i\}_{i\in \mathbf{N}}$ the corresponding set of eigenvalues (some of which may be coinciding) and $\{h_i\}$ the one-dimensional subspace of of a unit vector h_i. Then we have the decomposition of the quartet of subspaces over the degrees of freedom:

$$K_1 = \bigoplus_{i\in \mathbf{N}} \{h_i\}, \; K_2 = \bigoplus_{i\in \mathbf{N}} \{h_i + Y h_i\} ,$$

$$K_1^{\perp} = \bigoplus_{i\in \mathbf{N}} \{U h_i \}, \; K_2^{\perp} = \bigoplus_{i\in \mathbf{N}} \{U h_i - Y^{*} U h_i\} \tag{3.17}$$

(only the last formula here needs a short calculation). Hence, the standard continuity arguments entail

$$R_{\varphi,\pi} (B) = R (K_1 , K_2) = \{U (h_i), V (h_i + Y h_i), \; i \in \mathbf{N}\}'' \tag{3.18}$$

and similarly, taking account of (3.10),

$$R_{\varphi,\pi} (B^{\perp}) = R (K_2^{\perp} , K_1^{\perp}) = \{U (U h_i - Y^{*} U h_i), V (U h_i), \; i \in \mathbf{N}\}'' . \tag{3.19}$$

Infinite direct sums (3.17) in the one-particle space should imply representations of the Fock space $\mathcal{H}_F (K)$ and the factors $R_{\varphi,\pi} (B)$ in the form of ITP. However, it is not so simple to derive representations of the necessary form, for which criteria of belonging to definite types are known.

Let us introduce canonical variables (creation-annihilation operators) $\hat{a}_i^{\#} = (\hat{a} (h_i), \hat{a} (h_i)^{*})$ corresponding to a pair $U (h_i), V (h_i)$ of Weyl operators:

$$\hat{a} (h_i) + \hat{a} (h_i)^{*} = \underset{\lambda\to 0}{s\text{-lim}} \; \big(U \; (\lambda \, h_i) - I\big)/\lambda ,$$

$$\hat{a} (h_i) - \hat{a} (h_i)^{*} = \underset{\lambda\to 0}{s\text{-lim}} \; \big(V \; (\lambda \, h_i) - I\big)/\lambda .$$

Then operators $a_i^{\#}$ defined according to

$$a_i = \hat{a} \big(h_i + \frac{1}{2} Y h_i\big) - \frac{1}{2} \, \hat{a}^{*} \, (Y h_i) \tag{3.20}$$

correspond to the pair $U (h_i), V (h_i + Y h_i)$ and satisfy canonical commutation relations (CCR). This means that this pair of the Weyl operators implements an irreducible representation of CCR, the space of which we denote by $\mathcal{H}_i^{a}$. Similarly, it is easy to check that the operators $b_i^{\#}$ defined according to

$$b_i = \hat{a} \; \big(U h_i - \frac{1}{2} Y^{*} \, U h_i\big) - \frac{1}{2} \, \hat{a}^{*} \; (Y^{*} \, U h_i) \tag{3.21}$$

correspond to the pair $U(Uh_i - Y^* Uh_i)$, $V(Uh_i)$ and also satisfy CCR. Thus this pair implements an irreducible representation of CCR in a certain $\mathcal{H}_i^b$. Then in the space $\mathcal{H}_i^a \otimes \mathcal{H}_i^b$ there acts irreducibly the set of four Weyl operators (extended to this space by means of canonical isomorphism of the form $\tilde{U}(h_i) \in \mathcal{B}(\mathcal{H}_i^a) \to \hat{U}(h_i) \otimes I_i^b$). And in the ITP $\otimes_{i\in\mathbf{N}} (\mathcal{H}_i^a \otimes \mathcal{H}_i^b)$, if it can be defined correctly, the set of operators generating $R_{\varphi,\pi}(B)$ and $R_{\varphi,\pi}(B^{\perp})$ will act irreducibly, as it was acting in the initial $\mathcal{H}$ (remember that $\{R_{\varphi,\pi}(B) \cup R_{\varphi,\pi}(B^{\perp})\}'' = \mathcal{B}(\mathcal{H})$). For this reason we can identify $\mathcal{H}$ with the space $\otimes_{i\in\mathbf{N}} (\mathcal{H}_i^a \otimes \mathcal{H}_i^b)$, in which the Weyl operators are, obviously, realized in the form

$$U(h_i) = (I_1^a \otimes I_1^b) \otimes \ldots \otimes (\tilde{U}(h_i) \otimes I_i^b) \otimes I_{i+1}^a \otimes \ldots . \tag{3.22}$$

Since each Fock column (the set of which is dense in the state space) has only a finite number of nonzero components, correct definition of ITP $\otimes_{i\in\mathbf{N}}$ is the definition $\otimes_{i\in\mathbf{N}} = \otimes_{i\in\mathbf{N}}^{\Omega}$, where Ω is the Fock vacuum in $\mathcal{H}$. However $\mathcal{H}$ is realized now as the space of the CCR representation described by the canonical variables $\{a_i^{\#}, b_i^{\#}\}_{i\in\mathbf{N}}$, and this representation may not have the Fock vacuum. Indeed, for any finite number N of the degrees of freedom the representation $\{a_i^{\#}, b_i^{\#}\}_{i=1}^{N}$ is the Fock one and has the Fock vacuum Ω_N; but neither general theorems nor the formulae (3.21) ensure the Fock nature in case of the infinite system. For this reason an auxiliary representation of CCR is constructed in $\mathcal{H}$ for which it is not difficult to verify that it is a Fock representation. In this representation, like in the original one, there are for each degree of freedom $i \in \mathbf{N}$ two pairs of the Weyl operators U_i^A, V_i^A and U_i^B, V_i^B defined according to

$$U_i^A = U(h_i), \quad V_i^A = V(h_i + Y h_i - \lambda_i U h_i),$$
$$U_i^B = U(U h_i - Y^* U h_i + \lambda_i h_i), V_i^B = V(U h_i). \tag{3.23}$$

It is easy to see that this system corresponds to canonical variables

$$A_i = \hat{a}\left(h_i + \frac{1}{2} Y h_i - \frac{1}{2}\lambda_i U h_i\right) - \frac{1}{2}\hat{a}^*(Y h_i - \lambda_i U h_i),$$

$$B_i = \hat{a}\left(U h_i - \frac{1}{2} Y^* U h_i + \frac{1}{2}\lambda_i h_i\right) - \frac{1}{2}\hat{a}^*(Y^* U h_i - \lambda_i h_i),$$

which satisfy CCR but do not, in general. annihilate the original vacuum $\Omega \in \mathcal{H}$ for which

$$\hat{a}(h_i)\,\Omega = \hat{a}(U h_i)\,\Omega = 0, \qquad i = 1, 2, \ldots .$$

To prove that (3.23) is, nevertheless, a Fock representation, let us introduce an approximating Fock system in $\mathcal{H}$ which coincides with $\{A_i^{\#}, B_i^{\#}\}$ for the first N degrees of freedom:

$$A_i^N = \begin{cases} A_i, i \le N \\ \hat{a}(h_i), i > N \end{cases} \qquad B_i^N = \begin{cases} B_i, i \le N \\ \hat{a}(U h_i), i > N \end{cases} .$$

Clearly, this system implements an irreducible CCR representation in $\mathcal{H}$ which is Fock because so is the 'infinite part' of the system:

$$A_i^N \Omega = B_i^N \Omega = 0 \quad \forall\, i > N ,$$

and any 'finite part' is equivalent to a Fock representation. Hence there is in $\mathcal{H}$ the Fock vacuum Ω_N for this system and it suffices to prove that the sequence $\{\Omega_N\}_{N=1}^{\infty}$ is converging in $\mathcal{H}$ at $N \to \infty$. This property is equivalent to

$$\lim_{\substack{M,N\to\infty \\ N>M}} \| \Omega_N - \Omega_M \|^2 = 2 \lim_{\substack{M,N\to\infty \\ N>M}} \left(1 - \mathrm{Re}\,(\Omega_M , \Omega_N)\right) = 0 .$$

Calculations by the standard second quantization technique (cf. [135, Section 3]) give us that this condition is met if

$$\sum_{i=1}^{\infty} \| Y h_i - \lambda_i U h_i \|^2 < \infty \tag{3.24}$$

(where, unlike the preceding formula, $\| \cdot \|$ is the norm in the one-particle space). Since $Y = U\left(\alpha_d^{1/2} + \alpha_c^{1/2}\right)$ and $\lambda_i h_i = \alpha_d^{1/2} h_i$, one has $Y h_i - \lambda_i U h_i = U \alpha_c^{1/2} h_i$, and we see that (3.24) is fulfilled, because, by the von Neumann lemma, $\mathrm{Tr}\, \alpha_c < \infty$. The limiting vector $\Omega' = \lim_{N\to\infty} \Omega_N$ is the Fock vacuum for the system (3.23).

Now we construct the ITP $\widetilde{\mathcal{H}}$ of spaces $\widetilde{\mathcal{H}}_i = \mathcal{H}_i^A \otimes \mathcal{H}_i^B$ (where $\mathcal{H}_i^A$, $\mathcal{H}_i^B$ are the spaces of the CCR representations defined by the operators $A_i^{\#}$, $B_i^{\#}$, respectively) corresponding to Ω':

$$\widetilde{\mathcal{H}} = \bigotimes_{i\in \mathbf{N}}^{\Omega'} \widetilde{\mathcal{H}}_i ; \; \Omega' = \bigotimes_{i\in \mathbf{N}} \Omega'_i ; \; \Omega'_i = \Omega_{A_i} \otimes \Omega_{B_i} \tag{3.25}$$

Note that, by the definitions (3.23), the Weyl forms $U_i^{A,B}$, $V_i^{A,B}$ are functions of the Weyl forms $U_i^{a,b}$, $V_i^{a,b}$, corresponding to the operators $a_i^{\#}$, $b_i^{\#}$:

$$U_i^A = U_i^a ; \quad V_i^A = V_i^a \,(V_i^b)^{-\lambda_i} , \quad U_i^B = U_i^b \,(U_i^a)^{\lambda_i} , \quad V_i^B = V_i^b , \tag{3.26}$$

and, conversely, $U_i^{a,b}$, $V_i^{a,b}$ can be represented as functions of $U_i^{A,B}$, $V_i^{A,B}$ with the same index i. Hence, taking into account that all CCR representations with finite number of degrees of freedom are Fock, we can identify

$$\mathcal{H}_i^A \otimes \mathcal{H}_i^B = \mathcal{H}_i^a \otimes \mathcal{H}_i^b . \tag{3.27}$$

It means that it can be assumed $\widetilde{\mathcal{H}} = \mathcal{H}$ in (3.25) and this formula is the required realization of $\mathcal{H}$ in the form of an ITP adjusted to the decomposition $\mathbf{R}^3 = B \cup B^{\perp}$. By

(3.18), (3.19) and (3.22), the operators generating $R_{\varphi,\pi}(B)$ and $R_{\varphi,\pi}(B^{\perp})$ in this realization take the form:

$$U(h_i) = (I_1^a \otimes I_1^b) \otimes \ldots \otimes (U_i^a \otimes I_i^b) \otimes I_{i+1}^a \otimes \ldots$$

$$V(h_i + Yh_i) = I_1^a \otimes \ldots \otimes (V_i^a \otimes I_i^b) \otimes I_{i+1}^a \otimes \ldots$$

$$U(Uh_i - Yh_i) = I_1^a \otimes \ldots \otimes (I_i^a \otimes U_i^b) \otimes I_{i+1}^a \otimes \ldots$$

$$V(Uh_i) = I_1^a \otimes \ldots \otimes (I_i^a \otimes V_i^b) \otimes I_{i+1}^a \otimes \ldots$$

which entails the following representations for local algebras:

$$R_{\varphi,\pi}(B) = \bigotimes_{i\in\mathbf{N}}^{\Omega'} \mathcal{B}(\mathcal{H}_i^a) \otimes \mathcal{C}(\mathcal{H}_i^b) = \{\mathcal{C}(\mathcal{H}_1^a) \otimes \ldots$$

$$\ldots \otimes (\mathcal{B}(\mathcal{H}_i^a) \otimes \mathcal{C}(\mathcal{H}_i^b)) \otimes \mathcal{C}(\mathcal{H}_{i+1}^a) \otimes \ldots, i = 1, 2, \ldots\}'', \quad (3.28)$$

$$R_{\varphi,\pi}(B^{\perp}) = \bigotimes_{i\in\mathbf{N}}^{\Omega'} \mathcal{C}(\mathcal{H}_i^a) \otimes \mathcal{B}(\mathcal{H}_i^b) = \{\mathcal{C}(\mathcal{H}_1^a) \otimes \ldots$$

$$\ldots \otimes (\mathcal{C}(\mathcal{H}_i^a) \otimes \mathcal{B}(\mathcal{H}_i^b)) \otimes \mathcal{C}(\mathcal{H}_{i+1}^a) \otimes \ldots, \ i = 1, 2, \ldots\}'',$$

With the representations (3.28) we have achieved our main aim. They are extremely meaningful. First of all, they immediately produce duality which now coincides with the commutant theorem $((\otimes_i R_i)' = \otimes_i R'_i)$ proved for such algebras still by von Neumann himself [336]. Furthermore, they provide rich information about the structure of algebras. The algebras (3.28) are hyperfinite factors studied in a number of works [336, 226, 115] where, in particular, the necessary and sufficient conditions determining their type have been found. These conditions are formulated in terms of the sequence $\{d_i\}$ formed by the distances d_i of the generating vector Ω' in each component $\mathcal{H}_i^a \otimes \mathcal{H}_i^b$ from the nearest product vector $f_i \otimes g_i$:

$$d_i = \inf_{f_i \in \mathcal{H}_i^a \cap S_1,\ g_i \in \mathcal{H}_i^b \cap S_1} \|\Omega' - f_i \otimes g_i\|.$$

As proved in [336, 114], algebras of the form (3.28) are of type I iff $\Sigma_{i=1}^{\infty} d_i < \infty$ and type III in all the remaining cases. In our construction, as one easily sees, Ω' cannot be a product vector: by (3.23) $\Omega' = \otimes_{i\in\mathbf{N}} (\Omega_{A_i} \otimes \Omega_{B_i})$; on the other hand, by (3.26), U_i^A, V_i^A depend not only on U_i^a, V_i^a, but on V_i^b as well. Hence (3.27) cannot be enhanced to $\mathcal{H}_i^A = \mathcal{H}_i^a$, $\mathcal{H}_i^B = \mathcal{H}_i^b$. The calculation similar to the derivation of (3.24) gives the result

$$d_i = 2\left(1 - 1/\sqrt{1+\lambda_i^2}\,\right),$$

whence it follows due to $Y = U\left(\alpha_d^{1/2} + \alpha_c^{1/2}\right)$ and $\mathrm{Tr}\,\alpha_c < \infty$ that $\Sigma_{i=1}^{\infty} d_i < \infty$ iff Y is Hilbert–Schmidt. On the other hand, the antilocality of ω entails the following property of the pair of subspaces K_1, K_2: for any $\varepsilon > 0$ there exists a pair of vectors $\xi \in K_1$, $\eta \in K_2$ such that the angle between ξ and η is less than ε. Hence it follows that the domain of Y cannot be the whole of K_1. Thus Y is an unbounded operator and the hyperfinite factors $R_{\varphi,\pi}(B)$ and $R_{\varphi,\pi}(B^{\perp})$ are of type III.

For proving the type III_1 the shortest way is to use Longo's argument [311]. As we shall see below, local algebras change only isomorphically with the change of the mass of a free field in the interval $(0, \infty)$. Hence, the type of $R(D)$ is in our case the same as in the free massless field model. But in this model, due to its dilatation covariance, local algebras of double cones D, light cones V^+ and wedges W are all spatially isomorphic [245] and hence, of type III_1 by Proposition 1.3.12. Thus, in our case $R(D)$ is of type III_1 and so is $R_{\varphi,\pi}(B)$ for any bounded measurable $B \subset \mathbf{R}^3$; by duality this is true for unbounded regions as well. □

It should be remarked that duality can be proved by many different ways. In fact, as many as seven different proofs of it have been published! The original ones [39, 135] used the technique of infinite tensor products which seemed too complicated and not really necessary in the duality (in contrast to the type) problem. Hence, two other proofs have appeared, those by Osterwalder [341] and Landau [296]: they aimed mainly at simplicity and were based on expansions over the Wick products of canonical fields φ, π (such expansions, one of traditional tools of the 'old' quantum field theory can be made mathematically quite correct, see, e.g. [64]). On the other hand, the Tomita–Takesaki theory (still absent at the time of [39, 35]) suggested that the duality problem, like nearly any problem dealing with commutativity properties of von Neumann algebras, can be appropriately formulated and possibly solved by means of this theory. This suggestion has been realized successfully in the proofs by Eckmann and Osterwalder [177] and Rieffel [358] (to which we shall return soon, the modular structure of local algebras being of an independent interest). Finally, the most recent proof by Hislop [244] covers large classes of regions in M (the previous proofs concentrated on double cones) and both massive and massless fields, exploiting in the first case properties of local Sobolev spaces and in the second case dilatation covariance.

Furthermore, let us stress again that all the results proved for the canonical net give us the information about the covariant net only for diamonds $D = C(B)$, $B \subset S_0$, and their Poincaré transformations. Thus, on the whole, the structure of the covariant net still may be – and is, in a number of aspects – quite different from that of the canonical net. For instance, the following example by Araki [40, Section 7] shows that duality can be broken already for the simplest unions of double cones, $O = D_1 \cup D_2$. Let

$$T_{\pm} = \left\{x^0 \in (\pm t_1, \pm t_2)\right\}, \qquad T = \left\{x^0 \in (-t_2, t_2)\right\}, \qquad t^2 > t_1 > 0,$$

be open intervals on the time axis spanning the double cones $C(T_+) = V^+_{t_1} \cap V^-_{t_2}$,

$C\ (T_-) = V^-_{-t_1} \cap V^+_{-t_2}$, $C\ (T) = V^-_{t_2} \cap V^+_{t_1}$. Both the double cones $C\ (T_+)$, $C\ (T_-)$ are generated by the same ball in S_0:

$$B = \left\{ (x \in S_0 \,\middle|\, |x| < \frac{\sqrt{2}}{4}\ (t_2 - t_1) \right\} .$$

Hence it can be expected that the distinction between the nets $\{R_{\varphi,\pi}\ (B)\}$ and $\{R\ (O)\}$ will be reflected in the properties of the algebra $R\ (O_0)$, $O_0 \equiv C\ (T_+) \cup C\ (T_-)$.

PROPOSITION 3.3.5. *For the algebra $R\ (O_0)$ both duality and the diamond hypothesis are broken,*

$$R\ (O_0) \neq R\ (O_0')' ,\ R\ (O_0) \neq R\ (O_0'') .$$

Proof. By Definition 3.3.1, algebras $R\ (O)$ are a particular case of $R\ (H_1)$. We show first that the correspondence between subspaces $H_1 \subset \hat{L}$ and algebras $R\ (H_1)$ is one-to one:

$$R\ (H_1) = R\ (H_2) \Leftrightarrow H_1 = H_2 . \tag{3.29}$$

(Of course, it is not true for regions O_1, $O_2 \subset M$: owing to the dependence relations, $\hat{S}\ (O_1) = \hat{S}\ (O_2)$ or $R\ (O_1) = R\ (O_2)$ does not, in general, imply $O_1 = O_2$.) In (3.29) only the implication from left to right is nontrivial. It follows from the property, converse to isotony:

$$R\ (H_1) \subset R\ (H_2) \Rightarrow H_1 \subset H_2 . \tag{3.30}$$

If (3.30) is broken, then there exist a pair H_1, H_2 for which $R\ (H_1) \subset R\ (H_2)$ and a function $f \in H_2^{\perp}$ such that $f \notin H_1^{\perp}$ and hence, $(f_1 , f)_{\hat{L}} \neq 0$ for some $f_1 \in H_1$. Then, taking account of (3.2), we can find $\lambda \in \mathbf{R}$ such that $[W\ (\lambda f_1), W\ (\beta f)] = W\ (\beta f)\ W\ (\lambda f_1)\ \{\exp\ [-i\,\lambda\ (f_1 ,\ f)_{\hat{L}}\ ] - I\} \neq 0$. This means that $W\ (\beta f) \notin R\ (H_1)'$, though, clearly, $W\ (\beta f) \in R\ (H_2)'$. Thus $R\ (H_2)' \subset R\ (H_1)'$ which contradicts $R\ (H_1) \subset R\ (H_2)$. Hence, (3.30) and (3.29) follow.

Since $R\ (C\ (T))$ satisfies duality and $O_0' = C\ (T)'$, duality and the diamond hypothesis coincide for $R\ (O_0)$. Taking account of

$$R\ (O_0) = \{R\ (C\ (T_+)) \cup R\ (C\ (T_-))\}'' ,$$

we see that both duality and the diamond hypothesis for $R\ (O_0)$ are broken if

$$R\ (C\ (T)) \neq \{R\ (C\ (T_+)) \cup R\ (C\ (T_-))\}'' .$$

Due to (3.29), this inequality holds if

$$\hat{S}\ (C\ (T)) \neq \overline{L\ \{\hat{S}\ \ (C\ (T_+)) \cup \hat{S}\ \ (C\ (T_-))\}}^{\|\cdot\|_{\hat{L}}}$$

for which it suffices, in its turn, if there is a function g such that

$$g \in (\hat{S}\ (\ C\ (T_+)) \cup S\ (C\ (T_-)))^{\perp} , \qquad g \notin S\ (C\ (T))^{\perp} .$$

Consider the two-dimensional wave equation $f_{xx} - f_{tt} - m^2 f = 0$. For any boundary conditions $f_1(t) = f(t, 0)$, $f_2(t) = f_x(t, 0)$ from $\mathcal{D}(\mathbf{R})$ it has a unique C^∞ solution $f(t, x)$. By the well-known propagation properties of such solutions one has $f(t, x) = 0$ if $x^2 < (t - t')^2$ for all $t' \in (\operatorname{supp} f_1 \cup \operatorname{supp} f_2)$. Hence, there exists a C^∞ solution $f_0(t, x)$ such that $f_0(t, x) = 0$ in a neighbourhood of $C(T_+)$ and $C(T_-)$ while $f_0(t, x) \neq 0$ in a neighbourhood of some point $(t, 0)$ with $|t| < t_1$. Picking $h_1(\boldsymbol{x}) \in \mathcal{D}(\mathbf{R}^3)$, $h_1(\boldsymbol{x}) = 1$ for $|\boldsymbol{x}| \leq t_2$, put

$$g_1(\boldsymbol{x}) = f_0(0, x^1)\, h_1(\boldsymbol{x}), \qquad g_{2'}(\boldsymbol{x}) = (\dot{f}_0)_t\, (0, x^1)\, h_1(\boldsymbol{x}).$$

Then the solution of the Klein–Gordon equation $(\Box + m^2) F = 0$ with the initial data $F(0, \boldsymbol{x}) = g_1(\boldsymbol{x})$, $\dot{F}(0, \boldsymbol{x}) = g_2(\boldsymbol{x})$ coincides with $f_0(x^0, x^1)$ on $C(T)$. Since $g_1, g_2 \in \mathcal{D}(\mathbf{R}^3)$, there is $g(x) \in S(\mathbf{R}^4)$ such that $\delta_1 g_- = g_1$, $\delta_0 g_+ = -g_2$ (for instance, one can take $\tilde{g}(p) = -\tilde{g}_2(p)\, \alpha(p^2) + i\, \tilde{\omega}\, \tilde{g}_1(\vec{p})\, \alpha(p^2)\, \varepsilon(p^0)$ with a suitable choice of α). Then

$$\int d^4 y\, g(y)\, \Delta(x - y) = F(x).$$

But by construction $F(x)$ is equal to zero on $C(T_+) \cup C(T_-)$ and different from zero in some point of $C(T)$. Hence it follows that $\beta g \notin \hat{S}(C(T))^\perp$ and $\beta g \in \left(\hat{S}(C(T_+)) \cup \hat{S}(C(T_-))\right)^\perp$. ❐

Let us proceed to causal independence (Section 1.3.4). According to Theorem 1.3.30, this property is closely connected with a certain set of four equivalent structural properties ((1)–(4) in Theorem 1.3.30) being implied by any of them. They are also interesting by themselves and we shall show that they are all valid in our model (and so is causal independence, of course). Directly, we shall prove the condition (4) which asserts the existence of a normal product state for local algebras of regions $O_1 \times O_2$ or equivalently for algebras $R_{\varphi,\pi}(B)$ of regions $B_1 \cap B_2 = \emptyset$. It will be convenient to consider states on R as states on an abstract CCR algebra so that we could use the criteria of unitary and quasi-equivalence known for such states. There are several constructions of CCR algebra (see, e.g. [273, 316, 25, 10, Chapter 3]); following Buchholz [99] we shall use the 'self-dual CCR algebra' by Araki and Shiraishi [51] as one of the most adjusted to the Araki formalism for free field.

CCR algebra in the Araki–Shiraishi construction is generated by canonical fields

$$\Phi(F) = \varphi(h) + \pi(g) = \begin{pmatrix} \varphi \\ \pi \end{pmatrix} (h \oplus g)$$

defined on test functions $F = h \oplus g \in S(\mathbf{R}^3) \oplus S(\mathbf{R}^3) \equiv K$ and treated as elements of a free system of generators. This system determines on the test function space K the operator Γ of complex structure, by means of the formula

$$\Phi(F)^* = \Phi(\Gamma F), \Gamma F = \bar{F},$$

and the Hermitian form

$$\gamma(F, G) \equiv [\Phi(F)^*, \Phi(G)] = \left(F, \begin{pmatrix} 0 & -i \\ i & 0 \end{pmatrix} G\right),$$

where $(\cdot, \cdot)$ is the inner product on $L^2(\mathbf{R}^3) \oplus L^2(\mathbf{R}^3)$ (note that metrics on K defined by γ is indefinite). The set of all polynomals in generating elements $\Phi(F)$ related to the test function space K with the Hermitian form γ and the complex structure operator Γ is a unital *-algebra called 'self-dual CCR algebra' and denoted by $\mathfrak{A}(K, \gamma, \Gamma)$. In contrast to other CCR algebras, it is not normed and hence, states ψ on $\mathfrak{A}(K, \gamma, \Gamma)$ correspond to representations π_ψ, the images of which are Op^*-algebras (Section 3.1). The usual class of representations of CCR corresponds to the class of so-called regular states ψ on $\mathfrak{A}(K, \gamma, \Gamma)$ for which the operators $\overline{\pi_\psi(\Phi(F))}$ are essentially self-adjoint for all $F \in \operatorname{Re} K = \{G \in K \mid G = \Gamma G\}$ and the Weyl forms $W_\psi(F) = \exp[i\, \overline{\pi_\psi(\Phi(F))}]$ satisfy the Weyl–Segal commutation relations

$$W_\psi(F)\, W_\psi(G) = W_\psi(F + G) \exp\{\tfrac{1}{2}\gamma(F, G)\}.$$

$\mathfrak{A}(K, \gamma, \Gamma)$ can be endowed with a net of local subalgebras, the subalgebra for a region $B \subset \mathbf{R}^3$ being generated by the elements $\Phi(F)$ for which

$$F \in K(B) = \omega^{-1/2} S(B) \oplus \omega^{1/2} S(B).$$

Hence, it follows by continuity of Weyl operators and due to the definitions of spaces $L_\varphi(B), L_\pi(B)$ that if vacuum state ψ_0 on $\mathfrak{A}(K, \gamma, \Gamma)$ is regular, our algebras R and $R_{\varphi,\pi}(B)$ can be identified with the von Neumann algebras generated (by means of Weyl forms or, say, spectral projections) by the representation π_{ψ_0} of $\mathfrak{A}(K, \gamma, \Gamma)$ and its restriction to $\mathfrak{A}(K(B), \gamma, \Gamma)$, respectively. And the regularity of ψ_0 takes place because this state, corresponding to a free field theory, is, obviously, quasifree* and every quasifree state on $\mathfrak{A}(K, \gamma, \Gamma)$ is regular [51, Lemma 3.5].

With the help of this formalism we shall now prove the following proposition.

PROPOSITION 3.3.6 [99]. *For any pair of double cones D_1, D_2 such that $D_1 \times\!\!\times D_2$ there exists a normal product state for the algebras $R(D_1)$ and $R(D_2)$.*

Proof. By means of a Poincaré transformation the pair (D_1, D_2) can be transformed into a pair $C(B_1)$, $C(B_2)$ with $B_1, B_2 \subset S_0$, $d(B_1, B_2) > 0$. Denoting $O_i = C(B_i)$, $i = 1, 2$, one has by Proposition 3.3.2 that $R(O_i) = R_{\varphi,\pi}(B_i)$, $i = 1, 2$.

* Quasifree states introduced first by Robinson [367] constitute the usual and most studied class of states on CCR algebras. In broad terms, they can be defined as such functionals ψ on an algebra generated by a system of elements $A(h)$ or $W(h)$ which are determined uniquely by their 'two-point functions', i.e., objects of the form $\psi(A(h_1)\, A(h_2))$. General notion of quasifree state was suggested by quantum field theory where models of free and generalized free fields are characterized by the property that all their Wightman functions W_n are determined uniquely by the two-point function W_2. In the canonical formalism a quasifree state is characterized, obviously, by a bilinear form on one-particle space $\mathcal{H}_1$. Thus, their study amounts to studying objects in $\mathcal{H}_1$. In particular, criteria of quasi-equivalence of quasifree states [247, 128] are formulated as conditions for certain operators related to A to be Hilbert–Schmidt.

So we shall look for a product state for $R_{\varphi,\pi}(B_1)$ and $R_{\varphi,\pi}(B_2)$. If ψ_0 is a vacuum state on R, consider the state ψ_p on a *-algebra $\mathcal{A}_{12} \equiv R_{\varphi,\pi}(B_1) \vee R_{\varphi,\pi}(B_2)$ defined by the formula

$$\psi_p\left(\sum_{n=1}^{N} A_n B_n\right) = \sum_{n=1}^{N} \psi_0(A_n)\,\psi_0(B_n),\ A_n \in R_{\varphi,\pi}(B_1), B_n \in R_{\varphi,\pi}(B_2)\,. \tag{3.31}$$

To prove that ψ_p can be extended to a normal state on R, it suffices to prove that the corresponding representation π_p of $\mathcal{A}_{12}$ is equivalent to the representation π_0 corresponding to the restriction of ψ_0 to $\mathcal{A}_{12}$. To prove this equivalence, consider ψ_0 as a quasifree state on $\mathfrak{A}(K, \gamma, \Gamma)$. (Notions of (quasi) equivalence, factoriality, etc., for such states are defined with respect to von Neumann algebras generated by Weyl forms of essentially self-adjoint operators from corresponding regular representations of $\mathfrak{A}(K, \gamma, \Gamma)$.) Then its restriction to $\mathcal{A}_{12}$ as well as the state ψ_p (due to (3.31)) are quasifree states on the CCR algebra $\mathfrak{A}(K(B_1 \cup B_2), \gamma, \Gamma)$ and we can use the Araki condition [43] for the equivalence of such states.

To formulate this condition, introduce the main object characterizing quasifree state on CCR algebra. By means of the two-point function D_0 of the state ψ_0

$$D_0(F, G) \equiv \psi_0(\Phi(F)^* \Phi(G)) = \frac{1}{2}\left(F,\ \begin{pmatrix} 1 & -i \\ i & 1 \end{pmatrix} G\right) \tag{3.32}$$

define on $K(B)$, $B = B_1 \cup B_2$, the inner product

$$(F, G)_0 \equiv D_0(F, G) + D_0(\Gamma G, \Gamma \Phi)\,. \tag{3.33}$$

The set $K_0(B) \equiv \overline{K(B)}^{\|\cdot\|_0}$ is a closed subspace in $L^2(\mathbf{R}^3) \oplus L^2(\mathbf{R}^3)$ and the positive form D_0 on it is generated by a positive operator, the explicit form of which follows easily from (3.32), (3.33):

$$\forall_{F,\, G \in K_0(B)} \quad D_0(F, G) = (F, D_0 G)_0\,,\ D_0 = \frac{1}{2} E \begin{pmatrix} 1 & -i \\ i & 1 \end{pmatrix} E,\ 0 \le D_0 \le I, \tag{3.34}$$

where E is the projection on $K_0(B)$. Thus the restriction of ψ_0 to $\mathfrak{A}(K(B), \gamma, \Gamma)$ is completely determined by the operator D_0.

Similarly, one associates a positive operator D_p with the state ψ_p. Due to $d(B_1, B_2) > 0$, the decomposition

$$\forall_{F \in K(B)} \quad F = F_1 + F_2\,,\ F_i \in K(B_i)\,,\ i = 1, 2,$$

is determined uniquely and one has for the two-point function D_p of ψ_p:

$$D_p(F, G) \equiv \psi_p(\Phi(F)^* \Phi(G)) = D_0(F_1, G_1) + D_0(F_2, G_2)\,.$$

Define on $K(B)$ another inner product

$$(F, G)_p \equiv D_p(F, G) + D_p(\Gamma G, \Gamma F) = (F_1, G_1)_0 + (F_2, G_2)_0\,.$$

With the notation $K_p(B) = \overline{K(B)}^{\|\cdot\|_p}$ one sees that $K_p(B)$ is the orthogonal sum of $K_0(B_1)$ and $K_0(B_2)$ with respect to $(\cdot,\cdot)_p$:

$$K_p(B) = K_0(B_1) \oplus_p K_0(B_2) .$$

Denote by Θ_1 and Θ_2 the projections of this decomposition. Then the positive operator D_p generating the form D_p on $K_p(B)$:

$$D_p(F, G) = (F, D_p G) \ \forall F, G \in K_p(B) ,$$

can easily be expressed in terms of Θ_i and the projections E_i on $K_0(B_i)$, $i = 1, 2$, in $L^2(\mathbf{R}^3) \oplus L^2(\mathbf{R}^3)$:

$$D_p = \frac{1}{2} E_1 \begin{pmatrix} 1 & -i \\ i & 1 \end{pmatrix} E_1 \Theta_1 + \frac{1}{2} E_2 \begin{pmatrix} 1 & -i \\ i & 1 \end{pmatrix} E_2 \Theta_2 . \tag{3.35}$$

Now, the criterion of the equivalence of the representations π_0 and π_p of $\mathfrak{A}(K(B), \gamma, \Gamma)$ can be formulated [43, 99]: $\pi_0 \simeq \pi_p$, if the following conditions hold:

(1) 1/2 is not an eigenvalue for the operators D_0 on $K_0(B)$ and D_p on $K_p(B)$;

(2) the norms $\|\cdot\|_0$ and $\|\cdot\|_p$ on $K(B)$ are equivalent;

(3) the operators $A_1 = (D_0 - D_p)(I - 2D_0)^{-1}$ and

$$A_2 = (D_0\,(I - D_0))^{1/2} - (D_p\,(I - D_p))^{1/2}$$

belong to the class of Hilbert–Schmidt operators in $K_0(B)$ (here the spaces $K_0(B)$ and $K_p(B)$ are identified according to condition 2; the positive square root of $D_0(I - D_0)$ is understood with respect to $(\cdot,\cdot)_0$ and that of $D_p(I - D_p)$ with respect to $(\cdot,\cdot)_p$).

(4) 0 is not an eigenvalue for D_0 and D_p.

One must check that these conditions are met by the operators (3.34) and (3.35). By [43, Lemma 2.4] (1) is equivalent to the factoriality of π_0 and π_p. But the von Neumann algebra generated by π_0 is $R_{\varphi,\pi}(B)$ and that generated by π_p is isomorphic to $R_{\varphi,\pi}(B_1) \oplus R_{\varphi,\pi}(B_2)$, due to (3.31); by Proposition 3.3.3 both these algebras are factors. Furthermore, by [43, Lemma 6.2], conditions (1)–(3) ensure the quasi-equivalence of π_0 and π_p which in our case implies the unitary equivalence, the representations π_0 and π_p being factorial of type III [8, Section 5.6.6]. Hence, there is no need to check (4) and it only remains to check (2) and (3). This checking [99, Lemmas 3.2 and 3.3] is not difficult in principle but is rather cumbersome and so omitted here. We only point out that it uses again the separation of regions, $d(B_1, B_2) > 0$. □

As already said, actually Proposition 3.3.6 established for free fields all the properties in Theorem 1.3.30. By Corollary 2 to Proposition 1.3.31, Proposition 1.3.32 and (1.3.47), these properties, in their turn, imply all other causal properties discussed in Section 1.3.4.

COROLLARY. *For each pair of double cones* $D_1 \times\!\!\times D_2$ the local algebras $R(D_1)$ *and*

$R(D_2)$ are causal independent (in the sense of the condition $(\mathrm{CI})_N$) and enjoy the split property. The von Neumann algebra spanned by them is isomorphic to their W^*-tensor product,

$$\overline{R(D_1) \vee R(D_2)}^{\,w} \simeq R(D_1) \bar{\otimes} (D_2).$$

The pairs of the form $R(D_1)$, $R(D_2)$ as well as $R(D)$, $R(D')$, D being an arbitrary double cone, enjoy strict locality, the Schlieder property and extended locality.

Thus, the program of the study of local algebras which we have followed in Chapter 1 is fulfilled. For the free field model all the axioms and related physical properties of local quantum theory are verified and principal properties of the algebraic structure are established. Thus, the model can rightfully be considered as a concrete example proving the compatibility of the set of assumptions of local quantum theory. (Of course, dynamical triviality of this example is its essential shortcoming.) It can also serve as a testing ground for new properties and assumptions arising in algebraic quantum field theory.

In conclusion we discuss a few additional results and mention generalizations to other free fields. Technique of quasifree representations of CCR has also been used by Eckmann and Fröhlich [176] for the demonstration of the following fact: free scalar fields of different masses are locally unitarily equivalent, i.e., if the net $\{R(O, m)\}$ corresponds to a field of the mass m, then

$$R(D, m_1) \simeq R(D, m_2)$$

for all $m_1, m_2 \geq 0$ and any double cone D.

The proof is based on the following observation: if free fields are described by representations of CCR, the two representations corresponding to fields of different masses are connected with each other by a certain Bogoliubov transformation. Let us give its explicit form. A general Bogoliubov transformation is a linear orthogonal transformation in the one-particle space $\mathcal{H}_1$ of a canonical system which preserves CCR (or CAR) for the creation-annihilation operators $a^{\#}(h)$ and is defined by a pair $\beta = (\beta_+, \beta_-)$ of mappings of $\mathcal{H}_1$ into itself such that

$$\beta_+^* \beta_+ - \beta_-^* \beta_- = I, \quad \beta_+^* \beta_- = \beta_-^* \beta_+ .$$

The mappings $\beta_\pm$ are defined on dense domains $\mathcal{D}(\beta_\pm)$ in $\mathcal{H}_1$ and generate the canonical transformation to the following new variables $a_\beta^{\#}(h)$:

$$\forall_{h \in \mathcal{D}(\beta\pm)} \quad a_\beta(h) = a^*(\beta_- h) + a(\beta_+ h), \qquad a_\beta^*(h) = a^*(\beta_+ h) + a(\beta_- h). \quad (3.36)$$

If $a^{\#}(h)$, $h \in L^2(\mathbf{R}^3)$ correspond to a free field of mass m, they are connected with the canonical fields which we now denote by φ_m, π_m:

$$\varphi_m(\omega^{-1/2} h) = \frac{1}{\sqrt{2}} \left(a^*(\omega^{-1/2} h) + a(\omega^{-1/2} h)\right),$$

$$\pi_m(\omega^{1/2}h) = \frac{i}{\sqrt{2}}\left(a^*(\omega^{1/2}h) - a(\omega^{1/2}h)\right), \tag{3.37}$$

($\tilde{\omega} = |p|$ in the case of a massless field). Hence, if there are two fields with masses m_1, m_2 and the operators ω_1, ω_2 and the first of them has canonical variables $a^\#(h)$, then it follows from (3.36) and (3.37) that the second field has the canonical variables $a^\#_\beta(h)$ where

$$\beta_\pm = \frac{1}{2}\left(\omega_2^{-1/2}\ \omega_1^{1/2} \pm \omega_2^{1/2}\ \omega_1^{-1/2}\right). \tag{3.38}$$

In other words, the Bogoliubov transformation (3.38) induces the transformation of canonical fields $(\varphi_{m_1}, \varphi_{m_1}) \to (\varphi_{m_2}, \varphi_{m_2})$ and with it the transformations of Weyl forms and field algebras:

$$\tau_\beta : \mathcal{A}_{m_1} \to \mathcal{A}_{m_2}$$

($\mathcal{A}_{m_i}$, $i = 1, 2$, is algebra of all local observables for the field of mass m_i). τ_β is an invertible homomorphism which is a spatial isomorphism if β is a proper canonical transformation, i.e., β is the Hilbert–Schmidt operator, by the well known criterion [3]. The last property does not hold for (3.38). However, using the technique of constructive field theory, Eckmann and Fröhlich were able to prove that β_- is in a certain sense the Hilbert–Schmidt operator locally, i.e., when restricted to subspace $L^2(B)$; and this means exactly the local unitary equivalence of two fields.

Let us also give, following Eckmann and Osterwalder [177], some basic information about modular structure of free field. The cardinal fact here is that this structure is completely consistent with the second quantization formalism: all operators S, Δ, J, etc., of the modular theory for a local algebra $R_{\varphi,\pi}(B)$ in the Fock space have their 'one-particle prototypes', i.e., corresponding operators $\hat{S}$, $\hat{\Delta}$, $\hat{J}$, etc., in the one-particle space, of which they are obtained by means of the second quantization functor: $S = \oplus_{n=0}^{\infty} \hat{S}^{\otimes n}$, etc. Due to this property, study of modular theory objects reduces essentially to the study of their one-particle prototypes.

Since Ω is bicyclic for local algebras $R(K_1, K_2) = R_{\varphi,\pi}(B)$, we can introduce in the standard way the achieved generalized Hilbert algebra $\mathcal{U} = R(K_1 K_2)\Omega$ and operators S, F, Δ, J (Section 1.1). The following fact [177, Lemma 4] expresses the basic consistency between these objects and the Fock space structure:

> all the 'n-particle subspaces' $\mathcal{H}_n = \text{sym}\ \mathcal{H}_1^{\otimes n}$ of the Fock space $\mathcal{H}$ are invariant subspaces of S, F, Δ, J. This means that S commutes with the projections P_n on $\mathcal{H}_n$, i.e., $P_n S \subset S P_n$, and the restrictions $S_n = S|_{P_n \mathcal{D}^\#}$ are well-defined as closed operators in $P_n \mathcal{H}_n$ with the dense domains $P_n \mathcal{D}^\# = \mathcal{D}^\# \cap \mathcal{H}_n$; similar properties hold for F and Δ, while for J one has $[J, P_n] = 0$. Moreover, these operators leave invariant the real parts $\mathcal{H}_{n,R}$ of n-particle subspaces $\mathcal{H}_n = \mathcal{H}_{n,R} + i\,\mathcal{H}_{n,R}$.

Hence, the operators of modular structure are completely determined by their restrictions $\hat{S}$, $\hat{F}$, $\hat{\Delta}$, $\hat{J}$ to the real part $\mathcal{H}_{1,R}$ of the one-particle space which coincides with the test function space $K = L_\varphi$. The proof of this result gives also that $\mathcal{D}(S) = K_1 + K_2$.

Furthermore, it turns out that $\hat{S}$, $\hat{F}$, $\hat{\Delta}$, $\hat{J}$ can be easily calculated with the help of a certain transformation in K. Note that for any pair K_1, K_2 of subspaces in generic position in K there exist a Hilbert space $\tilde{\mathcal{H}}$ and a positive contracting operator T in $\tilde{\mathcal{H}}$ with the property $\ker T = \ker (I - T) = 0$ such that K_1, K_2 is unitarily equivalent to the pair of subspaces $\Gamma(T), \Gamma(-T)$ in $\mathcal{H} \otimes \mathcal{H}$ ($\Gamma(T)$ being the graph of T). Identifying K and $\tilde{\mathcal{H}} \otimes \tilde{\mathcal{H}}$, we obtain the following representations for the quartet of subspaces:

$$K_1 = \begin{pmatrix} h \\ Th \end{pmatrix}, \quad K_2 = \begin{pmatrix} h \\ -Th \end{pmatrix}, \quad K_1^{\perp} = \begin{pmatrix} -Th \\ h \end{pmatrix}, \quad K_2^{\perp} = \begin{pmatrix} Th \\ h \end{pmatrix}. \tag{3.39}$$

Using the formulae

$$\varphi(h)\,\Omega = \frac{1}{\sqrt{2}} h, \qquad i\pi(g)\,\Omega = \frac{1}{\sqrt{2}} g, \tag{3.40}$$

we find how $\hat{S}$ acts in this realization of K:

$$\hat{S} = \begin{pmatrix} 0 & T^{-1} \\ T & 0 \end{pmatrix}, \qquad \mathcal{D}^{\#} = \left\{ \begin{pmatrix} h_1 \\ Th_2 \end{pmatrix}, h_1, h_2 \in \tilde{\mathcal{H}} \right\},$$

hence, the formulae for other one-particle operators easily follow:

$$\hat{F} = \begin{pmatrix} 0 & T \\ T^{-1} & 0 \end{pmatrix}, \qquad \hat{\Delta} = \begin{pmatrix} T^2 & 0 \\ 0 & T^{-2} \end{pmatrix}, \qquad J = \begin{pmatrix} 0 & I \\ I & 0 \end{pmatrix}. \tag{3.41}$$

Now, using (3.40), we can get back to operators in the Fock space. In particular, due to $J\Omega = \Omega$, one easily finds

$$J\varphi(h)J\Omega = \varphi(\hat{J}h)\,\Omega, \qquad J\pi(g)J\Omega = \pi(\hat{J}g)\,\Omega,$$

whence it follows that

$$JU(h)J = U(\hat{J}h), \qquad JV(g)J = V(\hat{J}g) \qquad \forall h \in K_1, \quad g \in K_2.$$

Since it is clear from (3.39) and (3.41) that $\hat{J}K_1 = K_2^{\perp}$, $\hat{J}K_2 = K_1^{\perp}$, already these first elements of modular theory give us the 'abstract duality': $R(K_1, K_2)' = R\left(K_2^{\perp}, K_1^{\perp}\right)$. So far this formalism was not developed any further though possibly it would be of interest.

Finally, as mentioned in Section 1.2, the nuclearity condition has also been proved recently for free field [114, Appendix]. This fact, combined with another latest result: in general local quantum theory the nuclearity condition implies the split property [105, Theorem 2.1], immediately establishes for our model all causal properties in Section 1.3.4 and all structural properties in Proposition 1.3.30, in particular, making useless Proposition 3.3.6. However, the proofs of both these newest results are quite involved and so we did not adopt this way of the exposition.

As expected, the majority of the above results holds for other free fields as well. The ground for this is the possibility to formulate for these fields some or other version of the Araki formalism. In each concrete case such formulation is straightforward and we just mention briefly the results obtained.

(1) Scalar massless field. As said above, its local net is locally unitarily equivalent to that of the massive field and, therefore, the local algebras of bounded regions enjoy the same structural properties. According to [112], the nuclearity condition is also fulfilled which implies the split property, causal independence and other causal properties from Section 1.3.4. But there is, in addition, a number of properties reflecting physical distinctions of massless systems: their confo.mal symmetry and infrared effects. The first distinction produces the degeneration of vacuum [143, 390] and also proves to be very useful in structural studies: as already mentioned in Section 3.2.2, it implies spatial isomorphism between the local algebras of double cones, light cones and wedges, as well as provides the geometrical interpretation of modular structure of these algebras. The second distinction affects global structure. As discovered recently, quasilocal algebras of massless systems (free bosons [148], free fermions [162, 163, 97] and interacting particles[106]) typically possess so-called infrared representations which satisfy spectrum conditions but have no vacuum; the global von Neumann algebra of such a representation may be of any existing type.

(2) The Dirac field ($s = 1/2$, $m > 0$). Due to the presence of the gauge group $U(1)$ generated by the electric charge operator, this is the simplest model in which the fields are different from observables, $R(O) \subsetneqq F(O)$, and there is a superselection rule with countable set of abelian superselection sectors. For the equal-time field algebra the representation similar to (3.28) has been obtained [135] where the multipliers in the ITP are now type I_2 factors. Hence, it is proved that the algebra $F_{\varphi,\pi}(B)$ is of type I_∞ or III. As for duality, it obviously cannot hold in the usual form for fermion fields. It is well known, however, that there are certain transformations which bring the commutation structure of Fermi and/or para-field systems to the bosonic form [15, Chapter 5]. One of possible forms of such transformation is the twist transformation $F(O) \to F(O)^t$ introduced in [149] and defined by the condition that the original and transformed systems of fields are local relative to each other:

$$F(O') \subset F(O)^{t'}.$$

This condition is called the *twisted locality* and the corresponding transformation exists for any field with normal or para-statistics. Hence, it is obvious that for half-integer spin fields duality should take the form of the 'twisted duality':

$$F(O') = F(O)^{t'}.$$

As observed in [149], fulfillment of this condition for any half-integer spin field follows easily from the above-mentioned analogue of the representation (3.28). The split property for algebras $F(O)$ has been obtained by Summers [403] by means of the direct generalization of Buchholz's proof (Proposition 3.3.6).

Observable algebra for the Dirac field were studied in detail by Dadashyan and Horuzhy [125, 126, 248]. In [125] the covariant and equal-time nets are constructed, the correspondence between them similar to Proposition 3.3.2 is established and it is proved that they enjoy weak additivity though it can be violated for general Fermi systems. Type III for $R(D)$ and duality are proved by means of modular theory, without using ITP and much simpler. Causal independence is also given simpler proof than that in Proposition 3.3.6 for scalar field: we prove it first for the subalgebras $R_{\varphi}(B)$ generated by the forms $U(h)$ (such subalgebras are non-abelian in the formulation using complex test function spaces) and then transport to $R_{\varphi,\pi}(B)$ by means of constructing the faithful normal condition expectation $E : R_{\varphi,\pi}(B) \to R_{\varphi}(B)$ [126, 248].

(3) Free electromagnetic field. For this case there is a construction of the Araki formalism [62] and proofs of causal independence [126] and the nuclearity condition [112].

(4) Fields of arbitrary spins and masses were briefly considered in [135]. Equal-time nets and representations of their local algebras in terms of ITP, similar to (3.28), have been obtained. For integer spins these representations imply type III and duality for local algebras while for half-integer spins type I_{∞} is possible as well and duality holds in the form of twisted duality.

A kindred algebraic formalism is developed for the model of generalized free field (GFF). Although the fields and their time derivatives now, in general, cannot be correctly defined at $t = 0$, but still the c-number form of the commutation relations makes it possible to describe this popular model by means of a certain quasifree representation of CCR.

Recall or see, e.g. [15, Chapter II, Section 6; 4, Chapter 3, Section 4], that the GFF model can be defined as a Wightman theory satisfying one of the following equivalent conditions:

(1) the commutator of fields is a c-number, $[A(f), A(g)] = c(f, g) I$;

(2) only the functions W_1^{tr}, W_2^{tr} of all the truncated Wightman functions W_n^{tr} are different from zero;

(3) only a finite number of the truncated Wightman functions is different from zero;

(4) There is at least one spacelike point p_0 in momentum space which has an open neighbourhood $\mathcal{N}(p_0)$ lying entirely outside the support of the Fourier transform of the field, $\operatorname{supp} \tilde{A}(\tilde{f}) \cap \mathcal{N}(p_0) = \emptyset$ (there are a few more criteria, see, e.g. [309].

Putting the function W_1 equal to zero (without loss of generality) we see that the model is completely determined by the Wightman function W_2. It simplifies the reconstruction theorem and makes it the shortest way to correct formulation of the model taking account of its specific features. Wightman axioms imply the following representation for W_2:

$$W_2(x-y) = \int d\mu_+(p)\, e^{ip(x-y)}, \tag{3.42}$$

where $\mu_+(p)$ is a Lorentz-invariant measure of polynomial growth with the support in the forward light cone. In the Hilbert space reconstructed on the basis of the function

(3.42) a unitary continuous representation $U(a, \Lambda)$ of the Poincaré group and the Fourier transform $\tilde{A}(\tilde{f})$ of the field are naturally defined, $\tilde{A}(\tilde{f})$ being self-adjoint on a certain domain $\mathcal{D}_2$ such that $\mathcal{D} = \mathcal{D}_2 \cap \oplus_{n=0}^{\infty} S(\mathbf{R}^{4n})$ is a core for $\tilde{A}(\tilde{f})$. Since $\tilde{S}(\mathbf{R}^{4n}) = S(\mathbf{R}^{4n})$, it means that field $A(f) = \tilde{A}(\tilde{f})$ with $f \in S_R(\mathbf{R}^4)$ is essentially self-adjoint on $\mathcal{D}$.

Hence, it is already possible to introduce the Weyl forms and local algebras. We must, however, take into account that, $A(f)$ being defined via $\tilde{A}(\tilde{f})$, which, in its turn, is reconstructed on the basis of the measure $\mu_+(p)$, $A(f)$ should take the same values on all the test functions whose Fourier transforms differ outside the support of $\mu_+(p)$ only. It means that $A(f)$ actually depends only on the equivalence classes which form the quotient space $S_R(\mathbf{R}^4)/S_{\mu,0}(\mathbf{R}^4)$ where

$$S_{\mu,0}(\mathbf{R}^4) = \left\{ f \in S_R(\mathbf{R}^4) \mid \|f\|_\mu^2 \doteq \int |\tilde{f}(p)|^2 \, d\mu(p) = 0 \right\}$$

and, by definition, $\mu(p) = \mu_+(p) + \mu_+(-p)$. By the completion we obtain the real Hilbert space

$$\hat{L}_\mu(\mathbf{R}^4) = \overline{S_R(\mathbf{R}^4)/S_{\mu,0}(\mathbf{R}^4)}^{\|\cdot\|_\mu}$$

with the inner product

$$(f, g)_\mu = \int d\mu(p) \, \overline{\tilde{f}(p)} \, \tilde{g}(p) .$$

This space is completely similar to the space $\hat{L}(\mathbf{R}^4)$ in the Araki formalism for free fields. The complex structure operator β on $\hat{L}_\mu$ is defined as

$$\forall_{f \in \hat{L}_\mu} \quad (\widetilde{\beta f})(p) = i\,\varepsilon(p_0)\,\overline{\tilde{f}(-p)} . \tag{3.43}$$

It is easy to see that the operators $W(f) = e^{i\overline{A(f)}}$, $f \in \hat{L}_\mu$ implement the representation of CCR in the Segal form:

$$W(f)\,W(g) = W(f+g)\exp\left(\frac{i}{2}(f, g)_\mu\right),$$

and thus for any subspace $K \subset \hat{L}_\mu$ we can introduce a von Neumann algebra $R(K) = \{W(f), f \in K\}''$. Associating then with any region $O \subset M$ a certain subspace $K(O) \subset \hat{L}_\mu$ (in an obvious way), we obtain the definition of local algebras of the GFF: $R(O) \doteq R(K(O))$. It is completely similar to Definition 3.3.1; regions O may be unbounded as well as bounded.

The set $\{R(O)\}$ defines in $\mathcal{H} = \mathcal{H}_F(\hat{L}_\mu)$ the Haag–Araki theory satisfying the axioms I, IIa, III, V, VI: their fulfillment is verified exactly as in Proposition 3.3.1. The global algebra generated by the net $\{R(O)\}$ is irreducible, which can be checked by many ways. Note, for instance, that the characteristic functional of the system $\{W(f)\}$ can be easily calculated [395] with the answer

$$(\Omega, W(f)\,\Omega) = \exp\left(-\frac{1}{4}\|f\|_\mu^2\right),$$

corresponding to the Fock state and irreducible representation [3]. Furthermore, identically to the proof of the formula (3.25) we derive its analogue

$$R(K_1) = R(K_2) \Leftrightarrow K_1 = K_2 . \tag{3.44}$$

In fact, an even stronger result holds: the lattice of all subspaces K of the space $\hat{L}_\mu$ with the complement relation $K \to \beta K^\perp$ is isomorphic to the lattice of all the von Neumann algebras $R(K)$, since such an isomorphism has been proved by Araki [39, Theorem 1] for any system of von Neumann algebras $R(H_1/H)$ associated with the subspaces H_1 of the one-particle space H of any continuous representation of CCR. The said isomorphism is an isomorphism of structures, i.e., it includes the correspondence between operations in both lattices. Hence one has

$$R(\bigcap_i K_i) = \bigcap_i R(K_i), \quad R(\overline{L\{\bigcup_i K_i\}}^{\|\cdot\|_\mu}) = \{\bigcup_i R(K_i)\}'' , \tag{3.45}$$

and the complement $K \to \beta K^\perp$ in the lattice of subspaces corresponds to the commutant $R(K) \to R(K)'$ in the lattice of algebras. It means that in each regular Weyl–Segal system the 'abstract duality theorem'

$$R(K)' = R(\beta K^\perp) \tag{3.46}$$

is valid [39, (3.14)]. It still does not mean, however, the duality for local algebras since for the latter we need, in addition, the x-space identification: if a subspace K corresponds to the region $O \subset M$, to which region does $\beta K^\perp$ correspond? In the case of free fields such identification is proved by (3.10) and for GFF we shall obtain it in Proposition 3.3.9.

The set of structural properties found for the GFF model is less comprehensive than that for free fields; in particular, the results concerning the type and the nuclearity condition are not yet complete. The main bulk of results includes the facts about factoriality of local algebras, duality and some other causal properties. In their derivation it is convenient to use instead of the measure $\mu(p)$ the measure $\rho(m)$ which stands in the well known Källen–Lehmann representation for the two-point function:

$$W_2(x-y) = \frac{1}{i} \int_0^\infty \bar{\Delta}_{\sqrt{\lambda}}(x-y)\, d\rho(\lambda) ,$$

where $\bar{\Delta}_{\sqrt{\lambda}}$ is the negative frequency Pauli–Jordan function, supp $\rho \subset \mathbf{R}_+$ and $d\rho/d\lambda \in S'(\mathbf{R}_+)$. The connection between the two measures is as follows: at $p^2 = m^2$,

$$d\mu_+(p) = \tilde{\omega}^{-1}\, d^3p\, d\rho(m) .$$

In particular, in terms of the measure ρ (called the Källen–Lehmann spectral measure) one obtains a very useful criterion of the primitive causality fulfillment for GFF. (Recall that breaking of primitive causality for some GFF was already discovered in the pioneer work by Haag and Schroer [237] and has provided the independence proof for this axiom.)

For any $\delta > 0$ define the Hilbert space $\mathcal{H}(\delta)$ as follows: consider the space $Z((-\delta, \delta))$ of the Fourier transforms of all C^∞ functions with support in $(-\delta, \delta)$ as embedded into $L^2(\mathbf{R}_+, d\rho)$, form its quotient space by the subspace $Z((-\delta, \delta))_{\rho,0}$ of functions vanishing in the $L^2(\mathbf{R}_+, d\rho)$ norm and complete by this norm:

$$\mathcal{H}(\delta) = \overline{Z((-\delta, \delta))/Z((-\delta, \delta))_{\rho,0}}^{\|\cdot\|_\rho} .$$

PROPOSITION 3.3.7 [325, 395, 211]. GFF with the Källen–Lehmann spectral measure ρ satisfies the axiom IV if and only if $\mathcal{H}(\delta) = L^2(\mathbf{R}_+, d\rho)$ for any $\delta > 0$.

Proof. Necessity. One has $\mathcal{H}(\delta) \subset L^2(\mathbf{R}_+, d\rho)$ by construction. Take $f \in \mathcal{H}(\delta)^\perp$ and show that axiom IV entails $f = 0$. Exploit the already used (Proposition 1.3.35) method of representing the studied function by means of boundary values of five-dimensional wave equation and associate with $f(|p|)$ the distribution on $S(\mathbf{R}^5)$:

$$\Phi(x, s) = \int d\mu(p)\, \varepsilon(p_0)\, e^{ipx} f(\sqrt{p^2}) \cos(s\sqrt{p^2}), \tag{3.47}$$

satisfying the equation

$$\left(\Box_x - \frac{\partial^2}{\partial s^2}\right) \Phi(x, s) = 0 . \tag{3.48}$$

The wave operator being hypoelliptic in x_0, Φ and $\dot{\Phi}$ posses initial values at $x^0 = 0$ which belong to $S'(x, s)$ and can be easily calculated:

$$\Phi\big|_{x^0=0} = \int d\mu(p)\, \varepsilon(p_0)\, e^{-ipx} f(\sqrt{p^2}) \cos(s\sqrt{p^2}),$$

$$\dot{\Phi}\big|_{x^0=0} = \int 2i(2\pi)^3 \delta(x) \int d\rho(m) f(m) \cos(sm) .$$

Here the first integral vanishes identically, the measure in it being an even function and the integrand an odd one; for the same reason all the even x^0-derivatives of a Φ vanish at $x^0 = 0$, too. The lower integral vanishes if we smear it in the variable s with test functions from $\mathcal{D}((-\delta, \delta))$ since $f \in \mathcal{H}(\delta)^\perp$ and the integral in s is of the Fourier type; thus, $\dot{\Phi}(x, s) = 0$ at $x^0 = 0, |s| < \delta$. Acting on $\dot{\Phi}(x, s)$ by the operators $(\partial/\partial x^0)^n$, $n = 1, 2, \ldots$ we find this property to be valid for all odd x^0-derivatives of $\Phi(x, s)$. Thus, $\Phi(x, s)$ and all its x^0-derivatives vanish identically in the region $\{x^0 = 0, |s| < \delta\} \subset \mathbf{R}^5$. Then, due to the support properties of wave equation solutions (see, e.g. [30, Section 32]) $\Phi(x, s)$ should also vanish in the larger region $\{|x^0| + |s| < \delta\}$ and its boundary value at $s = 0$ (which is well defined in S' since the wave operator is hypoelliptic in s) vanishes in the time-slice $S_\delta = \{x \in M \mid |x^0| < \delta\}$. The aim of exploiting $\mathbf{R}^5$ is now achieved: we have obtained

$$\int d\mu(p)\, \varepsilon(p_0)\, e^{ipx} f(\sqrt{p^2}) = 0 \quad \forall\, x \in S_\delta .$$

Replacing x by $x + a$, $a = (0, \boldsymbol{a})$ and integrating over $\boldsymbol{a}$ with an arbitrary function

$\chi \in S(\mathbf{R}^3)$, $\chi = \chi(|a|)$, one has

$$\int d\mu(p)\, \varepsilon(p_0)\, e^{ipx} f(\sqrt{p^2})\, \chi(|\mathbf{p}|) = 0 \quad \forall x \in S_\delta .$$

Smearing this integral with a test function $u(x) \in S(\mathbf{R}^4)$ and taking account of (3.43) one obtains

$$(\mathring{\beta} u\,,\ \mathcal{F}^{-1}[f\chi])_\mu = 0\,,$$

where $\mathcal{F}^{-1}$ is the inverse Fourier transformation. It means that $\mathcal{F}^{-1}[f\chi] \in \beta K(S_\delta)^\perp$. But one has by axiom IV $R(K(S_\delta)) = R(\hat{L}_\mu)$. Due to (3.44) it implies $\beta K(S_\delta)^\perp = K(S_\delta)^\perp = \{0\}$ whence $f\chi = 0$. Taking $\chi(x) = e^{-x^2}$ one obtains $f = 0$ since $\tilde{\chi}$ has no zeros.

Sufficiency will be proved by the same method. At the same time we shall obtain a stronger result which will be useful below. Namely, for an arbitrary region $O \subset M$ symmetric with respect to S_0 the following properties will be proved:

(1) $f \in \beta K(O)^\perp$ entails $f_\pm \in \beta K(O)^\perp$, where $f_\pm$ are the components of f even and odd in x^0, respectively;

(2) if $\mathcal{H}(\delta) = L^2(\mathbf{R}_+, d\rho)$ for all $\delta > 0$, then in the sense of distributions in $\mathcal{D}'(O \cap S_0)$ and for almost all (in ρ) m,

$$\int d^3p\, e^{-ipx}\, \tilde{f}_+(\tilde{\omega}, \mathbf{p}) = \int d^3p\, e^{-ipx}\, \tilde{\omega}^{-1} \tilde{f}_-(\tilde{\omega}, p) = 0 . \tag{3.49}$$

Derivation of property (1) is left to the reader as an easy exercise. To derive (2) we introduce the distribution $\Phi'_+(x, s) \in S'(\mathbf{R}^5)$, replacing $f(\sqrt{p^2})$ by $\tilde{f}_+(p)$ in (3.47). This distribution also satisfies (3.48) and, as it is easy to find,

$$\dot{\Phi}_+(x, s)\big|_{x^0=0} = 2i \int d\rho(m)\, d^3p\, e^{-ipx}\, \tilde{f}_+(\tilde{\omega}, p)\, \cos(sm) . \tag{3.50}$$

It is equally easy to check that Φ_+ and its s, x^i- derivatives ($i = 1, 2, 3$) vanish for all $x \in O$ and $s = 0$: for odd s-derivatives it is implied by the presence of $\sin(s\sqrt{p^2})$ in the integrand and for the other distributions by $f_+ \in \beta K(O)^\perp$.

Pick up a point $(0, \mathbf{x}) \in O \cap S_0$. Since O is open, there is $\delta > 0$ such that the Cartesian product $U \equiv (-2\delta, 2\delta) \times B_{2\delta} \subset O$ (where $B_{2\delta}$ is an open ball of radius 2δ in S_0 and the interval $(-2\delta, 2\delta)$ is parallel to the x^0-axis, centres of both being in $(0, \mathbf{x})$) is in O. Using as above the support properties of the wave equation solutions one has

$$\Phi_+(x, s) = \frac{\partial^n \Phi_+}{\partial s^n}(x, s) = 0, \qquad n = 1, 2, \dots ; \quad s = 0, x \in U \Rightarrow$$

$$\Rightarrow \Phi_+(x, s) = \dot{\Phi}(x, s) = 0, \qquad |x| + |s| < 2\delta, \qquad \mathbf{x} \in B_{2\delta} .$$

It implies, in particular, $\dot{\Phi}_+(x, s\,|_{x^0=0} = 0$ for $|s| < 2\delta$. Hence integrating (3.50) over d^3x with a test function $g \in \mathcal{D}(B_{2\delta})$ produces

$$\int d\rho\,(m)\cos(sm)\,F_+(m,g)=0,\ |s|<2\delta\,, \tag{3.51}$$

where, by definition,

$$F_+(m,g)=\int d^3x\,g(\mathbf{x})\,F_+(m,\mathbf{x}),\ F_+(m,\mathbf{x})=\int d^3p\,e^{-i\mathbf{px}}\,\tilde{f}_+(\tilde{\omega},\mathbf{p})\,.$$

Smearing the integral (3.51) in s with the convolution (in t) of the functions $h\in\mathcal{D}((-\delta,\delta))$ and $\cos(sm)$ one obtains

$$\int d\rho\,(m)\cos(t\,m)\,\tilde{h}\,(m)\,F_+(m,g)=0,\ |t|<\delta\,.$$

$\mathcal{H}(\delta)=L^2(\mathbf{R}_+,d\rho)$ entails that $\tilde{h}\ F_+=0$ almost everywhere in ρ. But for any $m_0\in\mathbf{R}_+$ one can find $h\in\mathcal{D}((-\delta,\delta))$ such that $h(m_0)\neq 0$, and hence, $F_+(m,g)=0$ for almost all (in ρ) m. Since g is here an arbitrary function from $\mathcal{D}(B_\delta)$ and $F_+(m,g)$, as follows from its definition, is a continuous functional on $\mathcal{D}(B_\delta)$, one also has $F_+(m,\mathbf{x})=0$ as a distribution in $\mathcal{D}'(B_\delta)$. But the centre of B_δ is an arbitrary point in $O\cap S_0$ and hence, $F_+(m,\mathbf{x})=0$ as a distribution in $\mathcal{D}'(O\cap S_0)$ as well. This coincides with the condition (3.49) for f_+.

The proof of (3.49) for f_- is similar and can be obtained by replacing f_+ by f_- throughout. The only other difference is that now $\dot{\Phi}(x,s)|_{x^0=0}\equiv 0$ for parity reasons and so instead of (3.50) we must consider the distribution $\dot{\Phi}_-(x,s)|_{x^0=0}$, due to which the extra factor $\tilde{\omega}^{-1}$ appears in the final formula (3.49).

Finally, taking $O=S_\delta$ in properties 1,2 one obtains $K(S_\delta)=\hat{L}_\mu$, and by virtue of (3.44) the proof is finished. □

Since the equalities (3.49) are valid at a fixed mass value they open the possibility of using the free field technique and, first of all, the antilocal operator ω and the pair of subspaces $L_\varphi(B)$, $\omega L_\pi(B)$ in a generic position in $\mathcal{H}_1$. As we remember, such a structure in $\mathcal{H}_1$ implied many structural properties of local algebras, in particular, factoriality and duality.

PROPOSITION 3.3.8 [211]. *Let a* GFF *satisfy axioms* IV *and* VIa *and a region* $O\subset M$ *be symmetric with respect to* S_0 *and such that* $O\cap S_0=B_R$, $O'\cap S_0=\mathrm{int}\,B_R^\perp$ (B_R *is a ball of radius* R). *Then* $R(O)$ *is a factor.*

Proof. Due to (3.44)–(3.46), the factoriality of $R(O)$ is equivalent to $K(O)\cap\beta K(O)^\perp=\{0\}$. Picking a function f in the left-hand side one has by property (1) $f_\pm\in\beta K(O)^\perp$, and by property (2)

$$\int d^3p\,e^{-i\mathbf{px}}\,\tilde{f}_+(\tilde{\omega},\mathbf{p})=0,\ \mathbf{x}\in B_R\,,$$

$$\int d^3p\,e^{-i\mathbf{px}}\,\tilde{\omega}^{-1}\tilde{f}_-(\tilde{\omega},\mathbf{p})=0,\ \mathbf{x}\in B_R$$

for almost all (in the measure ρ) m. Denoting the left-hand sides of these equalities by

$F_+(x)$ and $F_-(x)$, respectively, and taking account of their support properties as well as the relation between the norms $\|\cdot\|_\mu$, $\|\cdot\|_\varphi$ and $\|\cdot\|_\pi$ one obtains for any fixed $m > 0$:

$$F_+ \in L_\varphi(B_R^\perp), \qquad F_- \in L_\pi(B_R^\perp).$$

On the other hand, locality of the net $\{R(O)\}$ together with (3.44) and (3.46) entail

$$f_\pm \in K(O) \Rightarrow f_\pm \in \beta K(O')^\perp,$$

and using again property (2) one has $F_+ \in L_\varphi(B_R)$, $F_- \in L_\pi(B_R)$. Thus,

$$F_+ \in L_\varphi(B_\rho) \cap L_\varphi(B_R^\perp) = \{0\}, \; F_- \in L_\pi(B_R) \cap L_\pi(B_R^\perp) = \{0\},$$

and since zero is an isolated point in the support of ρ (axiom VIa), by means of the Fourier transformation one obtains $f = f_+ + f_- = 0$. ❒

COROLLARY. *If a* GFF *satisfies axioms* IV *and* VIa, *its local algebras of double cones* D *and cylinders* Z *are factors.*

In the same way we can prove the following proposition.

PROPOSITION 3.3.9 [211]. *For an* GFF *satisfying axioms* VIa *the following properties are equivalent:*

(1) GFF *satisfies axiom* IV*;*

(2) *the diamond hypothesis holds, i.e.* $R(Z) = R(Z'')$ *for any cylinder* Z* *;*

(3) *duality for cylinders holds,* $R(Z)' = R(Z')$,

and imply

(4) *duality for double cones holds.*

Proof. According to the results in Section 1.3.3, the implications $3 \to 2 \to 1$ are valid already for general axiomatic theory; $3 \to 4$ is trivial since Z' is a double cone. Thus, it suffices to obtain $1 \to 3$. If Z is a S_0-symmetric cylinder, by Proposition 1.3.14, $R(Z) = R(Z_r)$, where Z_r is the spacelike hull of Z (See Figure 4). We shall prove the inclusion $R(Z') \supset R(Z_r)'$, which is equivalent to $K(Z') \supset \beta K(Z_r)^\perp$ due to (3.44)–(3.46). By property (1), $f \in \beta K(Z_r)^\perp$ implies $f_\pm \in \beta K(O)^\perp$. Note that $f_+ \in K(Z')$ if $(f_+, \beta g)_\mu = 0$ for all $g \in \beta K(Z')^\perp$, and for parity reasons $(f_+, \beta g_+)_\mu = 0$ for all $g_+ \in \hat{L}_\mu$. Hence, it suffices to prove $(f_+, \beta g_-)_\mu = 0$.
Applying property (2) to functions $f_\pm \in \beta K(Z_r)^\perp$ and $g_- \in \beta K(Z')^\perp$, one has

$$F_+(x) = \int d^3p \, e^{-ipx} \tilde{f}_+(\tilde{\omega}, p) = 0, \qquad |x| < R + T,$$

$$G_-(x) = \int d^3p \, e^{-ipx} \tilde{\omega}^{-1} \tilde{g}_-(\tilde{\omega}, p) = 0, \qquad |x| > R + T,$$

where R is the radius of Z and T its half-height (see Figure 4). At fixed $m > 0$ one

* We consider cylinders with timelike axes which can be assumed to be parallel to the x^0-axis without loss of generality.

obtains from (3.10), the definitions of F_+, G_- and their support properties:

$$F_+ \in L_\varphi\,(B^\perp_{R+T}) = \omega\, L_\pi\,(B_{R+T})^\perp\,, \qquad\qquad G_- \in L_\pi\,(B_{R+T})\,,$$

whence it follows that

$$(\omega^{-1} F_+ , G_-)_\pi = \int d^3 p\ \tilde{\omega}^{-1}\, \tilde{f}_+ (\tilde{\omega}, \boldsymbol{p})\, \tilde{g}_- (\tilde{\omega}, \boldsymbol{p}) = 0\,.$$

Integrating this relation with the measure ρ and taking account of axiom VIa one gets the required property:

$$\int d\,\mu_+ (p)\, \tilde{f}_+ (p)\, \tilde{g}_- (p) = \frac{1}{2} \int d\,\mu_+ (p)\, \varepsilon\, (p_0)\, \tilde{f}_+ (p)\, \tilde{g}_- (p) = (\tilde{f}_+ , \beta\, \tilde{g}_-)_\mu$$

$f_- \in K\,(Z')$ is derived identically. □

Different formalism for GFF which also makes it possible to use the free field technique has been developed by Landau [296]. The formalism is based on the fact that GFF can be considered as a free field of variable mass taking values in the support of ρ and so it can be described by means of one-particle space with functions depending on m in addition to x (or p):

$$\mathcal{H}_1 = L^2\,(\mathbf{R}^3) \otimes L^2\,(\mathbf{R}, d\,\rho\,(m)) \tag{3.52}$$

and canonical variables $a^\#\,(h)$, $h = h\,(p, m)$ in the Fock space $\mathcal{H}_F\,(\mathcal{H}_1)$:

$$\left[a\,(h_1), a^*\,(h_2)\right] = \int d^3 p\, d\,\rho\,(m)\, h_1\,(\boldsymbol{p}, m)\, h_2\,(\boldsymbol{p}, m)$$

(all the other commutators are trivial). In the usual way one can associate with the operators $a^\#\,(\boldsymbol{p}, m)$ relativistically covariant field $A\,(x, m)$ and canonical fields $\varphi\,(x, m) = A\,|_{x^0=0}$, $\pi\,(x, m) = \dot{A}\,|_{x^0=0}$. If one makes these heuristic relations rigorous (see (3.3) and below this formula), one obtains that the fields A, φ, π are defined on test functions from corresponding Schwartz spaces and essentially self-adjoint on a certain domain $\mathcal{D}$, when taken on real test functions. Hence, one can construct nets of local algebras:

$$R^{(m)}\,(O) = \left\{ e^{iA(\overline{F})} \,|\, F\,(x, m) \in S_R\,(\mathbf{R}^5),\ \operatorname{supp} F \subset O \times \mathbf{R} \right\}'' , \tag{3.53}$$

$$R^{(m)}_{\varphi,\pi}\,(B) =$$

$$\left\{ e^{i\varphi(\overline{H})}, e^{i\pi(\overline{G})} \,|\, H\,(x, m), G\,(x, m) \in S\,(\mathbf{R}^4),\ \operatorname{supp} H,\ \operatorname{supp} G \subset B \times \mathbf{R} \right\}'' .$$

Fields of variable mass enjoy a number of useful properties connected with the fact that their symmetric Fourier transformation with respect to mass

$$A\,(x, m) \to A\,(x, s) = \int d\,m\ \cos\,(s\, m)\, A\,(x, m)$$

is a solution (in the sense of S') of the five-dimensional wave equation

$$\left(\Box_x - \frac{\partial^2}{\partial s^2}\right) A\,(x,s) = 0,$$

satisfying the condition $\partial/\partial s\, A\,(x,s)|_{s=0} = 0$. Various implications of this fact are often derived with the help of the following observation. Taking an arbitrary bounded operator T and vectors $\Phi, \Psi \in \mathcal{D}$, consider the difference

$$\Delta\,(x,s) = (T^*\,\Phi, A\,(x,s)\,\Psi) - (A\,(x,s)\,\Phi, T\,\Psi)\,. \tag{3.54}$$

Then it is not difficult to verify that for any region O_5 in $\mathbf{R}^5$ such that $\Delta\,(x,s) = 0$ for all $(x,s) \in O_5$ and $\Phi, \Psi \in \mathcal{D}$ one has

$$T \in \left\{ e^{iA\,(\overline{\hat{F}})} \,|\, \hat{F} \in S_R\,(\mathbf{R}^5),\ \operatorname{supp}\,\hat{F}\,(x,s) \subset O_5 \right\}'.$$

Using this property together with the support properties discussed above of S'-solutions of the five-dimensional wave equation [30, Section 32], one derives easily the correspondence between the covariant and equal-time nets (3.53) similar to that in Proposition 3.3.2:

$$R^{(m)}\,(C\,(B)) = R^{(m)}_{\varphi,\pi}\,(B)\,. \tag{3.55}$$

It is easy to see that the usual operator of GFF is the boundary value of $A\,(x,s) : A\,(x) = A\,(x,s)|_{s=0}$. Returning to its local algebras, we can now prove the proposition.

PROPOSITION 3.3.10 [296]. *For any* GFF *essential duality* (*Definition* 1.3.2) *holds and, moreover,* $R\,(D')' = R^{(m)}\,(D)$.

Proof. It suffices to prove the following:

(1) $R^{(m)}\,(D')' = R^{(m)}\,(D)$;

(2) $R^{(m)}\,(D') = R\,(D')$.

Due to (3.55) duality for $R^{(m)}\,(D)$ is equivalent to that for $R^{(m)}_{\varphi,\pi}\,(B)$. We shall now use another good property of variable mass fields: their one-particle space (3.52) is such that the lattice of its subspaces is related to the lattice of local algebras $R^{(m)}_{\varphi,\pi}$ exactly in the same way as in the free field model. In other words, the following identification takes place:

$$R^{(m)}_{\varphi,\pi}\,(B) = R\,(K_1, K_2)\,, \qquad R^{(m)}_{\varphi,\pi}\,(B^\perp) = R\left(K_2^\perp, K_1^\perp\right). \tag{3.56}$$

It can be easily verified. Indeed, as subspaces K_1, K_2 we should now take $\overline{\omega^{\pm 1/2}\, S_R\,(B \times \mathbf{R})}^{\|\cdot\|_1}$, where $\|\cdot\|_1$ is the norm in the space (3.52). Using (3.10) and the explicit form of $\|\cdot\|_1$, we readily derive the required analogue of (3.10):

$$\overline{\omega^{-1/2}\, S_R\,(B \times \mathbf{R})}^{\|\cdot\|_1} = \left\{ \omega^{1/2}\, S_R\,(B^\perp \times \mathbf{R}) \right\}^\perp,$$

$$\overline{\omega^{-1/2}\, S_R\,(B \times \mathbf{R})}^{\|\cdot\|_1} = \left\{ \omega^{1/2}\, S_R\,(B^\perp \times \mathbf{R}) \right\}^\perp.$$

Due to (3.56), duality for $R^{(m)}_{\varphi,\pi}(B)$ follows from the abstract duality theorem in the form [39, (3.19)] : $R(K_1, K_2)' = R(K_1^{\perp}, K_2^{\perp})$. The proof of $R^{(m)}(D') = R(D')$ is similar to that of (3.55). Taking $T \in R(D')'$ in (3.54), we find that both $\Delta(x, s)$ and its normal derivative vanish on the hyperplane $s = 0$. Hence, by the properties of the wave equation solutions, one has $\Delta(x, s)|_{x^0=0} = \dot{\Delta}(x, s)|_{x^0=0} = 0$ for all s and $x \in B^{\perp}$ (if $D = C(B)$). But then the main property of $\Delta(x, s)$ gives us $T \in R^{(m)}_{\varphi,\pi}(B^{\perp})'$, which means $T \in R^{(m)}(D')'$, due to (3.55). ❐

Since every GFF is determined by its Källen–Lehmann measure $\rho(m)$, various types of GGF are appropriately classified in terms of the properties of $\rho(m)$ and, above all, its behaviour at $m \to \infty$. Structural properties of local algebras can frequently be ensured by certain conditions on this asymptotic behaviour.

To start with, a special class is formed by GFF which admit the equal-time formulation, i.e., the GFF $A(f)$ and its time-derivatives are well-defined at a fixed moment $t = t_0$. It is easy to see (already at the formal level one has $[A(x), \dot{A}(y)]|_{x^0=y^0} = i c\, \delta(x-y)$ $c = \int_0^{\infty} d\rho(m)$) that this class includes those and only those GFF for which all the moments of the Källen–Lehmann measure are finite,

$$c_n = \int_0^{\infty} m^n\, d\rho(m) < \infty, \qquad n = 0, 1, 2, \ldots . \tag{3.57}$$

It can be shown that primitive causality for GGF entails (3.57) [325], but the converse is in general not true [211]. Thus GFF for which primitive causality (and/or the diamond hypothesis and duality for cylinders) holds, form a smaller class with more restrictive conditions on the Källen–Lehmann measure. Several variants of such conditions are known because it is not so difficult to derive them using the condition $\mathcal{H}(\delta) = L^2(\mathbf{R}_+, d\rho)$ (Proposition 3.3.7) which is well studied due to its applications in probability theory (see, e.g. [1]). In particular, one sufficient condition is that the moment problem for the sequence $\{c_n\}$ has a unique solution, for which, in its turn, the following restriction on the growth of the sequence is sufficient: $\Sigma_{n=1}^{\infty}(c_n)^{-1/2n} = \infty$. The following condition is necessary and sufficient in the class of finite measures $\rho(m)$ on $\mathbf{R}_+$ which outside some finite interval $(0, M)$ on $\mathbf{R}_+$ are monotone decreasing and absolutely continuous with respect to Lebesque measure:

$$\int_M^{\infty} dm\, m^{-2} \ln \frac{d\rho}{dm} = -\infty .$$

Hence, it follows that if the Källen–Lehmann measure of some GFF has asymptotics of the form

$$d\rho(m) \underset{m \to \infty}{\sim} \exp(-m^{\alpha})\, dm ,$$

axiom IV and equivalent properties are valid when $\alpha \geq 1$ and broken when $\alpha < 1$.

Finally, in recent studies of the split property the subclasses of GFF which enjoy this property were also characterized in terms of Källen–Lehmann measure. According to

[156, Theorem 10.2], necessary condition of the split property is that this measure is concentrated on isolated points, $d\rho(m) = \Sigma_{n=1}^{\infty} c_i\, \delta(m - m_i)\, dm$, i.e. GFF in question has discrete mass spectrum and is, in fact, a system of a countable number of free fields. This condition is certainly not sufficient. Let us consider a weakening of the split property saying that a theory has a splitting distance $d(r) > 0$ if for its local algebras the inclusion $R(D_r) \subset R(D_{r+d})$ (where D_R is the double cone of base radius R centered at the origin) is split for $d > d(r)$ and not split for $d < d(r)$. Then it has recently been proved [129, Theorem 4.3] that for any $d_0 > 0$ there is a GFF with discrete mass spectrum and splitting distance $d(r)$ such that $d_0 \leq d(r) \leq 2\, d_0$ for all $r \geq 0$. Thus, to obtain sufficient conditions one should add some restrictions on the sequences $\{m_n\}$ and $\{c_n\}$; in particular, $\{m_n\}$ is allowed to have an accumulation point only at infinity [56]. As for the nuclearity condition, we know that it is stronger than the split property and so it also may hold in case of discrete mass spectrum only. The necessary and sufficient condition for it found in [114, 113] is as follows: $\Sigma_{n=1}^{\infty} \exp(-\beta\, m_i) < \infty$ for all $\beta > 0$. Obviously, it is also the sufficient condition for the split property, causal independence and other causal properties in Section 1.3.4.

Here we end the discussion of free and generalized free fields. There are a number of systems with nontrivial interaction for which the study of algebraic structure in both technique and results is not much different from free models: this is the case, for instance, for models with classical external fields (see, e.g. [67, 332, 342]). But it is more important that any rigorous study in relativistic quantum theory inevitably refers to results known for free systems and profits from methods developed for them. And it can rightfully be said that these methods together with general theorems of axiomatic approach are the two principal components on which algebraic quantum field theory is based.

Problems of Constructing Algebraic Gauge Quantum Field Theory

The extension of algebraic quantum field theory to gauge systems is perhaps its most topical problem nowadays. This problem was touched on briefly in Section 2.3, where it was stressed that the technique of local morphisms discussed there covers gauge models as well. However, if we want to analyze such basic features of gauge theory as gauge arbitrariness, potentials of gauge fields, field equations, etc., we inevitably need Wightman-type formulation or, in the algebraic approach, some formalism of field algebras. And here, we encounter a major obstacle: neither this formalism nor Wightman theory exist in their usual form in the case of a quantum system with gauge fields. The incompatibility between the structure of gauge theory and the system of Wightman axioms shows up in many facts, large and small, which for a long time remained 'part of folklore', until in the mid-seventies the main ones appeared in papers by F. Strocchi as rigorous theorems now somewhat conventionally referred to as Strocchi theorems. The first two of them demonstrate that the incompatibility is already present in the simplest theory of free electromagnetic field with potentials A_μ and field tensor $F_{\mu\nu}$. Let A_μ and $F_{\mu\nu}$ be defined as Wightman fields on test functions in corresponding Schwartz spaces. Then the following theorems hold.

The first Strocchi theorem [397]. If potentials A_μ are Lorentz-covariant and $\partial^\mu F_{\mu\nu} = 0$, then $F_{\mu\nu} \equiv 0$.

The second Strocchi theorem [397]. If the commutator of A_μ is local (even if only in a vacuum state) and $\partial^\mu F_{\mu\nu} = 0$, then $F_{\mu\nu} \equiv 0$.

The meaning of the theorems is obvious: validity of free Maxwell equations as operator identities in a Hilbert state space $\mathcal{H}$ excludes Lorentz-covariance as well as locality of the potential A_μ. The proofs do not use positivity of metric in $\mathcal{H}$ and hence, even in indefinite metric space locality and/or Lorentz-covariance of potentials imply that the field equations can hold only on a certain subspace.

The next theorem deals already with interacting quantum electrodynamics. Let $j_\mu(f^\mu)$ be the local electromagnetic current which is conserved ($\partial^\mu j_\mu = 0$) and determines the electric charge operator Q in the standard way:

$$Q_{R,d} = \int d^4x \; j_0(x^0, x) f_R(x) f_d(x^0),$$

where

$$f_R(x) = \begin{cases} 1, & |x| < R \\ 0, & |x| > R + \varepsilon \end{cases}$$

$$f_d(x^0) = 0, \; |x^0| > d \, ,$$

$$\int_{-\infty}^{\infty} f_d(t)\,dt = 1 \; . \qquad (A1)$$

For any field $\varphi(g)$ local relative to j_μ and any test function $g \in \mathcal{D}(\mathbf{R}^4)$ there exists the limit $\lim_{R\to\infty} [Q_{R,d}, \varphi(g)]$ which does not depend on f_d. Hence, the definition of charged field can be given: one says that a field φ carries the charge q if $\lim_{R\to\infty} [Q_{R,d}, \varphi(g)] = -q\,\varphi(g)$ on some dense domain in $\mathcal{H}$ and for all $g \in \mathcal{D}(\mathbf{R}^4)$, $Q_{R,d}$ generates locally (i.e. on fields in a bounded region dependent on R, d) the group $U(1)$ of gauge transformations of the first kind which acts nontrivially on a system of fields if and only if this system includes charged fields.

The third Strocchi theorem [189]. Let a Wightman theory be given with local conserved current j_μ, electromagnetic field $F_{\mu\nu}$ and an arbitrary set of other local fields. If j_μ generates nontrivial automorphisms of some local (polynomial) field algebra, the Maxwell equations $\partial^\nu F_{\mu\nu} = j_\mu$ do not hold.

Thus in the Wightman formulation of interacting quantum electrodynamics charged fields cannot be local.

Finally, the last theorem deals with Wightman formulation of a general gauge theory. Gauge character of the theory means that a certain local gauge symmetry is present. It can be described by a group of automorphisms of a polynomial field algebra, like symmetries of usual, nongauge fields; but parameters of the group are now real functions Λ_a, $a = 1, \dots, N$ (N is the dimension of the charge space) which can be taken from the class $O_M(\mathbf{R}^4)$ of multipliers of the Schwartz space. Infinitesimal local gauge transformations of fields

$$\varphi^{(i)}(g) \to \varphi^{(i)}(g) + \delta^{(\Lambda)}\varphi^{(i)}(g) \qquad (A2)$$

are generated by local generators $j^{\Lambda}_{\mu}(x)$:

$$\delta^{(\Lambda)}\varphi^{(i)}(g) = \lim_{R\to\infty} [Q^{(\Lambda)}_{R,d}, \varphi^{(i)}(g)]$$

$$= i \lim_{R\to\infty} \int d^4x \,[j^{(\Lambda)}_0(x), \varphi^{(i)}(g)]\, f_R(x) f_d(x^0) \, , \qquad (A3)$$

where $\partial^\mu j^{\Lambda}_{\mu} = 0$ and the functions f_R, f_d are the same as in ($A1$). The group of local gauge transformations includes the subgroup (internal symmetry group) of gauge transformations of the first kind obtained when $\Lambda^a = \text{const}\,\delta_{aa_0}$. The generators j^{μ}_{a} of this subgroup are local conserved currents with the following property (which indicates that they belong to a gauge theory): there is a skew-symmetric tensor $G^{\nu\mu}_{a}$ such that

$$(\Phi\,,(j_a^{\mu} - \partial_\nu G_a^{\nu\mu})\,\Psi) = 0 \qquad (A4)$$

for all vectors Φ, Ψ of physical states.

The fourth Strocchi theorem [400]. Let a Wightman theory be given with an internal symmetry group connected with local gauge transformations according to (A4) and generating nontrivial transformations of the field algebra $\mathcal{P}$. Then the metric in the state space of the theory cannot be positive definite.

Taken as a whole, this set of theorems shows a situation very familiar from the example of electrodynamics: for any gauge theory many different formulations (called gauges) are possible, but none of them can satisfy all the Wightman axioms including positivity of the metric. Different gauges turn out to be preferable in different problems depending on which parts of Wightman theory we are ready to sacrifice in each case. So in quantum electrodynamics one uses Coulomb gauge breaking locality and Lorentz-covariance but preserving the positive metric; Gupta–Bleuler gauge preserving locality and Lorentz-covariance but breaking positivity of the metric; etc. Obviously, aiming at rigorous formulation of a general gauge theory one should choose the formalism in which the fundamental axioms of locality and relativistic covariance are valid. By Theorem 4, in such formalism the state space $\mathcal{H}$ will have indefinite metric so that physical states will belong to a certain subspace of $\mathcal{H}$; by Theorems 1 and 2, field equations will also be satisfied only on a certain subspace.

Described formalism, a hybrid of Wightman theory and Gupta–Bleuler gauge, has originated in the paper [402] and is usually called the *Strocchi–Wightman theory*. It is determined by a system of fields $\varphi^{(i)}(g)$ defined as tempered distributions (in some models different test function spaces may be required, see [137]) and operators in a space with indefinite inner product $\langle\cdot\,,\,\cdot\rangle$. The following axioms are satisfied by the fields.

Relativistic covariance. A representation $U(a,\Lambda)$ of the proper Poincaré group is acting in $\mathcal{D}$; its operators leave invariant the form $\langle\cdot\,,\,\cdot\rangle$ and implement the usual transformation law (1.3.84) of field operators.

Locality. The fields $\varphi^{(i)}(g)$, according to their spins, are commuting or anti-commuting on every vector of $\mathcal{D}$ if the supports of their test functions are spacelike separated.

Uniqueness and cyclicity of vacuum. There is translationally invariant vector Ω in $\mathcal{D}$ which is positive ($\langle\Omega\,,\Omega\rangle > 0$) and such that $\mathcal{P}\Omega = \mathcal{D}$.

Spectrum condition. Fourier transforms $\widetilde{W}(q_1\,,\dots,q_{n-1})$ of vacuum expectation values $W(x_1 - x_2\,,\dots,x_{n-1} - x_n) = \langle\Omega\,,\varphi^{(i_1)}(x_1)\dots\varphi^{(i_n)}(x_n)\,\Omega\rangle$ in any of their variables have the support in $\overline{V^+}$ (translation operators lack spectral decomposition now which makes usual formulation of spectrum condition impossible).

One also assumes the existence of local gauge symmetries satisfying (A2)–(A4) and such that physical state vectors in (A4) span a certain positive subspace $\mathcal{D}' \subset \mathcal{D}$ for which $\Omega \in \mathcal{D}'$, $U(a,\Lambda)\,\mathcal{D}' \subset \mathcal{D}'$, $(j_a^{\mu} - \partial_\nu G_a^{\nu\mu})\,\mathcal{D}' \subset \mathcal{D}'$.

Like the usual Wightman theory, Strocchi-Wightman theory can be formulated as a certain representation of the Borchers algebra $\mathcal{A}$ (Section 3.1). Representations of $\mathcal{A}$ in indefinite inner product spaces can be constructed by means of a generalization of the

GNS construction for non-positive functionals on $\mathcal{A}$ [441, 327, 321]. Arbitrary functional W generates a representation Π which acts in the quotient space $\mathcal{D}$ of $\mathcal{A}$ by the ideal J_0 of all isotropic elements:

$$\mathcal{D} = \mathcal{A}/J_0 , \qquad J_0 = \{u \in \mathcal{A} \mid W(uv) = 0 \ \ \forall\, v \in \mathcal{A}\} .$$

The bilinear form $\langle \xi[u], \xi[v] \rangle \equiv W(v^* u)$ on $\mathcal{D}$ is indefinite, Hermitian and non-degenerate. Operators of the representation are defined in the standard way: $\Pi(u)\,\xi[v] \doteq \xi[uv]\ \forall u,\ v \in \mathcal{A}$.

(It is worth noting that the Borchers formalism can be generalized in yet another direction, thus opening the way to algebraic formulations of some other types of gauge theories and models. Namely, replacing the Schwartz space $S(\mathbf{R}^4)$ with the space of differential forms, we can construct the corresponding 'coordinateless' Borchers algebra and use it for describing theories on manifolds, including gravity, etc. This idea has been developed by Bongaarts [68], who has also noticed that it is useful to introduce the two Borchers algebras $\mathcal{A}^{(1)}$, $\mathcal{A}^{(2)}$ based on the spaces of 1-forms and 2-forms, respectively. Operators representing the first algebra are considered as potentials, those representing the second as field strengths (respectively, connection forms and curvature forms in the geometrical interpretation of gauge theory). The standard mapping of the space of 2-forms into that of 1-forms by means of the codifferentiation operator δ can be 'lifted' to the mapping $\Delta : \mathcal{A}^{(2)} \to \mathcal{A}^{(1)}$. Thus, we obtain the correspondence between potentials and field strengths and may use it to study problems connected with the gauge arbitrariness, i.e. the presence of different potentials corresponding to the same field tensor. In [67, 68] such a generalization of the Borchers formalism was considered in the case of quantum electrodynamics, while in the recent work [141] it was constructed for the Kaluza–Klein theory. In this way, the algebraic apparatus becomes combined with the differential geometry methods traditionally dominant in gauge theories.)

For both physical and technical reasons it is desirable to have on the indefinite metric space $\mathcal{D}$ the Hilbert space structure as well. One necessary and sufficient condition for it was pointed out by Yngvason [441]:

$$| W_{n+m} (\overset{*}{u}_m v_m) | \le p_n(u_n)\, p_m(v_m) \tag{A5}$$

where p_n is a certain sequence of Hilbert seminorms on $S(\mathbf{R}^{4n})$. This bound is not too restrictive and is easily verified in the case of free electromagnetic field. If it holds, there is a Hilbert inner product $(\cdot\,,\,\cdot)$ on $\mathcal{D}$ such that $\langle \cdot\,,\,\cdot \rangle = (\cdot\,, G\,\cdot)$ where G is a bounded Hermitian operator in the completion $\mathcal{H}$ of $\mathcal{D}$ with respect to the Hilbert norm. G is called the metric (or Gram) operator. If a Hilbert space structure on $\mathcal{D}$ exists, it is not unique. In particular, there can always be found [321] a structure for which G^{-1} exists and is a bounded operator; and for this structure there exists an equivalent topology for which G is Hermitian and unitary, i.e. it can be represented as a difference of two projection operators. In the last case it is usually denoted by $J : J = P_+ - P_-$, $P_+ + P_- = I$, and $\mathcal{H}$ is called the J-space or else Krein space.

Thus, the state space of Strocchi–Wightman theory satisfying ($A5$) can be realized as a

Krein space. Each field operator $\Pi(u)$ has then two adjoints, $\Pi(u)^0$ and $\Pi(u)^*$, with respect to $\langle\cdot,\cdot\rangle$ and $(\cdot,\cdot)$, respectively. If the algebra $\mathcal{P}\equiv\Pi(\mathcal{A})$ is invariant under the *-involution $\Pi(u)\to\Pi(u)^*$ it is an *Op**-algebra; but in general it is invariant under the '*J*-involution' $\Pi(u)\to\Pi(u)^0$ only. Such algebras called *OpJ* *-algebras in [127] have been little studied thus far (see [263]). Operators $U(a,\Lambda)$ leave invariant the form $\langle\cdot,\cdot\rangle$ on D. If they can be extended to $\mathcal{H}$ preserving this property, they are called *J*-unitary and are bounded operators; but they may, in general, be unbounded as well. Many attempts have been made to exploit this possibility for the description of confinement in nonabelian theories (see below). On the other hand, assuming $U(a,\Lambda)$ to be *J*-unitary, it is possible to extend to the net $\{\mathcal{P}(O)\}$ of *OpJ**-algebras the main results of the Bisognano–Wichmann formalism (except the results concerning modular structure of local algebras) and construct field algebras $F(O)$ of bounded operators [127]. The net $\{F(O)\}$ satisfies locality and other axioms, but little else can be said about it, since $F(O)$ are not *W** or *C** algebras anymore. They are weakly closed unital *B**-algebras with the *J*-involution called *WJ**-algebras and the study of such algebras is still at the rudimentary stage although the Tomita–Takeseki theory is already developed for them [127].

Development of Strocchi–Wightman theory and related algebraic formalism raises the problem of extending to gauge theory theorems and results of axiomatic approach. In this direction it has been proved in [400] that *CPT*-theorem and Reeh–Schlieder theorem remain valid and Wightman functions preserve the usual analytic properties. The main problems, however, lie not here but in studying new features and effects specific to gauge systems. They are, as a rule, more concrete than general Strocchi–Wightman formalism; in particular, we meet with different problems in abelian and nonabelian gauge theories. One important problem is, however, common to all indefinite metric formalism: it is the construction of physical state space which cannot now coincide with the space $\mathcal{H}$ obtained by the reconstruction theorem. According to the scheme outlined in [402], physical space is constructed as the subspace $\mathcal{H}'\subset\mathcal{H}$ satisfying the following conditions:

(1) $\mathcal{H}'$ is nonnegative (i.e. $\mathcal{H}'$ can have neutral subspaces and true physical state vectors are elements of the quotient $\overline{\mathcal{H}'/\mathcal{H}''}$ where $\mathcal{H}''$ is a maximal neutral subspace).

(2) Field equations of the (*A*4) type are satisfied on $\mathcal{H}'$. Since Wightman-type theory includes charged fields not only in neutral combinations but explicitly, physical states in $\mathcal{H}'$ should belong not only to a vacuum sector but to all charged sectors as well. Hence,

(3) $\mathcal{H}'$ includes states with nontrivial charge values.

Finally, operators of the Poincaré group and physical observables should leave physical states physical which implies

(4) $U(a,\Lambda)\,\mathcal{H}'\subset\mathcal{H}'$;

(5) $\mathfrak{A}\,\mathcal{H}'\subset\mathcal{H}'$, where $\mathfrak{A}$ is an algebra of observables (defined by the algebra $\mathcal{P}$ in some way).

Since field equations must be taken into account the construction of $\mathcal{H}'$ satisfying (1)–(5) (and even the proof of its existence) is also not an axiomatic problem but one depending on a concrete theory. In recent years much attention has been paid to this problem in quantum electrodynamics. The central fact which shows both the main difficulty in the problem and one way to overcome it is the theorem about the absence of local charged states [190]: no state vector of the form $F_q \Omega$, where F_q is a local field with the charge q, can lie in $\mathcal{H}'$, and hence, all physical states with nontrivial charge values are outside the Gårding domain $\mathcal{D} = \mathcal{P}\Omega$. Thus, physical charged states can only appear when a closure of $\mathcal{D}$ is taken. And since the closure operation is determined by the metric operator J which is highly nonunique, the problem is reduced to the following one: to find such a Hilbert topology on $\mathcal{D}$ the closure with respect to which gives vectors with nontrivial charge. Solutions of this problem based on technical assumptions have been obtained in [328].

One of the most topical problems to which Strocchi–Wightman formalism was applied is the rigorous description of charge confinement in nonabelian theories, above all, in quantum chromodynamics. This well known physical effect means the absence of experimentally observed (therefore, asymptotic) states with nonzero values of charges connected with nonabelian internal symmetry group. Clearly, for such an effect to take place, the usual Haag–Ruelle construction of asymptotic states [4, Chapter 4] must be invalid and hence, the confinement cannot be described in the standard Wightman theory. On the other hand, in Strocchi–Wightman theory, if the translation operators $T(a)$ do not commute with J, the assumptions of the Haag–Ruelle theory are broken and so its results may be broken as well. Some classes of unbounded representations of the translation group which make impossible the construction of asymptotic states for charged fields have been pointed out in [400, 327]. More careful analysis of group representations in Krein spaces [424] has lead in [423] to the choice of a different class of representations: it was claimed that the description of confinement is only possible with 'exotic' representations of the translation group which, in addition to the unboundedness, define a modified nonscalar transformation law of field operators. It should also be noted that there is no generally accepted choice for the rigorous criterion of confinement and different conditions are used, in terms of asymptotic states, coloured charge operators, etc.; moreover, these conditions and their verification often depend on the choice of physical subspace which is a problem in itself, as we have seen. So the confinement problem in Strocchi–Wightman theory, as in other existing approaches, cannot be considered as solved so far. Perhaps more detailed information is needed to obtain a conclusive solution. One promising idea in this direction is to use the BRS symmetry which appears if we complement the QCD formalism with the Faddeev–Popov ghost fields. This approach suggested in [292] and actively pursued within Lagrangian QCD had the two main advantages:

(1) it produces a well justified and very convenient choice of the physical subspace: $\mathcal{H}'$ is now the kernel of the BRS charge operator Q which generates the BRS symmetry and has special properties $Q = Q^0$, $Q^2 = 0$ simplifying the geometry of $\mathcal{H}$;

(2) it can be transferred to the string theory which also possesses ghost fields and BRS symmetry. Thus, it provides a unified mathematical formalism for both string theory and QCD. As numerous recent works demonstrate (see, e.g. [239]), this formalism, called the BRS quantization in the string theory, is now giving rise to considerable expectations.

References

BOOKS

[1] Akhiezer, N.I., *Vorlesungen über Approximationstheorie*, Akademie-Verlag, Berlin, 1967.

[2] Akhiezer, N.I., and Glasman, I.M., *Theory of Linear Operators in Hilbert Space*, Ungar, New York, 1961.

[3] Berezin, F.A., *The Method of Second Quantization*, Academic Press, New York, 1966.

[4] Bogoliubov, N.N., Logunov, A.A., and Todorov, I.T., *Introduction to Axiomatic Quantum Field Theory*, Benjamin, Reading, MA, 1975.

[5] Bratteli, O., and Robinson, D.W., *Operator Algebras and Quantum Statistical Mechanics*, Vol. 1, Springer–Verlag, New York, Heidelberg, Berlin, 1979.

[6] Courant, R., and Hilbert, D., *Methods of Mathematical Physics*, Vol. 2, Interscience, New York, London, 1962.

[7] Dixmier, J., *Les Algèbres d'Opérateurs dans l'Espace Hilbertien (Algèbres de von Neumann)*, 2nd edn., Gauthier–Villars, Paris, 1969.

[8] Dixmier, J., *Les C*-Algèbres et leurs Représentations*, 2nd edn., Gauthier–Villars, Paris, 1969.

[9] Dunford, N., and Schwartz, J.T., *Linear Operators. General Theory*, Interscience, New York, London, 1958.

[10] Emch, G.G., *Algebraic Methods in Statistical Mechanics and Quantum Field Theory*, Wiley–Interscience, New York, London, Sydney, Toronto, 1972.

[11] Gelfand, I.M., and Vilenkin, N.Ya., *Generalized Functions*, Vol. 4, Academic Press, New York, 1964.

[12] Grib, A.A., *The Problem of Noninvariant Vacuum in Quantum Field Theory*, Atomizdat, Moscow, 1978 (In Russian).

[13] Hille, E., and Fillips, R.S., *Functional Analysis and Semigroups*, AMS Colloq., Publ., Vol. 31, Providence R.I., 1957.

[14] Jauch, J.M., *Foundations of Quantum Mechanics*, Addison–Wesley, Reading, MA, 1968.

[15] Jost, R., *The General Theory of Quantized Fields*, Amer. Math. Soc., Providence, R.I., 1965.

[16] Kirillov, A.A., *Elements of The Theory of Representations*, Springer–Verlag, New York, Heidelberg, Berlin, 1976.

[17] Mackey, G.W., *The Mathematical Foundations of Quantum Mechanics*, Benjamin, New York, 1963.

[18] Naimark, M.A., *Normed Algebras*, Wolters–Noordhoff, Groningen, 1972.

[19] Pedersen, G.K., *C*-Algebras and their Automorphism Groups*, Academic Press, New York, San Fransisco, London, 1979.

[20] Reed, M., and Simon, B., *Methods of Modern Mathematical Physics*, Vol. 1, Academic Press, New York, London, 1972; Vol. 2, Ib., 1975.

[21] Riesz, F., and Sz.–Nagy, B., *Functional Analysis* , Ungar, New York,, 1960.

[22] Ruelle, *Statistical Mechanics. Rigorous Results*, Benjamin, New York, 1969.

[23] Sakai, S., *C*-Algebras and W*-algebras*, Springer–Verlag, New York, Heidelberg, Berlin, 1971.

[24] Sarymsakov, T.A., Ajupov S.A., Khadzhiev, J. and Chilin, V.I., *Ordered Algebras*, Fan, Tashkent, 1983 (In Russian).

[25] Segal, I.E., *Mathematical Problems of Relativistic Physics*, Amer. Math. Soc., Providence, R.I., 1963.

[26] Stein, E.M., *Singular Integrals and Differentiability Properties of Functions*, Princeton University Press, Princeton, N.J., 1970.

[27] Streater, R.F., and Wightman, A.S., *PCT, Spin and Statistics and all that*, Benjamin, New York, 1964.

[28] Takesaki, M., *Tomita's Theory of Modular Hilbert Algebras and its Applications. Lect. Notes In Math.*, Vol. 128, Springer–Verlag, New York, Heidelberg, Berlin, 1970.

[29] Takesaki, M., *Theory of Operator Algebras*, Vol. 1, Springer–Verlag, New York, Heidelberg, Berlin, 1970.

[30] Vladimirov, V.S., *Methods of Theory of Functions of Many Complex Variables*, MIT Press, Cambridge, MA, 1966.

[31] Yosida, K., *Functional Analysis*, 4th edn. Springer–Verlag, New York, Heidelberg, Berlin, 1974.

ARTICLES

[32] Amann, A., Broken symmetries and the generalization of classical observables in large systems, *Helv. Phys. Acta.*, 60, 384–393, (1987).

[33] Antoine, J.–P., Epifanio, G., and Trapani, C. Complete sets of unbounded observables, *Helv. Phys. Acta.*, **56**, 1175–1186, (1983).

[34] Antoine, J.–P., and Karwowski, W., Partial *-algebras of closed linear operators in Hilbert space, *Publ. RIMS Kyoto Univ.*, **21**, 205–236, (1985).

[35] Antoine, J.–P., and Mathot, F., Partial *-algebras of closed operators and their commutants. I. General structure, *Ann. Inst. H. Poincaré*, **46**, 299–324, (1987).

[36] Antoine, J.–P., and Mathot, F., Partial *-algebras of closed operators and their commutants. II. Commutants and bicommutants, *Ann. Inst. H. Poincaré*, **46**, 325–351, (1987).

[37] Araki, H., Einführung in die axiomatische Quantenfeldtheorie, *Lecture notes, ETH*, Zürich, (1961–1962).

[38] Araki, H., A generalization of Borchers theorem, *Helv. Phys. Acta.*, **36**, 132–139 (1963).

[39] Araki, H., A lattice of von Neumann algebras associated with the quantum theory of a free Bose field, *J. Math. Phys.*, **4**, 1343–1362, (1963).

[40] Araki, H., Von Neumann algebras of local observables for free scalar field, *Math. Phys.*, **5**, 1–13, (1964).

[41] Araki, H., On the algebra of all local observables, *Progr. Theor. Phys.*, **32**, 844–854 (1964); *Publ. RIMS Kyoto Univ.*, **5**, 1–16, (1964).

[42] Araki, H., Type of von Neumann algebras associated to the free field, *Progr. Theor. Phys.*, **32**, 956–961, (1964).

[43] Araki, H., On quasifree states of the canonical commutation relations. II., *Publ. RIMS Kyoto Univ.*, **7**, 121–152, (1972).

[44] Araki, H., Recent developments in the theory of operator algebras and their significance in theoretical physics, *Symp. Math.*, **20**, 395–424, (1976).

[45] Araki, H., On a characterization of the state space of quantum mechanics, *Comm. Math. Phys.*, **75**, 1–24, (1980).

[46] Araki, H., A remark on Machida–Namiki theory of measurement. *Progr. Theor. Phys.*, **64**, 719–730, (1980).

[47] Araki, H., and Haag, R., Collision cross-sections in terms of local observables, *Comm. Math. Phys.*, **4**, 77–91, (1967).

[48] Araki, H., Hepp, K., and Ruelle, D., On the asymptotic behaviour of Wightman functions in spacelike directions, *Helv. Phys. Acta*, **35**, 164, (1962).

[49] Araki, H., and Jurzak, J.–P., On a certain class of *-algebras of unbounded operators, *Publ. RIMS Kyoto Univ.*, **18**, 1013–1044, (1982).

[50] Araki, H., and Kishimoto, A., On clustering property, *Rep. Math. Phys.*, **10**, 275–281, (1976).

[51] Araki, H., and Shiraishi, M., On quasifree states of the canonical commutation relations. I. *Publ. RIMS Kyoto Univ.*, **7**, 105–120, (1972).

[52] Araki, H., and Woods, E.J. Representations of the canonical commutation relations describing a nonrelativistic infinite free Bose gas, *J. Math. Phys.*, **4**, 937–662, (1963).

[53] Ascoli, R., Epifanio, G., and Restivo, A., On the mathematical description of quantized fields, *Comm. Math. Phys.*, **18**, 291–300, (1970).

[54] Ascoli, R., Teppati, G., and Termini, S., Some remarks about particle-antiparticle superselection rules, *Lett. Nuovo Cim.*, **1** (4), 223–227, (1969).

[55] Ashtekar, A., and Sen, A., On the role of spacetime topology in quantum phenomena: superselection of charge and emergence of nontrivial vacua, *J. Math. Phys.*, **21**, 526–533, (1980).

[56] Avishai, Y., and Ekstein, H., Causal independence, *Found. Phys.*, **2**, 257–270, (1972).

[57] Avishai, Y., and Ekstein, H., H. Einstein's equivalence principle. An explicit statement and its derivation from special relativistic presymmetry, *Comm. Math. Phys.*, **37**, 193–220, (1974).

[58] Banai, M., Propositional systems in local field theories, *Int. J. Theor. Phys.*, **20**, 147–169, (1981).

[59] Bargmann, V., On unitary ray representations of continuous groups, *Ann. Math.*, **59**, 1–46, (1954).

[60] Baumgärtel, H., and Wollenberg, M., A class of nontrivial weakly local massive Wightman fields with interpolating properties, *Comm. Math. Phys.*, **94**, 331–352, (1984).

[61] Bell, J.S., Hughes, R.J., and Leinaas, J.M., The Unruh effect in extended thermometers, *Z. Phys. C.*, **28**, 75–80, (1985).

[62] Benfatto, G., and Nicolo, F., The local von Neumann algebras for the massless scalar free field and the free electromagnetic field, *J. Math. Phys.*, **19**, 653–660, (1978).

[63] Berezansky, Yu. M., On self-adjointness of field operators and integral representations of Wightman-type functionals, *Ukrain. Math. J.*, **18**, 3–12, (1966). (In Russian).

[64] Berzi, V., Arveson spectral theory and C^*algebraic Goldstone-type theorems, *Rep. Math. Phys.*, **16**, 293–302, (1972).

[65] Bisognano, J.J., and Wichmann, E.H., On the duality condition for a Hermitian scalar field, *J. Math. Phys.*, **16**, 985–1007, (1975).

[66] Bisognano, J.J., and Wichmann, E.H., On the duality condition for quantum fields, *J. Math. Phys.*, **17**, 303–321, (1976).

[67] Bongaarts, P.J.M., The electron-position field coupled to external electromagnetic potentials as an elementary C^*algebra theory, *Ann. Phys.*, **56**, 108–139, (1970).

[68] Bongaarts, P.J.M., Maxwell's equations in axiomatic quantum field theory. I. Field tensor and potentials, *J. Math. Phys.*, **18**, 1510–1516, (1977); II. Covariant and noncovariant gauges, *Ibid.*, **23**, 1818–1898, (1982).

[69] Bongaarts, P.J.M., Mathematical aspects of field quantization. Quantum electrodynamics, *Acta Phys. Pol.*, **B14**, 347–359, (1983).

[70] Bongaarts, P.J.M., Ruijsenaars, S.N.M., The Klein paradox as a many particle problem, *Ann. Phys.*, **101**, 289–318, (1976).

[71] Bonnard, J.–L., and Streater, R.F., Local gauge models predicting their own superselection rules, *Proc. V.A. Steklov Math. Inst.*, **135**, 81–86 (1975); *Helv. Phys. Acta*, **49**, 259–267, (1975).

[72] Borchers, H.J., Über die Vollständigkeit Lorentz-invarianten Felder in einer zeitartigen Röhre, *Nuovo Cim.*, **19**, 787–793, (1961).

[73] Borchers, H.J., On structure of the algebra of field operators, *Nuovo Cim.*, **24**, 214–236, (1962).

[74] Borchers, H.J., Field operators as C^∞ functions in spacelike directions, *Nuovo Cim.*, **33**, 1600–1613, (1964).

[75] Borchers, H.J., On the structure of the algebra of field operators, *Comm. Math. Phys.*, **1**, 49–56, (1965).

[76] Borchers, H.J., On the vacuum state in quantum field theory. II., *Comm. Math. Phys.*, **1**, 57–79, (1965).

[77] Borchers, H.J., Local rings and the connection of spin with statistics, *Comm. Math. Phys.*, **1**, 281–307, (1965).

[78] Borchers, H.J., Energy and momentum as observables in quantum field theory, *Comm. Math. Phys.*, **2**, 49–54, (1966).

[79] Borchers, H.J., A remark on a theorem of B. Misra, *Comm. Math. Phys.*, **4**, 315–323, (1967).

[80] Borchers, H.J., On the converse of the Reeh–Schlieder theorem, *Comm. Math. Phys.*, **10**, 269–273, (1968).

[81] Borchers, H.J., On groups of automorphisms with semibounded spectrum. In: Systèmes à un nombre infini des degrés de liberté. Colloque CNRS, mai 1969, **181**, 125–142.

[82] Borchers, H.J., On the implementability of automorphism groups, *Comm. Math. Phys.*, **14**, 305–314, (1969).

[83] Borchers, H.J., Über C^*-Algebren mit lokalkompakten Symmetriegruppen, *Nachr. Gött. Akad.*, **1**, 1–18, (1973).

[84] Borchers, H.J., Über Ableitungen von C^*-Algebren, *Nachr. Gött. Akad.*, **2**, 19, (1973).

[85] Borchers, H.J., C^*-algebras and automorphism groups, *Comm. Math. Phys.*, **88**, 95–103, (1983).

[86] Borchers, H.J., Algebras of unbounded operators in quantum field theory, *Physics*, **124A**, 127–144, (1984).

[87] Borchers, H.J., Translation groups and spectrum conditions, *Comm. Math. Phys.*, **96**, 1–13, (1984).

[88] Borchers, H.J., Locality and covariance of the spectrum, *Fizika*, **17**, 289–304, (1985).

[89] Borchers, H.J., Buchholz, D., The energy-momentum spectrum in local field theories with broken Lorentz symmetry, *Comm. Math. Phys.*, **97**, 169–186, (1985).

[90] Borchers, H.J., Haag, R., and Schroer, B., The vacuum state in quantum field theory, *Nuovo Cim.*, **29**, 148–162, (1963).

[91] Borchers, H.J., and Yngvason, J., On the algebra of field operators. The weak commutant and integral decomposition of states, *Comm. Math. Phys.*, **42**, 231–252, (1975).

[92] Borchers, H.J., and Yngvason, J., Integral representations for Schwinger functionals and the moment problem over nuclear spaces, *Comm. Math. Phys.*, **43**, 255–271, (1975).

[93] Borchers, H.J., and Yngvason, J., Necessary and sufficient conditions for integral representations of Wightman functionals at Schwinger points, *Comm. Math. Phys.*, **47**, 197–213, (1976).

[94] Borchers, H.J., and Zimmermann, W., On the self-adjointness of field operators, *Nuovo Cim.*, **31**, 1047–1050, (1964).

[95] Borek, R., Vacuum states and superselection rules in quantum field theory, *J. Math. Phys.*, **26**, 168–171, (1985).

[96] Borek, R. Representations of the current algebra of a charged massless Dirac field, *J. Math. Phys.*, **26**, 339–344, (1985).

[97] Borisov, N.V., Topological properties of the commutant of a field algebra representation, *Theor. Math. Phys.*, **11**, 313–315, (1972).

[98] Bratteli, O., Local norm convergence of the states on the zero time Bose fields, *Trans. Amer. Math. Soc.*, **188**, 269–289, (1974).

[99] Buchholz, D., Product states for local algebras, *Comm. Math. Phys.*, **36**, 287–304, (1974).

[100] Buchholz, D., Collision theory for massless fermions, *Comm. Math. Phys.*, **42**, 269–279, (1975).

[101] Buchholz, D., Collision theory for waves in two dimensions and a characterization of models with trivial S-matrix, *Comm. Math. Phys.*, **45**, 1–8, (1975).

[102] Buchholz, D., Collision theory for massless bosons, *Comm. Math. Phys.*, **52**, 147–173, (1977).

[103] Buchholz, D., On the structure of local quantum fields with nontrivial interaction, *Proc. Int. Conf. On Operator Algebras, Ideals and their Application in Theoretical Physics*, Leipzig: Teubner–Verlag, 146–153, (1978).

[104] Buchholz, D., The physical state space of quantum electrodynamics, *Comm. Math. Phys.*, **85**, 49–71, (1982).

[105] Buchholz, D., D'Antonio, C., and Fredenhagen, K., The universal structure of local algebras, *Comm. Math. Phys.*, **111**, 123–135, (1987).

[106] Buchholz, D., and Doplicher, S., Exotic infrared representations of interacting systems, *Ann. Inst. H. Poincaré*, **40**, 175–184, (1984).

[107] Buchholz, D., Doplicher, S., and Longo, R., On Noether's theorems in quantum field theory, *Ann. Phys.*, **170**, 1–17, (1986).

[108] Buchholz, D., and Epstein, H., Spin and statistics of quantum topological charges, *Fizika*, **17**, 329–343, (1985).

[109] Buchholz, D., and Fredenhagen, K., A note on the inverse scattering problem in quantum field theory, *Comm. Math. Phys.*, **56**, 91–99, (1977).

[110] Buchholz, D., and Fredenhagen, K., Charge screening and mass spectrum in abelian gauge theories, *Nucl. Phys.*, **154B**, 226–238, (1979).

[111] Buchholz, D., and Fredenhagen, K., Locality and structure of particle states, *Comm. Math. Phys.*, **84**, 1–54, (1982).

[112] Buchholz, D., and Jacobi, P., On the nuclearity condition for massless fields, *Lett. Math. Phys.*, **13**, 313–323, (1987).

[113] Buchholz, D., and Junglas, P., Local properties of equilibrium states and the particle spectrum in quantum field theory, *Lett. Math. Phys.*, **11**, 51–58, (1986).

[114] Buchholz, D., and Wichmann, E.H., Causal independence and the energy-level density of states in local quantum field theory, *Comm. Math. Phys.*, **106**, 321–344, (1986).

[115] Bures, D.J.C., Certain factors constructed as infinite tensor products, *Comp. Math.*, **15**, 169–191, (1983).

[116] Carey, A.L., and Hurst, C.A., A C^*-algebra approach to the Schwinger model, *Comm. Math. Phys.*, **80**, 1–21, (1981).

[117] Castell, L, A new superselection rule? *Lett. Nuovo Cim.*, **12**, 410–412, (1975).

[118] Chen, E., Facial aspect of superposition principle in algebraic quantum theory, *J. Math. Phys.*, **14**, 1462–1465, (1973).

[119] Chen, E., On the duality of local observables, *Ann. Inst. H. Poincaré*, **22**, 123–130, (1975).

[120] Cirelli, R., and Gallone, F., Algebra of observables and quantum logic, *Ann. Inst. H. Poincaré*, **22**, 297, (1973).

[121] Coleman, S., Quantum sine-Gordon equation as the massive Thirring model, *Phys. Rev.*, **11**, 2088–2097, (1975).

[122] Connes, A., Une classification des facteurs de type III, *Ann. Ec. Norm. Sup.*, **6**, 133–252, (1973).

[123] Connes, A., Classification des facteurs, *Proc. Symp. Pure Math.*, **38**, 43–110, (1982).

[124] Dadashyan, K.Yu., Causality properties in nonrelativistic algebraic models, *Theor. Math. Phys.*, **31**, 492–495, (1977).

[125] Dadashyan, K.Yu., and Horuzhy, S.S., Algebras of observables of the free Dirac field, *Theor. Math. Phys.*, **36**, 665–675, (1978).

[126] Dadashyan, K.Yu., and and Horuzhy, S.S., Conditional expectations on von Neumann algebras and the causal independence of quantum fields, *Theor. Math. Phys.*, **46**, 97–103, (1981).

[127] Dadashyan, K.Yu., and and Horuzhy, S.S., On field algebras in quantum theory with indefinite metric, *Theor. Math. Phys.*, **54**, 35–47, (1983); Formulation of modular theory in Pontryagin space, *Ibid.*, **62**, 21–30, (1985); Spectrum of the modular operator and Tomita's theorem, *Ibid.*, **70**, 125–133, (1987); *Ibid.*, **72**, 921–929, (1987).

[128] van Daele, A., Quasi-equivalence of quasifree states on the Weyl algebra, *Comm. Math. Phys.*, **21**, 171–191, (1971).

[129] D'Antonio, C., Doplicher, S., Fredenhagen, K., and Longo, R., Convergence of local charges and continuity properties of W^* inclusions, *Comm. Math. Phys.*, **110**, 325–348, (1987).

[130] D'Antonio, C., and Fredenhagen, K., Charges in spacelike cones, *Comm. Math. Phys.*, **94**, 537–544, (1984).

[131] D'Antonio, C., and Longo, R., Interpolation by type I factors and the split automorphism, *J. Funct. An.*, **51**, 361–371, (1983).

[132] De Facio, B., Causal independence in algebraic quantum field theory, *Found. Phys.*, **5**, 229–237, (1975).

[133] De Facio, B., and Taylor, D.C., Commutativity and causal independence, *Phys. Rev.*, **D8**, 2729–2731, (1973).

[134] Dell'Antonio, G.F., On some groups of automorphisms of physical observables, *Comm. Math. Phys.*, **2**, 384–397, (1966).

[135] Dell'Antonio, G.F., Structure of algebras of some free systems, *Comm. Math. Phys.*, **9**, 81–117, (1968).

[136] Dell'Antonio, G.F., Structure of algebras associated to free fields, In: *Axiomatic Approach to Quantum Field Theory and Many Body Problem*, Vol. IA, 52–97, Wroclaw, 1968.

[137] D'Emilio, E., and Minchev, M., Physical charged sectors in quantum electrodynamics. II. The charge operator, *Preprint* ISAS, Pisa, (1982).

[138] Derezinski, J., Boson free fields as a limit of fields of a more general type, *Rep. Math. Phys.*, **21**, 405–417, (1985).

[139] Dimock, J., Algebras of local observables on a manifold, *Comm. Math. Phys.*, **77**, 219–228, (1980).

[140] Dixmier, J., and Marechal, O., Vecteurs totalisateurs d'une algèbre de von Neumann, *Comm. Math. Phys.*, **22**, 44–50, (1972).

[141] Doebner, H.D., and Pasemann, F.B., Algebraic gauge QFT on generalized Kaluza-Klein spaces, In: *Conference on Differential Geometric Methods in Theoretical Physics* (Trieste 1981), World Sci. Publ. 1983, 154–162.

[142] Doplicher, S., An algebraic spectrum condition, *Comm. Math. Phys.*, **1**, 1–5, (1965).

[143] Doplicher, S., Fock representation and massless particles, *Comm. Math. Phys.*, **33**, 228–232, (1966).

[144] Doplicher, S., Fields, observables and gauge transformations, In: Systèmes à un nombre infini des degrés de liberté, Coll. CNRS, 1970, **181**, p. 113–123.

[145] Doplicher, S., Locality and particle statistics, *Proc. V.A. Steklov Math. Inst.*, **135**, 107–114, (1975).

[146] Doplicher, S., Quantum field theory and operator algebras of local observables, *Symposia Math.*, **20**, 109–115, (1976).

[147] Doplicher, S., Local aspects of superselection rules, *Comm. Math. Phys.*, **85**, 73–86, (1982).

[148] Doplicher, S., Figliolini, F., and Guido, D., Infrared representations of free Bose fields, *Ann. Inst. H. Poincaré*, **41**, 49–62, (1984).

[149] Doplicher, S., Haag, R., and Roberts, J.E., Fields, observables and gauge transformations, *Comm. Math. Phys.*, **13**, 1–23, (1969); *Ibid.*, **15**, 173–200, (1971).

[150] Doplicher, S., Haag, R., and Roberts, J.E., Local observables and particle statistics, *Comm. Math. Phys.*, **23**, 199–230, (1971), *Ibid.*, **35**, 49–85, (1974).

[151] Doplicher, S., Kadison, R.V., Kastler, D., and Robinson, D.W., Asymptotically abelian systems, *Comm. Math. Phys.*, **6**, 101–120, (1967).

[152] Doplicher, S., and Kastler, D., Ergodic states in a noncommutative ergodic theory, *Comm. Math. Phys.*, **7**, 1–20, (1968).

[153] Doplicher, S., Kastler, D., and Robinson, D.W., Covariance algebras in field theories and statistical mechanics, *Comm. Math. Phys.*, **3**, 1–28, (1966).

[154] Doplicher, S., Kastler, D., and Størmer, E., Invariant states and asymptotic abelianness, *J. Funct. An.*, **3**, 419–434, (1969).

[155] Doplicher, S., and Longo, R., Local aspects of superselection rules, *Comm. Math. Phys.*, **88**, 399–409, (1983).

[156] Doplicher, S., and Longo, R., Standard and split inclusions of von Neumann algebras, *Invent. Math.*, **75**, 493–536, (1984).

[157] Doplicher, S., Regge, T., and Singer, J.M., A geometrical model showing the independence of locality and positivity of the energy, *Comm. Math. Phys.*, **7**, 51–54, (1968).

[158] Doplicher, S., and Roberts, J.E., Fields, statistics and nonabelian gauge groups, *Comm. Math. Phys.*, **28**, 351–348, (1972).

[159] Doplicher, S., and Roberts, J.E., Compact Lie groups associated with endomorphisms of C^*-algebras, *Bull. Amer. Math. Soc.*, **11**, 333–338, (1984).

[160] Doplicher, S., and Roberts, J.E., Duals of compact Lie groups realized in the Cuntz algebras and their action on C^*-algebras, *J. Funct. An.*, **74**, 96–120, (1987).

[161] Doplicher, S., and Roberts, J.E., Compact group actions on C^*-algebras, Preprint Univ. di Roma, 'La Sapienza', (1987).

[162] Doplicher, S., and Spera, M., Representations obeying the spectrum condition, *Comm. Math. Phys.*, **84**, 505–513, (1984).

[163] Doplicher, S., and Spera, M., Local normality properties of some infrared representations, *Comm. Math. Phys.*, **89**, 19–25, (1983).

[164] Driessler, W., Comments on lightlike translations and applications in relativistic quantum field theory, *Comm. Math. Phys.*, **44**, 133–141, (1975).

[165] Driessler, W., On the type of local algebras in QFT, *Comm. Math. Phys.*, **53**, 295–297, (1977).

[166] Driessler, W., On the structure of fields and algebras on null planes, Local algebras, *Acta Phys. Austr.*, **46**, 63–96, (1977); Field structure, *Ibid.*, **46**, 163–196, (1977).

[167] Driessler, W., Duality and absence of locally generated superselection sectors for CCR-type algebras, *Comm. Phys. Math.*, **70**, 213–220, (1974).

[168] Driessler, W., and Fröhlich, J., Reconstruction of local observable algebras from the Euclidean Green's function of relativistic quantum field theory, *Ann. Inst. H. Poincaré*, **XXVII**, 221–236, (1977).

[169] Driessler, W., and Summers, S.J., Nonexistence of quantum fields associated with two-dimensional spacelike planes, *Comm. Math. Phys.*, **89**, 221–226, (1983).

[170] Driessler, W., and Summers, S.J., A dense set of cyclic vectors for quantum field polynomial algebras, *J. Math. Phys.*, **24**, 2809–2819, (1983).

[171] Driessler, W., and Summers, S.J., Central decomposition of Poincaré-invariant nets of local field algebras and absence of spontaneous breaking of the Lorentz group, *Ann. Inst. H. Poincaré*, **43**, 147–166, (1985).

[172] Driessler, W., and Summers, S.J., On the decomposition of relativistic quantum field theories into pure phases, *Helv. Phys. Acta*, **59**, 331–348, (1986).

[173] Driessler, W., Summers, S.J., and Wichmann, E.H., On the connection between quantum fields and von Neumann algebras of local operators, *Comm. Math. Phys.*, **105**, 49–84, (1986).

[174] Drühl, K., Haag, R., and Roberts, J.E., On parastatistics, *Comm. Math. Phys.*, **18**, 204–226, (1970).

[175] Dubois–Violette, M., A generalization of the classical moment problem of *-algebras with application to relativistic quantum theory, *Comm. Math. Phys.*, **43**, 225–254, (1975).

[176] Eckmann, J.P., and Fröhlich, J. Unitary equivalence of local algebras in the quasifree representation, *Ann. Inst. H. Poincaré*, **XX**, 201–209, (1974).

[177] Eckmann, J.P., and Osterwalder, K., An application of Tomita's theory of modular Hilbert algebras: duality for free Bose fields, *Funct. An.*, **13**, 1–12, (1973).

[178] Ekhaquere, G.O.S., The theory of superselection rules. I. A class of inequivalent irreducible *-representations of the CCR of the free electromagnetic field, *J. Math. Phys.*, **19**, 1751–1757, (1978).

[179] Ekhaquere, G.O.S., The theory of superselection rules. II. Sectors of the free electromagnetic field, *J. Math. Phys.*, **25**, 678–683, (1984).

[180] Ekstein, H., Relativistic particles in external fields, *Phys. Rev. D.*, **5**, 864–871, (1972).

[181] Englisch, H. Superselection rules, superposition and vector states, K.–Marx Universität (Leipzig), Sektion Mathematik, Preprint, (1976).

[182] Enss, V., Characterization of particles by means of local observables, *Comm Math. Phys.*, **45**, 35, (1975).

[183] Epifanio, G., and Trapani, C., V^*-algebras: a particular class of unbounded operator algebras, *J. Math. Phys.*, **25**, 2633–2637, (1984).

[184] Epstein, H., CPT-invariance of the S-matrix in a theory of local observables, *J. Math. Phys.*, **8**, 750–767, (1967).

[185] Epstein, H., Glaser, V., and Martin, A., Polynomial behaviour of scattering amplitudes at fixed momentum transfer in theories with local observables, *Comm Math. Phys.*, **13**, 257–316, (1969).

[186] Ezawa, Z.F., Quantum soliton operators for vortices and superselection rules, *Phys. Rev.*, **D18**, 2091–2099, (1978).

[187] Ezawa, Z.F., Quantum soliton operators for monopoles and fermionization mechanics of dyons, *Phys. Lett.*, **B81**, 325–328, (1979).

[188] Federbush, P., Local operator algebras in the presence of superselection rules, *J. Math. Phys.*, **9**, 1718–1721, (1968).

[189] Ferrari, R., Picasso, L.E., and Strocchi, F., Some remarks on local operators in quantum electrodynamics, *Comm Math. Phys.*, **35**, 25–38, (1974).

[190] Ferrari, R., Picasso, L.E., and Strocchi, F., Local operators and charged states in QED, *Nuovo Cim.*, **39A**, 1–8, (1977).

[191] Fidaleo, F., On the local implementations of gauge symmetries in local quantum theory, *Comm Math. Phys.*, **107**, 233–240, (1986).

[192] Fredenhagen, K., On the existence of antiparticles, *Comm Math. Phys.*, **79**, 141–151, (1981).

[193] Fredenhagen, K., On the modular structure of local algebras of observables, *Comm. Math. Phys.*, **97**, 79–90, (1985).

[194] Fredenhagen, K., A remark on the cluster theorem, *Comm. Math. Phys.*, **97**, 461–464, (1985).

[195] Fredenhagen, K., The structure of local algebras of observables, *Contemp. Math.*, **62**, 153–165, (1987).

[196] Fredenhagen, K., and Haag, R., Generally covariant quantum field theory and scaling limits, *Comm. Math. Phys.*, **108**, 91–115, (1987).

[197] Fredenhagen, K., and Hertel, J., Local algebras of observables and point-like localized fields, *Comm. Math. Phys.*, **80**, 555–561, (1981).

[198] Fredenhagen, K., and Marcu, M., Charged states in Z_2 gauge theories, *Comm. Math. Phys.*, **92**, 81–119, (1983).

[199] Fröhlich, J., On the infrared problem in a model of scalar electrons and massless scalar bosons, *Ann. Inst. H. Poincaré*, **A19**, 1–103, (1973).

[200] Fröhlich, J., New superselection sectors ('soliton states') in two-dimensional Bose quantum fields models, *Comm. Math. Phys.*, **47**, 269–310, (1976).

[201] Fröhlich, J., Phase transitions, Goldstone bosons and topological superselection rules, *Acta Phys. Austr. Suppl.*, **20**, 133–269, (1976).

[202] Fröhlich, J., The pure phases, the irreducible quantum fields, and dynamical symmetry breaking in Symanzik–Nelson positive quantum field theories, *Ann. Phys.*, **97**, 1–54, (1976).

[203] Fröhlich, J., Quantum theory of nonlinear invariant wave (field) equations or superselection sectors in constructive quantum field theory. In: *Invariant Wave Equations*, Lect. Notes in Phys., Vol. 73, 339–413, Springer–Verlag, New York, Heidelberg, Berlin, 1978.

[204] Fröhlich, J., The charged sectors of quantum electrodynamics in a framework of local observables, *Comm. Math. Phys.*, **66**, 223–265, (1979).

[205] Fröhlich, J., Morchio, G., and Strocchi, F., Infrared problem and spontaneous breaking of the Lorentz group in QED, *Phys. Lett.*, **B89**, 61–64, (1979).

[206] Fröhlich, J., Morchio, G., and Strocchi, F., Charged sectors and scattering states in quantum electrodynamics, *Ann. Phys.*, **119**, 241–284, (1979).

[207] Gachok, V.P., Self-adjointness of field operators and the problem of moments, *Ukrain. Math. J.*, **17**, 3–13, (In Russian), (1965).

[208] Gachok, V.P., Quasianalytic functionals and self-adjointness of field operators, *Nuovo Cim.*, **45**, 158, (1966).

[209] Gal–Ezer, E., and Reeh, H., Coleman and Coleman–Okubo theorems in relativistic quantum field theory, *Fortschr. Phys.*, **22**, 481, (1974).

[210] Gallone, F., Sparzani, A., Streater, R.F., and Ubertone, C., Twisted condensates of quantized fields, *J. Phys. A. Math. Gen.*, **19**, 241–258, (1986).

[211] Garber, W.D., The connexion of duality and causal properties for generalized free fields, *Comm. Math. Phys.*, **42**, 195–208, (1975).

[212] Galindo, A., Morales, A., and Nuñez–Lagos, R., Superselection principle and pure states of *n* identical particles, *J. Math. Phys.*, **3**, 324–328, (1962).

[213] Giles, R., Foundations for quantum mechanics, *J. Math. Phys.*, **11**, 2139–2160, (1970).

[214] Gilmore, T.Jr., and Park, J.L., Superselection rules in quantum theory. A new proposal for state restriction violation, *Found Phys.*, **9**, 537–556, (1979). Subensemble selection, *Ibid.*, **9**, 739–749, (1979).

[215] Glimm, J., and Jaffe, A., Boson quantum field models. In: *Mathematics of Contemporary Physics*. Academic Press, New York, 1972.

[216] Grundling, H., and Hurst, C.A., Algebraic structures of degenerate systems and indefinite metric., *J. Math. Phys.*, **28**, 559–572, (1987).

[217] Gudder, S., and Scruggs, W., Unbounded representations of *-algebras, *Pacific J. Math.*, **70**, 369–382, (1977).

[218] Guenin, M., Practical use of algebraic concepts in quantum field theory. In: *Axiomatic Approach to Quantum Field Theory and Many-Body Problem*, Vol. 1A, 185–217, Wroclaw, 1968.

[219] Guenin, M., Introduction to algebraic techniques in quantum field theory and statistical mechanics. In: *Lect. Theor. Phys.*, Vol. XII–C, 187–253, Gordon & Breach, New York, 1971.

[220] Guenin, M., and Misra, B., On the von Neumann algebras generated by field operators, *Nuovo Cim.*, **30**, 1272–1290, (1972).

[221] Guenin, M., and Misra, B., De la permutabilité des opérateurs non bornés, *Helv. Phys. Acta*, **37**, 233–240 (1964).

[222] Guenin, M., and Misra, B., Borchers classes and duality theorem, *Helv. Phys. Acta*, **37**, 267–269, (1964).

[223] Guenin, M., and Velo, G., Automorphisms and broken symmetries in algebraic quantum field theories, *Nuovo Cim.*, **47A**, 36, (1967).

[224] Guenin, and Velo, G., Mass renormalization as an automorphism of the algebra of field operators, *Helv. Phys. Acta*, **41**, 362–366, (1968).

[225] Guerra F., Local algebras in Euclidean quantum field theory, *Symp. Math.*, **20**, 13–26, (1976).

[226] Guichardet, A., Produits tensoriels infinis et représenations des relations d'anticommutation, *Ann. Ec. Norm. Sup.*, **83**, 1–52, (1966).

[227] Haag, R., Quantum field theories with composite particles and asymptotic conditions, *Phys. Rev.*, **112**, 669–673, (1958).

[228] Haag, R., Discussion des 'axiomes' et des proprietés asymptotiques d'une théorie des champs locaux avec particules composées. In: *Les problèmes mathématiques de la théorie quantique des champs*. Colloque CNRS, Vol. 83, 151–163, (mai 1957).

[229] Haag, R., Bemerkungen zum Nahwirkungsprinzip in der Quantenphysik, *Ann. der Physik*, **11**, 29–34, (1963).

[230] Haag, R., Structure analysis of quantum field theory. In: *Axiomatic Approach to Quantum Field Theory and Many Body Problem*, Vol. IA, 9–17, Wroclaw, 1968.

[231] Haag, R., Structural questions in quantum field theory, *Proc. Symp. Pure Math.*, **38**, 513–521, (1982).

[232] Haag, R., Mathematical structure of orthodox quantum theory and its relation to operationally definable physical principles. In: *Mathematical Problems in Theoretical Physics*. Lecture Notes in Physics, Vol. 153, 168–172, Springer–Verlag, New York, Heidelberg, Berlin, 1982.

[233] Haag, R., Local relativistic quantum physics, *Physica*, **124A**, 357–364, (1984).

[234] Haag, R., Kadison, R.V., and Kaster, D., Nets of C^*-algebras and classification of states, *Comm. Math. Phys.*, **16**, 81–104, (1970).

[235] Haag, R., and Kastler, D., An algebraic approach to quantum field theory, *J. Math. Phys.*, **5**, 848–861, (1964).

[236] Haag, R., Narnhofer, H., and Stein, U., On quantum field theory in gravitational background, *Comm. Math. Phys.*, **94**, 219–238, (1984).

[237] Haag, R., and Schroer, B., The postulates of quantum field theory, *J. Math. Phys.*, **3**, 248–256, (1962).

[238] Haag, R., and Swieca, J.A., When does a quantum theory describe particles? *Comm. Math. Phys.*, **1**, 308–320, (1965).

[239] Hata, H., Itoh, K., Kugo T., Kunitomo, H., and Ogawa, K., Covariant string field theory, *Phys. Rev. D.*, **34**, 2360–2429, (1986).

[240] Hegerfeldt, G.C., Extremal decomposition of Wightman functions and of states of nuclear *-algebras by Choquet theory, *Comm. Math. Phys.*, **45**, 133–135, (1975).

[241] Hegerfeldt, G.C., Kraus, K., and Wigner, E.P., Proof of the fermion superselection rule without the assumption of time-reversal invariance, *J. Math. Phys.*, **9**, 2029, 2031, (1968).

[242] Hepp, K., Quantum theory of measurement and macroscopic observables, *Helv. Phys. Acta.*, **45**, 237–248, (1972).

[243] Hermaszewski, Z.J., and Streater, R.F., Some local covariant representations of the canonical anticommutation relations, *J. Phys. A: Math. Gen.*, **16**, 2801–2810, (1983).

[244] Hislop, P.D., A simple proof of duality for local algebras in free quantum field theory, *J. Math. Phys.*, **27**, 2542–2550, (1986).

[245] Hislop, P.D., and Longo, R., Modular structure of local algebras associated with the free massless scalar field theory, *Comm. Math. Phys.*, **84**, 71–85, (1982).

[246] Hoffman, G., On the existence of quantum fields in spacetime dimension 4., *Rep. Math. Phys.*, **18**, 231–242, (1980).

[247] Holevo, A.S., Quasifree states of the C^*-algebra of CCR, *Theor. Math. Phys.*, **6**, 1–12, (1971); *Ibid.*, **6**, 103–107, (1971).

[248] Horuzhy, S.S., and Dadashyan, K.Yu., Causal independence of free quantum fields, *Czech. J. Phys.*, **B29**, 29–32, (1979).

[249] Horuzhy, S.S., Structure of local W^*-algebras in relativistic quantum theory, *Theor. Math. Phys.*, **2**, 253–260, (1970).

[250] Horuzhy, S.S., Fields and local algebras in axiomatic algebraic theory with superselection rules., *Theor. Math. Phys.*, **5**, 942–952, (1970).

[251] Horuzhy, S.S., Superposition principle in algebraic quantum theory, *Theor. Math. Phys.*, **23**, 413–421, (1975).

[252] Horuzhy, S.S., Fields and algebras of observables in models with superselection rules. Model with abelian gauge group, *Theor. Math. Phys.*, **25**, 1139–1148 (1975); Model with nonabelian

gauge group, *Ibid.*, **26**, 1–8, (1976).

[253] Horuzhy, S.S., and Voronin, A.V., Field algebras do not leave field domains invariant, *Comm. Math. Phys.*, **102**, 687–692, (1986).

[254] Horuzhy, S. S., and Voronin, A. V., *Op** versus *C** dynamical systems, *Theor. Math. Phys.*, **82**, N2, 3 (1989).

[255] Houtappel, R.R.M., Van Dam, H., and Wigner, E.P., The conceptual basis and use of the geometrical invariance principle, *Rev. Mod. Phys.*, **37**, 595–632, (1965).

[256] Hunziker, W., A note on symmetry operations in quantum mechanics, *Helv. Phys. Acta*, **45**, 233–236, (1972).

[257] Il'in V.A., Local nets of observable algebras for the free scalar charged field, *Vest. MGU, Ser.*, **3** (Physics, Astronomy), 16–21, (1979).

[258] Il'in V.A., and Slavnov, D.A., Observable algebras in the *S*-matrix approach, *Theor. Math. Phys.*, **36**, 32–41, (1978).

[259] Inoue, A., On a class of unbounded operator algebras, *Pacific J. Math.*, **65**, 77–96, (1977).

[260] Inoue, A., Ueda, H., and Yamauchi, T., Commutants and bicommutants of algebras of unbounded operators, *J. Math. Phys.*, **28**, 1–7, (1987).

[261] Jadczyk, A.Z., On spectrum of internal symmetries of the algebraic quantum field theory, *Comm. Math. Phys.*, **12**, 58–63, (1969).

[262] Jakobczyk, L., Borchers algebra formulation of an indefinite inner product quantum field theory, *J. Math. Phys.*, **25**, 617–622, (1984).

[263] Jakobczyk, L., Some algebras of unbounded operators in an indefinite inner product vector space, *J. Math. Phys.*, **27**, 116–123, (1986).

[264] Jancewicz, B., Remark on a theorem of Araki, *Acta Phys. Pol.*, **XXXIII**, 151–153, (1968).

[265] Jauch, J.M., and Misra, B., Supersymmetries and essential observables, *Helv. Phys. Acta*, **34**, 502, 699–709, (1961).

[266] Jordan, T.F., and Sudarshan, E.C.G., Reduction of operator rings and the irreducibility axiom in QFT, *J. Math. Phys.*, **3**, 587–594, (1962).

[267] Kadison, R.V., States and representations, *Trans. Amer. Math. Soc.*, **103**, 304–319, (1962).

[268] Kadison, R.V., Remarks on the type of von Neumann algebras of local observables in quantum field theory, *J. Math. Phys.*, **4**, 1511–1516, (1963).

[269] Kadison, R.V., Transformation of states in operator theory and dynamics, *Topology*, **3**, Suppl. 2, 177–198, (1965).

[270] Kadison, R.V., The energy-momentum spectrum of quantum fields, *Comm. Math. Phys.*, **4**, 258–260, (1967).

[271] Kadison, R.V., Some analytic methods in the theory of operator algebras, *Lect. Notes in Mathem.* **140**, 8–30, (1970).

[272] Karwowski, W., and Sznajder, N., Some remarks on the theory of several scalar fields in algebraic approach, *Rep. Math. Phys.*, **1**, 259–264, (1971).

[273] Kastler, D., The *C**-algebra of a free boson field. Discussion of the basic facts, *Comm. Math. Phys.*, **1**, 14–48, (1965).

[274] Kastler, D., Does ergodicity plus locality imply the Gibbs structure? *Proc. Symp. Pure Math.*, **38**, 467–489, (1982).

[275] Kastler, D., Mebkhout, M., Loupias, G., and Michel, L., Central decomposition of invariant states. Applications to the groups of time translations and of Euclidean transformations in algebraic field theory, *Comm. Math. Phys.*, **27**, 195–222, (1972).

[276] Kastler, D., and Robinson, D.W., Invariant states in statistical mechanics, *Comm. Math. Phys.*, **3**, 151–180, (1966).

[277] Kastler, D., Robinson, D.W., and Swieca, A., Conserved currents and associated symmetries: Goldstone theorem, *Comm. Math. Phys.*, **2**, 108–120, (1966).

[278] Kay, B.S., The double-wedge algebra for quantum fields on Schwarzschield and Minkowski space-times, *Comm. Math. Phys.*, **100**, 57–81, (1985).

[279] Kay, B.S., A uniqueness result for quasifree KMS states, *Helv. Phys. Acta*, **58**, 1017–1029, (1985).

[280] Kay, B.S., Purification of KMS states, *Helv. Phys. Acta*, **58**, 1030–1040, (1985).

[281] Kern, J.–D., Jaffe fields and algebras of local observables. In: *Topics in Quantum Field Theory and Spectral Theory*, 68–74, Berlin, 1986.

[282] Kharatyan, S.G., Von Neumann algebras of observables with nonabelian commutant and superselection rules, *Theor. Math. Phys.*, **14**, 227–232, (1973).

[283] Kharatyan, S.G., On connection between multiplicative symmetries and superselection rules in Wightman axiomatics, *Izv. AN Arm. SSSR, ser. Phys.*, **13**, 149–151, (1978).

[284] Kishimoto, A., External invariant states with a spectrum condition and Borchers' theorem, *Lett. Math. Phys.*, **3**, 193–196, (1979).

[285] Knight, J.M., Strict localization in quantum field theory, *J. Math. Phys.*, **2**, 459–471, (1961).

[286] Kovacs, I., and Szücs, J. Ergodic type theorems in von Neumann algebras, *Acta Sci. Math.* (Szeged), **27**, 233–246, (1966).

[287] Kraus, K., General quantum field theories and strict locality, *Z. Phys.*, **181**, 1–12, (1964).

[288] Kraus, K., Algebras of observables with continuous representations of symmetry groups, *Comm. Math. Phys.*, 7, 99–111, (1968).

[289] Kraus, K., An algebraic spectrum condition, *Comm. Math. Phys.*, **16**, 138–141, (1970).

[290] Kraus, K., and Streater, R.F., Some covariant representations of massless Fermi fields, *J. Phys., A: Math. Gen.*, **14**, 2467–2478, (1981).

[291] Kruszynski, P., and Napiorkowski, K., On the independence of local algebras, *Rep. Math. Phys.*, **4**, 303–306, (1973).

[292] Kugo, T., and Ojima, I., Local covariant operator formalism of nonabelian gauge theories and quark confinement problem, *Progr. Theor. Phys., Suppl.*, **66**, (1974).

[293] Lahti, P.J., A coherent superposition principle and the Hilbertian quantum theory, *Rep. Math. Phys.*, U22, **49–62**, (1985).

[294] Landau, L.J., A note on extended locality, *Comm. Math. Phys.*, **13**, 246–253, (1969).

[295] Landau, L.J., Asymptotic locality and the structure of local internal symmetries, *Commun. Math. Phys.*, **17**, 156–176, (1970).

[296] Landau, L.J., On local functions of fields, *Comm. Math. Phys.*, **39**, 49–62, (1974).

[297] Landau, L.J., On the violation of Bell's inequality in quantum theory, *Phys. Lett.*, **120**, 54–56, (1987).

[298] Landau, L.J., and Wichmann, E.H., On the translational invariance of local internal symmetries, *J. Math. Phys.*, **11**, 306–311, (1970).

[299] Lanford, O.E., and Ruelle, D., Observables at infinity and states with short correlations in statistical mechanics, *Comm. Math. Phys.*, **13**, *194–215*, (1969).

[300] Langerholc, J., and Schroer, B., On the structure of the von Neumann algebras generated by local functions of the free Bose field, *Comm. Math. Phys.*, **1**, 215–239, (1965).

[301] Lassner, G., Topological algebras of operators, *Rep. Math. Phys.*, **3**, 279–293, (1979).

[302] Lassner, G., Algebras of unbounded operators and quantum dynamics, *Physica*, **124A**, 471–480, (1984).

[303] Lassner, G., and Timmermann, W., On the essential self-adjointness of various algebras of field operators, *Theor. Math. Phys.*, **15**, 543–545, (1973).

[304] Lassner, G., and Uhlmann, A., On positive functionals on algebras of test-functions for quantum fields, *Comm. Math. Phys.*, **7**, 152–169, (1968).

[305] Leyland, P., and Roberts J.E, The cohomology of nets over Minkowski space, *Comm. Math. Phys.*, **62**, 173–189, (1978).

[306] Licht, A.L., Strict localization, *J. Math. Phys.*, **4**, 1443–1447, (1963).

[307] Licht, A.L., Local states, *J. Math. Phys.*, **7**, 1656–1669, (1966).

[308] Licht, A.L., Coherent superposition of local states, *J. Math. Phys.*, **9**, 1476–1480, (1968).

[309] Licht, A.L., and Toll, J.S., Two-point function and generalized free fields, *Nuovo Cim.*, **21**, 346, (1961).

[310] Longo, R, Notes on algebraic invariants for non-commutative dynamical systems, *Comm. Math. Phys.*, **69**, 195–207, (1979).

[311] Longo, R., Algebraic and modular structure of von Neumann algebras of physics, *Proc. Symp. Pure Math.*, **38**, 551–566, (1982).

[312] Longo, R., The joint modular structure of an inclusion of von Neumann algebras, *Contemp. Math.*, **62**, 529–538, (1987).

[313] Longo, R., and Peligrad, C., Noncommutative topological dynamics and compact actions on C^*-algebras, *J. Funct. An.*, **58**, 157–174, (1984).

[314] Maison, D., Eine Bemerkung zu Klustereigenschaften, *Comm. Math. Phys.*, ***10***, 48–51, (1968).

[315] Manuceau, J., Étude de quelques automorphisms de la C^*-algèbre du champ des bosons libres, *Ann. Inst. H. Poincaré*, **A8**, 117–138, (1968)

[316] Manuceau, J., C^*-algèbre des relations de commutation, *Ann. Inst. H. Poincaré*, **A8**, 139–161, (1968).

[317] Mathot, F., Topological properties of unbounded commutants, *J. Math. Phys.*, **26**, 1118–1124, (1985).

[318] Maurin, K., Mathematical structure of Wightman formulation of quantum field theory, *Bull. Ac. Pol. Sci. ser. Math.*, **11**, 115–120, (1963).

[319] Maurin, K., On some theorems of H.J. Borchers, *Bull. Ac. Pol. Sci. ser. Math.*, **11**, 121–123, (1963).

[320] Minchev, M., Quantization in indefinite metric, *J. Phys. A: Math. Gen.*, **13**, 1841–1859, (1980).

[321] Minchev, M., and D'Emilio, E., On the reconstruction of the physical Hilbert spaces for quantum field theories with indefinite metric, *J. Math. Phys.*, **22**, 1267–1271, (1981).

[322] Mirman, R., Nonexistence of superselection rules: definition of term 'frame of reference'? *Found. Phys.*, **9**, 283–299, (1979).

[323] Misra, B., On the representations of Haag fields, *Helv. Phys. Acta*, **38**, 189–206, (1965).

[324] Miyata, H., On clustering states, *Comm. Math. Phys.*, **34**, 1–6, (1973).

[325] Möllenhoff, C., Über die Vollständigkeit verallgemeinerter freien Felder in einer Zeitschicht, *Comm. Math. Phys.*, **11**, 227–232, (1968).

[326] Montvay, J., The representations of symmetry groups on the algebra of quasilocal observables, *Nuovo Cim.*, **40A**, 121, (1965).

[327] Morchio, G., and Strocchi, F., Infrared singularities, vacuum structure and pure phases in local quantum field theory, *Ann. Inst. H. Poincaré*, **A33**, 251–282, (1980).

[328] Morchio, G., and Strocchi, F., A nonperturbative approach to the infrared problem in QED, Construction of charged states, *Nucl. Phys.*, **B211**, 471, (1983).

[329] Morchio, G., and Strocchi, F., Spontaneous symmetry breaking and energy gap generated by variables at infinity, *Comm. Math. Phys.*, **99**, 153–175, (1985).

[330] Morchio, G., and Strocchi, F., Mathematical structure for long-range dynamics and symmetry breaking, *J. Math. Phys.*, **28**, 622–635, (1987).

[331] de Muynck, W.M., and van den Eijnde, J.P.H.W., A derivation of local commutativity from macrocausality using quantum-mechanical theory of measurements, *Found. Phys.*, **14**, 111–146, (1984).

[332] Nagamachi, S., and Mugibayashi, N., Neutral scalar field theory with a point source, *Progr. Theor. Phys.*, **49**, 329–340, (1973).

[333] Napiorkowski, K., On the independence of local algebras, *Rep. Math. Phys.*, **8**, 33–35,

[334] Neumann, H., and Werner, R., Causality between preparation and registration processes in relativistic quantum theory, *Int. J. Theor. Phys.*, **22**, 781–802, (1983).

[335] von Neumann, J., Zur Algebra der funktional Operationen und Theorie der normalen Operatoren, *Math. Ann.*, **102**, 370–427, (1930).

[336] von Neumann, J., On rings of operators III, *Ann. Math.*, **41**, 94–161, (1940).

[337] Nussbaum, A.E., Quasianalytic vectors, *Ark. Math.*, **6**, 179, (1965).

[338] Olesen, D., and Pedersen, G.K., Groups of automorphisms with spectrum condition and the lifting problem, *Comm. Math. Phys.*, **51**, 85–95, (1976).

[339] Orzalesi, C., Charges as generators of symmetry transformations in quantum field theory, *Rev. Mod. Phys.*, **42**, 381–409, (1970).

[340] Ostendorf, A., Feynmann rules for Wightman functions, *Ann. Inst. H. Poincaré*, **40**, 273–291, (1984).

[341] Osterwalder, K., Duality for free Bose field, *Comm. Math. Phys.*, **29**, 1–14, (1973).

[342] Palmer, J., Complex structures and external fields, *Arch. for Rat. Mech. and Anal.*, **75**, 31–49, (1980).

[343] Parenti, C., Strocchi, F., and Velo, G., Hilbert space sectors, dynamical charges and spontaneous symmetry breaking in classical field theory, *Phys. Lett.*, **B62**, 83–85, (1976).

[344] Parenti, C., Strocchi, F., and Velo, G., Hilbert space sectors for solutions of nonlinear relativistic field equations, *Comm. Math. Phys.*, **53**, 65–96, (1977).

[345] Pawlik, B., and Sadowski, P., On a construction of the maximal additive field of von Neumann algebras in quantum field theory, *Rep. Math. Phys.*, **4**, 223–225, (1973).

[346] Pedersen, G.K., and Takesaki, M., The Radon–Nikodym theorem for von Neumann algebras, *Acta Math.*, **130**, 53–87, (1973).

[347] Petrov, A., Asymptotic behaviour of the unitary representations of the Poincaré group, *Rep. Math. Phys.*, **4**, 183–188, (1973).

[348] Piron, C., Les règles de supersélection continuées, *Helv. Phys. Acta* **42**, 330–338, (1969).

[349] Pogrebkov, A.K., and Sushko, V.N, Quantum solitions and their connection with fermion fields for the $(\sin\psi)_2$ interaction, *Theor. Math. Phys.*, **26**, 286–289, (1976).

[350] Polivanov, M.K., Sushko, V.N., and Horuzhy, S.S., Axioms of observable algebras and the field concept, *Theor. Math. Phys.*, **16**, 629–641, (1973).

[351] Polley, L., Reents, G., and Streater, R.F., Some covariant representations of massless boson fields, *J. Phys. A: Math. Gen.*, **14**, 2479–2488, (1981).

[352] Powers, R.T., Self-adjoint algebras of unbounded operators, *Comm. Math. Phys.*, **21**, 85–124; *Trans. Amer. Math. Soc.*, **187**, 261–293, (1974).

[353] Reeh, H., Symmetry operations and spontaneously broken symmetries in relativistic quantum field theory, *Fortschr. d. Phys.*, **16**, 687–706, (1968).

[354] Reeh, H., and Schlieder, S., Bemerkungen zur Unitärequivalenz von Lorentzinvarianten Feldern, *Nuovo Cim.*, **22**, 1051–1068, (1961).

[355] Reeh, H., and Schlieder, S., Über den Zerfall der Feldoperator Algebra im Falle einer Vakuum Entartung, *Nuovo Cim.*, **26**, 32–42, (1962).

[356] Rehberg, J., and Wollenberg, M., Quantum fields as pointlike localized objects, *Math. Nachr.*, **125**, 259–274, (1986).

[357] Requardt, M., and Wreszinski, W.F., Temperature states, ground states and relativistic vacuum states in the context of symmetry breakdown, *J. Phys. A: Math. Gen.*, **18**, 705–712, (1985).

[358] Rieffel, M., A commutation theorem and duality for free Bose fields, *Comm. Math. Phys.*, **39**, 153–164, (1974).

[359] Rigotti, C., Remarks on the modular operator and local observables, *Comm. Math. Phys.*, **68**, 267–273, (1981).

[360] Roberts, J.E., Some applications of dilatation invariance to structural questions in the theory of local observables, *Comm. Math. Phys.*, **37**, 273–286, (1974).

[361] Roberts, J.E., Must there be a gauge group? *Proc. V.A. Steklov Math. Inst.*, **135**, 206–209, (1975).

[362] Roberts, J.E., Local cohomology and superselection structure, *Comm. Math. Phys.*, **51**, 107–119, (1976).

[363] Roberts, J.E., Localization in algebraic field theory, *Comm. Math. Phys.*, **85**, 87–98, (1982).

[364] Roberts, J.E., New light on the mathematical structure of algebraic field theory, *Proc. Symp. Pure Math.*, **38**, 523–550, (1982).

[365] Roberts, J.E., The search for quantum differential geometry. In: *Mathematical Problems in theoretical Physics*, Lecture Notes in Physics, **1153**, 374–379, Springer–Verlag, New York, Heidelberg, Berlin, (1982).

[366] Roberts, J.E., and Roepstorff, G., Some basic concepts of algebraic quantum theory, *Comm. Math. Phys.*, **11**, 321–338, (1969).

[367] Robinson, D.W., The ground state of the Bose gas, *Comm. Math. Phys.*, **1**, 159–171, (1965).

[368] Robinson, D.W., Algebraic aspects of relativistic quantum field theory. In: *Axiomatic Field Theory*, ed. M. Chretien, and S. Deser, 389–516. Gordon and Breach, New York, 1966.

[369] Robinson, D.W., and Ruelle, D., Extremal invariant states, *Ann. Inst. H. Poincaré*, **A6**, 299–310, (1967).

[370] Roos, H., Independence of local algebras in quantum field theory, *Comm. Math. Phys.*, **16**, 238–246, (1970).

[371] Ruelle, D., On the asymptotic condition in quantum field theory, *Helv. Phys. Acta*, **35**, 147–163, (1962).

[372] Ruelle, D., States of physical systems, *Comm. Math. Phys.*, **3**, 133–150, (1966).

[373] Sadowski, P., and Woronowicz, S.L., Total sets in quantum field theory, *Rep. Math. Phys.*, **2**, 113–120, (1971).

[374] Schlieder S., Some remarks about the localization of states in a quantum field theory, *Comm. Math. Phys.*, **1**, 265–280, (1965).

[375] Schlieder S., Einige Bemerkungen zur Zustandsänderung von relativistischen quantenmechanischen Systemen durch Messungen und zur Lokalitätsforderung, *Comm. Math. Phys.*, **7**, 305–331, (1968).

[376] Schlieder, S., Einige Bemerkungen über Projektionsoperatoren, *Comm. Math. Phys.*, **13**, 216–225, (1969).

[377] Schmüdgen, K., On trace representation of linear functionals on unbounded operator algebras, *Comm. Math. Phys.*, **63**, 113–130, (1978).

[378] Schoch, A., On the simplicity of Haag fields, *Int. J. Theor. Phys.*, **1**, 107–113, (1969).

[379] Segal, J.E., Postulates for general quantum mechanics, *Ann. Math.*, **48**, 930–948, (1947).

[380] Segal, J.E., A non-commutative extension of abstract integration, *Ann. Math.*, **57**, 401–457, (1953).

[381] Segal, J.E., and Goodman, R.W., Antilocality of certain Lorentz-invariant operators, *J. Math. Mech.*, **14**, 629–638, (1965).

[382] Sewell, G.L., Relativity of temperature and the Hawking effect, *Phys. Lett.*, **79A**, 23–24, (1980).

[383] Sewell, G.L., Quantum fields on manifolds, *Ann. Phys.*, **141**, 201–224, (1982).

[384] Snellman, H., Poincaré analytic vectors in axiomatic quantum field theory, *J. Math. Phys.*, **15**, 1054–1059, (1974).

[385] Steinmann, O., Haag–Kastler formulation of quantum electrodynamics, *Nucl. Phys.*, **107**, 339–364, (1976).

[386] Størmer, E., Large groups of automorphisms of C^*-algebras, *Comm. Math. Phys.*, **5**, 1–22, (1967).

[387] Størmer, E., Types of von Neumann algebras associated with extremal invariant states, *Comm. Math. Phys.*, **6**, 194–204, (1967).

[388] Størmer, E., Automorphisms and invariant states of operator algebras, *Acta Math.*, **127**, 1–9, (1971).

[389] Streater, R.F., Intensive observables in quantum theory, *J. Math. Phys.*, **5**, 581–590, (1964).

[390] Streater, R.F., Spontaneous breakdown of symmetry in axiomatic theory, *Proc. Roy. Soc.*, **A287**, 510–518, (1965).

[391] Streater, R.F., Gauge theories and superselection rules, *Acta Phys. Austr. Suppl.*, **XI**, 317–340, (1973).

[392] Streater, R.F., Charges and currents in the Thirring model. In: *Physical Reality and Mathematical Description*, 375–386; ed. Enz, Mehra; Reidel, Dordrecht, 1974.

[393] Streater, R.F., Dynamical selection rules, *Phys. Scripta*, **24**, 910–913, (1981).

[394] Streater, R.F., and Wilde, J.F., Fermion states of a boson field, *Nucl. Phys.*, **B24**, 561–575, (1970).

[395] Streit, L., Generalized free fields as cyclic representations of the canonical commutation relations, *Acta Phys. Austr.*, **32**, 107–112, (1970).

[396] Streit, L., Lightlike initial data for quantum field theory. In: *Recent Developments in Relativistic Quantum Field Theory and its Applications*. Proc. XIII Winter School of Theor. Phys., *Acta Univ. Wratisl.*, **N389**, 111–128, Karpacz, (1976).

[397] Strocchi, F., Gauge problem in quantum field theory, *Phys. Rev.*, **162**, 1429, (1967); Difficulties of combining Einstein equations and Wightman theory, *Ibid.*, **166**, 1302–1307, (1968); Quantization of Maxwell equations and weak local commutativity, *Ibid.*, **D2**, 2334–2340, (1970).

[398] Strocchi, F., Gauge groups in local field theory and superselection rules, *Proc. of the IV Intern. Colloquium on Group Theoret. Methods in Physics*, 87–92, June 23–27, Nijmegen, (1975).

[399] Strocchi, F., Superselection rules and infrared problems in Yang–Mills theories, *Lett. Nuovo Cim.*, **16**, 397–398, (1976).

[400] Strocchi, F., Local and covariant gauge quantum field theory. Clustering property, superselection rules and the infrared problem, *Phys. Rev.*, **D17**, 2020–2022, (1978).

[401] Strocchi, F., Gauss' law in quantum field theory. In: *Field theory, quantization and statistical physics*, 227–236, Reidel, Dordrecht, (1981).

[402] Strocchi, F., and Wightman, A.S., Proof of the charge superselection rule in local relativistic quantum field theory, *J. Math. Phys.*, **15**, 2198–2224, (1974).

[403] Summers, S.J., Normal product states and twisted duality for CCR- and CAR-type algebras with applications to the Yukawa_2 quantum field model, *Comm. Math. Phys.*, **86**, 111–141, (1982).

[404] Summers, S.J., and Werner, R., The vacuum violates Bell's inequalities, *Phys. Lett.*, **110A**, 257–259, (1985).

[405] Summers, S.J., and Werner, R., Maximal violation of Bell's inequalities is generic in quantum field theory, *Comm. Math. Phys.*, **110**, 247–259, (1987).

[406] Sushko, V.N., Fermionization of the $(\sin)_2$ interaction in a box, *Theor. Math. Phys.*, **37**, 949–968, (1978).

[407] Sushko, V.N., and Horuzhy, S.S., Vector states on algebras of observables and superselection rules. Vector states and Hilbert space, *Theor. Math. Phys.*, **4**, 758–774, (1970). Algebraic theory of superselection rules, *Ibid.*, **4**, 877–889, (1970).

[408] Sushko, V.N., and Horuzhy, S.S., Properties of $\mathcal{H}$-images of vector states on algebras of observables, *Theor. Math. Phys.*, **8**, 862–864, (1971).

[409] Sushko, V.N., and Horuzhy, S.S., Local and asymptotic structure of quantum systems with superselection rules, *Theor. Math. Phys.*, **13**, 1147–1160, (1972).

[410] Sushko, V.N., and Horuzhy, S.S., $\mathcal{H}$-images of vector states and causal properties of local algebras, *Theor. Math. Phys.*, **15**, 460–466, (1973).

[411] Takesue, K., Spatial theory for algebras of unbounded operators, *Rep. Math. Phys.*, **21**, 347–355, (1985).

[412] Todorov, T.S., On complete set of commuting observables, *Bull. Inst. Phys. Rech. Atom.*, **22**, 119–128, (1972).

[413] Todorov, T.S., A remark on the definition of superselection rules in terms of unbounded operators, *Int. J. Theor. Phys.*, **14**, 295–298, (1975).

[414] Todorov, T.S., On a generalization of the Reeh–Schlieder theorem, *Rep. Bulg. Ac. Sci.*, **34**, 491–492, (1981).

[415] Uhlmann, A., Über die Definition der Quantenfelder nach Wightman und Haag, *Wiss. Zs. Karl Marx–Univ.*, **11**, 213–217, (1962).

[416] Vassilev, A.N., La structure algébrique de la représentation et le problème de la symmetrie pour une classe de fonctionnelles de Wightman, *Comm. Math. Phys.*, **13**, 81–101, (1969).

[417] Vassilev, A.N., Sur la construction asymptotique et l'interprétation d'une théorie avec dégénérescence du vide, *Comm. Math. Phys.*, **13**, 102–118 (1969).

[418] Vassilev, A.N., Algebraic structure of local field theory with finite-fold vacuum degeneracy, *Theor. Math. Phys.*, **1**, 233–242, (1969).

[419] Vassilev, A.N., Theory of representations of a topological (non Banach) involutory algebra, *Theor. Math. Phys.*, **2**, 113–123, (1970).

[420] Vassilev, A.N., Algebraic aspects of Wightman axiomatics, *Theor. Math. Phys.*, **3**, 317–340, (1970).

[421] Vassilev, A.N., Criterion for the commutant of a quantized field to be algebraically closed, *Theor. Math. Phys.*, **5**, 1057–1060, (1970).

[422] Vassilev, A.N., and Borisov, N.V., Self-adjointness of some representations of field algebra, *Theor. Math. Phys.*, **11**, 306–312, (1972).

[423] Völkel, A.H., Confinement of quarks and gluons in gauge quantum field theory, *J. Math. Phys.*, **26**, 1057–1064, (1985).

[424] Völkel, A.H., Unbounded representations of symmetry groups in gauge quantum field theory. Confinement and differentiation, *J. Math. Phys.*, **26**, 2956–2973, (1985).

[425] Voronin, A.V., Discrete vacuum superselection in Wightman theory with essentially self-adjoint field operators, *Theor. Math. Phys.*, **66**, 13–29, (1986).

[426] Voronin, A.V., Sushko, V.N., and Horuzhy, S.S., Algebras of unbounded operators and vacuum superselection in quantum field theory. Some properties of Op^*-algebras and their vector states, *Theor. Math. Phys.*, **59**, 28–48, (1984); Mathematical structure of vacuum superselection, *Ibid.*, **60**, 323–343, (1984).

[427] Werner, R., Local preparability of states and the split property in quantum field theory, *Lett. Math. Phys.*, **13**, 325–329, (1987).

[428] Wichmann, E.H., On systems of local operators and the duality condition, *J. Math. Phys.*, **24**, 1633–1644, (1983).
[429] Wick, G.C., Wightman, A.S., and Wigner, E.P., The intrinsic parity of elementary particles, *Phys. Rev.*, **88**, 101–105, (1952).
[430] Wick, G.C., Wightman, A.S., and Wigner, E.P., Superselection rule for charge, *Phys. Rev.*, **D12**, 3267–3269, (1970).
[431] Wightman, A.S., Quantum field theory in terms of vacuum expectation values, *Phys. Rev.*, **101**, 860–873, (1956).
[432] Wightman, A.S., Relativistic invariance and quantum mechanics, *Nuovo Cim. Suppl.*, **14**, 81, (1959).
[433] Wightman, A.S., La théorie quantique locale et la théorie quantique des champs., *Ann. Inst. H. Poincaré*, **1**, 403–420, (1964).
[434] Wilde, I.F., A remark on the localization of a charged Bose field, *Comm. Math. Phys.*, **24**, 37–39, (1971).
[435] Wollenberg, M., On the relation between quantum fields and local algebras of observables, *Rep. Math. Phys.*, **22**, 409–417, (1985).
[436] Wollenberg, M., Quantum fields as pointlike-localized objects, *Math. Nachr.*, **128**, 287–298, (1986); Preprint Karl–Weierstrass Inst. für Mathematik, (1987).
[437] Woronowicz, S., A generalization of a theorem by Wightman, *Comm. Math. Phys.*, **9**, 142–145, (1968).
[438] Wreszinski, W.F., Charges and symmetries in quantum theories without locality, *Fortschr. d. Phys.*, **35**, 379–413, (1987).
[439] Wyss, W., The field algebra and its positive linear functionals, *Comm. Math. Phys.*, **27**, 223–234, (1972).
[440] Yngvason, J., On the algebra of test functions for field operators, *Comm. Math. Phys.*, **34**, 315–333, (1973).
[441] Yngvason, J., Remarks on the reconstruction theorem for field theories with indefinite metric, *Rep. Math. Phys.*, **12**, 57–64, (1977).
[442] Yngvason, J., On the decomposition of Wightman functionals in the Euclidean framework, *Rep. Math. Phys.*, **13**, 101–115, (1978).
[443] Yngvason, J., Translationally invariant states and the spectrum ideal in the algebra of test functions for quantum fields, *Comm. Math. Phys.*, **81**, 401–418, (1981).
[444] Yngvason, J., States, ideals and automorphisms of the algebra of test functions for quantum fields, *Physica*, **124A**, 621–628, (1984).
[445] Yngvason, J., On the locality ideal in the algebra of test functions for quantum fields, *Publ., RIMS Kyoto Univ.*, **20**, 1063–1081, (1984).
[446] Zinoviev, Yu.M., CPT-invariance in the theory of local observables, *Theor. Math. Phys.*, **16**, 642–650, (1973).
[447] Zinoviev, Yu.M., and Sushko, V.N., Physical symmetries in a theory of local observables, *Theor. Math. Phys.*, **18**, 9–18, (1974).

Subject Index